Plant Anatomy and Economic Botany

Plant Anatomy and Economic Botany

Editor:

Vikas Mishra, Ph.D.

Delve Publishing

www.delvepublishing.com

Plant Anatomy and Economic Botany

Editor: Vikas Mishra, Ph.D.

708 3rd Avenue, 6th Floor

New York, NY 10017

United States of America

www.delvepublishing.com

ISBN: 978-1-68095-761-7

Library of Congress Control Number: 2016951517

About the Editor

Vikas Mishra, Ph.D.

Vikas obtained his Ph.D. from King George's Medical University, India in 2010. His interests span from bio-resources, fermentation technology to medical virology and neuroscience research. He is currently working as a research associate at the University of British Columbia.

Contents

List of Contributors

John Charles Ryan
School of Communications and Arts, Edith Cowan University, 2 Bradford Street, Mount Lawley, Western Australia 6050, Australia

Luiz Rodrigo Saldanha Gazzaneo
Department of Biology, Botany Area, Applied Ethnobotany Laboratory, Federal Rural University of Pernambuco, Don Rua Manoel de Medeiros s/n, Two Brothers , Recife, Pernambuco, 52171-030 , Brazil

Reinaldo Farias Paiva de Lucena
Department of Biology, Botany Area, Applied Ethnobotany Laboratory, Federal Rural University of Pernambuco, Don Rua Manoel de Medeiros s/n, Two Brothers , Recife, Pernambuco, 52171-030 , Brazil

Ulysses Paulino de Albuquerque
Department of Biology, Botany Area, Applied Ethnobotany Laboratory, Federal Rural University of Pernambuco, Don Rua Manoel de Medeiros s/n, Two Brothers , Recife, Pernambuco, 52171-030 , Brazil

Aref Abu-Rabia
Department of Middle East Studies, Ben-Gurion University, Beer-Sheva, 84105, Israel

Andrea Pieroni
University of Gastronomic Sciences, Via Amedeo di Savoia 8, I-12060 Pollenzo/Bra, Italy

Maria Elena Giusti
Department of History of Art and Performing Arts, University of Florence, Via Gino Capponi, I- 50121 Florence, Italy

Fábio Ribeiro Araújo
Institute for Studies in Agricultural and Regional Development, Federal University of South and Southeast of Pará, Avenue Ipês - Ciudad Universitaria s/n, 68507-765, Maraba, PA, Brazil

Sol Elizabeth González-Pérez
Senior core Amazonian Studies, Federal University of Pará Augusto Corrêa Street, 01, 66075-110, Belém, PA, Brazil

Maria Aparecida Lopes
Institute of Biological Sciences , Federal University of Pará Augusto Corrêa Street, 01, 66075-110, Belém, PA, Brazil

Ismael de Jesus Matos Viégas
Federal Rural University of Amazonia, Rua Joao Pessoa, 113, 68700-030, Capanema, PA, Brazil

Renata Corrêa Martins
University of Brasília, Brasília, CP 4457, 70919-990 Brasilia, DF, Brazil

Tarciso de Sousa Filgueiras
Institute of Botany, São Paulo, CP 3005, 01061-950 São Paulo, SP, Brazil

Ulysses Paulino Albuquerque
Federal Rural University of Pernambuco, Laboratory of Applied and Theoretical Ethnobiology (LEA), Rua Dom Manoel de Medeiros, s/n, Dois Irmãos, 52171-900 Recife, PE, Brazil

A. Moteetee
Department of Botany and Plant Biotechnology, University of Johannesburg, South Africa

B-E. van Wyk
Department of Botany and Plant Biotechnology, University of Johannesburg, South Africa

S. Muruganandam
Post Graduate and Research Deportment of Plant Biology and Plant Biotechnology, Presidency College (Autonomous), Chennai-5

E. Kamaraj
Post Graduate and Research Deportment of Botany, Government Arts College for Men, (Autonomous), Nandanam, Chennai-35

V. Kadirvelmurugan
Post Graduate and Research Deportment of Plant Biology and Plant Biotechnology, Presidency College (Autonomous), Chennai-5

S. Ravikumar
Post Graduate and Research Deportment of Plant Biology and Plant Biotechnology, Presidency College (Autonomous), Chennai-5

Rainer W. Bussmann
William L. Brown Center, Missouri Botanical Garden, United States

Douglas Sharon
William L. Brown Center, Missouri Botanical Garden, United States

Ripu M Kunwar
Ethnobotanical Society of Nepal, G., Kathmandu, Nepal

Rainer W Bussmann
William L. Brown Center, Missouri Botanical Garden, St. Louis, MO, USA

Fakhruddin Ali Ahmed
Department of Botany, Jahangirnagar University, Savar, Dhaka-1342, Bangladesh

Anisur Rahman
Department of Botany, Jahangirnagar University, Savar, Dhaka-1342, Bangladesh

Sanzida Mubassara
Department of Botany, Jahangirnagar University, Savar, Dhaka-1342, Bangladesh

Gazi Mosharof Hossain
Department of Botany, Jahangirnagar University, Savar, Dhaka-1342, Bangladesh

Karolina Heyduk
2502 Miller Plant Sciences, Department of Plant Biology, University of Georgia, Athens, GA 30602, USA

Nia Burrell
2502 Miller Plant Sciences, Department of Plant Biology, University of Georgia, Athens, GA 30602, USA

Falak Lalani
2502 Miller Plant Sciences, Department of Plant Biology, University of Georgia, Athens, GA 30602, USA

Jim Leebens-Mack
2502 Miller Plant Sciences, Department of Plant Biology, University of Georgia, Athens, GA 30602, USA

Zhi-Yao Ma
Key Laboratory of Horticultural Plant Biology, College of Horticulture and Forestry Science, Huazhong Agricultural University, Ministry of Education, Wuhan, China

Jun Wen
Department of Botany, National Museum of Natural History, MRC166, Smithsonian Institution, Washington, DC, USA

Stefanie M. Ickert-Bond
UA Museum of the North Herbarium and Department of Biology and Wildlife, University of Alaska Fairbanks, Fairbanks, AK, USA

Long-Qing Chen
Key Laboratory of Horticultural Plant Biology, College of Horticulture and Forestry Science, Huazhong Agricultural University, Ministry of Education, Wuhan, China

Xiu-Qun Liu
Key Laboratory of Horticultural Plant Biology, College of Horticulture and Forestry Science, Huazhong Agricultural University, Ministry of Education, Wuhan, China

Franc Bavec
University of Maribor, Faculty of Agriculture and Life Sciences, Hoce, Maribor, Slovenia

Stephen N. Wegulo
Department of Plant Pathology, University of Nebraska-Lincoln, Lincoln, NE 68583, USA

Patricio Grassini
Department of Agronomy and Horticulture, University of Nebraska–Lincoln, Lincoln, Nebraska 68583-0915, USA.

Kent M. Eskridge
Department of Statistics, University of Nebraska–Lincoln, Lincoln, Nebraska 68583-0963, USA.

Kenneth G. Cassman
Department of Agronomy and Horticulture, University of Nebraska–Lincoln, Lincoln, Nebraska 68583-0915, USA.

Vessela Atanasova-Penichon
MycSA, National Institute for Agricultural Research, Villenave d'Ornon, France

Christian Barreau
MycSA, National Institute for Agricultural Research, Villenave d'Ornon, France

Florence Richard-Forget
MycSA, National Institute for Agricultural Research, Villenave d'Ornon, France

Gemechu Shale Ogato
Department of Rural Development and Agricultural Extension, Institute of Cooperatives and Development Studies, Ambo University, Ambo, Ethiopia

Boaresa K. Isaac
Department of Biotechnology, Faculty of Agriculture, University for Development Studies, Tamale, Ghana

Bawah Demuyakor
Department of Biotechnology, Faculty of Agriculture, University for Development Studies, Tamale, Ghana

Joseph A. Awuni
Department of Biotechnology, Faculty of Agriculture, University for Development Studies, Tamale, Ghana

Preface

The field of applied science, dealing with the practical or economic aspects of a subject, lends itself much better to such treatment than does the field of pure science. This is particularly true of botany. From earliest time plants 'have been intimately bound up with human existence. Not only have they played an important part in the everyday life of mankind, but they have had a profound influence on the course of history and civilization. A knowledge of the industrial, medicinal, and edible plants cannot fail to broaden one's outlook. Economic botany is the study of the relationship between people (individuals and cultures) and plants. Economic botany contributes significantly to many fields, including established disciplines such as agronomy, anthropology, archaeology, chemistry, economics, ethnobotany, ethnology, forestry, genetic resources, geography, geology, horticulture, medicine, microbiology, nutrition, pharmacognosy, and pharmacology. This link between botany and anthropology explores the ways humans use plants for food, shelter, medicines, textiles, and more. The average man is likely to consider himself as a being apart from the rest of the organic world, enabled by reason of hi:-; superior intellect to lead a self-sufficient and independent existence. He loses sight of the fact, or is ignorant of it, that he is absolutely dependent on other organisms for his very life, and his material happiness as well. His superior intelligence has made him more dependent rather than less so. Although various animal and mineral products contribute to his welfare, it is the plant kingdom that is most essential to man's well-being. Man's dependence on plants for the essentials of his existence has been of paramount importance in his life since the human race began. Human race cannot imagine to sustain in this world without plants. Plants have role in each and every aspect of human life. Besides photosynthesis and being a basic source of oxygen on the planet Earth, we need plants for food and feed, clothing, housing, medicinal and many more functions of basic human and animal sustenance. To the extent, that not very long ago the plants and grains were the basis for trade and economy of the world. In simple words, Economic botany is the interaction of human race with plants for sustainable solutions for some of their most pressing functions. It is very closely related to the ethnobotany, which in Greek means ethos, "race or cultural group" and botanikos, "herbs". The knowledge of ethnobotany has slowly evolved from the use of plants by the seemingly exotic cultures of the world, to systematic and quantitative analysis of plant use and benefits to our society.

Plant Anatomy and Economic Botany gives insight into the interactions of humans with various plants, which is based on the anatomy of the plants, for their economic and social uses. Primitive man probably had few needs other than food and a little shelter. Civilization, however, has brought with it an ever-increasing complexity, and has increased man's requirements to an amazing degree. The man of .today is no longer content merely to exist, with food and shelter as his only wants. He desires other commodities as well, and raw materials that can be converted into the many useful articles and products which contribute to his enjoyment of life, and which incidentally increase his debt to plants. Aside from their value as sources of food, drugs, and many of the raw materials of industrialism, plants are important to man in many other ways. The role of colorless plants in the economy of nature; the part that bacteria play in disease and many industries; and the effects of forests and other types of natural vegetation in controlling floods and erosion are but a few examples. The aesthetic value of plants has no small influence on man's enjoyment of life, as evidenced by the host of garden enthusiasts and flower lovers. The production and distribution of plant products have a profound influence on the economic and social life of the nations of the world, affecting both domestic conditions and international relations, and even changing the course of history.

Editor
Vikas Mishra, Ph.D.

INTRODUCTION

Plant anatomy can be defined as a branch of botany, which is concerned with the study of internal structure of plants. It is also called as Phytotomy. Plant anatomy is a Greek word, which means dissection. It completely deals with the structural organization of plants.

Plant anatomy is the study of the physical structure of plants. It is also known as phytoanatomy, with a practitioner of this scientific discipline being known as a phytoanatomist. As with animal anatomy, the goal is to learn more about how organisms are put together and how they work, with this information being used to obtain a deeper knowledge of how to care for plants and how to address diseases that affect them. Phytoanatomists work in a number of settings, including natural history museums, arboretums, and laboratories that develop new plants for agriculture and landscaping.

Phytoanatomists both study the structure of plants as a whole and dissect them to learn about their component parts. It can also take place on the microscopic level, with the anatomist examining plant cells to learn more about their function, and to differentiate between the various kinds of plant cells.

Plant anatomists are also interested in the development of plants, from their earliest stages as seeds through their maturity into adulthood.

By dissecting and studying plants, researchers can learn about the differences between various plants, which is an important part of plant taxonomy. Two plants may look very similar on the surface, for example, but be radically different when they are dissected and viewed under the microscope. These differences can be used to describe and categorize the plants so that they can be placed within a taxonomic system. Plant anatomy can also involve carefully studying newly discovered plants to confirm that they are unique and to gather data about them that can be used to categorize them.

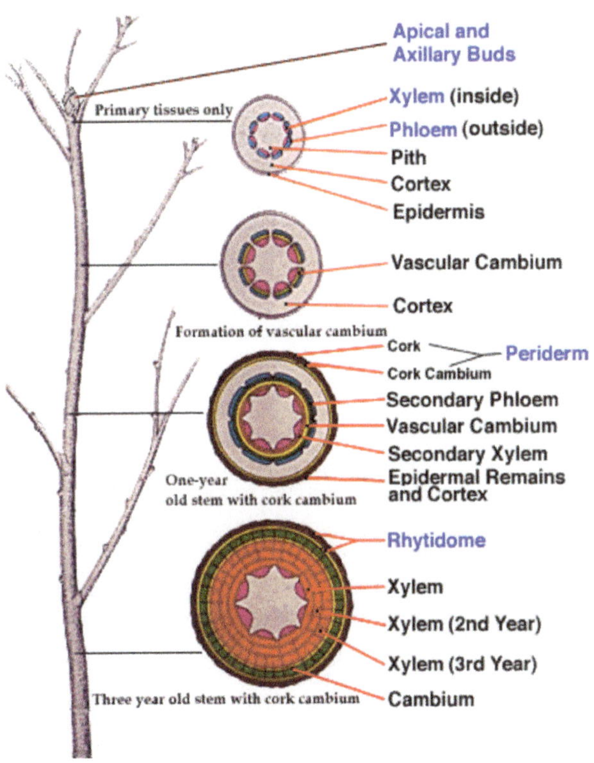

Increasingly, people are separating plant anatomy and morphology, with anatomy being concerned with the internal structure of plants, while morphology involves the exterior appearance of a plant. There are some overlaps between the fields, however. A flower, for example, can be examined by a morphologist and an anatomist, with both being interested in the exterior and internal structures of the flower.

The science of the structure of the organized plant body learned by dissection is called Plant Anatomy (anatomy-dissection). In general, Plant Anatomy refers to study of internal morphology, pertaining to different tissues. It is structure of Angiosperms, with emphasis on primary tissues.

Economic Botany is the interaction of people with plants. Economic botany is closely related to the field of ethnobotany that word is based on two Greek roots: ethnos (race: people: cultural group) and botanikos (of herbs) and can mean the plant lore of a race or people as well as the study of that lore.

Economic botanists are scientists who study the interactions between humans and plants. That makes the field of Economic Botany as far flung and diverse as both the human and plant life on our planet. Economic botanists study human-plant interactions from a variety of different angles. These skilled researchers rely on a variety of disciplines including archeology, sociology, and ecology in addition to basic botany to help them explain these interactions and their effects on plants, society and our dynamic planet.

Economic Botany sometimes focuses on the processes as well as the products involved in plant cultivation. Scientists ask questions about how knowledge of useful plants is acquired and transmitted between groups.

Tissue System

Plant body in Angiosperms is differentiated into root stem, leaf and flower. All these parts are made up of different types of tissues containing different cell types. A tissue is a mass of similar or dissimilar cells performing a common function. The body of a vascular plant is composed of dermal tissue, Ground tissue and Vascular tissue.

Dermal Tissue (Skin)

Dermal Tissue is protective in function. Basing on its origin, it is classified into two types – Epidermis and Periderm.

- Epidermis: This is the primary surface tissue of the entire plant. Epidermal cells are compactly and continuously arranged; the continuity is lost by the presence of Stomatal pores or breaks in the tissue. Covering the aerial epidermis, cutin (fatty substance) is present as an impregnation on cell wall. The cuticle can be separated from epidermis. The epidermis may produce unicellular or multicellular hairy outgrowths and other appendages. Epidermis provides mechanical protection, allows gaseous exchange through stomata, restricts transpiration with cuticle, and is also involved in storage, photosynthesis, secretion, absorption and perception to stimuli. Stomata are Pores, each guarded by two guard cells, which control the size of the pore. Cells surrounding guard cells, but differing from other epidermal cells, are called subsidiary cells. Guard cells are kidney shaped; their cell walls are thick on the inner surfaces. Guard cells contain many chloroplasts.
- Periderm: This is formed during secondary growth replacing primary epidermis.

Ground Tissue

This is inner to dermal tissue and is composed of simple tissues like parenchyma.

Vascular Tissue

Vascular tissue consists of conducting elements – xylem and phloem. Vascular tissue may be scattered in ground tissue or regularly arranged forming a ring. In the latter arrangement, ground tissue is differentiated into cortex (outer to vascular tissue) and pith (inner to vascular tissue). The ground tissue of leaves is called mesophyll, bound by upper and lower epidermis.

Tissue Types

In broad sense, tissues are classified as meristematic and permanent tissues.

Meristem (Meristos – divisible)

Initially all embryonic cells of an embryo have the capacity to divide and multiply but as the embryo develops into a plant body, this capacity for division is restricted to certain parts of the plant body called meristems which are active throughout the life of the plant body (unlike that of an animal body). When meristematic cells divide, a group of the daughter cells remain meristematic; the other daughter cells called derivatives differentiate into various tissue elements. Before the occurrence of any cell division, usually cells become enlarged accompanied with addition of protoplasmic and cell wall material.

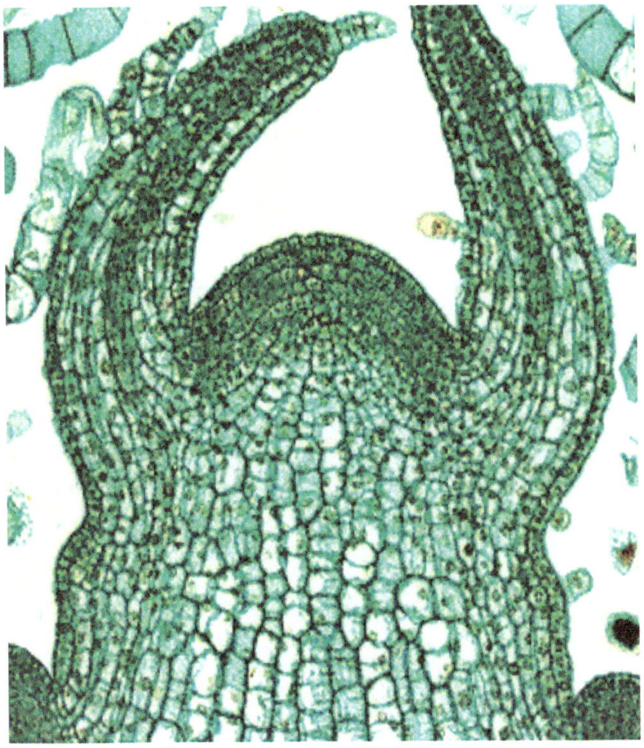

Meristematic cells are isodiametric, compactly arranged with dense cytoplasm, large nucleus, and small vacuoles or without vacuoles. Cell walls are thin.

Meristems which occur at the apices of stem, root and other branches are called apical meristems, which bring about primary growth of the plants, hence also called as primary meristems. In many plants in addition to apical meristems, lateral meristems like vascular cambium, cork cambium, intercalary meristems are found. Lateral meristems are arranged parallel to the sides of organs in which they occur.

Intercalary Meristem: This is also a primary meristem, found inserted between permanent tissues, in the bases of internodes and leaf sheaths of grasses. Sometimes, as in gynophore of groundnut, xylem may be present in intercalary meristem. Wherever stem is jointed, elongation of internodes is due to intercalary meristem. Example: Bamboos. Even prolonged growth of leaves, flowers and fruits may be regarded as an intercalary growth.

Vascular cambium and cork cambium: These are referred to as secondary meristems because they produce secondary tissues, and increase the thickness of the plant body. This process is called secondary growth, seen in dicotyledons and gymnosperms.

Permanent or Mature Tissues

Cells derived from meristems gradually change in their structure, metabolism and chemistry and acquire specialized characters by their various modes of differentiation. Not all the cells totally differ from the meristems. Some cells retain the power of division and others cannot divide. In a strict sense only cells which have lost the power for division must be regarded as permanent tissues, but in a broad sense, cells derived from meristem that have acquired a special function like photosynthesis, secretion, storage are treated as part of matured tissue. There are different types of mature tissues. Example: Parenchyma, Collenchyma, Sclerenchyma, Xylem and Phloem.

Types of Mature Tissues

Parenchyma (Para-beside, enchyma-In poured)

Parenchyma is the fundamental tissue of the plant body. It is found in every part of the plant body like pith and cortex of stem and root, mesophyll of leaves, flesh of fruits, floral parts and even in xylem and Phloem. Cells have thin primary walls and polyhedral shapes. The average number of faces of a polyhedron is 14 and is called as “tetra decahedron”. Cells are compactly arranged or more commonly spaciously arranged with intercellular spaces as in cortex and pith. Cells possess dense cytoplasm and are active metabolites.

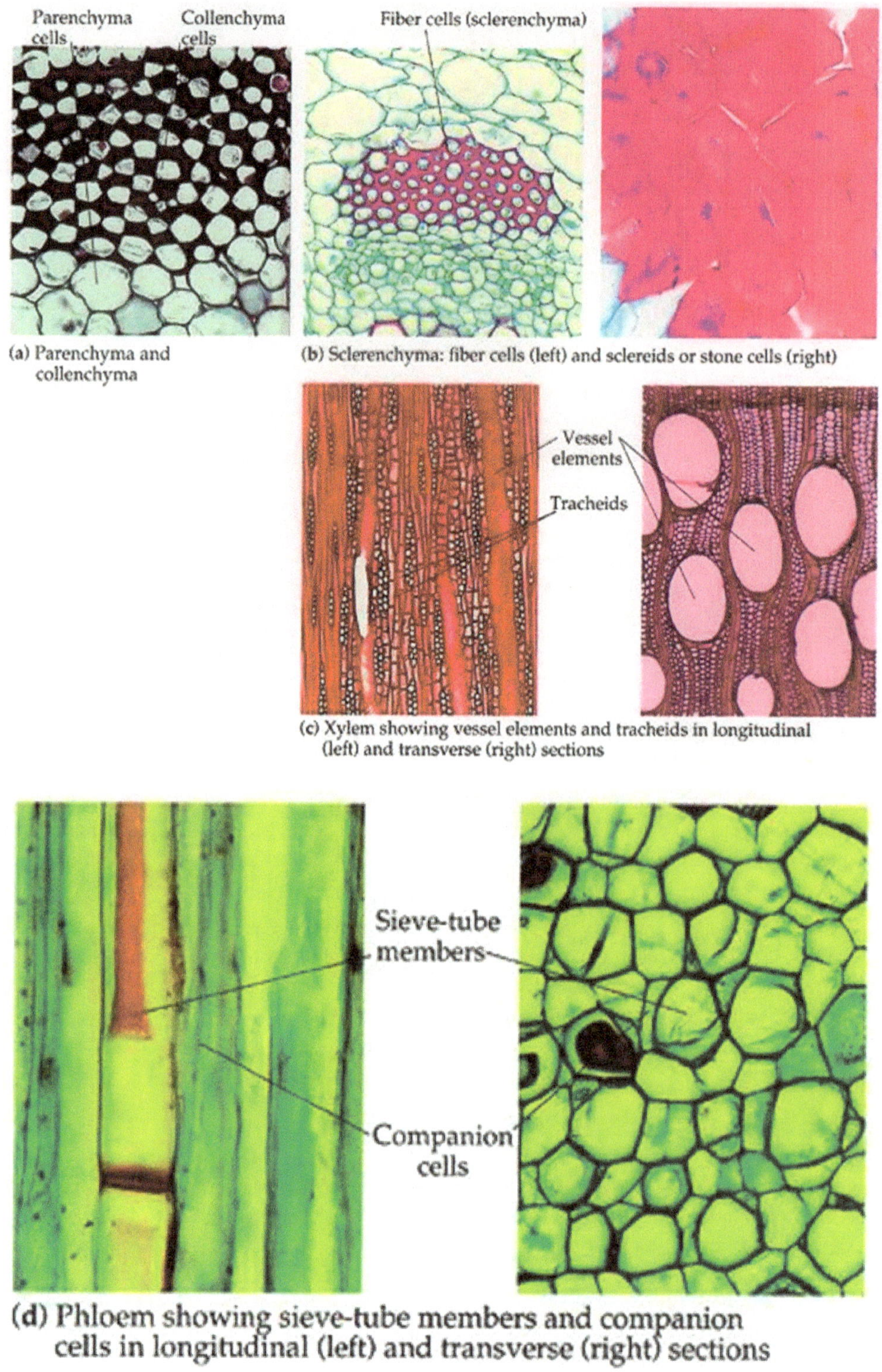

(a) Parenchyma and collenchyma

(b) Sclerenchyma: fiber cells (left) and sclereids or stone cells (right)

(c) Xylem showing vessel elements and tracheids in longitudinal (left) and transverse (right) sections

(d) Phloem showing sieve-tube members and companion cells in longitudinal (left) and transverse (right) sections

Generally cells of parenchyma are involved in storage of starch, sucrose, protein, water, phenol derivatives, many mineral substances, etc. Other metabolisms like respiration, protein synthesis etc., are active. Parenchymatous cells may also perform specialized functions and are structurally modified. The following are the different types of parenchyma.

Aerenchyma: Intercellular spaces filled with air, are large in size and many in number. Cells occupy a smaller area. Though the cells are smaller, they provide the required strength to the aquatic plants. In these plants air spaces are common, helping in aeration and buoyancy. Air spaces are also seen in roots of grasses, petioles of canna, aroids etc.

Chlorenchyma: Cells of photosynthetic parenchyma contain numerous chloroplasts. These cells are commonly seen in leaves, some times in young shoots. Cells of Chlorenchyma are of two types –

- Palisade cells that is elongated and compactly arranged.
- Spongy cells that are spaciously arranged and irregularly shaped.

Prosenchyma (Pros-to): Cells are elongated, have thick walls and serve as a supporting tissue. Example: Endosperm cells of seeds.

Armed Parenchyma: Cells are stellate in shape with inward projections of cell wall as arms. Example: Pinus leaf Mesophyll.

Collenchyma (Kolla – Glue)

Collenchyma is a simple, living tissue. Cell walls are thickened due to deposition of pectin. Collenchyma is the primary supporting tissue in stems, leaves and floral parts of dicots, where as in stems and leaves of monocots collenchyma is usually absent, (instead, sclerenchyma is present in monocots).

Collenchyma is usually hypodermal in position. Cells are more elongated and narrower than parenchyma. Like Parenchyma, collenchyma may also contain chloroplasts or may regain the thickening. Intercellular spaces may or may not be present. Three forms of collecnchyma are recognized based on the types of thickenings

- Thickening is on the tangential wall – lamellar collenchyma.
- Deposition of pectin is in the corners where several cells meet-angular collenchyma.
- Thickenings are around the intercellular spaces – lacunar collenchyma.
- Being plastic, collenchyma differs from elastic fibers of sclerenchyma.

Sclerenchyma (Scleros-hard)

Cells of sclerenchyma are thick walled and are usually lignified. The thickness is due to formation of secondary wall. At least initial the secondary wall is free from primary wall. At maturity, usually the cells are devoid of protoplast. The cell wall encloses a cavity lumen. On the cell wall, pits are usually present (simple type). Commonly sclerenchyma cells are classified into fibers and sclereids. The primary function is mechanical. (Pits-thin areas without secondary wall material deposition).

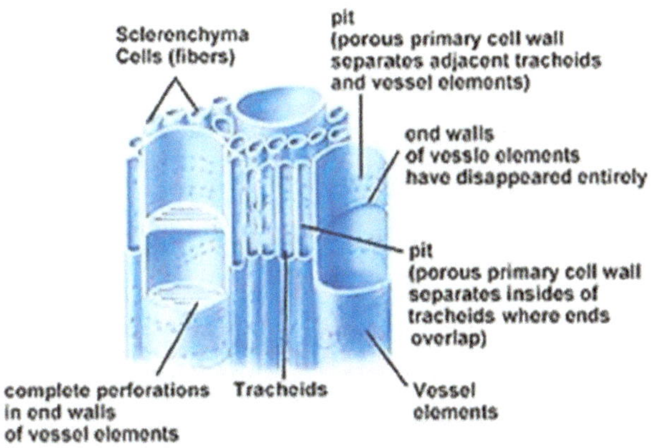

Fibers: These are usually long, spindle shaped structures, with tapering or blunt ends. Longest fiber is seen in Boehmeria nivea (55 cms). They are arranged in groups. Secondary thickening may account for 90% of the area of the cell; the lumen is narrow. Fibers are grouped into xylary fibers and extraxylary fibers. Xylary fibers, also called wood fibers are parts of xylem and are longest among xylem elements. Extraxylary fibers are classified as – Bast or phloem fibers, cortical fibers and perivascular fibers (peripheral to the vascular bundles). These are lignified or non lignified. On account of their elasticity, fibers enable the plant body to withstand various strains. Commercial fibers like jute, flax, ramie are extraxylary fibers.

Sclereids: These are shorter than fibers. Sclereids occur singly or in groups. Sclereids are commonly found in fruit wall, seed coat, epidermal scales, and occasionally found in cortex, pith, mesophyll and petiole of submerged aquatics.

There are many types of sclereids:

- Asterosclereids are star shaped
- Macrosclereids are similar to palisade cells
- Osteosclereids are bone like that are enlarged at their ends
- Brachysclereids are isodiametric like parenchyma

Plant Growth

Plant growth is a phenomenon different from animal growth.

Animas exhibit a growth pattern called determinate growth.

- After fertilization, the zygote cells are rapidly dividing, undifferentiated cells
- However, after a certain critical stage, the cells differentiate and form tissues.
    - From this point onward, their developmental fate is sealed
    - There are exceptions to this (i.e. stem cells in bone marrow)

- Most animals have a pre-programmed body plan (i.e. barring mutation or accident, most humans have 10 fingers and toes, two eyes, a heart with four chambers, etc..)
- Most animals quit growing after a certain age

Plants, however, exhibit a growth pattern called indeterminate growth

- The plant retains areas where rapidly dividing, undifferentiated cells remain all through the life of the plant
- These areas are called meristems
    - Meristematic tissue continues to rapidly divide producing undifferentiated cells which may eventually differentiate to form the tissue and cell types discussed above
- Plants do not have a pre-programmed body plan
    - There are constants like leaf shape and branching patters (opposite, alternate, etc.) but you can never predict where a new branch will come about on a tree.
- Plants continue to grow throughout their life

Meristems

The pattern of plant growth depends upon the location of meristems

Apical meristems

- located at the tips of roots and shoots
- supply cells for the plant to increase in length (grow up for shoots and down for roots)
    - growth in this direction is known as primary growth
    - primary growth found in herbaceous and woody plants
    - primary growth found in monocots and dicots

Lateral meristems

- located near the periphery of the plant, usually in a cylinder
- supply cells for the plant to increase in girth
    - growth in this direction is known as secondary growth
    - found in all woody and some herbaceous plants
    - lateral meristems and secondary growth found only in dicots

CHAPTER 1 PASSIVE FLORA? RECONSIDERING NATURE`S AGENCY THROUGH HUMAN-PLANT STUDIES (HPS)

John Charles Ryan

Keywords: Plants, society, environmental philosophy, human-animal studies

INTRODUCTION

> *What else than the living organism allows one to see and sense true time? For a plant, a form is equivalent to an age—form is linked to size. Time is inextricably and correlatively tied to its life. A year is a node, a layer, a body separated from its surroundings and annexed, added onto, raised, directed, appointed, placed, built upon.*
>
> —French poet and philosopher Paul Valéry [1]

As agents of healing, sources of food and fiber, objects of ornamentation, symbols of inspiration and images of beauty, plants have been integrally connected for ages to human societies worldwide [2]. As the ethnobotanists Michael Balick and Paul Cox assert in *Plants, People, and Culture: The Science of Ethnobotany*, "the very course of human culture has been deeply influenced by plants, particularly plants that have been used by indigenous peoples around the world" [3]. Economic, aesthetic, and medicinal values are only a few of many associated with plants, both in indigenous and non-indigenous societies. Cultivated plants, in particular, have had multiple aesthetic purposes throughout history. In seventeenth-century Holland, for example, the fervid love of flowers galvanized the social and economic furore over tulips known as "tulipmania" [4]. In Australia, according to herbalist and botanical researcher Cheryll Williams, "few truly appreciate the immense contribution that our native plants have already made to our welfare—nor their extraordinary future potential and the need for their conservation" [5].

Citation: John Charles Ryan. Passive Flora? Reconsidering Nature`s Agency through Human-Plant Studies (HPS), Societies, 2012; 2(3), 101-121. DOI:10.3390/soc2030101. ISSN 2075-4698.

As the study of people, plants, and values, ethnobotany has been positioned traditionally in the social sciences as an interdiscipline linking botany and anthropology with an emphasis on indigenous uses of plants [6,7]. Economic botany focuses on the material role of plants across societies as foods, fibers, and objects of commerce. As defined by G.E. Wickens, economic botany is "the study of plants, fungi, algae and bacteria that directly or indirectly, postively or adversely affect man [*sic*], his livestock, and the maintenance of the environment. The effects may be domestic, commercial, environmental, or purely aesthetic; their use may belong to the past, the present or the future" [8]. In contrast, medical botany examines specifically the influence of plant medicines on human well-being [9]. However, in addition to ethno-, economic, and medical botany, there are recent noteworthy developments in the discipline of geography as "human-plant geographies" [10,11]; philosophy and theology as "philosophical botany" [12]; critical theory and literary studies as "critical plant studies" [13]; and cultural studies and transdisciplinarity as "cultural botany" [14].

By and large, utilitarianism is one of the dominant social values associated with plants: as food, fodder, fiber, and medicine; as substances or active ingredients; as the raw material of the natural world acted upon by the social world in a myriad of ways. Non-utilitarian social values associated with plants include aesthetic (e.g., the planting of kangaroo paws at botanical gardens or the use of herbs, such as parsley, as a soup garnish); symbolic (e.g., the use of trees and wheat in heraldry or the long-standing association between roses and love); linguistic and metaphorical (e.g., expressions such as "solid as an oak" and "flexible as a willow"); nationalistic (e.g., the wattle on the Coat of Arms of Australia, as shown in Figure 1); and cosmological and mythological (e.g., the Norse *Yggdrasil* or Tree of Life). For instance, plant creatures figure prominently in J.R.R. Tolkien's Middle-earth fantasy writings, including *The Hobbit*, *The Lord of the Rings* and *The Silmarillion*. For example, in *The Lord of the Rings*, Frodo Baggins is put under a spell of sleep by the wise Old Man Willow. Other plants in the Tolkien legendarium are represented as embodiments of humans, often with magical or healing capabilities [15].

However, in contemporary thought, the view of plants as active and intentional beings is mostly limited to fantasy, mythological, and childrens' literature. Admittedly, such myth-based associations between plants and human life, prosperity, and death do not provide an adequate basis for establishing plant intelligences, although these kinds of myths do affirm the cultural relevance of plants across diverse societies. However, human-plant geographies, critical plant studies, philosophical botany, and cultural botany, as four significant developments in the way in which plants are regarded and researched socially and culturally, offer a way forward for contemporary perceptions of plants. These four approaches to plants differ markedly from ethno-, economic, and medical botany in their renewed conceptualization of plants as autonomous, agentic beings, rather than as mute materials or mere messages. In *Environmental Culture*, the Australian environmental philosopher Val Plumwood proposes a new model for thinking about and studying the non-human world, one which would regard nature as an *agent in*—rather than an *object for*—the production of knowledge and socioecological practices. The paradigm envisioned by Plumwood points to an ethics of dialogue between a researcher and an autonomous subject of research. Plumwood's approach would necessitate "sensitive listening and attentive observation [and] an open stance that has not already closed itself off by stereotyping the other that is studied in reductionist terms as mindless and voiceless" ([16], p. 56).

Figure 1. The Coat of Arms of Australia is the recognized symbol of the country. Initially granted to Australia by King Edward VII in 1908, this version was granted in 1912 by King George V. The Coat consists of major floral and faunal symbols of Australian nationhood, including the Golden Wattle (*Acacia pycnantha*) on which the kangaroo and emu are posed. This is one example of the symbolic use of plants to represent national identity. (Image source: WikiMedia Commons).

As Plumwood's argument goes, a dialogical model of knowledge-making is a response to the subject-object hierarchies—promulgated by conventional science-based research including ethno-, economic and medical botany—between a disengaged knower and a passive known (or a yet-to-be-known). Identifying this subject-subject knowledge paradigm as "an uncompleted task for modernist science," Plumwood posits an urgent ecological and intellectual need to "recast the nature-devaluing aspect of rationalism" ([16], p. 55). According to Plumwood, an ethical science—evolving from a reconceptualization of the subject of research—would at the same time help to redress the hyper-separation between the humanities and the sciences, thus bringing forward an "ecological humanities" ([16], pp. 50–56; [17,18,19,20,21]).

Plumwood's call is perhaps nowhere more relevant than to the study of plants—organisms conventionally regarded as automatons; as voiceless objects worked over by animate agents such as animals; as mindless constituents of the biosphere or landscape; and as mere material for physical and aesthetic consumption [11,12,22,23,24]. As Matthew Hall argues trenchantly in *Plants as Persons: A Philosophical Botany* (2011), plants are regarded ordinarily in scientific and social scientific discourses as passive, and are assumed to lack intelligence, consciousness, and behavior—those qualities defining animality [12,25,26,27]. As living beings, plants continually transact with their socioecological milieus, as the histories of agriculture, horticulture, silviculture, and biological invasions attest. Indeed, plants constitute certain social practices and customs as well as the ethics surrounding them. However, the agency of plants and their particular intelligences have yet to figure fully into the study of social practices and knowledge production. As suggested above, ethnobotany and its affiliates are dominated by a view of plants as food, fodder, fibers, and medicines. Of course, societies *act upon* plants: as weeds to expunge, as forests to harvest, as crops to trade, as seeds to plant, as ornamentals to accompany, as backdrops to ongoing social dramas. Houle asserts that the "backgrounding of herbality—indeed of ecology—is directly linked to the foregrounding of animality. It is a gestalt operation" [28]. While plants are more active and autopoietic than they have been depicted as in literature, art, and science, there are obvious fundamental differences between plants and animals—especially in terms of mobility, psychology, and emotions—that ought to be recognized. However, as recent research in the plant sciences argues, the differences between plants and creatures normally regarded as sentient are neither as profound as previously thought nor as substantial as purported for maintaining the predominant attitude towards plants as disposable materials, ecological automatons, or static backdrops to human desires.

What if we were to consider how plants *act upon* us, contributing to the co-generation of our cultural practices, values, perceptions, relations, artifacts, and all else through their volitions in the *umwelt* of which all living beings are part? The outlandishness of the question belies a deeply engrained assumption that, of course, plants do not produce *anything* but are shrewdly used by intelligent beings to produce the things of culture. The apparent foolishness of the question reaffirms the socially and scientifically indoctrinated position—the anthropocentric and zoocentric assumption—that plants cannot bear agency because they do not have brains. In his popular account of the reciprocality between plants and people, *The Botany of Desire: A Plant's-Eye View of the World*, Michael Pollan comments on the language used to describe human-vegetal relationships: "Even our grammar makes the terms of this relationship perfectly clear: *I choose the plants, I pull the weeds, I harvest the crops.* We divide the world into subjects and objects, and here in the garden, as in nature generally, we humans are the subjects" ([29], p. xiv). I extend Pollan's argument that, rather than a condition of people *acting upon* plants, it is more productive to consider global phenomena, such as agriculture, horticulture, and silviculture, as co-generated by people and plants in a shared *umwelt* reflecting a non-individuated, non-zoocentric form of intelligence.

The observation that people assimilate plants to social practices and customs reflects only one side of the reciprocality that characterizes human-vegetal relationships; it entails only the single condition of humans *acting upon* plants, or in Pollan's words, "I choose the plants, I pull the weeds." Human-plant studies (HPS) would consider the obverse condition as well, that of plants *acting upon* humans to co-generate our milieus of sustenance, the places in which our everyday lives run their courses. The recognition of this dual aspect of the human-vegetal dynamic lays the groundwork for the more contentious discussion of intelligence(s). However, considered to lack intelligence and volition, plants have been construed as inactive objects—decorative, aesthetic, and utilitarian (see Figure 2). As some theorists argue, the narrow ascription of intelligence to animals can be attributed to "brain chauvinism," the assumption that thinking, volition, and communication are the qualities of "brained" organisms ([30], pp. 6–11). As Hall and Hallé argue, this zoocentric bias predicates intelligence on animal definitions ([12]; [25], p. 29). Indeed, the study of flora, across the humanities, social sciences, and biological sciences, reflects the assumption that plants are *acted upon* constituents—rather than *acting* partners—in socioecological networks. As a consequence, the notion that plants could possess ethical standing—as an outcome of their autonomy as demonstrated by recent scientific insights—is "downright silly" according to a recent article in *Nature* [31].

Figure 2. *Passiflora laurifolia* by Francisco Manuel Blanco (1880–1883). A species in the family Passifloraceae, *Passiflora laurifolia* is also known as Jamaican honeysuckle. An invasive plant in many parts of the world, it is native to Central and South America. Although often visually captivating, botanical illustration tends to communicate to a viewer only one phase in the life cycle of a plant and often neglects to depict the complex biological and cultural environments of which the plant is part. Representations such as these tend to fix living beings in space and time, reinforcing the perception of plants as static objects or "passive flora". (Image source: WikiMedia Commons).

The object-making discourses intrinsic to the humanities, social sciences, and biological sciences devalue the agencies of plants. In light of this, it is imperative to interrogate our social, cultural, and scientific assumptions about the botanical world. Despite their marginalized and passive position in Western worldviews, plants constitute "more than 99% of the earth's biomass" ([32], p. 463). Plant physiologist Anthony Trewavas, a prominent proponent of floristic intelligence, observes the capacity of plants for adaptive behavior; they "possess a fine ability to adjust and optimally exploit the local environment" [33]. Moreover, a recent body of writings characterizes plants as agentive, autonomous, and intelligent subjects and considers the wide-ranging implications of research into plant intelligence and consciousness, pioneered most notably by Charles Darwin in the 19th century [12,26,27,34,35,36].

The aim of this article is to collect these embryonic pieces of theory, science, and historical exegesis—scattered across the humanities, social sciences, and biological sciences—under the heading of human-plant studies (HPS), an interdisciplinary research framework which would regard plants as intelligent, conscious, and capable subjects, rather than mechanical accessories—as "passive flora." Building upon the conceptual positions of environmental philosopher Val Plumwood, anthropologist Deborah Bird Rose, philosopher and botanist Matthew Hall, and plant scientists such as Anthony Trewavas, as well as the precedent of human-animal studies (HAS), the basis for the ongoing development of human-plant studies through various frameworks will be posited. The implication of this framework is the reconsideration of plants in all forms of research, including the role of plants in society and their representations in the humanities. I will address the following keystones of human-plant studies: (1) plants are intelligent and volitional organisms; (2) plants are integral to socioecological networks and practices; (3) plant intelligences are viable exemplars for societies, cultures, and communities; (4) the roles of plants in society are best articulated through interdisciplinary research that considers art, literature, philosophy, Indigenous knowledges, and science; (5) the HPS framework will complement existing paradigms of ethnobotany and its affiliates.

THE PARADOX OF ACTIVE ANIMALS AND PASSIVE PLANTS

Although biologically alive, plants have been conceptualized as unthinking assemblies which only react when acted upon. Indeed, many of the dominant cultural metaphors for passivity are botanical. For example, to plant one's feet; the set down roots; and to be vegetative imply a limited capacity for movement and intelligence. A "vegetable" is someone trapped in a mindless state—verging on death—without the ability to reason or feel. Trewavas asks rhetorically, "Do plants exhibit intelligent behaviour? The use of the term 'vegetable' to describe unthinking or brain-dead human beings perhaps indicates the general attitude" ([24], p. 1).

In the same regard, geographers Russell Hitchings and Verity Jones note a paradox. Although represented as passive elements of the environment, plants are also the quintessence of dynamic nature, epitomizing the ancient notion *physis*: "Vegetation is something passive in contemporary understanding: to be in a vegetative state is to be without mind. Yet the root meaning of the word 'vegetative' is associated with activity and enlivened animation" ([11], p. 15). Indeed, the late 14th century word "vegetative" suggested "endowed with the power of growth," but it has been associated pathologically with brain death only since 1893 [37].

The assumption that intelligence is an attribute of animals and that plants are "mindless and voiceless" ([16], p. 56; [38]) is not endemic to the social or biological sciences. In works of art, plants are often cast as passive objects of aesthetics; wildflowers in particular have widespread popularity—though limited agency and intelligence—in many forms of art and literature [39] (see Figure 1 and Figure 2). D.H. Lawrence and Mollie Skinner's depiction of the Western Australian red gum or marri (*Corymbia callophylla*) in *The Boy in the Bush* (1924) presents one example: "Red gum everywhere! Fringed leaves dappling, the glowing new sun coming through, the large, feathery, honey-sweet blossoms flowering in clumps, the hard, rough-marked, red-bronze trunks rising like pillars of burnt copper" [40]. The passage depicts living trees in static terms of color and form to create aesthetic objects—"pillars of burnt copper".

While recent scholarly work in literary criticism and animal studies has produced "zoocriticism" to research the linkages between animals, literature, and culture [41], an analogous humanities-based field has yet to emerge for plants. (And would it be called *phytocriticism*?) How are the humanities and social sciences to couple new scientific understandings of flora to the criticism and creation of art, literature, and social practices involving plants? How can the humanities and social sciences help to resolve a widespread societal perspective "that regards plants basically as automatons" ([24], p. 2).

As a tentative response to these questions, I offer some background. In critical theory, which has affected greatly the course of contemporary humanities and social sciences, the rhizome (of plants) has become a significant metaphor for lateral thinking that cross-cuts subject-object binarisms (see Figure 3). For Deleuze and Guattari, the rhizome encapsulates postmodern yearnings because it "has no beginning or end; it is always in the middle, between things, interbeing, intermezzo. The tree is filiation, but the rhizome is alliance, uniquely alliance" [42]. The rhizome counters the growth-focused "arborescent" tropes of modernity by "burrowing through substance, fragmenting into simultaneous sprouts, moving with a certain stealth, powerful in its dispersion" [43].

Figure 3. The Mangles Kangaroo Paw (*Anigozanthos manglesii*) is the floral emblem of Western Australia and has adapted to its environment through an vigorous rhizomal system that enables the plant to endure extreme climatic fluctuations including drought and fire. The interest in the rhizome, as a potent intellectual metaphor in critical theory and postmodernism, is well-suited to the Mangles Kangaroo Paw, an enduring expression of environmental beauty and resilience in Western Australia. (Image source: Author).

The rhizome signifies the broad, subtly perceptible time scales over which plants evolve and exist. Valéry intimates this with his statement "time is inextricably and correlatively tied to [a plant's] life" ([1], p. 107). Carl Jung, as well, commented that "life has always seemed to me like a plant that lives on its rhizome. Its true life is invisible, hidden in the rhizome. The part that appears above ground only lasts a single summer…What we see is the blossom, which passes. The rhizome remains" [44]. In contrast to animals, plants exhibit liminal temporality—an interbeing, *intermezzo*—between immediately mobile mammality and relatively immobile geology. The difference of time scales, observes Trewavas, "frequently makes plants *seem unmoving* [italics added]" ([24], p. 1).

Research into plant communication and behavior—which counters the notion of "passive flora"—commenced in North America in the early 1980s ([28], p. 98). This scientific research increasingly characterizes plants as autonomous agentic beings. The notion of intelligent plant behavior, however, can be traced back to Charles Darwin. In *The Power of Movements in Plants* (1880), Darwin and his son Francis described the embryonic root or the plant radicle as a brain:

> *It is hardly an exaggeration to say that the tip...acts like the brain of one of the lower animals; the brain being seated within the anterior end of the body, receiving impressions from the sense organs, and directing the several movements.*
>
> [45]

Charles Darwin and his son noted that the root tips perceived environmental stimuli and responded in particular to touch, moisture, gravity, and light. Furthermore, the radicle could discrimate between stimuli by selecting a growth response that would most benefit the whole plant. Plant scientist Peter Barlow writes "the complete set of tips endows the plant with a collective brain, diffused over a large area, gathering, as the root system grows and develops, information important for plant nutrition and survival" [46]. Rather than mechanical tropic movement towards stimuli as in phototropism or geotropism, the radicle exhibits "root nutation," which for Darwin described the volition of the radicle brain, signified by its discriminatory ability. The plant brain is located diffusely in its "circumnutating root," a system which can identify self and non-self roots, further suggesting the idea of volition, intention, and self-directedness in plants ([46], p. 48).

Drawing from the observations of the Darwins, plant neurobiology has recently been established as a discipline within the plant sciences. The premise of plant neurobiology is that plants are intelligent organisms, endowed with the capacity to learn, to communicate, and to make key decisions regarding their survival in the face of particular environmental constraints. Research increasingly confirms that plants use a variety of volatile compounds to thwart insects and other predators, but also to communicate with other plants within an ecosystem [47]. In fact, the immobility of plants relative to animals has generated a battery of adaptive strategies deployed as deliberate survival choices. These strategies, moreover, include the capacity for remembering environmental stressors and factoring these memories into future decisions [48]. Trewavas defines plant intelligence as "adaptively variable growth and development during the lifetime of the individual" ([24], p. 1). He characterizes the particular intelligence of plants as "mindless mastery" to intimate that plant intelligence exists without a central brain or mind, but rather as a diffuse system of intelligence located loosely in the roots (and signified by the phenomenon of root nutation, as mentioned above) [27]. An extensive review of current research in plant intelligence is out of the scope of the present discussion, but can be found in key works such as *Communication in Plants: Neuronal Aspects of Plant Life* (editors Baluška, Mancuso, and Volkmann). However, the argument that plant intelligence is located in the roots reflects a zoocentric gravitas amongst plant physiologists to seek a brain analogue in plants or, in other words, a single organ or physical repository of intelligence governing the volitions of the whole organism. The kind of botanical intelligence I advocate is, indeed, diffuse, but is not the property of certain plant tissue located in discrete anatomical regions and manifested through pheromonal mechanisms. Drawing from biosemiotics, I will look towards the*umwelt* of the plant in defining intelligent behavior in relation to the condition of reciprocality with people and environment. It is only through this form of intelligence that plants have adapted to their environmental constraints over remarkable expanses of space and time.

Only recently have social and cultural studies of plants considered the scientific advances of the last thirty years. Of significance is Matthew Hall's *Plants as Persons* (2011) [12]. After tracing the notion of plant "personhood," or the moral consideration of plants in social practices, in various traditions from Christianity to Neopaganism, Hall synthesizes scientific research in plant sensation, movement, signalling, communication, intelligence, and learning to build a well-grounded, interdisciplinary argument for the reconsideration of plants ([12], pp. 137–156). As Hall argues elsewhere, the implications of intelligence (here used to include the capacity for sensation, signalling, learning, and communication) are far-reaching; plants are "sensitive, active, self-governing organisms" ([22], p. 173), "intentional, intelligent agents which actively try to maintain and perpetuate their own existence" ([12], p. 156). Similarly, Trewavas pinpoints a turning point "sweeping away the detritus of passivity, replacing it with an exciting dynamic—the investigation of plant intelligence" [27]. The intelligence and self-directedness of plants could catalyze a comparable revolution in the way in which we study environment and society.

Moreover, the intelligence of plants manifests in their transactions with other organisms as part of self-governing socioecological systems, or what Hoffmeyer calls in the context of biosemiotics "this infinite swarm of swarming swarms" or the "inner semiosphere," relating individual bodies and physiologies to the collective intelligences produced by multiple individuals working in symphony over durations of time in expanses of space ([49], p. 125). Hoffmeyer defines the semiosphere as:

> *Like the atmosphere, the hydrosphere, and the biosphere. It penetrates to every corner of these outer spheres, incorporating all forms of communication: sounds, smells, movements, colors, shapes, electrical fields, thermal radiation, waves of all kinds, chemical signals, touching, and so on. In short, signs of life.*
>
> [49], p. vii

As part of the semiosphere, plants can be seen to demonstrate highly integrated forms of intelligent behavior without exhibiting brain physiology analogous to animals. Not only is the biological model of intelligence predicated on zoological rather than botanical definitions, it further assumes a strongly individuated organism with the capacity to exploit its environment and manipulate other species in the Darwinian sense of survival fitness. It is most optimal, then, to regard the particular intelligence of plants as a dual mode comprising individual expression—the volition, behavior, and adaptation of herbs, shrubs, and trees on the level of single organisms—coupled to the manifestation of these same traits collectively by communities and populations of plants in ecosystems. The consideration of this acutely expressed dual mode of plant intelligence can liberate human-plant studies (HPS) from the erroneous assumption that, in order to argue for intelligence in the botanical world, researchers only need to identify the "brain-like" or neuronal features of plant anatomy. Recent plant research shows that, not only have neuronal molecules been identified in plants, but also that some of the same neurobiological processes in animal brains are present in plants.

In countering the notion of plants as passive and somewhat fixed objects of science and culture that lack brains, botanical research across disciplines could consider principles of process and their implications for human understandings of plants as non-passive, as agentic and active participants in socioecological systems. The plant morphologist Rolf Sattler identifies how stasis—stillness, inertia, synchronicity—figures into contemporary understandings of plants. He transfigures the binary between stasis and dynamism by characterizing plant structures themselves as processes in perpetual action rather than fixed material manifestations:

> *Structure tends to be considered static, whereas process is dynamic. If we mistake the map for the territory, we conclude that plants consist of structures within which processes occur. On closer inspection we learn, however, that what appears static is in fact also dynamic.*
>
> [50]

Extending Sattler's ideas here, one could say that plants inhabit the world dynamically and liminally at the margins between being and becoming; stasis and process; synchronicity and diachronicity; and visual representation and bodily experience. Sustained sensory awareness of flora in a place implies that the "map"—the static appearance of the greenness of plants in botanical imagery, for example—is not the "territory"—the place of bodily interaction, the social milieu where people transact with plants in multiple, commonplace ways. In contrast to an atomistic philosophy of nature as an aggregation of stable things, a process perspective elicits "temporality, historicity, change, and passage as fundamental facts" [51]. Rather than biochemical extracts or anatomized parts, plants can be defined by their connectivities to dynamic events and other mutable beings. As plants change before our senses, so too do our bodies and social practices in relation to the plants. Process as a guiding principle in botanical research taps into the capacity of plants for complex and dynamic behaviors independent of the zoocentric model of intelligence.

Positively for the view of plants as agentic and autopoietic, Hoffmeyer's interpretation of biosemiotics rests on the assertion that "living creatures are self-referential, they have a history, they react selectively to their surroundings, and they participate in the evolutionary incorporation of the present in the future" ([49], p. 51). Hoffmeyer argues for "a theory on organisms as subjects" and, in doing so, stresses the importance of Jakob von Uexküll's *umwelt* or lifeworld theory. The*umwelt* is where interpenetration between an organism and its environment happens as

> *The conquest of vital aspects of events and phenomena in the world around [the creature], inasmuch as these aspects are continually being turned—by way of the senses—into an integral part of the creature. The umwelt is the representation of the surrounding world within the creature.*
>
> [49], p. 58

Importantly, the *umwelt* "allows the creature to become a part of the semiotic network found in that particular ecosystem" ([49], p. 58), including the sounds, smells and sensations of the environment as these stimuli impact reciprocally upon individual organisms and intelligent systems. A population of living things inherently dwells in a "semiotic niche" ([49], p. 59). Significantly, biosemiotics, according to Hoffmeyer, permits interpretations of intelligence independent of zoocentric models. Biosemiotics:

> *Releases the genie of reason from the well-guarded bottle which we know as the human brain and accords it an immanent position in the natural history fairy tale. This move enables us to unite the two separate spheres: Cultural history runs parallel to natural history; at one time they were one and the same.*
>
> [49], p. 95

Notably, Hoffmeyer describes "swarm intelligence," implying that "rather than the brain being pre-programmed to produce intelligence, intelligence seems to swarm out of it" ([49], p. 114). Insofar as the model can get past the location of intelligence in a brain, Hoffmeyer's biosemiotics appears to offer a way through the predominant paradigm of animal intelligence that has proved the basis for judging, and subsequently dismissing, plant intelligence: "Only animals have nervous systems and brains; these have never been found in plants—and from the dawn of evolution their purpose has been to control bodily actions, behavior" ([49], p. 47). On the diffuse intelligence of plants, Marder comments:

> *Each plant in its singularity is a collective being, a loose and disorganized assemblage, and, hence, a community of plants that do not comprise a unified whole, do not constitute either an individual or an organism. In vegetal beings, life is de-centered—not, as some might think, concentrated in the vital 'organ', the root, but dispersed and disseminated throughout the body of plant communities.*
>
> [52]

HUMAN-PLANT STUDIES (HPS): ADDRESSING THE QUESTION OF THE PLANT

Drawing from intellectual currents such as biosemiotics, which reconfigures the separation between nature and culture, human-plant studies (HPS) would become the site for the revolution—an interdisciplinary framework that would suffuse the arts, humanities, and social sciences with scientific awareness of plants as intelligent beings, as outlined in the previous section. In an issue of the journal *Antennae* titled "Why Look at Plants?" Giovanni Aloi asks provocatively:

> *Would there be productive opportunities in attempting to understand plants from different perspectives, just as the field of human-animal studies has proved possible with animals? What contributions to our understanding of animals could a focus on plants make? Could we even envisage that, in the near future, we may have a field of human-plant studies?*
>
> [53]

A precedent for HPS exists in human-animal studies (HAS), focusing on animals and human-animal relations and differing to the anthropological subdisciplines ethnobiology and ethnozoology. Armstrong and Simmons identify "the cultural, philosophical, economic and social means by which humans and animals interact" as central to HAS research [54]. In similar terms, Shapiro and DeMello note the "question of the animal" that has been at the centre of HAS since its inception in the early 1990s: "Why do we think about animals in the ways that we do?" [55]. A corresponding question would guide human-plant studies: "Why do we think about plants in the ways that we do?"

Definitions of HAS emphasize its interdisciplinarity, akin to other "studies" with radical political origins—environmental, women's, and cultural, for instance ([55], p. 6). HAS is "a rapidly growing interdisciplinary field that examines the complex and multidimensional relationships between humans and other animals. [It comprises] psychology, sociology, anthropology, political science, history, literary criticism and other disciplines in the humanities" [56]. HARN, the Human-Animal Research Network, characterizes HAS as critical of the history of "human" and "animal." HAS addresses, for example, "why some animals have protective legislation and others do not; why some of us eat particular animals but not others; why being 'an animal' is akin to denigration, and what cruelty to animals says 'about us'" [57].

As a common subject of analysis, animals and the question of animality and society define HAS. The field addresses the "lack of scholarly attention given to nonhuman animals and to the relationships between human and nonhuman" ([55], p. 2). HAS investigates human relations to animals and the experiences of animals as autonomous creatures, not merely as "cultural artifacts, symbols, models, or commodities in a largely human-centred world" [58]. According to recent HAS theorists, in order to make progress as a scholarly field, HAS will need to identify "ways of understanding animals and human-animal relations that are not constrained by traditional disciplinary boundaries and methods" [59].

The strength of HAS is its receptivity to scientific perspectives—biology, zoology, and ethology in particular—without its possible constriction by empirical methods solely ([59], p. 3). Freeman and Leane advocate a transition within HAS from multidisciplinarity—in which disciplines comingle without significant integration of knowledges and methods—to interdisciplinarity where "scientists, social scientists, and scholars in the humanities collaborate intellectually" [60]. The steady growth of human-animal studies in universities and scholarly journals suggests the timeliness of these approaches to the question of the animal ([55], p. 3). An April 2012 article in *The Washington Post* concludes that HAS is in transition from a fringe activist movement to an increasingly established scholarly field [61].

Extending the HAS precedent, human-plant studies (HPS) would redress the lack of scholarly focus given to plant intelligences, as well as secular or sacred human-plant interactions. Comprising "more than 99% of the earth's biomass" [32], plants are integral to culture, society, literature, and art—not only ecology and scientific awareness. As an example, in the biodiverse South-West of Western Australia, botanical imagery proliferates in visual culture—contemporary wildflower photography [62]; endemic symbols of statehood [63] (see Figure 2); and the works of botanical artists [64,65,66]. HPS would identify innovative ways of understanding plants and people-plant relations. It would do so through knowledge of empirical advances but without the limitations of disciplinary parameters, as HAS strives to do with increasing recognition and influence. HPS would engage an interdisciplinary dialogue with botany, conservation biology, forestry, agronomy, horticulture, sociology, cultural studies, medical herbalism, ethnobotany, and landscape architecture, for example.

In light of advances in scientific understandings, HPS would regard the autonomy of plants and their bearing in social practices—not merely their relevance as cultural symbols or aesthetic accessories. The further negotiation of the moral standing of plants would also be included within HPS [12,22,67,68]. What constitutes cruelty to herbs, shrubs, and trees? What does the unbridled overuse of flora for human physical and aesthetic consumption say "about us?" How should we reconsider plant and human interactions through the prism of recent advances in conceptualizing plants? Moreover, HPS would re-examine kinship and connectivity between plants and society, including the paradigm of plants as teachers [69]. In Fremantle, Western Australia in 1833, to offer a colonial example again from the biodiverse corner of Australia, the Austrian botanist Baron Charles von Hügel remarked in his journal "I could not wait to get out into Nature, into my *kindred* world of plants, which had so often held my gaze when bowed down by deep sorrow" [italics added] [70]. A feeling of kindredness, reflected here by von Hügel, implies that plants are much more than inert backdrops to human affairs, as Matthew Hall, Anthony Trewavas and other scholars in various fields have already argued in convincing detail.

THE WEST AUSTRALIAN CHRISTMAS TREE THROUGH A HUMAN-PLANT STUDIES LENS

In order to glimpse how the HPS framework can be applied to actual plants, I offer a brief example comprising traditional Indigenous Australian Dreaming knowledges and contemporary scientific understandings. According to ethnobotanist Philip Clarke, "the 'Dreaming' is an Aboriginal English term used to embrace indigenous religious beliefs about the Creation and the laying down of Aboriginal Law" ([7], p. 23). Agency in the botanical world is perhaps best exemplified through the Aboriginal concept of Dreaming plants, which facilitate strong, long-standing ontological links between plants and people. Sacred plants are often recognized as ancestral beings themselves or as deeply implicated with the spiritual passage of human beings to the after world. As Clarke goes on the explain, "even when plants do not represent the actual Ancestors, they may be seen as having been involved with Dreaming activities in the Creation" ([7], pp. 23–24). Some plants have clear utilitarian roles as foods, fibers and medicines in Aboriginal societies (and have thus been used intelligently and sustainably by those societies). Others have *perceived* agency as Creation beings actively contributing to the generation of cultural beliefs, practices, and traditions; and/or *actual* agency in their resilient modes of adapation to the socioecological landscapes of which they are part. Indeed, some plants are known as powerful, sacred, and vital plants to Aboriginal societies: agentic, autonomous, and volitional Creation beings, qualities of the botanical world that have only recently been corroborated by plant science. A prominent example is the West Australian Christmas Tree—known as *mudjar* in the Aboriginal Australian language Nyoongar and *Nuytsia floribunda* to scientists—is endemic to the South-West corner of Western Australia. The plant has numerous significant ecocultural meanings [71]. Modern botanical science classifies *Nuytsia*as a mistletoe endemic to the South-West region. As a root and rhizome hemi-parasite, the plants parasitizes a number of hosts but also has the ability to make its own food through photosynthesis [72]. The parasitizing rootlets coming off the main roots of the *Nuytsia* are so tenacious that they have been known to burrow into underground utility lines.

In the 1930s, the ethnographer Daisy Bates [73] noted the close associations between Nyoongar (or Bibbulmun) spirituality and the Christmas Tree:

> *The tree-Moojarr, or Moodurt...was to the Bibbulmun [Nyoongar] the 'Kaanya Tree' 'the tree of the souls of the newly dead'. From time immemorial the soul of every Bibbulmun rested on the branches of this tree on leaving its mortal body for its heavenly home, Kurannup, the home of the Bibbulmun dead which lay beyond the western sea.*

Nuytsia facilitated the passage of souls to the after world, but as Bates claimed, the tree was feared and avoided. Other early ethnographers recorded the use of *Nuytsia* as food, water, and decoration, suggesting that within the South-West, there have been variable spiritual beliefs and cultural practices surrounding the tree [74].

The colonial diarist and settler George Fletcher Moore described the Christmas Tree as "*Mut-yal*, s.-*Nuytsia floribunda*; colonially, cabbage-tree. The only loranthus or parasite that grows by itself. Another anomaly in this land of contradictions. It bears a splendid orange flower" [75]. As the world's largest parasitic plant, the Christmas Tree epitomized the baffling growth habits of New World species in the eyes of European botanists. It represented the departure of the Australian landscape from Old World norms. Even the name "Christmas Tree," flowering in yellow as it does in the heat of the late spring and early summer months of November and December, runs contrary to the image of the evergreen Christmas tree brought indoors from the cold and deep snow of the English countryside.

Early European colonists to Western Australia reported mixed admiration and disdain for *Nuytsia*. The tree in flower was first recorded by the crew of Dutch explorer Pieter Nuyts's vessel Gulden Zeepaard in 1627 [74]. *Nuytsia* was assigned as its scientific name in 1831 by the botanist Robert Brown and the tree was referred to as "Fire Tree" [76] amongst the colonists. In the journals of Western Australian surveyors Alfred Hillman and Septimus Roe, *Nuytsia* indicated infertile country and was described disparagingly as part of the intolerable scrubbiness of the landscape [71]. In 1880, peripatetic artist Marianne North painted "Study of the West Australian Flame-tree or Fire-tree," now part of the botanical art collection at Kew Gardens in England. However, the tree depicted in the painting appears more like a European elm tree with its pleasant vase-like form than most Christmas Trees found in the Western Australian wild (see Figure 4).

Figure 4. Marianne North's "Study of the West Australian Flame-tree or Fire-tree" (1880). *Nuytsia floribunda*, or the West Australian Christmas Tree, is represented here as a European elm with a fluted canopy and fairly symmetrical overall form. Human-Plant Studies, as a framework for researching *Nuytsia* and other plants, would invoke interdisciplinary perspectives on the species in question, while also exploring the agency of plants in contributing to the formation and cohesion of human societies and cultural practices: Indigenous, colonial and post-colonial. (Image source: Reproduced with the kind permission of the Director and the Board of Trustees, Royal Botanic Gardens, Kew).

For some Aboriginal Australian people, the Christmas Tree is considered a sacred, Dreaming plant linked to the souls of the deceased. The writings of early settlers and ethnographers point to some of the spiritual beliefs and material practices surrounding *Nuytsia*. In the 1930s, Bates reported the view of *Nuytsia* as a home for disembodied spirits when she wrote that the Nyoongar people "did not fear the tree; they loved it, but held it sacred for its spiritual memories. The souls of all their forbears had rested on the spirit tree on their way to Kurannup" ([73], p. 153). However, writing in the 1880s, diarist and settler Ethel Hassell reported the use of *Nuytsia* root as a candy:

> *They gave me one of the roots to taste, telling me it was called mungah. The outer skin was pale yellow but easily stripped off leaving a most brittle centre tasting very like sugar candy.*
>
> [77], p. 26

A ghoulish creature called a *gnolum*, in the form of a very tall, very thin man, enticed boys away by offering them the roots of the *mungah* tree ([77], p. 65). In traditional Nyoongar thought, the *mungah tree* is literally ensouled, bearing the spirits of the deceased and facilitating their passage to the next life. The tree is not simply the passive material or medium for the forging of cultural artifacts or practices. Indeed, the sacred tree is *acted upon* as a sugar candy eaten or as decoration worn for ceremonial purposes. However, in the tree's capacity to liaise with the spiritual world, to invoke fear and respect, and to associate with other-worldly figures, it manifests some degree of agency synchronized to the active processes of its social, spiritual, and ecological milieu—its broader *umwelt*. Hence, human-plant studies reflects on Indigenous knowledges of plants in conjunction with scientific and Western cultural knowledges to identify points of commonality and *cross-pollination*.

RECONSIDERING THE ROLE OF PLANTS IN SOCIETY THROUGH HPS

What intellectual niche would human-plant studies serve and what can recent developments in the humanities and ecocultural theory contribute? For an answer, I return to environmental philosopher Val Plumwood's *Environmental Culture*. HPS would need to reconstitute subject-object and disciplinary dichotomies that posit plants as passive and the study of plants as empirical only. To this effect, Plumwood advocates an "ecologically-integrated humanities knowledge field" ([16], p. 51). For Plumwood, the "ecological humanities" bridge "the great split ...between nature (science) on the one hand and culture (philosophy and the humanities, cultural studies) on the other" ([16], p. 51). In Plumwood's view, plant research, most commonly associated with the sciences (botany) or social sciences (anthropology and ethnobotany), is restricted to "the sphere of the 'objective'...where the subject/object constructions reign supreme" ([16], p. 52).

Echoing Plumwood, anthropologist Deborah Bird Rose and historian Libby Robin argue that the ecological humanities bridge dichotomies intrinsic to Western thinking about nature through an "ontology of connectivity" [21]. Canonical works in the ecological humanities, such as Carolyn Merchant's *The Death of Nature*, portray nature as "an *active partner* rather than*passive accessory* in the unfolding of human communities [italics added]" [78]. Through the premise of connectivity between plants and society, HPS would provide a framework for conceptualizing plants as active partners in knowledge production and cultural practices, "as social beings with agentive efficacy" ([69], p. 183). HPS would investigate plants and everyday human-plant interactions—for example, South African botanical species in Western Australia (see Figure 5)—towards a reconsideration of "planthood." Hence, HPS would align closely with the theoretical advances of the ecological humanities, as well as the methodological precedents of ecocriticism [41]; ecocultural studies [17]; human-animal studies [58]; human-plant geographies [10,11]; multispecies ethnography [69]; and biosemiotics [49,79].

Figure 5. Exotic *Gladiola* spp. plants alongside a railway line near Bunbury, Western Australia. Termed "invasive" and "opportunistic," weeds often exhibit adaptive group behaviors that enable them to establish vigorous populations around the world. Should weeds like this beautiful *Gladiola* be known as obnoxious pests or intelligent colonizers? (Image source: Author).

Importantly, this framework would look towards Indigenous knowledges in reconceptualizing these divides [7,80]. Intrinsic to Aboriginal Australian relations to non-human beings, kinship (involving totemism and custodianship) expresses a "connectivity ontology" between plants and people ([7], p. 58; [21]). Philosopher Mary Graham outlines two axioms of Aboriginal philosophy: (a) the land is the law and (b) you are not alone in the world, the former signifying that "meaning comes from land" and the latter, "a kinship system which extends into land." The stories of different clans describe "Creator Beings," including plant beings, that enabled sleeping "proto-humans to become fully human" [80].

The interdisciplinary focus of HPS would bring about novel research approaches—steeped in ancient ontologies—in which plants and human-plant interactions and social relations might be better understood. HPS would conceptualize botanical intelligence broadly as a syncretic quality emerging from intelligent systems and encompassing the potentiality for plants to become "persons," as central to particular Indigenous worldviews. Most crucially for the development of human-plant studies is the differentiation between botanical intelligence—plants *acting upon* people to co-orchestrate cultural beliefs and practices in the *umwelt* of living organisms—and the intelligent use of plants—people *acting upon* plants in utilitarian and potentially exploitative ways that posit "intelligent" animals against "passive" plants. The ecological humanities afford a strong basis for this further articulation of human-plant studies. Above all else—as Hall, Plumwood, Rose, and Robin suggest—HPS implies an ethics of research involving care and connectivity with plants. Such dialogical approaches—to be advanced further within HPS—regard nature as an agent in knowledge formation and "articulate ethical and social engagement with respect for what is studied" ([16], pp. 55–56). Extending the animal and society precedent, human-plant studies would interrogate the history of "human" and "plant" (and their association in Western ontology), the latter laden with connotations of passivity and the former with presuppositions of intelligence. This call for interdisciplinarity posits plants, their intelligences, and their social and cultural relations as productive areas for research to be explored further from a myriad of perspectives. As Marder asks, "and what would it mean to write and think in a vegetal—if not a vegetative—state, having left one's head behind or walking on one's head? What is the outcome of our approximating the locus of vegetal being?" ([13], p. 474).

ACKNOWLEDGMENTS

The author is indebted to the Centre for Research in Entertainment, Arts, Technology, Education and Communications (CREATEC) and the School of Communications and Arts at Edith Cowan University (Australia) for their support during this research. He also thanks the three anonymous referees whose suggestions greatly enhanced the article. The author finally wishes to acknowledge that the concepts and general argument underpinning this article borrow extensively from the recent work of Australian environmental philosopher Matthew Hall and his landmark publication *Plants as Persons: A Philosophical Botany* (2011).

REFERENCES AND NOTES

1. Paul Valéry quoted in Hallé, F. *In Praise of Plants*; Timber Press: Portland, OR, USA, 2002; p. 107.
2. Goody, J. *The Culture of Flowers*; Cambridge University

Press: Cambridge, UK, 1993.

3. Balick, M.; Cox, P. Preface. In *Plants, People, and Culture: The Science of Ethnobotany*; Scientific American Library: New York, NY, USA, 1996; p. vii.

4. Goldgar, A. *Tulipmania: Money, Honor, and Knowledge in the Dutch Golden Age*; University of Chicago Press: Chicago, IL, USA, 2007.

5. Williams, C. *Medicinal Plants in Australia: Bush Pharmacy*; Rosenberg Publishing: Dural Delivery Centre, NSW, Australia, 2010; Volume 1, p. 8.

6. Balick, M.; Cox, P. *Plants, People, and Culture: The Science of Ethnobotany*; Scientific American Library: New York, NY, USA, 1996.

7. Clarke, P. *Aboriginal People and Their Plants*; Rosenberg Publishing: Dural Delivery Centre, NSW, Australia, 2011.

8. Wickens, G.E. Economic botany. In *Economic Botany: Principles and Practices*; Wickens, G.E., Ed.; Kluwer Academic Publishers: Dordrecht, The Netherlands, 2001; p. 2.

9. Lewis, W.; Elvin-Lewis, M. *Medical Botany: Plants Affecting Human Health*, 2nd ed; Wiley: New York, NY, USA, 2003.

10. Head, L.; Atchison, J. Cultural ecology: Emerging human-plant geographies. *Progr. Hum. Geogr.* 2009, *33*, 236–245.

11. Hitchings, R.; Jones, V. Living with plants and the exploration of botanical encounter in human geography research.*Ethics Place Environ.* 2004, *7*, 3–18.

12. Hall, M. *Plants as Persons: A Philosophical Botany*; SUNY Press: Albany, NY, USA, 2011.

13. Marder, M. Vegetal anti-metaphysics: Learning from plants. *Continent. Philos. Rev.* 2011, *44*, 469–489.

14. Ryan, J. Cultural botany: Towards a model of transdisciplinary, embodied and poetic research into plants. *Nat. Cult.*2011, *6*, 123–148.

15. Hazell, D. *The Plants of Middle-Earth: Botany and Sub-creation*; Kent State University Press: Kent, OH, USA, 2007.

16. Plumwood, V. *Environmental Culture: The Ecological Crisis of Reason*; Routledge: London, UK, 2002; p. 56.

17. Conley, V. Eco-subjects. In *Rethinking Technologies*; Conley, V., Ed.; University of Minnesota Press: Minneapolis, MN, USA, 1993; pp. 77–91.

18. Giblett, R. *People and Places of Nature and Culture*; Intellect Press: Bristol, UK, 2011.

19. Griffiths, T. The humanities and an environmentally sustainable Australia. *Australian Humanities Review*, 2007, 43. available online: http://www.australianhumanitiesreview.org/archive/Issue-December-2007/EcoHumanities/EcoGriffiths.html (accessed on 1 May 2012).

20. Haraway, D. *When Species Meet*; University of Minnesota Press: Minneapolis, MN, USA, 2008.

21. Rose, D.; Robin, L. The ecological humanities in action: An invitation. *Australian Humanities Review*, 2004, 31–32. available online: http://www.australianhumanitiesreview.org/archive/Issue-April-2004/rose.html (accessed on 1 May 2012).

22. Hall, M. Plant autonomy and human-plant ethics. *Environ. Ethics* 2009, *31*, 169–181.

23. Ryan,J.'Plantsthatperformforyou'?Fromfloralaesthetics to floraesthesis in the Southwest of Western Australia. *Aust. Humanit. Rev.* 2009, 47, pp. 117–140. Available online: http://www.australianhumanitiesreview.org/archive/Issue-November-2009/ryan.html (accessed on 1 May 2012).

24. Trewavas, A. Aspects of plant intelligence. *Ann. Bot.* 2003, *92*, 1–20.

25. Hallé, F. *In Praise of Plants*; Timber Press: Portland, OR, USA, 2002.

26. Trewavas, A. The green plant as an intelligent organism. In *Communication in Plants: Neuronal Aspects of Plant Life*; Baluška, F., Mancuso, S., Volkmann, D., Eds.; Springer-Verlag: Berlin, Germany, 2006; pp. 1–18.

27. Trewavas, A. Mindless mastery. *Nature* 2002, *415*, 841.

28. Houle, K. Animal, vegetable, mineral: Ethics as extension or becoming? The case of becoming-plant. *JCAS* 2011, *1/2*, 89–116.

29. Pollan, M. *The Botany of Desire: A Plant's-Eye View of the World*; Random House: New York, NY, USA, 2002; p. xiv.

30. Vertosick, F. *The Genius Within: Discovering the Intelligence of Every Living Thing*; Harcourt: New York, NY, USA, 2002; pp. 6–11.

31. Editorial. Open to interpretation. *Nature* 2008, *453*, 824.

32. Breckle, S. *Walter's Vegetation of the Earth: The Ecological Systems of the Geo-Biosphere*, 4th ed; Springer-Verlag: Berlin, Germany, 2002; p. 463.

33. See ref. 24.

34. Baluška, F.; Volkmann, D.; Hlavacka, A.; Mancuso, S.; Barlow, P. Neurobiological view of plants and their body plan. In *Communication in Plants: Neuronal Aspects of Plant Life*; Baluška, F., Mancuso, S., Volkmann, D., Eds.; Springer-Verlag: Berlin, Germany, 2006; pp. 19–35.

35. Barlow, P. Charles Darwin and the plant root apex: Closing a gap in living systems theory as applied to plants. In*Communication in Plants: Neuronal Aspects of Plant Life*; Baluška, F., Mancuso, S., Volkmann, D., Eds.; Springer-Verlag: Berlin, Germany, 2006; pp. 37–51.

36. Karban, R. Plant behaviour and communication. *Ecol. Lett.* 2008, *11*, 727–739.

37. Harper, D. Vegetative. *Online Etymological Dictionary*, 2012. Available online: http://www.etymonline.com/index.php?term=vegetative (accessed on 1 May 2012).

38. See also page 84 in Marder, M. Plant-soul: The elusive meanings of vegetative life. *Environ. Philos.* 2011, *8*, 83–89.

39. Ryan, J. Plants as objects: Challenges for an aesthetics of flora. *Philos. Stud.* 2011, *1*, 222–236.

40. Lawrence, D.H.; Skinner, M. *The Boy in the Bush*; Eggert, P., Ed.; Cambridge University Press: Cambridge, UK, 2002; pp. 92–93.

41. Huggan, G.; Tiffin, H. *Postcolonial Ecocriticism:*

Literature, Animals, Environment; Routledge: New York, NY, USA, 2010; pp. 133–202.

42. Deleuze, G.; Guattari, F. *A Thousand Plateaus: Capitalism and Schizophrenia*; Massumi, B., Translator; Continuum: London, UK, 2004; p. 27.
43. Kaplan, C. *Questions of Travel: Postmodern Discourses of Displacement*; Duke University Press: Durham, NC, USA, 1996; p. 87.
44. Jung, C. *The Earth Has a Soul: The Nature Writings of C.G. Jung*; North Atlantic Books: Berkeley, CA, USA, 2002; p. 37.
45. Darwin, C.; Darwin, F. *The Power of Movements in Plants*; Murray: London, UK, 1880; p. 573.
46. Barlow, P. Charles Darwin and the plant root apex: Closing a gap in living systems theory as applied to plants. In*Communication in Plants: Neuronal Aspects of Plant Life*; Baluška, F., Mancuso, S., Volkmann, D., Eds.; Springer-Verlag: Berlin, Germany, 2006; p. 39.
47. Baluška, F.; Volkmann, D.; Hlavacka, A.; Mancuso, S.; Barlow, P. Neurobiological view of plants and their body plan. In *Communication in Plants: Neuronal Aspects of Plant Life*; Baluška, F., Mancuso, S., Volkmann, D., Eds.; Springer-Verlag: Berlin, Germany, 2006; p. 21.
48. Goh, C.; Nam, H.; Park, Y. Stress memory in plants: A negative regulation of stomatal response and transient induction of rd22 gene to light in abscisic acid-entrained *Arabidopsis* plants. *Plant J.* 2003, *36*, 240–255.
49. Hoffmeyer, J. *Signs of Meaning in the Universe*; Haveland, B., Translator; Indiana University Press: Bloomington and Indianapolis, IN, USA, 1996; p. 125.
50. Sattler, R. Homology, homeosis, and process morphology in plants. In *Homology: The Hierarchical Basis of Comparative Biology*; Hall, B., Ed.; Academic Press: San Diego, CA, USA, 1994; p. 451.
51. Rescher, N. *Process Philosophy: A Survey of Basic Issues*; University of Pittsburgh Press: Pittsburgh, PA, USA, 2000; p. 3.
52. Marder, M. Resist like a plant! On the vegetal life of political movements. *Peace Stud. J.* 2012, *5*, 24–32.
53. Aloi, G. Editorial. *Antennae* 2011, *17*, 3–4.
54. Armstrong, P.; Simmons, L. Beastiary: An introduction. In *Knowing Animals*; Simmons, L., Armstrong, P., Eds.; Brill: Leiden, The Netherlands, 2007; p. 1.
55. Shapiro, K.; DeMello, M. The state of human-animal studies. *Soc. Anim.* 2010, *18*, 2–17.
56. Animals and Society Institute, Inc. Human–Animal Studies. Available online: http://www.animalsandsociety.org/pages/human-animal-studies (accessed on 1 May 2012).
57. The University of Sydney. HARN: About us. HARN: Human Animal Research Network 2012. Available online: http://sydney.edu.au/arts/research/harn/about/index.shtml (accessed on 1 May 2012).
58. Shapiro, K. Editor's introduction: The state of human-animal studies: Solid, at the margin! *Soc. Anim.* 2002, *10*, 331–337.
59. Potts, A.; Armstrong, P. Hybrid vigor: Interbreeding cultural studies and human-animal studies. In *Teaching the Animal: Human-Animal Studies Across the Disciplines*; DeMello, M., Ed.; Lantern Books: Brooklyn, NY, USA, 2010; p. 3.
60. Freeman, C.; Leane, E. Introduction. In *Considering Animals: Contemporary Studies in Human-Animal Relations*; Freeman, C., Leane, E., Watt, Y., Eds.; Ashgate: Surrey, UK, 2011; p. 4.
61. Joseph, M. The growing field of animal studies. *The Washington Post*, 2012. Available online: http://www.washingtonpost.com/lifestyle/magazine/the-growing-field-of-animal-studies/2012/04/10/gIQA9AvjCT_gallery.html (accessed on 1 May 2012).
62. Breeden, S.; Breeden, K. *Wildflower Country: Discovering Biodiversity in Australia's Southwest*; Fremantle Press: Fremantle, Australia, 2010.
63. Hopper, S. *Kangaroo Paws and Catspaws: A Natural History and Field Guide*; Department of Conservation and Land Management: Como, Australia, 1993.
64. Nikulinsky, P.; Hopper, S. *Life on the Rocks: The Art of Survival*; Fremantle Press: Fremantle, Australia, 1999.
65. Nikulinsky, P.; Hopper, S. *Soul of the Desert*; Fremantle Press: Fremantle, Australia, 2005.
66. Pelloe, E. *Wildflowers of Western Australia*; C.J. De Garis: Melbourne, Australia, 1921.
67. Kohák, E. Speaking to trees. *Crit. Rev.* 1993, *6*, 371–388.
68. Stone, C. *Should Trees Have Standing? Law, Morality, and the Environment*, 3rd ed; Oxford University Press: New York, NY, USA, 2010.
69. Kirksey, S.; Helmreich, S. The emergence of multispecies ethnography. *Cult. Anthropol.* 2010, *25*, 545–576.
70. Von Hügel, B. *New Holland Journal: November 1833–October 1834*; Clark, D., Translator; Melbourne University Press at the Miegunyah Press in Association with the State Library of New South Wales: Melbourne, Australia, 1994; p. 23.
71. Hopper, S. Nuytsia floribunda. *Curtis's Botanical Magazine* 2010, *26*, 333–368.
72. Paczkowska, G.; Chapman, A. *The Western Australian Flora: A Descriptive Catalogue*; Wildflower Society of Western Australia, Inc., the Western Australian Herbarium, CALM and the Botanic Gardens & Parks Authority: Perth, Australia, 2000.
73. Bates, D. *Aboriginal Perth Bibbulmun Biographies and Legends*; Bridge, S., Ed.; Hesperian Press: Victoria Park, Australia, 1992; p. 86.
74. Cunningham, I. *The Land of Flowers: An Australian Environment on the Brink*; Otford Press: Caringbah, Australia, 2005.
75. Moore, G. *Diary of Ten Years Eventful Life of an Early Settler in Western Australia*; University of Western Australia Press: Nedlands, Australia, 1884; p. 80.
76. Lindley, J. *A Sketch of the Vegetation of the Swan River Colony*; James Ridgway: London, UK, 1840; p. xxxix.
77. Hassell, E. *My Dusky Friends*; C.W. Hassell: Dalkeith, Australia, 1975; p. 26.
78. Eckersley, R. The death of nature and the birth of the

ecological humanities. *Organ. Environ.* 1998, *11*, 183–185.

79. Hoffmeyer, J. *Biosemiotics: An Examination into the Signs of Life and the Life of Signs*; Favareau, D., Ed.; Hoffmeyer, J.; Favareau, D., Translators; University of Scranton Press: Scranton, PA, USA, 2008.

80. Graham, M. Some thoughts about the philosophical underpinnings of Aboriginal worldviews. *Aust. Humanit. Rev.*2008, p. 45. Available online: http://www.australianhumanitiesreview.org/archive/Issue-November-2008/graham.html(accessed on 1 May 2012).

CHAPTER 2

KNOWLEDGE AND USE OF MEDICINAL PLANTS BY LOCAL SPECIALISTS IN AN REGION OF ATLANTIC FOREST IN THE STATE OF PERNAMBUCO (NORTHEASTERN BRAZIL)

Luiz Rodrigo Saldanha Gazzaneo, Reinaldo Farias Paiva de Lucena, and Ulysses Paulino de Albuquerque

Keywords: Ethnobotany, Atlantic Forest, medicinal plants, traditional knowledge

INTRODUCTION

Fifteen percent of Brazil was once covered by Atlantic Forest, and although less than 5% of the original forest remains today [1], it is still one of the highest biodiversity areas on the planet [2]. This forest also demonstrates extremely high rates of endemism: up to 74.4% for bromeliads, 55% for trees, and 64% for palms [1, 3], and we can infer that there is yet much to be studied and discovered within this ecosystem, especially in terms of its useful resources.

The Atlantic Forest ecosystem has been the victim of successive economic cycles since colonial times (brazilwood, sugarcane, mining, coffee, and cattle), especially in areas very near the coast. Today, large cities (50% of the Brazilian population) are located in areas once covered by these forests, and 80% of the GNP is produced there [1]. For these reasons, innumerous species have disappeared since Brazil's discovery, and continue to disappear without having been recorded; and with them, important information related to ecology, pharmacology, botany, and other fields, may vanish completely.

Rural communities are considered to be the most neglected area in terms of ethnobotanical studies [4]. Despite the availability of modern medicines, most agro-cultural communities (either by choice or for lack of economic resources) still use and detain an extensive pharmacopoeia of native plants [4].

The study of local knowledge about natural resources is becoming increasingly important in defining strategies and actions for conservation or recuperation of residual forests. This study therefore sought to: collect information from local populations concerning the use of Atlantic Forest medicinal plants; verify the sources of medicinal plants used; determine the relative importance of the species surveyed, and; calculate the informant consensus factor in relation to medicinal plant use.

MATERIALS AND METHODS

The Study Area and Its Inhabitants

This field study was undertaken with the aid of the "Três Ladeiras" community in the municipality of Igarassu, Pernambuco State, in northeastern Brazil. Igarassu is located in the mesoregion of the Recife metropolitan area (the state capital) and in the Itamaracá microregion (Figure 1). The municipal seat is located 26 km from the state capital [5] at 20 m above sea level (7° 50' 20"S and 35° 00' 10"W). The municipality has a total area of 304.2 km^2, a population of 72,990 (219.9 inhabitants/km^2), and it is 74.9% urbanized. The regional climate is humid, with a mean annual temperature of 27°C. Average annual rainfall is 2,000 mm, with a moderate water deficit in the summer [6].

Citation: Luiz Rodrigo Saldanha Gazzaneo, Reinaldo Farias Paiva de Lucena, and Ulysses Paulino de Albuquerque Knowledge and use of medicinal plants by local specialists in an region of Atlantic Forest in the state of Pernambuco (Northeastern Brazil), Journal of Ethnobiology and Ethnomedicine, 2005; 1:9. DOI:10.1186/1746-4269-1-9.

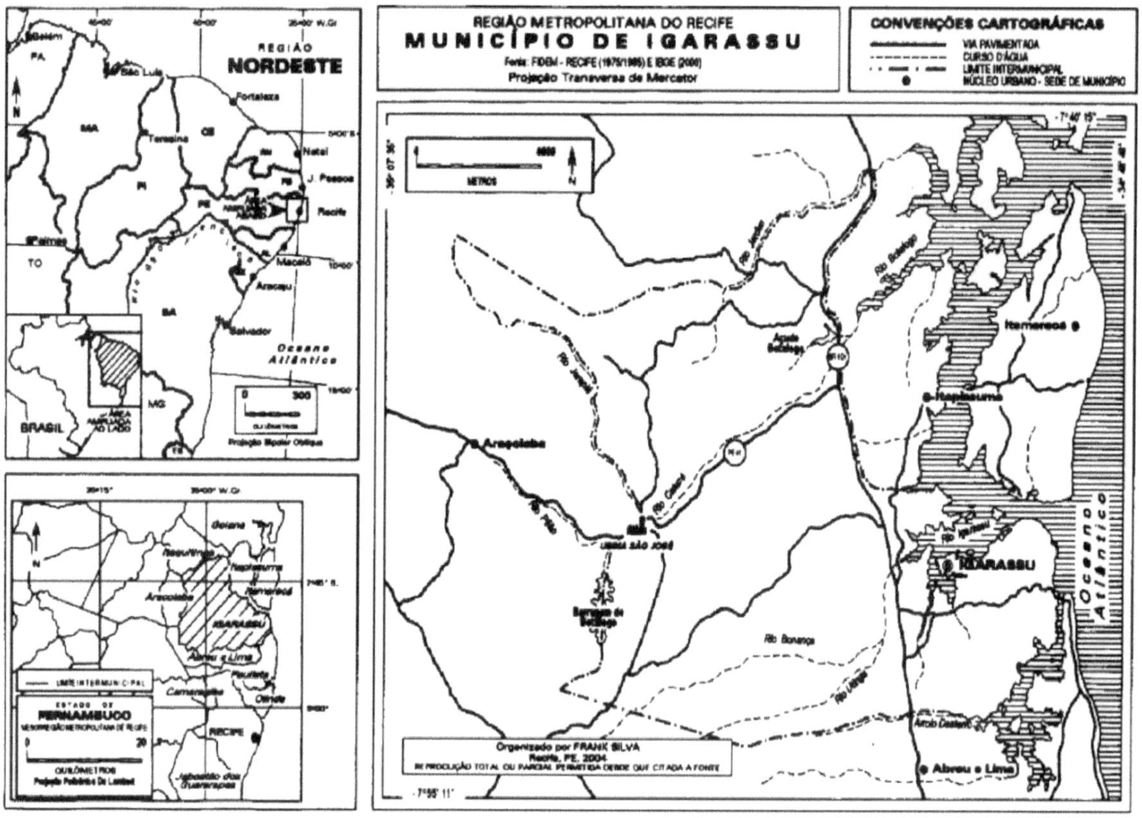

Figure 1. Place of ethnobotanical data collection of the medicinal plants cited by the population in the municipality of Igarassu (Northeast Brazil).

The Rural Inhabitants

The community studied is part of the "Usina São José" sugarcane refinery, which employs most of the male population. The community is surrounded by Atlantic Forest fragments that belong to the refinery's ecologic reserve [7] (Fig. 1). The fragments occupy a total area of 210 hectares, with altitudes varying from 50 to 140 m. The number of people employed by the refinery oscillates during the year, increasing and decreasing according to the seasons and the periods of land preparation, planting, and harvesting. It is common to find families cultivating small plantations that serve as a nutritional and/or economic reserve between periods of employment activity. There is a 323.3 ha forest southeast of the refinery, which represents 0.80% of the total area of the municipality. The forest is part of the Botafogo River Basin source protection area, according to law N° 9860, dated August 12, 1986. The purpose of this reserve is to protect the landscape, soil, and the water basin [8].

The community views itself as a coherent group, and is formed basically by active or retired workers of the local sugar cane refineries. The women of the community carry out almost all of the activities related to caring for the family and their property. There are two public schools and a small Peruvian cherry ("*acerola*") plantation that produces fruit pulp and provides job opportunities for youths during the harvest season. There is no basic sanitation, and a single health care center treats only simple health problems; patients requiring more intensive care are removed to hospitals in Igarassu.

Data Collection

Information on the use of medicinal plants was collected for one year (2003) with the help of rural dwellers of the "Três Ladeiras" community. During this period, door-to-door visits were made in order to attempt to identify local people with a specialized knowledge of medicinal plant use [9]. As such, the sampling was intentionally non-random [10], under the assumption that local specialists would provide more specific and higher quality information concerning medicinal plants. Data was obtained using semi-structured forms to record the interviewee's personal information and topics related to the medicinal use of specific plants. During the first contacts with the local population, a specialist was identified by the inhabitants themselves. A specialist is defined as "a person recognized by the community as having deep knowledge about the uses of native and/or introduced plants in manufacturing remedies and in promoting cures". Using the snowball method [11], names of other specialists were then obtained. Six local specialists were eventually interviewed (three men and three women), aged 51–102. All the people identified as specialists had been living in the community for at least 30 years and were frequently sough out by other community members for advice on the use of plants. The men had all been woodsmen or hunters, although these activities are currently prohibited. The women were housewives who had acquired a broad knowledge of medicinal plants by way of experimentation and trading information with relatives or people from the surrounding communities.

Data Analysis

The first step employed in the data analysis calculating the informant consensus factor (ICF) [12]. ICF values will be low (near 0) if plants are chosen randomly, or if informants do not exchange information about their use. Values will be high (near 1) if there is a well-defined selection criterion in the community and/or if information is exchanged between informants.

The ICF is calculated as follows: number of use citations in each category (n_{ur}) minus the number of species used (n_t), divided by the number of use citations in each category minus one:

$$ICF = \frac{n_{ur} - n_t}{n_{ur} - 1}$$

All citations were placed into one of 14 categories: undefined pains or illnesses; skin and subcutaneous tissues; diseases of the endocrine glands, metabolism, and nutrition; blood and hematopoietic organs; skeletal, muscle, and connective tissues; infectious and parasite-related diseases; neoplasies; problems of the circulatory system; problems of the digestive system; problems of the genitourinary system; problems of the nervous system; problems of the respiratory system; problems of the sensorial system – ear; and problems of the sensorial system – eye.

The use value (adapted from the proposal of Phillips et al. [13]), a quantitative method that demonstrates the relative importance of species known locally, was also calculated:

$$UV = \Sigma U/n$$

where: UV = use value of a species; U = number of citations per species; n = number of informants

All of the material collected was processed, identified with the aid of specialists, and subsequently deposited in the PEUFR herbarium of the Biology Department of the Federal Rural University of Pernambuco. All material was collected with the help of local informants.

RESULTS

Local Specialists and Medicinal Plants

The plants collected represent 125 plants, distributed among 61 botanical families. Lamiaceae (12 spp.) was best represented in terms of the number of species, followed by Asteraceae (6 spp.), Myrtaceae (5 spp.), Arecaceae (5 spp.), and Poaceae (5 spp.). When analyzing the number of citations for the plant parts used to prepare local remedies, a preference for the use of leaves becomes noticeable. The use of teas had the highest relative value (42%), followed by the use of syrups (20%), and *in natura* (16%).

Plants cited are listed by collection locality (anthropogenic zones or forest) and degree of management (cultivated or non-cultivated). It was observed that 54.5% of the species used are cultivated, while native and/or weed plants composed the non-cultivated category. In terms of where the plants were collected, 82.7% were from areas backyards and small farms, while only 17.3% were gathered from inside the forest. Most of the non-cultivated plants were weed, with little participation of native plants.

The species with the highest use-value was *Pithecellobium cochliocarpum* (Gomez) Macbr, popularly known as "barbatenon", with an UV of 1.8. Its main importance resides in its attributed healing effects on wounds. It is used externally, in the form of an alcoholic bark extract. *Alpinia zerumbet* (Pers.) Burt. ex R. M. Smith ("colônia") had the second highest UV (1.6). This species is frequently used as an ornamental plant in home gardens. Its main medicinal uses are to treat coughs (flower, leaf, and/or a root syrup), and headaches (a lightly heated leaf is applied to the forehead).

Two species shared the third highest use value (1.3): *Hymenaea martiana* Hayne ("jatobá") and *Aeolanthus suaveolens* Mart. ex Spreng. ("macassá"). *H. martiana* was cited for the treatment of blows to the body, inflammations, and rheumatism. It is principally prepared as an alcoholic extract of the fruit, using the local distilled drink ("cachaça") or wine. *A. suaveolens* is well known among the interviewees as being capable of treating general pains, especially earaches (for which it is applied *in natura*, as an earplug).

Cited for the treatment of inflammations, *Schinus terebinthifolius* Raddi ("aroeira"), *Boerhavia diffusa* L ("pega-pinto"), and *Borreria verticillata* G. F. W. Meyer ("vassoura de botão") all had a use value of 1.1. *Protium heptaphyllum* (Aubl.) March. ("amescla") is used to treat tooth and headaches, and *Solanum panicultum* L. ("jurubeba branca") to treat several ailments (such as indigestion, anemia, hepatitis, kidney, and spleen problems); both had use-values of 1.

Consensus Factor Among Specialists

The highest ICF values were linked to problems related to the blood and to hematopoietic organs (1.0), and for problems of the sensorial system – ear (0.60). The use category with the lowest ICF value was "diseases of the endocrine glands, metabolism, and nutrition" (0.14). The ICF for infectious and parasite-related diseases was strikingly low (0.27) – the result of the use of a number of species that was almost as varied as the number of citations. A more detailed description of each category follows (see Table 1).

Table 1. Informant Consensus Factor by corporal systems categories or diseases.

Category	Species	(%) All Species	Use citations	(%) All use citations	ICF
Undefined pains or illnesses	42	32.3	68	19.88	0.38
Skin and subcutaneous tissues	16	12.3	34	9.9	0.54
Diseases of the endocrine glands, metabolism and nutrition	7	5.3	8	2.3	0.14
Blood and hematopoietic organs	1	0.7	5	1.4	1.00
Skeletal, muscle and connective tissues	13	10.0	23	6.7	0.45
Infectious and parasite-related diseases	14	10.7	19	5.5	0.27
Neoplasies	1	0.7	1	0.2	0.00
Problems of the circulatory system	7	5.3	10	2.9	0.33
Problems of the digestive system	36	27.6	60	17.5	0.40
Problems of the genitourinary system	24	18.4	34	9.9	0.30
Problems of the nervous system	3	2.3	4	1.16	0.33
Problems of the respiratory system	38	29.2	65	19.0	0.42
Problems of the sensorial system (ear)	3	2.3	6	1.7	0.60
Problems of the sensorial system (eye)	3	2.3	5	1.4	0.50

Blood and Hematopoietic Organs

This category had the highest ICF value (1.0). There were five citations for a single species, demonstrating total agreement on selection criteria among the informants. The main problem treated within this category was hemorrhaging, often caused by accidents involving the perforating or cutting instruments used in sugarcane culture and in subsistence farming. Blood loss is diminished by applying a cloth soaked in a tincture made from the bark of *Pithecellobium cochliocarpum* (Gomez) Macbr ("barbatenon").

Problems of the Sensorial System – Ear

This group obtained the second highest ICF value (0.60). Of all of the plants cited by informants (127), 2.3% (three species) are used to treat medical problems within this group. Only one of the interviewees did not cite an ear-related disease. Earaches were cited by all of the specialists, and there was general agreement in the citations of *Aeolanthus suaveolens* Mart. ex Spreng. as the main remedy. There was also agreement on the use of *Plectranthus sp.* ("hortelã grande") and *Ruta graveolens* L ("arruda"). In all cases, the manner of usage was an earplug made from the leaves (*in natura*).

Skin and Subcutaneous Tissues

With a total of 34 citations (10% of the total), this category had an ICF of 0.54. Its main representative – as in the first category cited – was *Pithecellobium cochliocarpum* (Gomez) Macbr ("barbatenon"), a plant of great importance to the local population.

Problems of the Sensorial System – Eye

This is a heterogeneous group, with an ICF of 0.50. Of the five use-citations, three species are recommended for "tired eyesight" and conjunctivitis. The most cited phytotherapeutic resource was *Ocimum basilicum* L. ("manjericão"), with three citations. Recipes mention the use of an aqueous extract of this plant (obtained as a tea) as an eyewash.

Skeletal, Muscle, and Connective Tissues

With an ICF of 0.45, the 23 citations of this group included 10% (13 species) of the total number of plants cited in this study. The most important representatives of this group were *Hymenaea martiana* Hayne ("jatobá"), with four citations, and *Caesalpinia ferrea* Mart. ex Tul. ("jucá"), with three citations. It is interesting to note that in both species the fruit is used, and it is prepared in a very similar fashion – wine is employed to obtain an alcoholic solution for internal use.

Problems of the Respiratory System

This category had the second highest number of citations (65), only slightly less than the group of undefined pains and illnesses (68). The ICF of this category was 0.42. It is important to note that approximately 32% (42) of the total number of species cited were included within this group, demonstrating a great diversity in the knowledge of medicinal plants for the treatment of respiratory problems. The remedies used are mainly administered in the form of syrups to treat coughing, breathlessness, asthma, bronchitis, and the common cold – maladies that especially affect children, either due to the fragility of their immunological system, or to problems related to their immature respiratory system.

According to the data collected, sugar is often added to the syrup preparation – for in addition to its preservative properties (it increases the osmolality of the solution), its sweet taste masks the bitter and unpalatable taste of some of the herbs.

The documented use of a wide variety of species is in great part due to the practice of using innumerous herbs (sometimes over 15) in recipes for a single syrup, and by the belief that the greater the number of herbs (always in odd numbers) the better the syrup. Another reason for the large number of citations is the frequency of cases of breathlessness and asthmatic problems when the sugarcane fields surrounding the community are burned during the pre-harvest preparations.

Problems of the Digestive System

This category includes all problems related to organs directly or indirectly linked to digestion, including the teeth and gums. The ICF value for this group was 0.40, considered low. However, if we consider only the citations for toothaches, the ICF value would be nearly 1.0, since there was great agreement on the use of*Aeolanthus suaveolens* Mart. ex Spreng. ("macassá") to treat this complaint.

Undefined Pains or Illnesses

This category includes all citations for undefined pains, and for diseases with unspecific symptoms. The ICF value was 0.38. This group had the highest number of use-citations (68 of a total of 342) as well as species used (42). This result is compatible with the category's diversity of problems.

Once again, *Aeolanthus suaveolens* Mart. ex Spreng. ("macassá") stands out, followed by "verga morta" (an undetermined species) and *Protium heptaphillum* (Aubl.) March. ("amescla"). All of these plants have analgesic uses. *Pfaffia glomerata* (Spreng.) Pederson ("acônito") had the highest number of citations among plants used to treat fevers. *Ruta graveolens* L. ("arruda") stood out for its importance as an anti-inflammatory agent.

Problems of the Circulatory System

Heart diseases and blood pressure are examples of problems within this category. A total of seven species were cited, with 10 uses. The ICF for this category was low (0.33). Strokes received great attention from the interviewees, totaling 60% of the citations. *Argemone mexicana* L. ("cardo santo") was the species most cited.

Problems of the Nervous System

An ICF of 0.33 was observed for this category. Fainting and insomnia were the most frequent citations. Half of the citations referred to the calming effects attributed to a tea made from orange tree flowers or leaves (*Citrus sinensis* (L.) Burmann), which is also frequently used to treat insomnia.

Problems of the Genitourinary System

With 34 citations, this category comprises 24 plant species, with an ICF of 0.30. Women in the community commonly use sitting baths made with tea from the bark of *Boerhavia diffusa* L ("pega-pinto") or*Pithecellobium cochliocarpum* (Gomez) Macbr ("barbatenon").

Infectious and Parasite-Related Diseases

The ICF for this category was 0.27. The common cold was frequently cited, followed by worms and hepatitis. A total of 19 disease citations were recorded, and 14 plants species were indicated as cures for these diseases, or as treatments for their symptoms.

Diseases of the Endocrine Glands, Metabolism, and Nutrition

The most cited diseases were diabetes (metabolism) and anemia (nutrition). Seven plants were cited for eight diseases, and the ICF was correspondingly very low (0.14).

Neoplasies

As only a single citation was obtained (for cancer) and therefore this category's ICF appears as 0, as this index requires citations from at least two informants.

DISCUSSION

There seems to be a tendency for a few plant families to stand out in any pharmacopoeia. In a study by Almeida & Albuquerque [14], the family Lamiaceae was classified as the richest in species citations. The families Lamiaceae and Asteraceae stood out in the studies of Bennett & Prance [15], for together they represented 21% of the 216 species surveyed. Hanazaki *et al.* [2] likewise found that these two families had the greatest number of species cited by fishing communities and in other coastal areas of Atlantic Forest in São Paulo State [16]. These plants are normally herbaceous species that can either be cultivated or occur as weeds. Preference for their use may be related as much to their ready availability, for they are common in disturbed areas [17], as to factors related to their biological activity. Based on evidence and availability theory, Stepp & Moerman [18] suggest that these plants concentrate very active biological compounds as a function of their habit or of their life strategies. Much evidence has accumulated indicating that chemical and ecological factors orient the selection and use of medicinal plants in local communities in all parts of the world [cf. [19,20]].

Another interesting factor is that the cited species generally are indicated to the same illnesses in other localities [2, 3, 14, 17, 23, 25, 28, 29]. Moreover, observing the lists of species of other studies, it can be evidenced that the most important species vary of region for region [14, 20, 23, 27], although to verify a set of species, in special the cultivated ones, common in several works, as: *Plectranthus* spp., *Mentha* spp. and*Ocimum* spp.

Another relevant observation relates to the tendency of communities living near humid forests to use plant leaves. In a study undertaken in Barra do Piraí (state of Rio de Janeiro) with merchants and traditional medicine users, Parente & Rosa [21] observed that leaves were the most used category of plant parts. In this study, leaves were used in 42% of the time – over twice the number of citations for the second most used part (the root, at 18%). Similar observations had already been recorded for other communities near forested areas, where vegetation is always green and leaves are abundant [16, 22]. On the other hand, communities in dry regions tend to focus their attention on plant parts that are continuously available, such as the bark [23, 14]. As the plants in these localities are regularly exposed to long periods of drought and lose their leaves, bark and roots are more often used. This observed difference in usage of plant parts in different areas should be more closely investigated. Zschocke *et al.* [24] suggested, for example, that the use of bark is much more common in tropical regions than in temperate zones. However, as noted above, even in tropical regions the use of bark seems to be most strongly associated with arid and semi-arid zones, reflecting the resources most readily available to these communities. Additionally, the preference towards leaves places more emphasis on the preparation of medicinal teas. Parente & Rosa [21] also found high percentages for teas (51%), baths (39%), and other forms (liquid extracts, infusions, and *in natura* – 10%) due to the great preference for leaves in preparing remedies.

The emphasis on cultivated plants observed in this study was not unexpected. However, it might also have been expected that a larger proportion of the plants cited would be native species in view of the fact that the informants are all specialists having an excellent knowledge of the native resources. Nonetheless, the proportion of non-cultivated to cultivated species used varies among different communities that have access to the resources in the Atlantic Forest [2, 25, 16]. Amorozo [22] observed a similar situation in the community of Barcarena (state of Pará), where 57.2% of the plants used were cultivated, while only 42.8% were not. Yet, in the same study, Amorozo [22] demonstrated that in the Santo Antonio do Leverger community (state of Mato Grosso) the percentages invert themselves, with non-cultivated plants totaling 56.2% and cultivated plants 43.8%. This result may be due to the fact that Santo Antonio de Leverger is undergoing a process of modernization, resulting in additional anthropogenic alterations in the natural areas where many medicinal plants grow, and causing the devaluation of local culture as well as the loss of traditional practices such as plant cultivation.

In regards to the population studied here, their preference for using plants from anthropogenic zones may be explained by the establishment of the "Usina São José" forest reserve, which does not permit plant collection. This would force the local specialists to rely more on weed and cultivated plants. Local specialists informed us that, in function of the prohibition, they prevent to collect plants in the fragment forest. In this sense, they "opted" to using non-native plants. The ample use of plants derived from disturbed areas is quite common in many parts of the world [26], including areas of Atlantic Forest [17]. This may be explained by easy access to these species, especially when there is little cultural resistance to using such plants. Albuquerque *et al.* [27] counters this idea when he cites the preference of a rural northeastern Brazilian community for native species even when they have easy and rapid access to substitutes in disturbed areas.

In general, the results presented here give rise to two questions. First, what are the implications of the local use of a large number of non-native species? A mixture of native and introduced species in the pharmacopoeias of tropical zone communities is quite common [15], and in view of the complexity of the local knowledge, cannot be simply explained. On one hand, Begossi *et al.* [25] view the question from a conservationist's point of view, as the use of cultivated species reduces pressure on native forest products. On the other hand, in the case of the present study, the use of exotic and cultivated plants is a result of the need for alternatives to native (prohibited) resources in spite of their great importance to this community. This suggests that a contextual analysis of this phenomenon must take into account not only the proportion of species used (native as well as introduced), but also local preferences.

The second question relates to the transmission of local knowledge. The enforced restriction of the access to native plants may result, in the long run, to the abandonment of their use, and thus a loss of local knowledge. This is particularly serious in the present study, as the most knowledgeable people are very old and the young have not shown much interest in this accumulated knowledge. This will surely affect the resilience of the local knowledge [25]. Additionally, the opportunities for these specialists to invent and experiment with new native resources will be lost, as well as the possibility for scientists to investigate new bio-active products. In agreement with observations in other Atlantic Forest communities [28, 29, 30], we consider that "the process of household Forest uses, of economic alternatives, and of environmental education may be closely tied and may serve as a vehicle for conservation, and key individuals from each community should be included in management programs" [25].

CONCLUSION

This study demonstrated that local specialists in the Atlantic Forest community studied tend to agree with each other in terms of the plants used to treat blood-related problems, but cite a much more diverse group of plants to treat problems related to the respiratory and digestive systems – two important categories in studies undertaken in different parts of the world. In the case of the community examined here, the extensive repertory of plants employed to treat respiratory problems seems to be a response to problems related to the extensive use of fire in local agricultural practices. The local medicinal flora is largely based on plants that are either cultivated or obtained from anthropogenic zones, possibly due to the use and access restrictions of the legally protected neighboring forest. Despite these restrictions, the species with the highest use-value by this community was *Pithecellobium cochliocarpum* (Gomez) Macb., a native plant of the Atlantic Forest.

REFERENCES

1. Consórcio Mata Atlântica, Universidade Estadual de Campinas: Reserva da Biosfera da Mata Atlântica. 1992, Plano de Ação. Referências Básicas
2. Hanazaki N, Tamashiro JY, Leitão-Filho HF, Begossi A: Diversity of plant uses in two Caiçaras communities from the Atlantic Forest coast, Brazil. Biodiversity and Conservation. 2000, 9: 597-615. 10.1023/A:1008920301824.
3. Rossato SC, Leitão-Filho H, Begossi A: Ethnobotany of caiçaras of the Atlantic Forest Coast (Brasil). Economic Botany. 1999, 53: 387-395.
4. Prance GT: What is ethnobotany today?. Journal of Ethnopharmacology. 1991, 32: 209-216. 10.1016/0378-8741(91)90120-3.
5. FIAM – Fundação de Desenvolvimento Municipal do Interior de Pernambuco: Perfil Municipal de Igarassu, Pernambuco. 1997
6. FIDEM: Diagnóstico das Reservas Ecológicas: Região Metropolitana do Grande Recife. Recife. 2001
7. SECTAM: Diagnósticos das Reservas Ecológicas da Região Metropolitana do Recife. Recife. 2001
8. FIDEM: Monitoramento das Reservas Ecológicas da

Região Metropolitana do Grande Recife. Recife. 1993

9. Davis A, Wagner JR: Who Knows? On the Importance of Identifying "Experts" When Researching Local Ecological Knowledge. Human Ecology. 2003, 31: 463-489. 10.1023/A:1025075923297.

10. Albuquerque UP, Lucena RFP: Seleção e escolha dos informantes. Métodos e técnicas na pesquisa etnobotânica. Edited by: Albuquerque UP, Lucena RFP. 2004, Recife: Editora NUPEEA, 19-35.

11. Bailey K: Methods of social research. 1994, New York: The Free Press, 4

12. Heinrich M, Ankli A, Frei B, Weimann C, Sticher O: Medicinal plants in Mexico: Healers'consensus and cultural importance. Social Science and Medicine. 1998, 47: 1863-1875.

13. Phillips O, Gentry AH, Reynel C, Wilki P, Gávez-Durand CB: Quantitative ethnobotany and Amazonian conservation. Conservation Biology. 1994, 8: 225-248. 10.1046/j.1523-1739.1994.08010225.x.

14. Almeida CFCBR, Albuquerque UP: Uso e conservação de plantas e animais medicinais no Estado de Pernambuco (Nordeste do Brasil): um estudo de caso. Interciencia. 2002, 27: 276-285.

15. Bennett BC, Prance GT: Introduced plants in the indigenous pharmacopoeia of Northern South America. Economic Botany. 2000, 54: 90-102.

16. Di Stasi LC, Oliveira GP, Carvalhaes MA, Queiroz-Junior M, Tien OS, Kakinami SH, Reis MS: Medicinal plants popularly used in the Brazilian Tropical Atlantic Forest. Fitoterapia. 2002, 73: 69-91. 10.1016/S0367-326X(01)00362-8.

17. Voeks RA: Tropical Forest healers and habitat preference. Economic Botany. 1996, 50: 382-400.

18. Stepp JR, Moerman DE: The importance of weeds in ethnopharmacology. Journal of Ethnopharmacology. 2001, 75: 19-23. 10.1016/S0378-8741(00)00385-8.

19. Stepp JR: The role of weeds as sources of pharmaceuticals. Journal of Ethnopharmacology. 2004, 92: 163-166. 10.1016/j.jep.2004.03.002.

20. Almeida CFCBR, Silva TCL, Amorim ELC, Albuquerque UP: Life strategy and chemical composition as predictors of the selection of medicinal plants from the *caatinga* (Northeast Brazil). Journal of Arid Environments. 2005, 62: 127-142. 10.1016/j.jaridenv.2004.09.020.

21. Parente CET, Rosa MMT: Plantas comercializadas como medicinais no Município de Barra do Piraí, RJ. Rodriguésia. 2001, 52: 47-59.

22. Amorozo MCM: Uso e diversidade de plantas medicinais em Santo Antônio de Leverger, MT, Brasil. Acta Botânica Brasílica. 2002, 16: 189-203.

23. Albuquerque UP, Andrade LHC: Uso de recursos vegetais da caatinga: o caso do agreste do Estado de Pernambuco (Nordeste do Brasil). Interciencia. 2002, 27: 336-345.

24. Zschocke S, Rabe T, Staden J: Plant part substitution – a way to conserve endangered medicinal plants?. Journal of Ethnopharmacology. 2000, 71: 281-292.18. 10.1016/S0378-8741(00)00186-0.

25. Begossi A, Hanazaki N, Tamashiro JY: Medicinal Plants in the Atlantic Forest (Brasil): Knowledge, Use and Conservation. Human Ecology. 2002, 30: 281-299. 10.1023/A:1016564217719.

26. Anklin A, Sticher O, Heinrich M: Medical ethnobotany of the Yucatec Maya: Healter's consensus as a quantitative criterion. Economic Botany. 1999, 53: 144-160.

27. Albuquerque UP, Andrade LHC, Silva ACO: Use of plant resources in a seasonal dry forest (northeastern Brazil). Acta Botânica Brasilic. 2005, 19: 27-38.

28. Figueiredo GM, Leitão-Filho HF, Begossi A: Ethnobotany of Atlantic Forest coastal communities: diversity of plant uses in Gamboa (Itacuruçá Island, Brazil). Human Ecology. 1993, 21: 419-430. 10.1007/BF00891142.

29. Figueiredo GM, Leitão-Filho HF, Begossi A: Ethnobotany of Atlantic Forest Coastal Communities: II. Diversity of plant uses at Sepetiba Bay (SE Brasil). Human Ecology. 1997, 25: 353-360. 10.1023/A:1021934408466.

30. Hanazaki N, Leitão-Filho HF, Begossi A: Uso de recursos na Mata Atlântica: o caso da Ponta do Almada (Ubatuda, Brasil). Interciencia. 1996, 21: 268-276.

CHAPTER 3 URINARY DISEASES AND ETHNOBOTANY AMONG PASTORAL NOMADS

Aref Abu-Rabia

INTRODUCTION

People have been using traditional medicine including ethno-botany for several thousand years. Ancient Arabic medicine was influenced by the ancient medicinal practices of Mesopotamia, Greece, Rome, Persia and India. The Greco-Roman system of medicine was developed based primarily on the writings of Hippocrates (460-360 B.C.), Dioscorides (circa 54 to 68 AD) and Galen (130–201 AD). A combination of political and religious factors caused many Greek and Syriac-speaking scholars to move eastward to Persia and to establish centers of learning there. The city of Gundishapur in southwest Iran also became a center of learning, with a well-known medical school, in the sixth century AD [1, 2]. One of the Arab physicians during the time of the Prophet Muhammad (571–632 AD) was al-Harith ibn Kalada (d. 634), one of the most prominent physicians of his time, who traveled to Gundishapur in Persia and studied medicine prior to the establishment of Islam. Another renowned Arab physician was Ibn Abi Rimtha. The sayings (*Hadith*) of the Prophet Muhammad on health and illness were systemized and became known as *The Medicine of the Prophet* (*al-Tibb al-Nabawi*) [3, 2]. During the Umayyad rule (from 661–750 in the East, based in Damascus), many ancient medical works began to be translated. For five centuries (750–1258) the Abbasids, based in Baghdad, dominated the socio-political life of the greater part of the Muslim world. Countless manuscripts, particularly those written in Greek, were collected and stored in *Bayt al-hikmah* (The House of Wisdom, established in 830, by the Caliph al-Ma'mun), where scholars worked to translate them into Arabic [4, 5].

Within a century, Muslim physicians and scientists were writing original contributions to medical and botanical knowledge. One of the greatest and most famous Islamic doctors was Ibn Sina (Avicenna 980–1037), author of*The Canon of Medicine* (*Kitab al-Qanun fi al-Tibb*), the epitome of Islamic medicine. This work is the culmination and masterpiece of the Arab systematization of medical science, and includes many descriptions of the uses of medicinal plants [6]. Other Arabic philosopher-physicians were al-Razi (Rhazes 865–923) who wrote *The Comprehensive Book on Medicine* (*Kitab al-Hawi fi al-Tibb*). The material written by al-Hawi is arranged under headings of different diseases, with separate sections on pharmacological topics. Ibn Sina's and al-Razi's works were later translated into Latin, and continued to influence medical science well into the nineteenth century [7, 8, 9].

In the western part of the Islamic empire, the Umayyads of Andalus (Islamic Spain) made their capital at Cordoba. Areas of Cordoba and Granada became centers of learning. The richness and diversity of the flora of Spain were major contributing factors to the development of medical botany. The majority of physicians were herbalists and vice versa. The physician Ibn al-Baytar (1197–1248), authored *The Compendium of Simple Drugs and Food* (*al-jami' li-mufradat al-adwiya wa'l-aghdhiya*), in which he described more than 1400 medicinal drugs, 300 of which had not previously been described, recording them alphabetically and discussing them with great clarity and detail. The work specified the names of herbs and remedies in various languages, thus providing a first

Citation: Aref Abu-Rabia. Urinary diseases and ethnobotany among pastoral nomads, Journal of Ethnobiology and Ethnomedicine, 2005; 1:4. DOI: 10.1186/1746-4269-1-4.

class tool for the comparative research of medicinal plants. Other well-known physicians who wrote on plant uses were: Ibn Juljul, al-Ghafiqi, Ibn Bajjah, Ibn Samajun, and Abu'l-Hassan al-Andalusi [10, 7]. Traditional medical information grounded in the Arab medicine of the Middle Ages was gradually transferred to traditional healers and to the general public [11]. The use of herbal medicine is still widespread throughout the populations of the Middle East, including the pastoral nomadic tribes [12, 13, 14,15, 16, 17, 18, 19].

Among the pastoral Bedouin, hundreds of species of trees and shrubs are employed as analgesics, astringent, diuretics, emetics, purgatives, poultices, salves, and tonics. Some of these herbs are aimed at cleansing the pastoralist's body of polluting influences, bad spirits, *jinns*, and the negative effects of sorcery and/or witchcraft. The pastoral nomadic tribes depend on their local healers and traditional medicine as recorded in Table 1. (see Additional file 1).

METHODOLOGY

The data for this paper are derived from a broad twenty-year study of ethnobotany and folk medicine among the pastoral nomadic Bedouin tribes in the Negev, Jordan and Sinai deserts, carried out from (1984–2004). The paper is based on interviews with healers and patients. All the material was recorded in field logs, and some was tape-recorded. Unstructured interviews and the observation of participants were carried out in the informants' homes (120 men and 120 women), as well as in the homes of traditional healers (15 men and 10 women). Most of the healers were in the age range of forty to eighty. All the informants were married and over thirty. There were five males from each desert, and four female healers from the Negev, three from Sinai and three from Jordan. The informants were divided into two groups of forty men and forty women from each desert. The collected information was used to construct a list of the indigenous ethnobotanic medicine. Samples from all the plants were collected and identified by healers, patients and university botanists.

RESULTS AND DISCUSSION

This paper describes the treatment of diseases and disorders of the urinary tract by traditional herbalists among the pastoral nomadic Bedouin tribes in the Middle East. In this study, we identified eighty-five plant species, which belong to thirty-six families.

The use of traditional medicine by the pastoral nomads, and the appeal to traditional healers over the course of many centuries established a psychological-therapeutic dependence of the pastoral nomadic tribes upon these healers. The rich variety of approaches employed by pastoral nomadic healers to treat disorders and diseases of the urinary tract is indicative of the depth and breadth of indigenous medicine practiced among the pastoral nomads in the twentieth century. The analysis of my collected data, together with the information extracted from the literature on herbal and ethnobotanic medicine of countries in the Middle East [14, 20, 15,21, 18], yielded Table 1 (see Additional file 1). This table includes eighty-five plants with medicinal potential which have been used among the pastoral nomadic Bedouin tribes in the Middle East from generation to generation as reported by my informants.

Table 1 presents information on which parts of the plants are used and in what manner. It should be noted that for some plants, the uses in different countries of the Middle East are similar [22, 16]. However, dissimilar uses were also observed for certain plants in different countries/tribes in the Middle East [23, 15, 19, 24]. The important information gathered in this study will help to preserve the heritage and knowledge of ethnobotanic and folk medicine of the indigenous pastoral tribes of the Middle East. This study will generate awareness in the region concerning the potential for conserving plant resources in medicine, food, nutrition and folk heritage. It is of the utmost importance to preserve this heritage, which relates to the traditional, economic and medicinal uses of available plant resources in the countries of the Middle East.

The many medicinal substances which we were able to identify as used in traditional medicine included various plants species. The analysis of the findings shows that the three deserts where I conducted my research served as the geographic origin of the medicinal substances. These plants were available because they grew as wild and cultivated plants and were part of the natural flora of these deserts. The pastoral nomads used these plants as food and as medicine, by eating different parts of the plants including the following: leaves, flowers, barks, stems, stalks, roots, rhizomes, bulbs, pith, fruit, corms, inflorescenes, shells, berries, seeds, stones/pits (in fruit), soft seed pods, buds, and shoots.

It should be noted that wild desert plants also contain a host of other biologically active compounds besides nutrients. The physiological effects of these other compounds in relation to plant nutrients are not well known, but could affect nutrient and medical utilization or other functions. These topics are of relevance for future research in terms of improving our understanding of human nutritional and medical requirements of the pastoral nomads in the Middle East.

ACKNOWLEDGEMENTS

Part of my research was supported by a grant from the British Council in Tel-Aviv; The Faculty of Humanities and Social Sciences at the Ben-Gurion University of the Negev; The Natural Medicine Research Unit at Hadassah Jerusalem; The National Council for Research and Development in the Ministry of Science. To all of these I owe a debt of gratitude. I am indebted to Prof. Clinton Bailey and Dr. Lauren Basson for their reading, advice and comments. Special thanks to Prof. Allan Witztum, Nissim Krispil and Yusef Or, for identifying the plants. Finally, my great thanks go to the pastoral nomadic healers and the many people I have not mentioned individually by name, who kindly provided me with invaluable information, without whose help and kindness I would not have managed to conduct this research.

ADDITIONAL MATERIAL

Additional File 1

Table 1: Urinary Diseases and Ethnobotany among Pastoral Nomads in the Middle East

Scientific family	Scientific name	Local name	Plant parts	Preparation	Ethno-botanical uses
Adiantaceae	Adiantum capillus-veneris L	*Kuzbarat al-Beir, Sha'arat al-Ghulih*	leaves and roots	boil in water and drink	Colds, diuretic, urinary tract infections
Anacardiaceae	Rhus coriaria L	*Summaq*	fruit	drink extracted liquid from the fruit	urinary tract infections and urinary retention
Boraginaceae	Anchusa strigosa Banks et Sol.	*Ihmim, lisan al-thur*	leaves and roots	crush, soak in water, and drink	urinary tract infections
Caesalpiniaceae	Cassia italica (Mill.) Lam. Ex F.W. Andrews	*Kasiah, Sinamakky*	leaves	boil in water and drink	urinary retention
	Cercis siliquastrum L.	*'Arus al-Ghaba*	leaves and flowers	boil in water and drink	urinary retention and infections
Capparaceae	Capparis spinosa L.	*Qubbar, Lasaf*	bulbs	soak in water and drink	diuretic, urinary tract infections and retention
Caryophyllaceae	Paronychia argentea Lam.	*Rijil al-Hamamih, Rijil al-'Asfur*	leaves and flowers	boil in water and drink	diuretic, urinary retention and urinary tract infections
Chenopodiaceae	Beta vulgaris L.	*Shamandar, salq, banjar*	leaves and roots	raw	urinary tract infections
	Chenopodium album L.	*Fisa al-Kalb*	the whole herb	soak in water and drink	urinary tract infections and retention
Asteraceae	Cichorium pumilum Jacq.	*'Iltt, 'Ilk, hindabah*	green leaves	eat, or soak in water and drink	diuretic, urinary tract infections and retentions
	Cnicus benedictus L.	*Kassuba*	leaves, stalks and roots	boil in water and drink	urinary retention and urinary tract infections, prostate illness
	Inula viscose (L.) Ait.	*'Irq al-Tayun, Tayun*	leaves	boil in water and drink	urinary retention and urinary tract infections
	Matricaria aurea (Loefl.) Sch. Bip.	*Babunaj, Rabil*	leaves and flowers	boil in water and drink	urinary retention, urinary tract infections, kidney pains, testicle pain
	Phagnalon rupestre (L.) DC.	*Qadih,Sufan Qray'i*	leaves and stalks	boil in water and drink	urinary retention and urinary tract infections
	Scolymus maculatus L.	*Sinariya Hawliya*),	stems, stalks	boil in water	urinary tract infections
	Taraxacum megalorrhizon (Forssk.) Hand.-Mazz.	*Salatat al-Ruhban*	Flowers and leaves	eat as raw salad	urinary retention and urinary tract infections
	Varthemia iphionoides Boiss. et Bl.	*Slimaniya, Ktilih*	Leaves and stalks	boil in water and drink	urinary retention and urinary tract infections

Brassicaceae	Anastatica hierochuntica L.	*Kaf al-Rahaman, Kaff Maryam*	dried fruit	crush the fruit, soak in water and drink	urinary retention and urinary tract infections, female sterility, male impotency
	Brassica nigra (L.) Koch	*Left*	stems, seeds	eat or soak in water and drink	urinary retention and urinary tract infections, stomach ache
	Capsella bursa-pastoris L. Medik.	*Kis al-Ra'y*	dried leaves	boil in water and drink	urinary retention and urinary tract infections
	Lepidium sativum L	*Habb al-Rashad, Hurf*	seeds	boil in water and drink	urinary retention and urinary tract infection, prostate illnesses
	Raphanus raphanistrum L.	*Fijil barri, Fijjilih*	leaves	boil the leaves in water and drink	urinary tract infections and urinary retention
	Sisymbrium irio L.	*Kibs*	leaves and flowers	eat as raw salad	urinary retention and urinary tract infections
Cucurbitaceae	Citrulus vulgaris Schirad	*Battykh*	fruit	eat	urinary tract infections
	Cucumis melo L.	*Shimmam*	fruit	eat	urinary retention and urinary tract infections, kidney pains, constipation
	Cucumis sativa L.	*Khiar, Khyar*	fruit	eat	urinary retention and urinary tract infections
	Ecbalium elaterium (L.) A. Rich.	*Qitha al-Hameir, fagus al-Hameir*	fruit, seeds	squeeze or soak in water and drink	urinary retention and urinary tract infections, yellow fever, liver infections
Cupressaceae	Juniperus communis, Juniperus oxycedrus L. Juniperus phoenicea L.	*'Ar'ar, 'Itran, Qitran*	berries	soak in water and drink	diuretic, urinary retention and urinary tract infections, kidney pains, prostate, impotency, female sterility, vaginal infection.
Globulariaceae	Globularia arabica Jaub. et Spach	*'Aynun*	leaves, flowers	soak in water and drink	urinary incontinence and urinary tract infections, kidney pains
Poaceae	Elymus repens (L.) Gould	*Nejil, Njil*	leaves and roots	boil in water and drink	urinary tract infections, and cystitis
Poaceae	Hordeum sativum Hordeum spontaneum C. Koch. Avena sterilis L.	*Sha'ir* *khafour, shufan*	seeds, bran Leaves, seeds	eat as pita, boil crushed seeds in water and drink eat bread or prepare porridge	urinary tract infections and retention urinary retention and infections, diabetes, increasing breast milk and strengthening mother body after childbirth
	Triticum aestivum L.	*Qamh*, hintta	Leaves, seeds, bran	eat as pita, soak seeds or bran in water and drink	urinary retention and prostate illnesses, impotency, diabetes, high blood pressure

	Zea mays L.	*dhrah, dhurah, dhurah shamiya*	Maize fruit	soak in water and drink	Diuretic, urinary retention, urinary tract infections and cystitis, diabetes
Iridaceae	Iris pallida Lam. Iris palaestina (Back.) Boiss.	*Sauwsan*	Leaves, rhizome	boil in water and drink	urinary tract infections and retention
Labiatae	Coridothymus capitatus (L.) Reichb.	*Zahayfy, Za'tar Farsy*	leaves and flowers	boil leaves and flowers in water and drink	urinary tract infections and retention, prostate, testicles pains
	Mentha longifolia L.	*Na'na' barri, Habaq*	leaves	boil in water and drink	blood in the urine, urinary tract infections and retention, vomiting, nausea
	Micromeria myrtifolia Boiss. et Hohen.	*Zufih*	leaves	boil in water and drink	urinary retention and urinary tract infection.
	Rosmarinus officinalis L.	*Iklil al-Jabal*	leaves	boil the leaves in water and drink	urinary tract infections and retention, vagina infection and sterility
Liliaceae	Allium ampeloprasum L.	*Thum barri*	fruit, leaves and stalks	mix with olive oil and eat	urinary tract infections and retention, kidney pain
	Asparagus aphyllus L.	*Halyun, Hiliun*	leaves	eat as raw salad	diuretic, urinary retention and urinary tract infections
	Asphodelus aestivus Brot.	*Swai, 'Unsol, 'Onsol*	roots	boil in water and drink	urinary retention and urinary tract infection
	Colchicum ritchii R. Br.	*Wada', 'Akna, 'Ukna, Farj al-Ard*	corms	apply corms in hot aches, peel and eat it	kidney infections, urine pains, and prostate
	Ruscus aculeatus L.	*Ass Barri, Khizana, 'Urf al-Deik*	roots, leaves	boil in water and drink	urinary tract infections and retention
Linaceae	Linum pubescens Banks et Sol.	*Kittan*	seeds	boil in water and drink	urinary tract infections, urinary pains, and prostate illnesses
Malvaceae	Althaea officinalis L.	*khitmiya, khutmiya, Khutmiy*	leaves	eat as raw salad	urinary tract infections and the incontinence
	Lavatera trimestris L.	*Khubbizat al-Far*	leaves, seeds	eat as raw salad, or cook	urinary tract infections, retention and prostate illnesses
	Malva sylvestris L.	*Khubiza, Khubaizih*	leaves	eat as raw salad, or cook, or boil in water and drink	urinary tract infections, retention and prostate illnesses, vaginal infection and skin irritation
Mimosaceae	Acacia raddiana Savi	*Sayal*	resin	boil in water or milk and drink	urinary retentions and urinary tract infections, and prostate
	Prosopis farcta (Banks et Sol.)	*Yanbut*	leaves, fruit and roots	boil in water and drink	urinary tract infections and retention, diabetes, high blood pressure

Oleaceae	Olea europaea L.	*Zaytun, Zeit*	oil	drink	urinary retention and urinary tract infection, prostate, kidney pains, skin irritation, constipation
Oxalidaceae	Oxalis corniculata L.	*Hummaydh*	whole plant	soak in water and drink	urinary retention and urinary tract infection
Fabaceae	Alhagi maurorum Medik	*'Aqul, 'uqul, Shuk al-Jimal*	rhizome	boil in water and drink	urinary retention and urinary tract infection
	Lupinus albus L.	*Turmus*	seeds, whole herb	soak in water and drink	diuretic, urinary tract infections and retention
	Ononis antiquorum L.	*Shibriq Shaik*	leaves, flowers and dried roots	boil in water and drink	diuretic, urinary tract infections and retention
	Scorpiurus muricatus L.	*lisan al-Kalb*	leaves, stems and stalks	eat as raw salad	urinary tract infections
	Trigonella foenum-graecum L.	*Hilbih*	seeds	boil seeds in water and drink	urinary tract infections and retention, diabetes, and high blood pressure
Pinaceae	Pinus halepensis Mill.	*Snubar, Qriesh*	leaves	boil in water and drink	urinary retentions, cystitis and prostate illnesses
Plantaginaceae	Plantago lagopus L.	*Yanama, 'Ishbit al-Baraghith*	leaves	boil in water and drink	urinary tract infections
Polygonaceae	Polygonum equisetiforme Sm.	*Qudhdhab*	leaves	boil in water and drink	urinary retention and urinary tract infections, blood in the urine.
	Rheum palaestinum Feinbr.	*'Uttrafan*	leaves and roots	boil in water and drink	urinary retention and urinary tract infections.
	Rumex cyprius Murb.	*Hummidhah, Hummidh*	leaves	eat as raw salad	urinary tract infections and retentions, blood in the urine
Portulacaceae	Portulaca oleracea L.	*Farfahina, Rijlih*	leaves and stalks	crush, soak in water and drink, or eat as raw salad	diuretic, urinary tract infections and retention, cystitis
Ranunculaceae	Adonis aestivalis L.	*'Ain al-Deik*	leaves	soak in water and drink	urinary tract infections, prostate, impotency
	Nigella sativa L.	*Habbit al-Barakah, Habbih Suda, Qazhih*	seeds, or extracted oil	boil the seeds in water and drink, or drink the oil	urinary tract infections and retention, prostate illnesses, blood in urine, diabetes, impotency, sterility
Rhamnaceae	Zizyphus spina-christi (L.) Desf.	*Sidr, Dowm*	fruit, or leaves	eat the fruit, soak leaves in water and drink	urinary tract infections and prostate
Rosaceae	Rubus sanctus Schreb.	*'Ullayq*	berries, or leaves	eat the berries, soak leaves in water and drink	urinary retention and prostate illnesses
Rubiaceae	Rubia tenuifolia D'Urv.	*Fuwwa*	roots	boil in water and drink	diuretic, urinary retentions, skin irritation

Rutaceae	Citrus limon L.	*lamun*	fruit	drink with water or tea	urinary tract infections and retention, stomach ach, diarrhea, upset, nausea
Solanaceae	Mandragora autumnalis Bertol.	*Mjininih, Yabruh, Tuffah al-Majanin*	ripe fruit	eat	urinary tract infections, sterility, vaginal infections, increasing breast milk, evil eye
Scrophulariaceae	Linaria cymbalaria Mill.	*Hashishat al-Shuquq*	leaves	boil in water and drink	diuretic, urinary tract infections and retention, breast infection
Umbelliferae	Anisum vulgare L.	*Yansun*	seeds	soak or boiled in water and drink	diuretic, urinary tract infections and retention
	Apium graveolens L.	*Karafs*	leaves, roots	boil in water and drink	urinary tract infections and pains, incontinence and retention
Umbelliferae	Conium maculatum L.	*Shiqran, Sikran, shoukaran*	dried leaves	soak in water and drink	urinary tract infections and prostate illnesses
	Petroselinum crispum/ Petroselinum sativum	*Baqdunis*	leaves	soak in water and drink	urinary tract infections and retention, and prostate illnesses
	Pituranthos tortuosus (Desf) Benth.ex Aschers. et Schweinf.	*Zaqquh, Zagguh*	stems and flowers	boil in water and drink	blood in the urine, urinary retention
Urticaceae	Parietaria judaica L.	*Kuzziza*	leaves	Boil in water and drink	urinary tract infections, urinary retention, vaginal infection
	Urtica pilulifera/ Urtica urens	*Qurris, Hurriq*	leaves	Boil in water and drink	urinary tract infection and retention
Zygophyllaceae	Tribulus terrestris L.	*Draiseh, Qatab*	dried leaves	Boil in water and drink	urinary tract infections and retention

REFERENCES

1. Mursi: 'Arab: *Dirasat fi'l-Shi'un al-tibbiyah al-'arabiyah* = Studies on Arab Medical Affairs. Al-Iskandariya: al-Ma'arif (Arabic). 1966, 7-12.
2. Savage-Smith, Emilie: Medicine. Encyclopedia of the History of Arab Science. Edited by: Roshdi Rashid, Regis Morelon. 1996, London and New York: Routledge, 3: 907-908.
3. Hawting GR: The Development of the Biography of al-Harith ibn Kalada and the Relationship between Medicine and Islam. The Islamic World, from Classical to Modern Times. Edited by: Bosworth CE, Charles Issawi, Roger Savory, Udovitch AL Princeton. 1989, New Jersey: The Darwin Press, Inc, 127-137.
4. Hitti, Philip: History of the Arabs, From the Earliest Times to the Present. 1952, London: Macmillan & Co, 311-312.
5. Ullmann, Manfred: Islamic Surveys: Islamic Medicine. 1978, Edinburgh: University Press, 7-40.
6. Al-Said, Mansour Solyman: Medicine in Islam. Encyclopedia of the History of Science, Technology, and Medicine in Non-Western Cultures. Edited by: Helaine Selin. 1997, Dordrect/Boston/ London: Klewer Academic Publishers, 695-698.
7. Johnstone, Penelope: Prophet Medicine. Ibn Qayyim al-Jawziyya. Medicine of the Prophet. Edited by: Penelope Johnstone. 1998, Cambridge: The Islamic Texts Society, xxx-xxxii.
8. Murad, Amnah: *Lamhat min tarikh al-tibb al-qadim* = Glimpses From the History of Early Medicine. al-Qahira. Maktabat al-Nasr al-Haditha (Arabic). 1966, 1-32.
9. Al-Shatti, Ahmad: *al-'Arab wa-al-tibb = The Arabs and Medicine*. Dimashq: Manshurat Wazarat al-Thaqafa (Arabic). 1970, 12-65.
10. Al-Najjar, Amer: Fi Tarikh al-Ttib fi al-Dawlah al-Islamyya: History of Medicine in the Islamic Empire. Al-Qahira: Dar al-Ma'aref (Arabic). 1994, 159-229.
11. Lev, Efraim, Zohar Amar: Ethnopharmacological survey

of traditional drugs sold in Israel at the end of the 20th century. Journal of Ethopharmacology. 2000, 72: 191-205. 10.1016/S0378-8741(00)00230-0.

12. Abu-Khusa, Ahmad: Beer al-Saba' wal-Hayah al-Badawiya. Amman: Matabi' al-Mu'assasa al-Sahafiya al-Urdunniya (Arabic). 1976, 70–71: 106-113.
13. Abu-Rabia, Aref: Traditional Bedouin Medicine. 1999, Tel-Aviv: Mod Publishing, (Hebrew), 35-92.
14. Bailey, Clinton, Danin A: Bedouin plant utilization in Sinai and the Negev. Economic Botany. 1981, 35 (2): 145-162.
15. Jabbur, Jibrail: The Bedouins and the Desert: Aspects of Nomadic Life in the Arab East. 1995, Albany: State University of New York Press, 64-81.
16. Palevitch, Dan, Zohara Yaniv: Medicinal Plants of the Holy Land. 2000, Tel-Aviv: Modan Publishing House, 9-88.
17. Pillsbury, Barbara: Traditional Health Care in the Near East. 1978, A report prepared for the U.S. Agency for International Development, Washington D.C, 1-25.
18. Shuqayr N: Tarikh Sina. Cairo: Matba'at al-Ma'arif (Arabic). 1916, 340-375.
19. Tal, Pnina: Medicinal Plants. 1981, Tel-Aviv: Reshafim (Hebrew), 15-17.
20. Burckhardt, John Lewis: Travels in Syria and the Holy Land. 1992, London: Darf Publishers Ltd, 12-97.
21. Musil, Alois: The Manners and Customs of the Rwala Bedouins. 1928, New York: American Geographical Society, 90-98.
22. Boulos, Loutfy: Medicinal Plants of North Africa. 1983, Algonac: Reference Publications, Inc, 23-67.
23. Ghazanfar, Shahina: Handbook of Arabian Medicinal Plants. 1994, London: CRC Press, 44-87.
24. Abu-Rabia, Aref: Bedouin Health Services in Mandated Palestine. Middle Eastern Studies. 2005, 41 (3): 421-429.

CHAPTER 4

ALPINE ETHNOBOTANY: TRADITIONAL KNOWLEDGE OF GASTRONOMIC AND MEDICINAL PLANTS

Andrea Pieroni and Maria Elena Giusti

BACKGROUND

In recent years, only a few ethnobotanical researches in the Alps have focused on the interaction between plant resources and human societies within ethnic minority groups [1, 2, 3, 4, 5, 6]. This is regrettable because comparative and/or quantitative analyses aimed at increasing the understanding of how Traditional Knowledge (TK) of botanicals changes across cultures and over time and space are essential if we are to form strategies aimed at sustaining bio-cultural diversity and unique experiences of interactions between nature and cultures, and their transmission mechanisms along generations [7, 8, 9, 10, 11, 12, 13].

Furthermore, the Alps represent one of the most interesting but least studied regions in Europe, particularly with regard to the exploration of tangible and intangible cultural heritage related to TK of plants, even though the potential outputs of such researches could be important in sustainability projects focused on organic farming, home gardens, local foods, eco-tourism, eco-gastronomy, and eco-museology; and even though climate change may have a tremendous impact on Alpine biodiversity and related ethnobotanical resources [14].

Although a number of ethnobotanical inventories were compiled during the past 50 years in a few other Occitan [15, 16, 17, 18, 19, 20], Franco-Provencal and Walser Alpine [21, 22, 23] valleys in Piedmont and the surrounding areas [24, 25, 26], no field ethnobotanical studies had been conducted in the Varaita valley. Instead only a few linguistic [27] and ethnographic researches [28, 29, 30, 31, 32] took place.

The aims of this study were therefore:

- To record the food and medical ethnobotanical knowledge of the Occitans living in the upper Varaita valley (Chianale and Bellino/Blins);
- To compare the collected data with those available in the ethnobotanical literature of other Alpine valleys in Piedmont and surrounding areas (Aosta Valley, and the French and Swiss sides of the Alps), and to point out the eventual occurrence of a specific "Occitan" ethnobotany;
- To suggest local plant resources of particular interest for sustainable small-scale agricultural/gathering activities and eco-tourism activities.

METHODS

Ethnographic, Historical, and Environmental Context

The presence of a single "Occitan" culture/language is one of the most disputed issues in European ethnolinguistics [33], given the considerable heterogeneity of the vast area covered by all linguistic varieties linked under the "Occitan" umbrella. In Italy, "Occitan" languages are spoken in a dozen of valleys in the Western Alps in the Piedmont region [34], where the interest of the media has been concentrated in recent years on tracing this disappearing culture [35], and within a small

Citation: Andrea Pieroni and Maria Elena Giusti. Alpine ethnobotany: traditional knowledge of gastronomic and medicinal plants, Journal of Ethnobiology and Ethnomedicine, 2009; 5:32. DOI:10.1186/1746-4269-5-32.

enclave in Southern Italy (i.e. Guardia Piemontese in the Calabria region).

The focus of this study, the Varaita valley, is located in the Western Alps in the Piedmont region, Cuneo province, in North-Western Italy (figure 1). The valley is crossed by an homonymous alpine torrent, which is 75 km long and springs from the slopes of the mountain, Monviso, in the Cottian Alps near the French border, and enters the Po River near Casalgrasso. From the beginning of the 12th century until 1601, most of the Varaita valley was part of the small but influential marquisate of Saluzzo. The upper Varaita valley, called Chastelada, had a different fate however. For four centuries, from 1343 until 1713, it belonged to an independent federation known as Escartons, which was based in Briançon in France. This federation was a good example of enlightened and progressive self-government [28]. After 1713, the upper valley had a quite complex history due to its location on the border of Italy and France. It was the locus of continuous military incursions in the 18th Century from the French side, and in the 19th and most of the 20th century it saw much intensive smuggling activity.

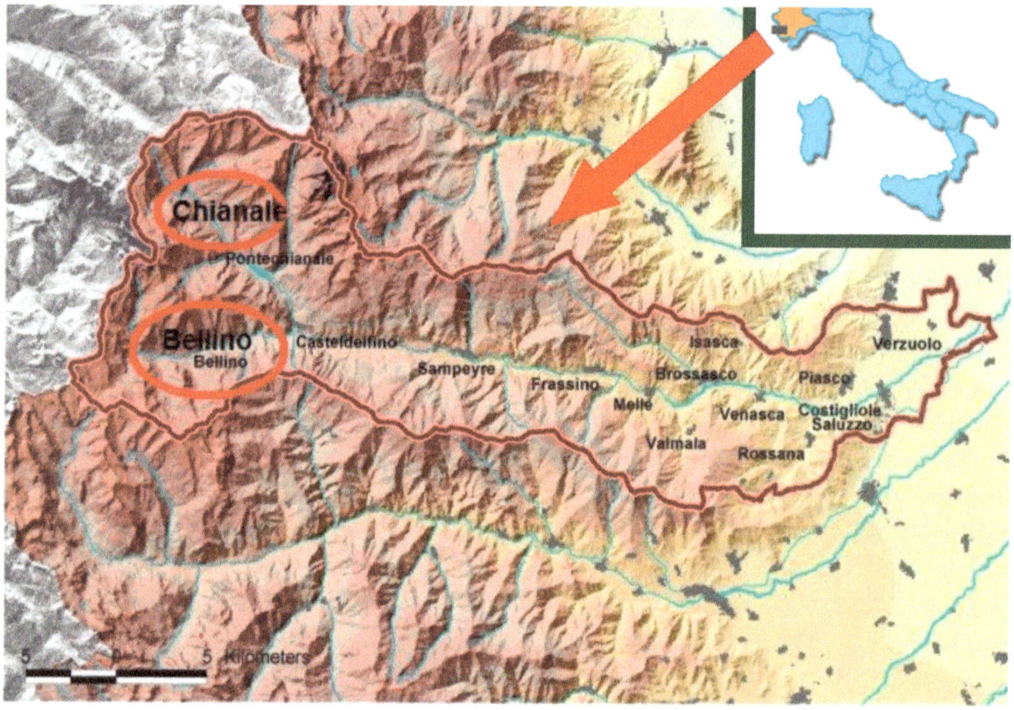

Figure 1. Location of the study area.

At present, only mainly elderly persons still live year round in the upper part of the Valley. Bellino (1480 m a.s.l.) includes the hamlets of Çelle, Chiazale, Chiesa Fontanile, Pleyne, Ribiera, and Sant›Anna, and has the status of a municipality. Chianale (1797 m a.s.l., figure 2) is a hamlet in the municipality of Pontechianale. The hamlets of Bellino and Chianale are among the smallest Alpine villages in Italy (numbering approx. 15-30 permanent, mainly elderly inhabitants in each village).

Figure 2. Chianale.

Regarding vegetation, the upper Varaita valley presents mainly the association *Rhododendro-Vaccinietum cembretosum* (characterized by *Rhododendron ferrugineum, Vaccinium myrtillus* and, to a lesser extent, *V. vitis-idaea, V. uliginosum, Lonicera coerulea, L. alpigena, L. nigra, Juniperus nana, Rosa pendulina, Daphne mezereum, Clematis alpina, Sedum anacampseros, Luzula sieberi, Festuca flavescens and Homogine alpine*), and the subassociation *Junipero-Arctostaphyletum cembretosum* (characterized by *Juniperus nana, Arctostaphylos uva-ursi, Cotoneaster integerrimus, Avenella flexuosa, Vaccinium myrtillus, Minuartia laricifolia, Festuca flavescens, Galium obliquum and Brachypodium caespitosum*). The total rainfall is 854 mm/yr (Casteldelfino, 1296 m a.s.l.); the climate is sub-littoral alpine, with some continental characteristics [36].

The local economy of the upper valley is based on traditional agro-pastoral activities (cow breeding), but this is carried out in only a minor way. There is some tourism, especially in the summer (alpine walking and trekking). Almost all of the few middle-aged and younger members of the community are employed in the lower part of the valley. Traditional cultivation of local staples (rye, barley, buckwheat) had disappeared by the 1970's and home gardens these days are managed as a secondary activity only.

Field Study

Ethnobotanical investigations of the upper Varaita Valley (figure 1), in the communities of Bellino (Celle, Prafauchier, Chiazale, Chiesa, Pleyne, Ribiera, and Sant'Anna, and Bals) and Chianale (figure 2) were carried out over four weeks in July 2008. Local informants (n = 67) were asked about local food and medicinal plants via open and semi-structured interviews (figure 3). Detailed information was sought about the plants' vernacular names, and particularly their past and present uses.

Figure 3. Documentation of a few interviews conducted during the field study (photo: Nicola Robecchi).

Prior informed consent (PIC) was obtained verbally before commencing each interview, and ethical guidelines adopted by International Society of Ethnobiology [37] were followed. The quoted botanical species were identified using Pignatti's *Flora d'Italia* [38] (with the help of a recent standard work on Alpine flora [39]), and systematically framed as suggested by the most recent checklist of Italian vascular flora [40]. Voucher specimens of the wild taxa were collected, and they are now deposited at the Herbarium of the University of Gastronomic Sciences.

Data Analysis

The data collected during the field study were reported using Microsoft® Excel, and a descriptive comparison was conducted with ethnobotanical data available in the scientific literature (e.g. field studies previously conducted in Piedmont and other surrounding alpine areas [15, 16, 17, 18, 19, 20, 21, 22, 23, 24, 25, 26]). However, given the tremendous heterogeneity of the methodological approaches adopted in all these previous studies (in which the number of informants was rarely reported, and in a few cases even the names of the villages visited were omitted), it was impossible to carry out a proper statistical analysis aimed at capturing similarities between our data and the data occurring in the ethnobotanical literature (i.e.: correspondence analysis, Jaccard index).

RESULTS

All the quoted botanical taxa (n = 88), including their related recorded local names and traditional gastronomic and/or medicinal uses, are reported below. We have included in brackets the voucher specimen codes and the quotation index (+++: taxa quoted by at least 40% of the informants; ++: taxa quoted by more than 10% and less than 40% of the informants; +: taxa quoted by fewer than 10% of the informants).

Achillea herba-rotta All. (Asteraceae). Wild. Canamìo, Rua. Dried flowering aerial parts (figure 4): tea or macerated in alcohol as a digestive, for treating toothache and headaches, as a mild tranquillizer, or an emmenagogue. Topical application with butter and heated paper, which is then covered with a piece of woollen cloth for treating bad coughs. (CHI-CAN) +++

Figure 4. Dried flowering tops of *Achillea herba-rotta*.

Achillea millefolium L. (Asteraceae). Wild. Ervo taiouro. Dried flowering aerial parts: tea as an anti-haemorrhoidal (humans) or an anti-diarrheic (cattle). Fresh aerial parts: topically applied as a cicatrizing agent. (CHI-ETA) ++

Alchemilla vulgaris (Rosaceae). Wild. Gunnellettes (plural). Dried leaves: use not recalled. Traded in the past. +

Allium cepa L.(Alliaceae). Cultivated. Seoulo. Fresh bulbs: consumed in salads and as a seasoning in the local cuisine. +++

Allium porrum L. (Alliaceae). Cultivated. Por. Fresh whole plant: traditionally consumed mixed in a kind of maize polenta. +

Allium sativum L. (Alliaceae). Cultivated. Ai. Fresh bulbs: as a seasoning in the local cuisine (rare). In the past made into necklaces and worn by children as a treatment for worms. ++

Allium schoenoprasum L. (Alliaceae). Wild. Poureto. Fresh leaves: added in pancakes (*bignes*) as a seasoning. To be avoided by cows and sheep, as the milk takes on a horrible taste. (CHI-POU) +++

Antennaria dioica (L.) Gaertn. (Asteraceae). Wild. Piot de chat. Dried flowering aerial parts: tea as a digestive. Traded in the past. +

Arnica montana L. (Asteraceae). Wild. Strugnarelo. Dried flowers: macerated in alcohol or in oil and applied externally for treating bruises. Traded now and in the past. (CHI-STR) +++

Artemisia absinthium L. (Asteraceae). Wild. Erbio bioncho, Ousenzo. Dried aerial parts: put in cattle stables and in wardrobes as an anti-parasitic or to drive away moths. Tea used as an anti-hypertensive, digestive, and an antihelminthic (for children). (CHI-BIO) +++

Artemisa genipi Weber (Asteraceae) (Figure 5). Wild. Genepì maschi, Genepì nero. Dried flowering aerial parts: tea as a panacea (especially against coughs and as a digestive). Macerated in alcohol as a digestive. Traded in the past and now (illegally). (CHI-GEM) +++

Figure 5. *Artemisia genipi.*

Artemisia glacialis L. (Asteraceae) (Figure 6). Wild. Genepì fumelo, Bava dei ghiacciai. Dried flowering aerial parts: tea as a panacea (rare). Macerated in alcohol as a digestive. Traded now and in the past. (CHI-GEF1) +++

Figure 6. *Artemisia glacialis.*

Artemisia umbelliformis Lam. (Asteraceae) (Figure 7). Wild and cultivated. Genepì fumelo. Dried flowering aerial parts: macerated in alcohol as a digestive. Traded in the past and now (illegally). (CHI-GEF2) ++

Figure 7. *Artemisia umbelliformis.*

Artemisia vulgaris (Asteraceae). Wild. Artumiso. Dried aerial parts: tea used as an emmenagogue (women) and to facilitate calving (cows). (CHI-ART) +++

Avena sativa L. (Poaceae). Cultivated in the past. Aveno. Fruit used as fodder for animals. +

Boletus spp. (Boletaceae). Wild. Boulet. Fruiting body: consumed (rare). +

Brassica oleracea L. (Brassicaceae). Cultivated. Chaoul. Fresh aerial parts: consumed in soups and as fodder for animals. Often harvested during the snow season and stewed with anchovies (*sancraou*). ++

Bunium bulbocastanum L. (Apiaceae). Wild. Gravaioun. Fresh tubers: eaten raw, or boiled, also in the dough of the *ravioles* (small dumplings made with flour, potatoes and alpine hut cheese, *tumo*). These uses are recalled as occurring only in the past. +

Carlina acaulis L. (Asteraceae). Wild. Chardousso. Fresh flower receptacle as a snack. Dried flowering aerial parts are hung on the outside door year round and used to forecast weather. (CHI-CHA1) ++

Carum carvi L. (Apiaceae). Wild. Chimmel, Serieis. Dried fruits: tea or macerated in alcohol as a digestive. Added to food as a seasoning (especially meat and poultry, particularly rabbit, or *sarass, a* local fermented and spiced ricotta cheese). (CHI-SER) +++

Cetraria islandica (L.) Acharius (Parmeliaceae). Wild. Gratacoul. Dried thallus: tea as a bechic. Traded in the past. (CHI-GRA) ++

Chelidonium majus L. (Papaveraceae). Wild. Erbo de Santo Mario. Fresh latex: topically applied for treating warts. (CHI-ESM) +

Chenopodium bonus-henricus (Chenopodiaceae). Wild. Bies, Zerbes, Orle (plural). Fresh young leaves: consumed boiled in soups and in omelettes. Also preserved in salt. Fodder for cows (reputed to be good as a "cleansing" agent). Topically applied for treating skin inflammation caused by stinging nettle. (CHI-BIE) +++

Cirsium acaule Scop. (Asteraceae). Wild. Vazanel. Fresh flower receptacle, consumed raw as a snack. +

Cirsium arvense (Asteraceae). Wild. Choussio. Fresh young leaves: boiled in soups. (CHI-CHO) +

Cirsium spinosissimum (L.) Scop. (Asteraceae). Wild. Chardousso. Fresh flower receptacles as a snack. Dried flowering aerial parts are hung on an outside door year round and used to forecast weather. (CHI-CHA2) ++

Cynodon dactylon (L.) Pers.(Poaceae). Wild. Gramoun. Roots: tea as a depurative. (CHI-GRA) +

Echium italicum L. (Borraginaceae). Wild. Erbo vieio. Fresh young leaves: boiled in soups. +

Fagopyrum esculentum Moench. (Polygonaceae). Cultivated in the past. Furmetin, Pignoulet. Fruits: ground into flour and used as the main ingredient in the local polenta, often prepared by alternating layers of polenta with layers of local cheese (*tumo*) and butter, or sometimes even adding potatoes and leeks. +++

Fragaria vesca L. (Rosaceae). Wild. Ambourses (plural), Mourso. Fruits: consumed as a snack. (CHI-MOU) ++

Gentiana acaulis L. (Gentianaceae). Wild. Chaousso de cucu. Dried leaves: tea as a digestive and anti-hypertensive. Dried flowers: macerated in alcohol or wine, as a digestive. (CHI-CHC) +++

Gentiana lutea L. (Gentianaceae). Wild. Ergensano. Dried roots (*reis*): macerated in wine, as a digestive, for "cleansing blood "/depurative, or "to make the blood thinner". Decoction used to enhance rumination in cows. (CHI-ERG) +++

Heracleum sphodylium L. (Apiaceae). Wild. Turmel. Fresh aerial parts: to be avoided as fodder (hay would remain humid and be of a very poor quality). (CHI-TUR) +

Hordeum vulgare L. (Poaceae). Cultivated in the past. Uerge.

Folk species: Brins, Brens. Fruits: crushed and used as fodder for hens; the bran used as fodder for cattle. Traditionally consumed as main ingredient in a particular kind of *polenta* (*pouties*). Folk species: Eplà, Pelá. Fruits: roasted and ground, in decoction, as a "coffee substitute". ++

Hypericum perforatum L. (Hypericaceae). Wild. Erbo di San Juan. Fresh flowering aerial parts: macerated in oil for treating burns. (CHI-ESJ) +

Juglans regia L. (Juglandaceae). Semi-domesticated. Nounzal (tree). Kernels: pressed to obtain the oil, which was commonly used in the past in the local cuisine. +

Juniperus communis L. (Cupressaceae). Wild. Chais. Dried fruits: used as a seasoning or macerated in grappa as a digestive. (CHI-CHI) ++

Laburnum alpinum (Mill.) Bercht. & J. Presl (Leguminosae). Wild. Albourn. Fresh flowering aerial parts (*bindareles*): put in the hen house to deter lice and flies. (CHI-ALB) ++

Larix decidua Mill. (Pinaceae). Wild. Melze, Merze. Resin: topically applied as a cicatrizing agent. (CHI-MEL) +++

Lathyrus sp. (Leguminosae). Wild. Lioum. Seeds: ground and the flour used in the past for preparing a kind of polenta. +

Lens culinaris Medik. (Leguminosae). Cultivated in the past. Lentie (plural). Seeds: in soups, or traditionally consumed as main ingredient in a particular kind of *polenta* (*pouties*). ++

Linum usitatissimum L. (Linaceae). Cultivated in the past. Lin. Seeds: as fodder for calves. Mixed with water, heated and externally applied on the chest for treating colds and coughs. +

Malva neglecta Wallr. and *M. sylvestris* L. (Malvaceae). Wild. Rioundelo. Fresh leaves: tea, also used in gargles, as an anti-inflammatory. Boiled in soups. (CHI-RIU) +++

Mentha spicata L. (Lamiaceae). Cultivated in home gardens. Mento. Fresh leaves: tea as a digestive. Also as a seasoning for pancakes (*bignes*). ++

Myrrhis odorata (L.) Scop. (Apiaceae). Wild. Charvei. Aerial parts: to be avoided as they are "poisonous". (CHI-CHV) +

Narcissus poeticus L. (Liliaceae s.l.). Wild. Susarelo, Joes de mel, Fior del mel. Flowers: sucked as a snack (especially by children in the past). (CHI-SUS) +

Nasturtium officinale R. Br. (Brassicaceae). Wild. Creissentin. Fresh leaves: consumed in salads. (CHI-CRE) +

Nepeta cataria L. (Lamiaceae). Wild. Erbo da chat. Dried aerial parts: tea as a mild tranquillizer. Traded in the past. +

Onobrychis montana DC. (Leguminosae). Wild. Sparsei. Fresh and dried aerial parts: highly valued fodder for animals. (CHI-SPA) +

Petasites hybridus (L.) P. Gaertn., B. Mey & Scherb. (Asteraceae). Wild. Chapus. Fresh leaves: fodder for hens. (CHI-CHP) ++

Peucedanum ostruthium (L.) W.D.J. Koch (Asteraceae). Wild. Algrot. Fresh root (*reis*): decoction or directly administered crushed and mixed with salt as a digestive (for cattle and sheep). (CHI-ALG) +++

Pinus cembra L. (Pinaceae). Wild. Elvio. Seeds: pressed to obtain oil (in the past). Cones: macerated in alcohol as a digestive. (CHI-ELV) +++

Piper nigrum L. (Piperaceae). Not local; bought in shops. Pepe. Dried fermented fruits: boiled in milk and butter for treating coughs. +

Plantago lanceolata L. and *P. major* L. (Plantaginaceae). Wild. Piantai. Dried leaves: tea as an appetizer and digestive (for calves only). Fresh leaves: topically applied with cream from cows' milk and bread or clay as a suppurative (also for cows, especially for treating inflamed hooves). (CHI-PIA1/ CHI-PIA2) ++

Polygonum bistorta L. (Polygonaceae). Wild. Zonbuines (plural), Lingo boino, Linboina. Fresh young leaves: consumed in soups. (CHI-ZON) +++

Prunus avium L. (Rosaceae). Semi-cultivated. Serieses (fruits, pl.). Fresh fruits: consumed raw as a snack. +

Prunus domestica L. (Rosaceae). Semi-cultivated. Brigne (fruits, pl.). Fresh fruits: consumed raw. +

Rosa canina L. (Rosaceae). Wild. Bouchier, Bouchiet. Dried pseudofruits: tea as an anti-diarrhoeic. Ground and added to bread dough in the past. (CHI-BUC) ++

Rubus ideaus L. (Rosaceae). Wild. Amurse (plural). Fresh fruits: eaten raw or processed into home-made jams. +++

Rumex acetosa L. and *R. acetosella* L. (Polygonaceae). Wild. Setou. Fresh leaves. eaten raw as a snack. (CHI-SET1 and CHI-SET2) +

Rumex alpinus L. (Polygonaceae). Wild. Gravasso, Ravasso. Fresh leaves: topically applied as a cicatrizing agent. Fresh leaves and stems: fodder for cattle. Fresh stems: eaten raw with salt as a snack or boiled like asparagus; occasionally used in soups (especially in the past). (CHI-GRA) ++

Sambucus nigra L. (Caprifoliaceae). Wild. Sambuc. Fresh fruits: wine and jams (only in the lower part of the studied area). +

Secale cereale L. (Poaceae). Cultivated in the past. Seil. Fruits: flour. Widely used in the past as main ingredient for bread, and also for the local home-made noodles, *courzeitin*, and a porridge-like preparation based on milk and butter, *panado*. +++

Sedum album L. (Crassulaceae). Wild. Salabron. Fresh aerial parts: topically applied for treating skin inflammation caused by stinging nettle. (CHI-SAL) +

Solanum tuberosum L. (Solanaceae). Cultivated. Trufes (plural). Fresh tubers: consumed boiled; traditionally added to the dough of the local dumplings (*ravioles*). +++

Sorbus aria (L.) Cratz. (Rosaceae). Wild. Alier. Fresh fruits (*alìe*, pl.): eaten after natural fermentation on straw. Once dried, ground and added to bread dough (in the past). +

Sorbus aucuparia L. (Rosaceae). Wild. Pizzera. Fresh fruits: to be avoided by humans but appreciated by wild birds. +

Tanacetum vulgare L. (Asteraceae). Wild and semi-domesticated. Archebuse. Fresh leaves: macerated in alcohol as digestive. (CHI-ARC) +++

Taraxacum officinale Weber (Asteraceae). Wild. Mourioun. Fresh leaves: consumed in salads or boiled. (CHI-MOU) +++

Thymus pulegioides L. s.l. (Lamiaceae). Wild. Serpour, Serpoul. Dried leaves: tea as a digestive, for treating sore-throats, or as a diuretic. As a seasoning in the local cuisine (especially the local, home-made, fermented and spiced ricotta: *sarass*). (CHI-SER) ++

Tragopogon pratensis L. (Asteraceae). Wild. Barbouch, Barbabouch. Fresh flower buds and young leaves: consumed raw or boiled. (CHI-BAR) +++

Tussilago farfara L. (Asteraceae). Wild. Pupettes. Dried flowers: tea as a bechic. Traded in the past. (CHI-PUP) +

Urtica dioica L. and other *U.* species (Urticaceae). Wild. Urtio. Fresh leaves: consumed in soups or frittata. Tea as a diuretic (mainly for women; men should take it in limited amounts only.). (CHI-URT) +++

Vaccinium myrtillus L. (Ericaceae). Wild. Aize. Fresh fruits: consumed raw or processed in jams or cordials. +++

Vaccinium uliginosum L. (Ericaceae). Wild. Poumarette (plural). Fresh fruits: often mixed with *V. myrtillus* fruits, consumed raw or processed in jams or cordials (less popular than *V. myrtillus* fruits) ++

Vaccinium vitis-idaea L. (Ericaceae). Wild. Petemerlet. Fresh fruits: raw as a snack. ++

Valerianella sp. (Valerianaceae). Wild. Salzet. Fresh young leaves: consumed in salads. (CHI-SAL) +

Veratrum album L. (Liliaceae s.l.). Wild. Varaire. Roots: topically applied as an anti-lice treatment for calves. (CHI-VAR) +

Verbascum nigrum L. (Schrophulariaceae). Wild. Nevioun. Flower: tea as a bechic. (CHI-NEV) +

Veronica allionii Vill. (Schrophulariaceae). Wild. Jaspertero. Dried flowering, aerial parts: tea (use not recalled). Decoction given to cows (use not recalled). Traded in the past. (CHI-JAS) +++

Veronica beccabunga L. (Schrophulariaceae). Wild. Seiraset. Fresh leaves: consumed raw in salads. (CHI-SER) +

Vicia faba L. (Leguminosae). Cultivated in the past. Faves (plural). Seeds: traditionally consumed as main ingredient of a particular kind of *polenta* (*poouties*). +

Viola odorata L. (Violaceae). Wild. Vioulette (plural). Dried leaves: tea for treating colds and coughs (low dosage). Fresh leaves: boiled in soups. Traded in the past. (CHI-VIO) ++

Vitis vinifera L. (Vitaceae). Vinegar. Purchased in the surrounding lower areas. Poultice mixed with clay: externally as a suppurative. ++

Zea mais L. (Poaceae). Cultivated. Meliga. Seeds: ground into flour and used as main ingredient for a particular kind of local polenta. +++

Unidentified taxon. Wild. Erbo dousso. Fresh aerial parts. Boiled in soups. +

Unidentified taxon. Wild. Gialinettes, Gialinettes grasses (plural). Fresh aerial parts: boiled in soups. ++

Unidentified taxon. Wild. Ontu. Root: decoction given to calves to treat digestive problems. +

DISCUSSION

Is the Occitan TK of Gastronomic and Medicinal Plants Different from Other Western Alpine TK Systems?

Table 1 shows a comparison between the medico-ethnobotanical data collected in the study area and data gathered from previous ethnobotanical studies conducted in other Alpine valleys in Piedmont and surrounding regions (figure 8) [15, 16, 17, 18, 19, 20, 21, 22, 23, 24, 25, 26]. The data available in the literature on Piedmontese ethnobotany are not methodologically consistent, and the field methods used in early studies have only rarely been properly described, nevertheless the descriptive comparison shows that approximately one fourth of the quoted taxa in our study occur in the TK systems of other Occitan valleys. One exception, however, is represented by the ethnobotanical data of two Occitan valleys whose inhabitants are Waldensians. This group shows an ethnobotanical knowledge that seems to be very distinct from that of the bulk of all other Occitan areas.

Table 1. Comparison between the folk medicinal taxa quoted in the upper Varaita valley and those quoted in other ethnobotanical studies conducted in the past in the Western Alps (see Figure 8).

Area	Year of publication of the field ethnobotanical study	Ethnic/cultural group(s)	Recorded folk medicinal taxa	Medicinal taxa also used in the Upper Val Varaita (VAR)
Ubaye Valley (1)	1990	OC	112	25
Maira Valley (2)	1982	OC	127	24
Sangone Valley (6)	1977	FP	72	24
Stura Valley (1)	2006	OC	102	23
Susa Valley (7)	1977	OC and FP	99	20
Mastallone Valley (Rimella) (11)	1957	WA	25	15
Les Allues (8)	1986	FP	27	13
Formazza Valley (17)	1955	WA and PE	27	12
Grande Valley (9)	1957	WA and PE	23	11
Sermenza Valley (10)	1957	WA and PE	21	11
Strona Valley (12)	1957	PE	25	11
Mastallone Valley (Forbello) (11)	1957	PE	32	10
Anzasca Valley (13)	1955	WA and PE	27	10
Bognanco Valley (15)	1955	PE	14	8
Vigezzo Valley (18)	1955	PE and LO	20	8
Antrona Valley (14)	1955	PE and LO	13	7
Valtournanche (16)	1983	FP	64	6
Chisone Valley (5)	1984	OC*	35	2
Germanasca Valley (4)	1984	OC*	19	0

Numbers in the first column correspond to those used in Figure 2.

LO: Lombard. FP: Franco-Provençal. OC: Occitan. OC*: Occitan and Waldensian (as per religion). PE: Piedmontese. WA: Walser

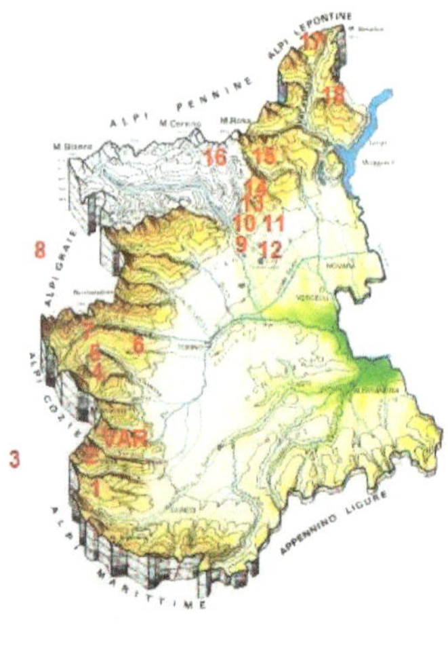

Figure 8. Location of the Western Alpine sites, where previous ethnobotanical studies have been carried out (numbers correspond to those used in the first column of Table 1; VAR: study area - upper Varaita valley).

Genepì: From an Herbal Tea to a Gastronomic Specialty

The term genepì/génépì (synonyms: génepi, génipi) is applied in the Western Alps (in Occitan, Franco-Provencal, and also in French and Italian too) to diverse species of locally growing Alpine wormwood (esp.*Artemisia genipi, A. glacialis, A. umbelliformis*). This term has probably Savoyard origins, and is derived from the Latin *Dianae spicum* (*épi de Diane* in French, *Diana's ear* in English) [41].

The ethnobotany of *genepì* in the Western Alps has never been recorded specifically or systematically, although scattered notes have occurred in Alpine ethnobotanical studies over the last few decades (Table 2). From these brief ethnographic reports we can conclude that genepí was originally used for preparing herbal teas, and shepherds in the mountainous areas frequently used to decoct the aerial parts in milk and butter instead of water. Following the introduction of wine and grappa (purchased from farmers living in the hills and on the plains), the herb probably was macerated in alcohol as its use as a liqueur began to appear over a century ago in different gastronomic and commodity treatises in France and Italy [42, 43, 44, 45].

Table 2. Folk medical uses of the aerial parts of Alpine wormwood (*Artemisia genipi, A. glacialis, and A. umbelliformis*) in diverse North-Western Alpine areas (See Figure 8 to located most of these areas on the map).

Area(s)	Ethnic/cultural group(s)	Preparation(s)	Folk medical use(s)
Anniviers Valley, Switzerland	FP	Macerated in alcohol; made into home-made distillate with chopped apples and sugar.	Depurative, digestive; high doses are considered dangerous (inducing circulatory disturbances in eyes and nervous system)
Aosta Valley, Italy	FP	Infusion in milk with honey	Anti-tussive, anti-cold
Grande and Formazza valleys, Italy	WA and PE	Infusion in water; decoction in milk or wine; macerated in alcohol	Anti-tussive, tonic, digestive
Les Allues, France	FP	Infusion in water or milk and butter (with sugar added); on rare occasions macerated in alcohol	Anti-cold, panacea; high doses are considered dangerous
Sangone Valley, Italy	FP	Infusion in water	Digestive
Susa Valley, Italy	OC and FP	Infusion in water	Digestive
Germanasca and Chisone valleys, Italy	OC*	Infusion in water	Digestive
Maira Valley, Italy	OC	Infusion in water; macerated in alcohol	Diaphoretic, to prevent mountain sickness and syncopal attacks

FP: Franco-Provençal. OC: Occitan. OC*: Occitan and Waldensian (according to religion).

PE: Piedmontese. WA: Walser

This seems to concur with what has been hypothesized in other Italian Alpine regions regarding herbal macerates in *grappa* [46], which they were considered sometimes traditional "medicinal" preparations. Preparation of liqueurs began to be relevant especially in the 1960's, when Italy began to enjoy better economic conditions which enabled isolated rural communities to have at their disposal larger amounts of grappa and/or industrial alcohol.

This shift is quite interesting since it demonstrates, not only the overlapping between food and medicine, but also how the preponderant role of one of the two domains can be the result of dynamic historical, social, and economic changes.

Nowadays, *A. umbelliformis* is widely cultivated in the Western Alps to supply the local distilleries that produce the alcoholic macerate of *genepì*, which widely traded all over the world. Because of this economic interest, local institutions in Piedmont are very active in putting in place measures for improving the quality and yield of cultivated Alpine wormwood [47].

All our informants, however, were very keen to underline the difference in taste between the industrial genepì (produced from the aerial parts of *A. umbelliformis* only - the only *Artemisia* species of the three that has been and can be cultivated with substantial agronomic results) and the home-made liqueur (which generally always contain *A. genipi* or sometimes a mix of *A. genipi* and *A. glacialis*). According to locals in the upper Varaita valley, the specific occurrence in the recipe of *A. genepì* (locally defined as the "male genepì") is crucial for achieving a superb tasting alcoholic beverage.

Collection of Local Herbs from the Wild: Sustainable Perspectives for the Future

A. genipi grows in the study area only between 2600-2700 meter a.s.l., and it is very difficult to find, which is why it is so highly valued. Although the species is endangered and its collected is prohibited in many Alpine areas, in the Piedmontese legislation the plant is not included in the regional Red List and aerial parts corresponding to up to 5 plants per person can be gathered [48].

Locals in the upper Varaita valley continue to gather the flowering tops (locally named "ears") of *A. genipi*during July and August, to dry them, and them macerate them in alcohol for at least a month and a half (the alcoholic macerate is then diluted with sugar and water, and filtered). In the past this species was probably heavily exploited by the locals in Chianale, who used to sell it (generally dried) to local distilleries both on the plain of Piedmont and on the French side (often smuggled across secondary pathways over the Alps that were passable with donkeys only). Informants in Chianale recalled how this activity provided the inhabitants with their most important income from a few decades after World War II. There is probably still some illegal gathering of this species that provides additional cash to the people of the mountain. A few informants told us that the income that can be derived from gathering genepì (*A. genipi* and - to lesser extent - *A. glacialis*) can reach 15-20 thousand Euros every summer for one single household.

In table 3 we report on the current legal framework [48, 49, 50], the estimated ecological availability and potentialities for the local market of the most quoted, traditionally used wild herbs in the study area.

Table 3. Local market potential of the most quoted wild gastronomic and medicinal herbs (excluding wild berries) in the upper Varaita valley

Species	Estimated ecological availability in the upper Varaita valley	Gathering permitted in the Cuneo Province	Estimated potential for the local market (details)
***Achillea herba-rotta*(herb)**	Medium	Aerial parts corresponding to five plant samples; up to 1 kg (dried aerial parts) if special permission is obtained from local authorities.	**High** **(dried herb to be sold as a tea).**
Arnica montana **(flowers)**	Medium	Aerial parts corresponding to five plant samples; up to 5 kg (dried flowers/roots) if special permission is obtained from local authorities.	**High** **(dried flowers to be sold as a tea).**
Artemisia absinthium (herb)	High	Aerial parts corresponding to five plant samples; up to 2 kg (dried aerial parts). if special permission is obtained from local authorities.	Low (dried herbs to be sold as a tea); Medium (ingredient of home-made digestives).
Artemisia genipi, A. glacialis, A. umbelliformis (aerial parts and flowers)	Low	Aerial parts corresponding to five plant samples; up to 1 kg of each species (dried aerial parts). if special permission is obtained from local authorities.	Very low, given the limited ecological availability.
Carum carvi **(fruits)**	Medium-high	Not clear: aerial parts corresponding to five plant samples or unlimited gathering. Unlimited gathering, due to the fact that the plant can be considered "commonly consumed" in the study area.	**High (dried fruits as a tea or seasoning).**
***Chenopodium bonus-henricus*(leaves)**	Very high	Unlimited gathering, due to the fact that the plant can be considered "commonly consumed" in the study area.	**High (as a preserved vegetable?)**
Gentiana acaulis (leaves and flowers)	Medium-low	Aerial parts corresponding to five plant samples	Low (as an ingredient for digestives) due to the limited availability in the area.
Gentiana lutea **(roots)**	High	Up to 10 kg (dried roots), if special permission is obtained from local authorities.	**High (dried root as a tea or ingredient for digestives).**
Malva spp.	High	Unlimited gathering, due to the fact that the plant can be considered "commonly consumed" in the area.	Medium (as a tea or as a preserved vegetable).
Peucedanum ostruthium (roots)	Medium-low	Up to 2 kg (dried roots), if special permission is obtained from local authorities.	Low (as a tea).
Pinus cembra (cones)	Low	Theoretically unlimited.	Low (as an ingredient for digestives) due to the limited availability.

Polygonum bistorta (leaves)	High	Unlimited gathering, due to the fact that the plant can be considered "commonly consumed" in the area.	Medium (as a preserved vegetable).
Rumex alpinus (leaves/stalks)	Medium	Unlimited gathering, due to the fact that the plant can be considered "commonly consumed" in the area.	Low (as a preserved vegetable).
Tanacetum vulgare **(leaves)**	Medium	Aerial parts corresponding to five plant samples; up to 5 kg (dried aerial parts) if special permission is obtained from local authorities.	**High (as a tea or as an ingredient for digestives).**
Taraxacum officinale **(leaves)**	High	Unlimited gathering, due to the fact that the plant can be considered "commonly consumed" in the area.	**High (as a preserved vegetable or a tea).**
Tragopogon pratensis **(leaves and flower buds)**	Medium	Unlimited gathering, due to the fact that the plant can be considered "commonly consumed" in the area.	**High (as a preserved vegetable).**
Urtica spp. (leaves)	High	Unlimited gathering, due to the fact that the plant can be considered "commonly consumed" in the area.	Low (as a preserved vegetable).
Veronica allionii (flowering aerial parts)	Low	Aerial parts corresponding to five plant samples.	Low (as a tea) due to the limited availability in the area.

It is clear that further exploitation of the *genepì* species cannot be recommended, however sustainable gathering of other local plant resources (wild vegetables and herbal teas) that are widely available in the area, could be a very interesting way to complement the eco-touristic activity, that is quite exclusively concentrated during the summer months, thereby providing an important source of income for the few remaining inhabitants.

CONCLUSION

Traditional food and medical uses of 88 local plants were recorded in the upper part of the Varaita valley. Analytical comparisons with other ethnobotanical studies previously carried out in other Piemontese and surrounding areas have shown that: a) traditional knowledge has been heavily eroded in the Varaita valley, and b) approximately one fourth of the quoted botanical taxa were also recorded in earlier surveys conducted in other Occitan valleys in the Italian Alps. Furthermore, the continued widespread local collection from the wild of the aerial parts of Alpine wormwood for the preparation of liqueurs should be seriously reconsidered in terms of sustainability, given the limited availability of these species even though this collection is culturally salient in the entire study areas, On the other hand interesting potentialities for the local market could be tested for other widely available wild species, such as *Achillea herba-rotta, Arnica montana, Carum carvi, Chenopodium bonus-henricus, Tanacetum vulgare, Taraxacum officinale*, and *Tragopogon pratensis*.

ACKNOWLEDGEMENTS

Special thanks are due to all the Occitans of the upper *Varacho*, for agreeing to share their knowledge with the authors; to Professor Alberto Capatti (food historian, University of Gastronomic Sciences, Pollenzo, Italy) for providing us with invaluable historical sources on the liquoristic traditions of *genepì*; to Nicola Robecchi for logistic assistance and photographic services; to the Region Piedmont, for having sponsored the project.

This article is dedicated to the unforgettable colleagues Nina L. Etkin (University of Hawai'I, USA), medical anthropologist, and Miguel Martinez, (UNAM, Mexico), ethnobotanist, who passed away a few months ago.

AUTHORS' CONTRIBUTIONS

AP and MEG have collected the data in the field; AP has analyzed the data and drafted the conclusions. All authors read and approved the final manuscript.

REFERENCES

1. Coassini Lokar L, Poldini L: Herbal remedies in the traditional medicine of the Venezia Giulia region (North East Italy). J Ethnopharmacol. 1998, 22: 231-278. 10.1016/0378-8741(88)90238-3.
2. Zampiva F: Le principali erbe della farmacopea cimbra. Cimbri-Tzimbar Vita Cultural Comunità Cimbre. 1998, X: 135-146.
3. Zampiva F: Róasan (Flora cimbra). Verona (Italy). 2000, Edizioni Curatorium Cimbricum Veronense
4. Barbini S, Tarascio M, Sacchetti G, Bruni A: Studio preliminare sulla etnofarmacologia delle comunità ladine dolomitiche. Inf Bot Ital. 1999, 31: 181-182.
5. Chiovenda-Bensi C: Tradizioni e usi fitoterapici popolari. La Valsesia. Atti Accademia Ligure Scienze

Lettere. 1957, 13: 190-205.

6. Pieroni A, Giusti ME, Münz H, Lenzarini C, Turkovic G, Turkovic A: Ethnobotanical knowledge of the Istro-Romanians of Žejane in Croatia. Fitoterapia. 2003, 74: 710-719. 10.1016/j.fitote.2003.06.002.

7. Balick M, Kronenberg F, Ososki A, Reiff M, Fugh-Berman A, O›Connor B, Roble M, Lohr P, Atha D: Medicinal plants used by Latino healers for women›s health conditions in New York City. Econ Bot. 2000, 54: 344-357.

8. Pieroni A, Münz H, Akbulut M, Baser KH, Durmushahya C: Traditional phytotherapy and trans-cultural pharmacy among Turkish migrants living in Cologne, Germany. J Ethnopharmacol. 2005, 101: 258-270. 10.1016/j.jep.2005.04.028.

9. Pieroni A, Vandebroek I, Eds: Traveling plants and cultures. The Ethnobiology and ethnopharmacy of human migrations. 2007, Oxford: Berghahn

10. Pieroni A, Houlihan L, Ansari N, Hussain B, Aslam S: Medicinal perceptions of vegetables traditionally consumed by South-Asian migrants living in Bradford, Northern England. J Ethnopharmacol. 2007, 113: 100-110. 10.1016/j.jep.2007.05.009.

11. Waldstein A: Mexican migrant ethnopharmacology: pharmacopoeia, classification of medicines and explanations of efficacy. J Ethnopharmacol. 2006, 108: 299-310. 10.1016/j.jep.2006.07.011.

12. Pieroni A, Gray C: Herbal and food folk medicines of the Russlanddeutschen living in Künzelsau/Talächer, South-Western Germany. Phytother Res. 2008, 22: 889-901. 10.1002/ptr.2410.

13. Pieroni A, Sheikh QZ, Ali W, Torry B: Traditional medicines used by Pakistani migrants from Mirpur living in Bradford, Northern England. Compl Ther Med. 2008, 16: 81-86. 10.1016/j.ctim.2007.03.005.

14. Pieroni A, Torry B: Does the taste matter? Taste and medicinal perceptions associated with five selected herbal drugs among three ethnic groups in West Yorkshire, Northern England. J Ethnobiol Ethnomedicine. 2007, 3: 21-10.1186/1746-4269-3-21.

15. Volpato G, Godínez D, Beyra A: Migration and ethnobotanical practices: The case of Tifey among Haitian immigrants in Cuba. Hum Ecol. 2009, 37: 43-10.1007/s10745-008-9211-4.

16. Grabherr G: Biodiversity in the high ranges of the Alps: Ethnobotanical and climate change perspectives. Global Environ Change. 2009, 19: 167-10.1016/j.gloenvcha.2009.01.007.

17. Caramiello Lomagno R, Rovera L, Lomagno PA, Piervittori R: La fitoterapia popolare nella Val Maira. Annali Facoltà Scienze Agrarie Università di Torino. 1982, XII: 217-275.

18. Caramiello Lomagno R, Piervittori R, Lomagno PA, Rolando C: Fitoterapia popolare nelle Valli Chisone e Germanasca. Annali Facoltà Scienze Agrarie Università di Torino. 1984, XIII: 259-298.

19. Musset D, Dore D: La muave et l›erba bianca. 2006, Salagon (France): Musée départmental ethnologique de Haute-Provence

20. Novaretti R, Lemordant D: Plants in the traditional medicine of the Ubaye valley. J Ethnopharmacol. 1990, 30: 1-34. 10.1016/0378-8741(90)90014-K.

21. Meilleur BA: Alluetain ethnoecology and traditional economy: the procurement and production of plant resources in the Northern French Alps. PhD Thesis. 1986, University of Washington, Department of Anthropology

22. Lomagno P, Lomagno Caramiello R: La fitoterapia popolare nella Valle di Susa. Allionia. 1970, 16: 166-174.

23. Lomagno P, Lomagno Caramiello R: La fitoterapia popolare nella Valle del Sangone. Boll SIFO. 1977, 23: 445-452.

24. Chiovenda-Bensi C: Tradizioni e usi fitoterapici popolari. La Valsesia. Atti Accademia Ligure Scienze Lettere. 1957, 13: 190-205.

25. Chiovenda-Bensi C: Piante medicinali nell›uso tradizionale della Valle d›Ossola. Atti Accademia Ligure Scienze Lettere. 1955, 11: 32-53.

26. Chimenti Signorini R, Fumagalli M: Indagine etnofarmacobotanica in Valtournanche (Valle d›Aosta). Webbia. 1983, 27: 69-94.

27. Denarier N: Venti erbe per stare bene. Piante medicinali della Valle d›Aosta. 2006, Gressan (Italy): Edizioni Vida

28. Brünschweiler S: Plantes et saviors des Alpes. L›example du val d›Anniviers. 1999, Sierre (Switzerland): Editions Monographic

29. Bernard G: Lou saber. Dizionario enciclopedico dell›occitano di Blins. 1996, Venasca (Italy). Edizioni Ousitanio Vivo

30. Anonymus: Popolamento e spopolamento di una vallata alpina. Ricerche antropo-ecologiche nell›alta Val Varaita e testimonianze di cultura occitana. Archivio Antropologia Etnologia. 1976, CVI (Suppl): 1-290.

31. Bernard JL: Nosto modo. Testimonianza di civiltà provenzale alpina a Blins (Bellino). 1992, Coumboscuro (Italy): Centre Provençal

32. Anonymus: Bén minjà bén begü. Alimentazione e cucina tradizionale a Chianale. Bra (Italy). 1988, Edizioni del Comitato per San Lorenzo

33. De Matteis L: Blins. L›abitare di una comunità rurale delle Alpi occitane. 1993, Torino: Priuli e Verlucca

34. Para E: Un carnevale alpino. Il lupo di Chianale (University Thesis). Torino: Università degli Studi

35. Toso F: Lingue d›Europa. La pluralità linguistica dei Paesi europei fra passato e presente. 2006, Milan: Baldini Castoldi Dalai

36. Allasino E, Ferrier C, Scamuzzi S, Telmon T: Le lingue del Piemonte. 2007, Torino: Istituto di Ricerche Economico Sociali del Piemonte

37. Diritti G: Il vento fa il suo giro (DVD). 2006, Bologna: Arancia Film

38. Motta R, Nola P: Growth trends and dynamics in sub-alpine forest stands in the Varaita Valley (Piedmont, Italy) and their relationships with human activities and global change. J Veg Sci. 2001, 12 (219): 230.

39. Code of Ethics of the International Society of Ethnobiology. [http://ise.arts.ubc.ca/global_coalition/ethics.php]
40. Pignatti S: Flora d›Italia. 1982, Bologna: Edagricole
41. Aeschimann D, Lauber K, Moser DM, Theurillat JP, Eds: Flora alpina. 2004, Bern: Haupt Verlag, 3:
42. Conti F, Abbate G, Alessandrini A, Blasi C: An annotated checklist of the Italian vascular flora. 2005, Rome: Palombi Editori
43. *Le trésor de la langue française informatisé.* [http://atilf.atilf.fr/tlf.htm]
44. Roques J: Nouveau traité des plantes usuelles spécialement appliqué à la médecine domestique et au régime alimentaire de l›homme sain ou malade. Paris: Dufart; 1837-1838
45. Seigneurie A, Seigneurie H, Bertin J, de Borsat X: Dictionnaire encyclopédique de l›épicerie et des industries annexes. 1904, Paris: L›Épicier
46. Rossi A: Manuale del liquorista. 1450 ricette pratiche. 1904, Milan: Hoepli
47. Volpato G: Etnobotanica delle grappe medicinali nel Veneto (Italia). 2004, Montebelluna (Italy):De Rerum Natura: Quaderni del Museo di Storia Naturale e Archeologia di Montebelluna
48. Soster M: Il genepì, l›agricoltura delle vette. Quaderni Regione Piemonte Agricoltura. 2008, 58: 27-31.
49. Norme per la conservazione del patrimonio naturale e dell›assetto ambientale. Legge Regionale 2 Novembre n. 32. [http://arianna.consiglioregionale.piemonte.it/ariaint/TESTO?LAYOUT=PRESENTAZIONE&TIPODOC=LEGGI&LEGGEANNO=1982&LEGGE=032]
50. Elenco delle piante dichiarate officinali. Regio Decreto 26 Maggio n. 772. [http://www.ismea.it/flex/AppData/Redational/Normative/20020528000200001.pdf]
51. Disciplina della coltivazione, raccolta e commercio delle piante officinali. Legge 6 Gennaio n. 99. [http://www.pianteofficinali.org/main/Leggi/Legge_99_1931.pdf]

CHAPTER 5

ETHNOBOTANY OF BABASSU PALM (ATTALEA SPECIOSA MART.) IN THE TUCURUÍ LAKE

Fábio Ribeiro Araújo, Sol Elizabeth González-Pérez, Maria Aparecida Lopes, and Ismael de Jesus Matos Viégas

Keywords: Local knowledge, non-timber forest resources, palm uses, riverine community, Tocantins river

INTRODUCTION

Considering the human populations living within or near conservation areas, local ecological knowledge can be of great value in management proposals based on the principles of local participation and sustainability (Sillitoe 1998; Huntington 2000). Ethnobotanical information can clarify the level of dependency of a community in relation to local plant resources, and provide information about the consequences of certain types of resource exploitation (Phillips 1996).

Palm trees of the family Arecaceae are one of the major non-timber forest resources for local populations. Their importance to local communities has been demonstrated in numerous studies in the Neotropics (Rocha & Silva 2005; Albán *et al.* 2008; Nascimento 2010; Araújo & Lopes 2012). This plant family is possibly the most commonly used by Amerindians and rural populations in the Amazon (Plotkin & Balick 1984; Borchsenius *et al.* 1998; Macía *et al.* 2011; Paniagua-Zambrana *et al.* 2015) and has a key role in the subsistence strategies of many communities (Anderson 1991;Byg & Balslev 2004). Research on the local perception of the importance of palm trees is of great interest because it can support successful practices for the conservation of the biological environment (Byg & Balslev 2001).

Socioeconomic factors affect the distribution of ethnobotanical knowledge of palms used by informants. Knowledge about these relationships has a practical importance for the sustainable use and conservation of forest biodiversity, because it clarifies which social groups depend most on these natural resources and also underscores the mechanisms that cause their exploitation (Byg & Balslev 2004).

Among the species of the family, the babassu (*Attalea speciosa*) stands out as one of the most abundant palm trees in the Amazon. It is distributed widely in the southern part of the region, from the Atlantic Ocean to Bolivia (Balick & Pinheiro 2000), especially in the eastern part of the Amazon, in the transition zones between the rainforest biome and the semi-arid of northeastern Brazil (May *et al.* 1985; Meirelles 2004). It can occur in low densities, in mature forests, or in high densities, in large degraded areas where it is considered a pioneer and dominant species (Anderson *et al.* 1991). The fact that all palm parts are useful is what makes this such an important palm tree for the livelihood of many traditional communities (Lima *et al.* 2003; Campos *et al.* 2015). The leaves are used for thatch to support roof structures (Rezende *et al.* 2010), in various crafts and household items (Corrêa *et al.* 2010), walls of homes and animal facilities, and as a fertilizer (Nascimento *et al.* 2009). The fruits are used as food, raw material for cosmetics, drugs and as a fuel source (Lima *et al.* 2003).

Given the importance of the family, it is vital, from a scientific point of view as well as an economic and conservationist one, to learn more about the local needs which are being met by palm products, how these are used, which social groups are more dependent on palm resources, or how palms are extracted (Rufino *et al.* 2008).

This study aims to explore the ethnobotanical knowledge of the babassu, by characterizing its uses regarding diversity

Citation: Fábio Ribeiro Araújo, Sol Elizabeth González-Pérez, Maria Aparecida Lopes, and Ismael de Jesus MatosViégas. Ethnobotany of babassu palm (Attalea speciosa Mart.) in the Tucuruí Lake, Acta Bot. Bras., 2016; vol.30, no.2, pp.193-204. ISSN:0102-3306. http://dx.doi.org/10.1590/0102-33062015abb0290.

and purpose, describing the forms of use and exploitation, and evaluating how socioeconomic factors affect the distribution of knowledge in riverine populations inhabiting the Tucuruí Lake Protected Areas Mosaic - eastern Amazon.

We expect, as a result of this study, that most of the uses will be concentrated and more diverse in the categories of utensils and tools and construction, with most intense use of leaves and seeds, respectively. In addition, we predict that the knowledge of babassu use will be more influenced by the economic activity of the informants.

MATERIALS AND METHODS

Study Area

The Tucuruí hydroelectric station is located on the Tocantins River in Pará state, eastern Brazilian Amazonia. The river was dammed in 1985 and about 2875 km^2 of forest was flooded, causing changes in regional ecosystems and landscapes, including the formation of the actual reservoir and various islands made from higher non-flooded ground.

Since 1984, several Protected Areas have been created by the state government in the area of influence of the reservoir, composing in 2002 the Lake Tucuruí Protected Areas Mosaic (PAM) (Jatobá 2006). The PAM includes areas with different levels of protection of the biota: an Environmental Protection Area (EPA) and two Sustainable Development Reserves (SDR), where sustainable resource use is encouraged, and two zones of Wildlife Conservation (ZWC), which are strictly Protected Areas (Fig. 1). The original vegetation in the Lake Tucuruí PAM was predominantly open rainforest, however, even before the river was dammed, several areas of human influence already existed, which included fallow fields (secondary forest), cultivation, bare soil and pastures (Ohashi *et al.* 2004).

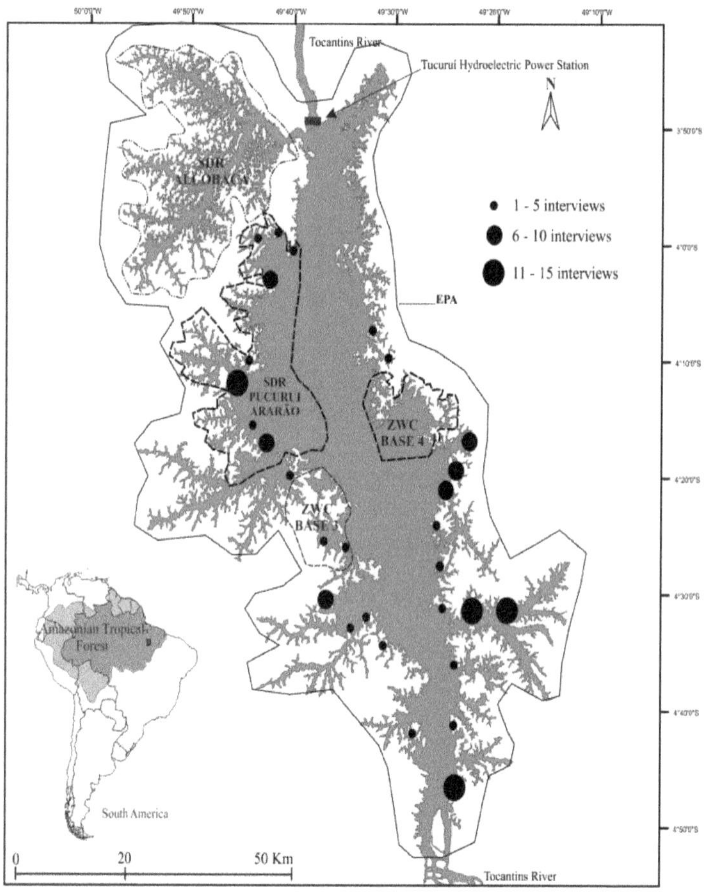

Figure 1. Map showing the location and boundaries of the Tucuruí Lake Protected Areas Mosaic. Black dots indicate locations where interviews were conducted and the number of interviews in each; EPA - Environmental Protection Area; SDR - Sustainable Development Reserve; ZWC - Zone of Wildlife Conservation.

The region is occupied by groups of riverine populations, which may be dispersed or in communities, and rural properties of various sizes along the extension of the EPA (Araújo 2009). The population is estimated to be approximately 6,500 inhabitants, with two regions with the highest concentration of occupation of the islands: SDR Alcobaça (2,677 inhabitants) e SDR Pucuruí-Ararão (1,389 inhabitants) (SECTAM 2002). The activities performed by these populations are directly related to subsistence activities such as extraction of natural resources, agriculture and fishing (Jatobá 2006). Most of the residents do not have any formal education and do not benefit from essential public services such as water supply and drainage or universal services such as electricity, education and health (Ravena *et al.* 2009).

Ethnobotanical Data and Socio-economic Variables

We based the study on interviews that were made with household heads who were present in their homes throughout the study in an accidental manner (Albuquerque *et al.* 2008). The interviews were conducted from February to March 2010 and in June 2014, with 193 families that inhabit the Tucuruí Lake Protected Areas Mosaic. In 2010, 37 interviews were lost in a boat accident on the lake, which were only restored in 2014 after again interviewing the same household heads for which the information had been lost.

We tried to distribute the interviews along the whole reservoir and on both banks, and included all types of Protected Areas, particularly those that allow the use of natural resources (SDR and EPA). The research was authorized by the relevant government agencies (Environmental Department of the state of Pará, authorization n. 018/2009) and all respondents agreed to participate in the study after being informed of its goal.

The questionnaire consisted of semi-structured interviews about the knowledge and use of babassu with an introductory part containing the informant's personal data and socioeconomic characteristics (gender, age, federal state of birth, productive activity), and another part with questions related to aspects of the knowledge and use of palms in the study area (number of known uses, use purpose, plant parts used and form of exploitation).

The species was photographed and identified in the field by consulting the specialized literature (Henderson 1995; Lorenzi *et al.* 2010) and later confirmed by experts of the graduate course in Botany at the Federal Rural University of Amazonia and Herbarium João Murça Pires of the Museu Paraense Emílio Goeldi (MG), Belém, Pará (Fig. 2).

Figure 2. Photographic documentation of the species in the field. A) babassu palm; B) cluster of babassu palms; C) external aspect of the babassu fruit; D) internal aspect of the babassu fruit.

Data Analyses

Uses reported by informants were classified into use categories and subcategories according to Macía *et al.* (2011) that in a review article on the use of palm trees on the western portion of South America, classified the ethnobotanical uses in ten categories each divided into subcategories, with ethnobotanical information unclassified in previous subcategories inserted into 'other'.

To determine how the species is used and which aspects of use contributes to importance according to informants of the Tucuruí Lake PAM we calculated the Use-diversity Value (UDV) index, that measures the importance of an use category and how each category contributes to the total use value, obtained from the number of recorded indications for a category divided by the total number of indications for all categories; Purpose Consensus Value (PCs) which measures the degree of agreement among informants concerning the use purpose, both based on Byg & Balslev (2001), and Useful Plant Part Values (PPV) index, indicating which plant parts are more intensively used, obtained from the total number of citations reported for each part of the plant divided by the total number of citations for all plant parts, according to Gomez-Beloz (2002).

We used the Kolmogorov-Smirnov test to assess normality of distribution of knowledge of the babassu among the informants, and we conducted multiple regressions to evaluate the effect of socioeconomic factors on the knowledge of the use of the babassu as cited by the informants. In the regressions, we considered the socioeconomic data as independent variables and the measures of knowledge as dependent variables. We also performed a nonparametric test (Kruskal-Wallis) between the socioeconomic variables that showed greater effect on the knowledge in the regressions, to specify the socioeconomic variable that most influences the accumulation of knowledge of palms, from the comparison of sample means using Dunn's method. All tests were performed in BioEstat 5.3 software.

RESULTS AND DISCUSSION

Socioeconomic Aspects of the Informants

The set of informants included 77 women and 116 men and the age varied between 17 and 79 years, the majority varying from 26 to 55 years. In most families, fishing is the only productive activity (66%), but some families combine fishing and agriculture (25%) and others practice only agriculture (9%). The origin of the interviewees is diverse: 45% are from the state of Pará, 44% from the Brazilian northeast region, especially from the neighboring Maranhão state, and 11% from other parts of the country. A total of 75% of the interviewees have been living in the study area from six to 25 years, 16% from one to five years, and 9% have been living on the margins of the Tocantins River from 26 to 55 years. Ravena *et al.* (2009) and Piratoba (2014) recorded intense social mobility in the community at Lake Tucuruí PAM, resulting in a population that has been formed only recently, with fishing as the main current economic attraction.

Diversity of Babassu Use

We recorded 1,226 citations about the use of babassu, grouped into 60 different uses, which were classified into nine use categories: utensils and tools, environmental uses, construction, human food, animal food, fuel, medicinal, cultural uses and other uses (Tab.1). Other studies in different regions of the Amazon attest to the use of this palm tree in these categories, such as Arruda *et al.* (2014) in quilombos, González-Pérez *et al.* (2012) in the *Terra Indígena Las Casas* Mebêngôkre-Kayapó in the southwest and southeast of the Brazilian Amazon, respectively, and Paniagua-Zambrana (2005) in the Madidi region in Bolivia.

Table 1. Types of babassu uses classified into nine categories of use indicating the subcategory of use, plant part used, the number of citations for each different use, in descending order of the number of citations per category.

Use category	Use subcategory	Types of babassu uses	Plant Part	Number of citations
Utensils and Tools	Domestic Utensils	Basket	Leaves	98
		Fans	Leaves	95
		Mats	Leaves	53
		Sieves	Leaves	16
		Broom	Leaves	8
		Hamper	Leaves	2
		Feeder to animals	Spathe	2
		Shelf	Stem	1
	Hunting and Fishing Tools	*Matapi* (trap for shrimp)	Leaves	57
		Camaroeira (nurseries for shrimp)	Leaves	27
	Wrappers	Bowl to food	Spathe	1
		Insect repellent	Fruit (endocarp)	30
		Soap	Fruit (seed oil)	26
		Nests of domestic birds	Leaves	19
	Other	Birdcage	Leaves	1
Environmental Uses	Soil Improvers	Organic fertilizer	Stem	11
Construction	Houses	Walls	Leaves	85
		Walls	Stem	36
		Wall for animal installations	Stem	7
		Floor	Stem	2
	Thatch	Thatched roof	Leaves	165
		Ridge	Leaves	1
		Roof to henhouse	Leaves	1
	Other	Palisade wall	Leaves	16
		Washbasin	Leaves	1
Human Food	Oil	Oil	Fruit (seed)	118
	Food	Coconut milk	Fruit (seed)	111
		Coconut *in natura*	Fruit (seed)	18
		Heart of palm	Meristem	16
		Mesocarp triturated in milk	Fruit (mesocarp)	12
		Porridge	Fruit (mesocarp)	4
		Coconut candy	Fruit (seed)	3
		Seed bagasse for *beiju*	Fruit (seed)	1
Animal Food	Fodder	Feed for animals	Fruit (seed)	2
	Fish Bait	Bait shrimp	Fruit (seed)	1
Fuel	Firewood	Cooking food	Fruit (endocarp)	93

Medicinal	Blood and Cardio-vascular System	Blood coagulation	Leaves	19
		Hemorrhoids	Fruit (seed oil)	5
	Dental Health	Toothache	Fruit (endocarp)	24
	Digestive System	Gastritis	Fruit (mesocarp)	6
		Laxative	Fruit (seed oil)	5
		Hepatitis	Fruit (mesocarp)	1
	Infections and Infestations	Vermifuge	Fruit (seed oil)	3
	Muscular-Skeletal System	Osteoporosis	Fruit (mesocarp)	1
	Poisoning	Bite of spiders	Fruit (seed)	1
	Reproductive System and Reproductive Health	Female genital inflammation	Fruit (seed oil)	1
	Sensory System	Eye infection	Fruit (liquid albumen)	1
	Skin and Subcutaneous Tissue	Myiasis	Fruit (seed oil)	2
		Skin infection	Fruit (seed oil)	2
		Mycosis	Fruit (seed oil)	1
	Not Specified	Analgesic	Fruit (endocarp)	1
		Anti-inflammatory	Fruit (endocarp)	1
		Anti-inflammatory	Fruit (seed oil)	1
		Antithermic	Fruit (endocarp)	1
Cultural Uses	Cosmetics	Hair moisturizing	Fruit (seed oil)	1
Other Uses	Miscellaneous	Sale	Fruit (seed oil)	4
		Beetle larvae*	Larva*	2
		Earache	Oil of larva*	1
		Bait for fish	Larvae of beetle*	1
		Hair moisturizing	Oil of larva*	1

The most prominent use categories were utensils and tools, evidenced by the higher UDV, followed by construction and human food (Tab. 2). Campos *et al.* (2015), while studying the babassu oil extractive communities in northeastern Brazil, also registered the highest UDV in all three categories in the same order of importance presented here, however, the uses that we classified as utensils and tools were presented as handicrafts in Campos *et al.* (2015).

Table 2. Use-diversity values (UDV) for use categories of babassu indicated by informants of the Tucuruí Lake PAM.

Categories	N° different uses	UDV
Utensils and tools	15	0.126
Construction	9	0.066
Human Food	8	0.053
Fuel	1	0.006
Medicinal	18	0.004
Environmental uses	1	0.001
Other Use	5	0.001
Animal food	2	0.001
Cultural Use	1	0.001

All respondents knew and used the species with an average of six listed uses, ranging from one to 19 uses. The most important use indicated by the degree of agreement among respondents on the use purpose is thatch, which obtained the highest PCs (Tab. 3). Campos *et al.* (2015) recorded thatch as the most common use in two rural communities of babassu extrativists in northeastern Brazil.

Table 3. Purpose consensus value (PCs) for the ten main babassu uses determined by informants of the Tucuruí Lake PAM.

Use	PCs
Thatch	0.0181
Oil	0.0093
Coconut milk	0.0082
Basket	0.0064
Fans	0.0060
Cooking food	0.0058
Walls	0.0048
Matapi (trap for shrimp)	0.0022
Mats	0.0019
Walls	0.0009

Utensils and Tools

The category with the most use citations was utensils and tools (436), and leaves were the main part of the plant used, corresponding to 86% of the citations. Leaves are used to make baskets with different shapes and purposes, such as for transporting products collected in the forest, produced on the farm and fished, or to store food. They are also used for making bird nests, fans, sieves and mats, which often replace mattresses.

Babassu utensils are also important in fishing, the main economic activity of the riverine community in the Tucuruí Lake Protected Areas Mosaic. Ninety-four percent of respondents subsist from fishing, either exclusively or in combination with other activities. Fishing for Amazonian shrimp (*Macrobrachium amazonicum*) is an important activity for the economy and livelihood of the inhabitants of the Tucuruí Lake PAM. The capture is performed with a trap, called '*matapi*', made from the babassu petiole. It has a cylindrical shape closed at the ends with cones. After capturing the shrimps, these are deposited in nurseries, called *'camaroeiras'*, made from the same material as the matapi.

Araújo *et al.* (2014), while researching fishing and shrimp capture procedures in the Amazon downstream from the Tucuruí hydroelectric station, demonstrated that the matapi stands out as the main utensil used: of 126 respondents, 72% reported using only the matapi for shrimp capture.

Construction

The most common use in this category is thatched roof. Eighty-five percent of respondents live in houses with roofs made from babassu leaves, 37% in homes with walls made of leaves and 15% in mud houses with internal structures made of slats taken from the stem.

In the Brazilian Amazon, babassu has been pointed out by several studies as the main building resource for traditional populations and forest dwellers and for rural communities. In the eastern part of the biome, these include indigenous communities such as the Apinajé and Guajajara (Balick 1988), the Guajá (Forline 2000), the Xicrin (Fuerst 2006), the Mebêngôkre-Kayapó (González-Pérez *et al.* 2012), the Krahò (Nascimento *et al.* 2009), and non-indigenous rural populations, such as the coconut-breaking women from Maranhão (Shiraishi 1999). Other examples include the Saterê-Mawé (Coelho-Ferreira 1994) in the west and the quilombo communities of Mato Grosso do Sul (Arruda *et al.* 2014).

For legal, economic and cultural reasons, babassu is the primary building resource of the inhabitants of the Tucuruí Lake PAM. Logging is illegal in almost all of the Protected Areas of the Tucuruí Lake Protected Areas Mosaic (Pará 2002). Logging is costly and demands a minimal infrastructure, including equipment, such as chainsaws, and fuel consumption. On the other hand, the absence of specific legal restrictions and inspection of the exploration of babassu in the Tucuruí Lake PAM is possibly another factor that influences the choice of this palm as the main construction resource by the inhabitants. Most of the population (44%) comes from the Brazilian northeast, especially from the states of Maranhão and Piauí, where there are oligarchic formations of this palm tree, strongly linked to the livelihood and culture of the people (Barbieri 2004).

Human Food

The fruit is the main part of the palm used for food, and oil is the main product derived from the seed, with 50.8% of the respondents confirming its everyday use in cooking. Babassu is widely used in the Amazon and northeastern Brazil for this purpose and is considered the world's leading source of wild oil for domestic use (Lorenzi *et al.* 1996).

All respondents attest that the fruit collection is independent of gender, while oil extraction is carried out almost exclusively by women. This division of labor is repeated in other cultures in the Amazon (Shiraishi 1999; González-Pérez *et al.* 2012; Arruda *et al.* 2014). The seed is also consumed when processed with water. The mixture, called 'coconut milk', is used in the preparation of cooked meals, mainly fish, with 47.8% of respondents stating that this is part of their weekly meals. Other uses of the seed are less frequent, such as its direct raw consumption or when it is crushed to prepare confections, such as coconut candy.

The dry mesocarp is ground and sieved, originating the "babassu powder" that is mixed with milk as a drink or used as a substitute for wheat flour in the formulation of cakes and porridge. Another type of food use cited is the consumption of the apical meristem, called the palm heart. This form of consumption, common in the Amazon and in the northeast (Posey 1987; Rufino *et al.* 2008), involves the death of the plant. However, respondents claim that the palm trees are not felled for this purpose, and the use of meristem for food occurs when the palm is felled for other purposes, such as using the stem for buildings. They also state that the consumption of palm heart is sporadic and its frequency has diminished, because its extraction is a very labor intensive activity. This decrease was also recorded in quilombo communities of Mato Grosso (Arruda *et al.* 2014).

Fuel

The charcoal made from woody babassu endocarp is the only source of fuel for 40% of informants. Mainly used for cooking food, it helps to reduce excessive gathering of wood for these purposes in the forests of the Tucuruí Lake PAM. Several studies attest to the use of this important energy source by rural communities in the regions in which it occurs (May *et al.* 1985; González-Pérez *et al.* 2012; Rufino *et al.* 2008; Arruda *et al.* 2014). According to Albiero *et al.* (2007), babassu coal is the only source of fuel in various regions of northeastern Brazil.

The quality of babassu coal, added to the high price of coal from the wood of plants, and legal restrictions on the use of the latter, attracted the interest of steel companies based in southeastern Brazilian Amazonia who used babassu coal instead of coal from native trees. Between 1999 and 2007 the marketing of babassu coal for the steel industry was a supplementary income generating activity for approximately 33% of the informants.

Medicinal

The largest number of different uses was cited in the medicinal category; altogether 18 different uses with 76 citations. The use of babassu in folk medicine has been recorded by numerous studies that have indicated the use of many products derived from this plant by various local populations (Plotkin & Balick 1984; Forline 2000; López *et al.*2011; Sousa *et al.* 2011; Arruda *et al.* 2014). In addition, several studies of preclinical analysis contributed to the validation of this knowledge, attesting its anti-inflammatory action (Silva & Parente 2001), immunomodulatory action (Nascimento *et al.* 2006), antitumor activity (Fortes *et al.* 2009; Souza *et al.* 2011), and its use for stimulating the production of lymphocytes (Barroqueiro 2001) as well as its analgesic effect (Pinheiro *et al.* 2012).

The fruit was the most important plant part for medicinal purposes, resin endocarp being its main by-product with the highest number of citations (27) being the relief of toothache its most quoted use. The ingestion of the triturated mesocarp, dried and mixed with water, was cited in combating gastritis, hepatitis and osteoporosis, and in the form of a poultice in cases of mycosis. Sousa *et al.* (2011) also reported the use of mesocarp powder for various therapeutic purposes in six rural communities of coconut breakers in Maranhão. The main uses were to treat gastritis, general inflammation and leukorrhea. Studies have demonstrated the effectiveness of the aqueous layer of babassu mesocarp for gastric (Batista *et al.* 2006) and skin healing (Maciel *et al.* 2007), supporting their therapeutic use.

The oil extracted from the seed is used after being refined as well as the residue called *'borra'* which, according to respondents, is stronger and toxic. The refined form is used in the treatment of hemorrhoids, skin infections and as a general anti-inflammatory drug, and the *'borra'* is used as a laxative, worm medicine, and for combating myiasis and female genital inflammation, such as leukorrhea. Sousa *et al.* (2011) reported the use of oil and *'borra'* in the treatment of female genital tract inflammations and various skin infections. Balick (1984) also reported the use of oil as worm medicine and as an expectorant by traditional communities in Bolivia.

The liquid albumen, taken from the immature fruit, was quoted as being used to treat eye infections and the juice of the unripe petiole had the second highest use citation in the medicinal category, being used to stop hemorrhages. The seed bagasse is also used for spider bites, with direct application to the bite site.

Animal Food, Cultural and Environmental Use

Babassu is also used as animal food, for which the only use cited is feeding domestic birds. Its use to attract the shrimp into the *matapi* was also recorded - for this, babassu seed crumbs are used, wrapped in cupuassu (*Theobroma grandiflorum*), cacao (*Theobroma cacao*) or banana (*Musa* sp.) leaves.

Regarding the cultural use of babassu by the riverine communities, the removal of babassu oil from the fruit is cited for hair moisturizing purposes. Babassu oil is applied directly to the hair. González-Pérez *et al.* (2012) show that for the Kayapó indigenous people the most cited use of babassu oil is as a cosmetic; the Kayapós spread it over the skin so that it becomes "beautiful and fragrant."

Referring to environmental use, after decomposition on the forest floor the stipe is used as an organic fertilizer called "*paú*" that is mixed with the soil of small vegetable grown for family consumption. The practice is not widespread among the informants; only eleven citations of this use were recorded. This use was also recorded with few citations in studies in other parts of the country, such as in Campos *et al.* (2015) for extractive communities in the northeast and Nascimento *et al.* (2009) for indigenous people in the central region.

Other Uses

The trade of babassu derivatives is insignificant in the Tucuruí Lake Protected Areas Mosaic, with the sporadic sale of babassu oil having only been reported for four families. The sale happens when there are opportunities to go to neighboring villages and towns and the amount traded is small; four to six liters are transported each time, generally once a month. This data is corroborated by Jatobá (2006), who reported that non-timber extraction is poorly practiced in the Tucuruí Lake PAM and is designed mainly for family consumption, being commercialized only sporadically. This shows that the palm has an almost exclusive subsistence role for the inhabitants of the surveyed area.

Indirect babassu uses also occur, such as the use of larvae of the beetle *Pachymerus nucleorum* that develops inside the fruit, instead of the seed. It can be used directly in human food or used as bait for fish, and the oil removed from the larvae by heating has medicinal (analgesic for earaches) and aesthetic (capillary moisture) uses.

Resources Extraction and Management

In the Tucuruí Lake PAM, the plant parts most intensively used are the fruits and leaves, evidenced by the higher PPV values (Tab. 4). Other studies on babassu use in Brazil also recorded a use concentration in fruits and leaves showing a high PPV, in western quilombos (Martins *et al.* 2014); an indigenous group in the north (González-Pérez *et al.* 2012) and extractivist communities in the northeast (Campos *et al.* 2015).

Table 4. Useful Plant Part Values (PPV) for babassu determined by the uses of the inhabitants of the Tucuruí Lake PAM.

Plant part	PPV
Fruit	0.554
Leaves	0.304
Stipe	0.089
Spata	0.036
Meristem	0.018

Eighty-two percent of the informants state that extraction does not meet management criteria, and all claim to be unaware of technical guidelines for good non-timber forest management practices for the species, such as how many and which leaves should be removed per plant, the minimum number of palm trees that should remain per hectare or how much fruit should be left in the forest for recruitment of the species.

Construction is probably the use category that most increases the exploration pressure on the species in the study area, and thatch is the category with most intensive use, as already described, with widespread use and the highest PCs value among the informants. In the study area, each family uses 200-350 babassu leaves per thatch, which are replaced every two or three years, creating a continuous demand. The leaves are removed from both mature and juvenile palms.

According to the respondents, in regions where the occurrence of the plant is low and there are large individuals, it is common to overthrow the whole plant in order to make complete use of the leaves. The respondents claim that the resource is scarce or non-existent in many parts of the study area; 31 of these, all residents of the SDR Pucuruí-Ararão stated that they traveled up to 25 km in boats in search of babassu elsewhere.

The information that the distribution of babassu is not uniform across the Tucuruí Lake Protected Areas Mosaic is supported by Kahn (1986), who carried out inventories of palm trees in eight forest areas spread over 200 km on both banks of the Tocantins River, in areas currently submerged by the reservoir. The species was recorded in three of the eight surveyed areas, all in the southern part of the study area, more concentrated on the left bank (upstream to downstream direction).

Distribution of Knowledge among the Informants

The distribution of knowledge about babassu, in terms of number of uses known by the informants, showed a non-normal distribution (Kolmogorov - Smirnov d: 0.172, $p < 0.01$), which indicates that knowledge of palms is not homogeneously distributed among the residents of the study area, where most of the respondents know a higher or lower number than the average listed uses.

Socioeconomic factors that most influence the knowledge of babassu were gender and economic activity of the responders. Other factors also contributed but with lower significance, such as the amount of time living in the study area and age of the informant (Tab. 5).

The time living in the study area and the age of the informant were directly related with the knowledge of babassu, indicating that older informants know more uses than young residents. This difference may be related to the gradual accumulation of knowledge throughout life and the erosion of knowledge (Phillips & Gentry 1993).

Table 5. Relationship between knowledge of the informants (Number of known uses, number of uses per category) and the socioeconomic characteristics of the informants. We included in the table only results that were statistically significant.

	Gender	Age	Time	Origin	Activity	P	R^2
Number of known uses	-***				+****	< 0.0001	0.1664
Number of uses p/ category							
Utensils and tools					+****	0.0124	0.0356
Construction	+**		+*		+****	< 0.0001	0.1081
Human Food	-**				+****	< 0.0001	0.2562
Fuel	+*	+*				0.0015	0.0583
Medicinal	-****				+*	< 0.0001	0.4684
Environmental uses						0.5424	0.0078
Cultural Use	-**		+**			0.5555	0.0063
Animal food					+***	< 0.0001	0.1559
Other Use						0.6475	0.0125

+: Positive correlation, -: negative correlation; significance levels: * $0.01 < P \leq 0.05$, ** $0.001 < P \leq 0.01$, *** $0.0001 < P \leq 0.001$, **** $P \leq 0.0001$

The economic activity of the informant was also the main factor of influence in all categories of use, particularly in utensils and tools and human food (Tab. 5). Farmers showed greater means related to number of uses known than fishermen (Fig. 3). Byg & Balslev (2001), in Madagascar, found a direct positive relationship between agricultural diversification and knowledge of use of palms and considered that this trend could be related to the direct action that the farmer has on the floristic environment.

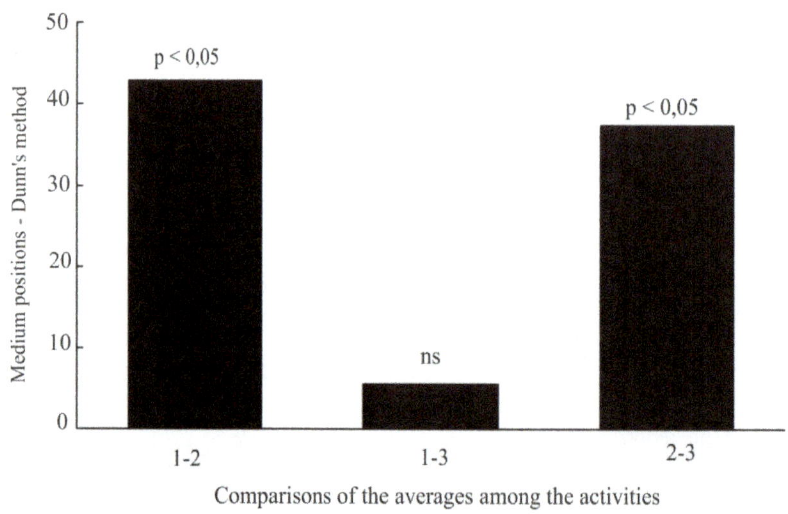

Figure 3. Comparison of the number of uses known per informant according to the economic activity (Kruskal-Wallis non-parametric tests) using Dunn's method for comparing means: 1 only agriculture, 2 only fishing, 3 agriculture and fishing; ns: not significant.

Socioeconomic factors did not show a statistically significant influence on the categories: environmental uses, cultural uses and other uses (Tab. 5)

In a general manner, women know more about babassu uses than men; they particularly have greater medicinal knowledge about the species (Tab. 6). Differences in knowledge of men and women are related to differences in their areas of responsibilities and daily activities: women often know more uses related to medicinal and food categories, while men know more about construction and commercially valuable species (Hanazaki *et al.* 2000; Luoga*et al.* 2000; Byg & Balslev 2004).

Table 6. Comparison of means related to knowledge of General use and Medicinal use of babassu between men and women (Kruskal-Wallis test) using Dunn's method for comparison of means.

Parameters	General use	Medicinal use
H =	15.75	49.59
Degrees of freedom	1	1
(P)	< 0.0001	< 0.0001
Women (medium position)	123.40	144.97
Men (medium position)	86.58	79.69
Comparison (Dunn's method)	Difference of positions	
Medium position men and women	36.81	65.28

CONCLUSION

The daily dependence that the inhabitants of the Tucuruí Lake Protected Areas Mosaic have on this palm tree, mainly in the utensils and tools, for construction, human food and fuel, shows that babassu is an important resource for their livelihood strategies.

Although informants have shown strong sociocultural connection with the species, there is a lack of appropriate management of a more sustainable exploration by most of them.

Distribution surveys and current babassu abundance throughout the area should subsidize the exploitation strategies, because exploration that decimates the local population of babassu will put at risk the guarantee of shelter and food and energy security for much of the population. Instead, it will also encourage an increased exploitation pressure on timber forest resources.

The study of knowledge distribution patterns among the informants revealed the role of agriculture and the female gender as the main factors determining the detention of knowledge about the use of babassu.

ACKNOWLEDGEMENTS

We thank The Tucuruí PAM communities. The Federal Rural University of Amazonia, and CNPQ (Conselho Nacional de Desenvolvimento Científico e Tecnológico), Eletrobrás - Eletronorte SA. Antônio Elielson Sousa Rocha for confirmation of species identification. The sharing of the results of this research with local communities will be in the form of lectures where we will discuss the sustainable use of the resource and good use practices and management of babassu. Text and grammar Review by Magda Alves de Sousa Gobira.

REFERENCES

1. Albiero D, Maciel J, López C, Mello C, Gamero CA. 2007. Proposta de uma máquina para colheita mecanizada de babaçu (*Orbignya phalerata* Mart.) para agricultura familiar nas regiões de matas de transição da Amazônia. Acta Amazonica 37: 337-346.
2. Albuquerque UP, Lucena RFP, Cunha LVFC. 2008. Métodos e técnicas na pesquisa etnobotânica. 2nd. edn. Recife, Comugraf.
3. Anderson AB. 1991. Forest management strategies by rural inhabitants in the Amazon estuary. In: Goméz-Pompa A, Whitmore TC, Hadley M. (eds.) Rain forest regeneration and management. Paris, United Nations Educational, Scientific and Cultural Organizations. p. 351-360.
4. Anderson AB, May P, Balick M. 1991. The subsidy from nature: Palm forests, peasantry, and development on an Amazon frontier. New York, Columbia University Press.
5. Araújo ARO. 2009. Os territórios protegidos e a Eletronorte na área de influência da UHE Tucuruí/PA. MSc Thesis, Universidade Federal do Pará, Brazil.
6. Araújo FR, Lopes MA. 2012. Diversity of use and local knowledge of palms (Arecaceae) in eastern Amazonia. Biodiversity and Conservation 21: 487-501.
7. Araújo MVLF, Silva KCA, Silva BB, Ferreira ILS, Cintra IHA. 2014. Pesca e procedimentos de captura do camarão-da-Amazônia à jusante de uma usina hidrelétrica na Amazônia brasileira. Biota Amazônia 4: 102-112.
8. Arruda JC, Sander NL, Silva CJ. 2014. Conhecimento e Uso do Babaçu (*Attalea Speciosa* Mart.) por Quilombolas em Mato Grosso. Fragmentos de Cultura 24: 239-252.
9. Balick MJ. 1984. Ethnobotany of palms in the Neotropics. Advances in Economic Botany 1: 9-23.
10. Balick MJ. 1988. The use of palms by the Apinajé and Guajará Indians of Northeastern Brazil. Advances in Economic Botany 6: 65-90.
11. Balick MJ, Pinheiro C. 2000. Babaçu *Orbignya phalerata*. In: Clay JSP, Clement C. (eds.) Biodiversidade Amazônica: Exemplos e Estratégias de Utilização. Óleos Industriais. Manaus, Instituto Nacional de Pesquisa da Amazônia, SEBRAE. p. 255-263.
12. Barbieri ALC. 2004. Desenvolvimento recente do Maranhão: uma análise do crescimento do PIB e expectativas. MSc Thesis, Universidade Federal do Pernambuco, Brazil.

13. Barroqueiro ESB. 2001. Efeito do mesocarpo de babaçu sobre a produção de anticorpos auto-reativos. MSc Thesis, Universidade Federal do Maranhão, Brazil

14. Batista CP, Torres OJM, Matias JEF, et al 2006. Effect of aqueous extract of *Orbignya phalerata* (babassu) in the gastric healing in rats: morphologic and tensiometric study. Acta Cirúrgica Brasileira 21: 26-31.

15. Borchsenius F, Borgtoft PH, Balslev H. 1998. Manual to the Palms of Ecuador. Aarhus, Denmark: Department of Systematic Botany, Aarhus University-Pontificia Universidad Católica del Ecuador, Quito. Association of American Universities Reports 37.

16. Byg A, Balslev H. 2001. Diversity and use of palms in Zahamena, eastern Madagascar. Biodiversity and Conservation 10: 951-970.

17. Byg A, Balslev H. 2004. Factors affecting local knowledge of palms in Nangaritza Valley in South-Eastern Ecuador. Journal of Etnhnobiology 24: 255-278.

18. Campos JLA, Silva TLL, Albuquerque UP, Peroni N, Araújo EL. 2015. Knowledge, Use, and Management of the Babassu Palm (*Attalea speciosa* Mart. ex Spreng) in the Araripe Region (Northeastern Brazil). Economic Botany 69: 240-250.

19. Coelho-Ferreira M. 1994. Essai sur l´ethnopharmacologie et la gestion des ressources naturelles chez les Saterê-Mawé du fleuve Marau, Moyen Amazonas, Brésil. Paris, Muséum National d´Histoire Naturel, Rapport de Mission.

20. Corrêa DAG, Ming LC, Vasques MP. 2010. Manejo de fibras vegetais utilizadas em artesanatos por comunidades tradicionais do parque estadual e turístico do Alto Ribeira, e seu entorno, Iporanga, SP. In: Silva VA, Almeida ALS, Albuquerque UP. (eds.) Etnobiologia e Etnoecologia: Pessoas & natureza na América Latina. 1st. edn. Recife. Nupeea. p. 173-208.

21. Forline LC. 2000. Using and sustaining resources: the Guajá Indians and the babassu palm (*Attalea speciosa*). Indigenous Knowledge and Development Monitor 8: 3-7.

22. Fortes TS, Azevedo APS, Barroqueiro ES, et al 2009. Desenvolvimento do Tumor de Ehrlich em Camundongos após incubação in vitro com mesocarpo de babaçu. Revista de Ciências da Saúde 11: 11-18.

23. Fuerst R. 2006. Xicrin hommes oiseaux d'Amazonie. Milan, 5 Continents Editions.

24. Gomez-Beloz A. 2002. Plant use knowledge of the Winikina Warao: the case for questionnaires in ethnobotany. Economic Botany 56: 231-241.

25. González-Pérez SE, Coelho-Ferreira M, Robert P, Garcés CLL. 2012. Conhecimento e usos do babaçu (*Attalea speciosa* Mart. e *Attalea eichleri* (Drude)) entre os Mebêngôkre-Kayapó da Terra Indígena Las Casas, estado do Pará, Brasil. Acta Botanica Brasilica 26: 295-308.

26. Hanazaki N, Tamashiro JY, Leitão-Filho HF, Begossi A. 2000. Diversity of plant uses in two Caiçara communities form the Atlantic forest coast, Brazil. Biodiversity and Conservation 9: 597-615.

27. Henderson A. 1995. The Palms of the Amazon. New York, Oxford University Press.

28. Huntington HP. 2000. Using traditional ecological knowledge in science: methods and applications. Ecological Applications 10: 1270-1274.

29. Jatobá SUS. 2006. Gestão do território e a produção da socionatureza nas ilhas do Lago de Tucuruí na Amazônia brasileira. MSc Thesis, Universidade de Brasília, Brazil.

30. Kahn F. 1986. Les palmiers des forêts tropicales humides du Bas Tocantins (Amazonie brésilienne). Revue d'Ecologie: La Terre et La Vie 41: 3-13.

31. Lima ES, Felfili JM, Marimon BS, Scariot A. 2003. Diversidade, estrutura e distribuição espacial de palmeiras em um cerrado sensu stricto no Brasil Central - DF. Revista Brasileira de Botânica 263: 361-370.

32. López C, Gonzalez-Pérez S, Coelho-Ferreira M, et al. 2011. Laboratório de práticas sustentáveis em Terras Indígenas próximas ao Arco de Desmatamento Terra Indígena Las Casas - PA/Terra Indígena Kayapó-Moikarakô e Terra Indígena Alto Turiaçu aldeias Xiepihu-rena e Paracui-rena. Belém, INCT-MPEG - Relatório anual.

33. Lorenzi H. 1996. Palmeiras no Brasil nativas e exóticas. Nova Odessa, Editora Plantarum.

34. Lorenzi H, Kahn F, Noblick LR, Ferreira E. 2010. Flora Brasileira Lorenzi: Arecaceae (Palmeiras). Nova Odessa, Editora Plantarum .

35. Luoga EJ, Witkowski ETF, Balkwill K. 2000. Differential utilization and ethnobotany of trees in Kitulanghalo forest reserve and surrounding communal lands, eastern Tanzania. Economic Botany 54: 328-343.

36. Macía MJ, Armesilla PJ, Cámara-Leret R, et al. 2011. Palm Uses in Northwestern South America: A Quantitative Review. The Botanical Review 77: 462-570.

37. Maciel MCG, Barroqueiro ESB, Silva LA, Guerra RNM, Nascimento FR. 2007. Avaliação do Efeito Cicatrizante do Mesocarpo de Babaçu (*Orbignya phalerata*) em Feridas de Pele Induzidas em Camundongos. Revista de Ciências da Saúde 9: 1-4.

38. Martins RC, Filgueiras TS, Albuquerque UP. 2014. Use and diversity of palm (Arecaceae) resources in central western Brazil. The Scientific World Journal 14: 1-14.

39. May PH, Anderson AB, Balick MJ, Frazão JM. 1985. Subsistence benefits from the Babassu palm (*Orbignya martiana*). Economic Botany 39: 113-129.

40. Meirelles JF. 2004. O Livro de Ouro da Amazônia: Mitos e verdades sobre a região mais cobiçada do planeta. 4th. edn. São Paulo, Ediouro.

41. Nascimento AR. 2010. Riqueza e etnobotânica de palmeiras no território indígena Krahô, Tocantins, Brasil. Floresta 40: 209-220.

42. Nascimento AR, Santos A, Martins R, Borges T. 2009. Comunidade de palmeiras no território indígena Krahò, Tocantins, Brasil: biodiversidade e aspectos etnobotânicos. Interciência 34: 182-184.

43. Nascimento FRF, Barroqueiro ESB, Azevedo APS, et al 2006. Macrophage activation induced by *Orbigna*

*phalerata*Mart. Journal of ethnopharmacology 103: 53-58.

44. Ohashi ST, Vieira ICG, Leão NVM. 2004. Fitossociologia e estrutura de uma floresta remanescente da área de soltura 4 da UHE - Tucuruí. Belém, Relatório Técnico Eletronorte.
45. Paniagua-Zambrana NY. 2005. Diversidad, densidad, distribución y uso de las palmas en la región del Madidi, noreste del departamento de La Paz (Bolivia). Ecología em Bolivia 40: 265-280.
46. Paniagua-Zambrana NY, Cámara-Leret R, Macía MJ. 2015. Patterns of medicinal use of palms across northwestern South America. The Botanical Review 81: 317-415.
47. Pará. 2002. Lei N° 6.451, de 8 de abril de 2002. Cria Unidades de Conservação da Natureza na região do Lago de Tucuruí no território sob jurisdição do Estado do Pará. Belém, Assembléia Legislativa do Estado do Pará.
48. Phillips O, Gentry AH. 1993. The useful plants of Tambopata, Peru: II. Additional hypothesis testing in quantitative ethnobotany. Economic Botany 47: 33-43.
49. Phillips O. 1996. Some quantitative methods for analyzing ethnobotanical knowledge. In: Alexiades M. (ed.) Selected guidelines for ethnobotanical research: a field manual. New York, New York Botanical Garden. p. 171-197.
50. Pinheiro MMG, Boylan F, Fernandes PD. 2012. Antinociceptive effect of the *Orbignya speciosa* Mart. (Babassu) leaves: evidence for the involvement of apigenin. Life Sciences 91: 293-300.
51. Piratoba MD. 2014. Dinâmica Temporal da Paisagem: mudanças, percepções e dificuldades de recuperação na RDS Alcobaça, área de influência da UHE Tucuruí/PA. MSc Thesis, Universidade Federal do Pará, Brazil
52. Plotkin MJ, Balick MJ. 1984. Medicinal uses of South American palms. Journal of ethnopharmacology 10: 157-179.
53. Posey D. 1987. Manejo da floresta secundaria, capoeira, campos e cerrados (Kayapó). In: Ribeiro B. (ed.) Suma Etnológica Brasileira. Vol. 1. 2nd. edn. Petrópolis, Vozes. p. 173-185.
54. Ravena N, Cañete VR, Souza CL, Cañete TMR. 2009. Lições não aprendidas: Hidroelétricas, impactos ambientais e políticas de recursos hídricos. Paper NAEA 239.
55. Rezende JBA, Lima J, Meira FM, Lima P, Tenório G 2010. Manejo de Caraná no Alto Tiquié. In: Cabalzar A. (ed.) Manejo do mundo: conhecimento e práticas dos povos indígenas do Rio Negro, noroeste Amazônico. São Paulo, FOIRN & ISA.
56. Rocha AES, Silva MFF. 2005. Aspectos fitossociológicos, florísticos e etnobotânicos das palmeiras (Arecaceae) de floresta secundária no município de Bragança, PA, Brasil. Acta Botanica Brasilica 19: 657-667.
57. Rufino MUL, Costa JTM, Silva VA, Andrade LHC. 2008. Conhecimento e uso do Ouricuri (*Syagrus coronata)* e do babaçu (*Orbignya phalerata*) em Buíque, PE, Brasil. Acta Botanica Brasilica 22: 1141-1149.
58. SECTAM - Secretaria Executiva de Ciência, Tecnologia e Meio Ambiente do Estado do Pará. 2002. Levantamento Socioeconômico das Ilhas do Lago de Tucuruí. Belém, Governo do Estado do Pará.
59. Shiraishi NJ. 1999. As quebradeiras de coco no meio norte. Paper NAEA 121.
60. Sillitoe P. 1998. The development of indigenous knowledge. Current Anthropology 39: 223-252.
61. Silva BP, Parente JP. 2001. An anti-inflammatory and immunomodulatory polysaccharide from Orbignya phalerata. Fitoterapia 72: 887-93.
62. Sousa MHSL, Monteiro CA, Figueredo PMS, Nascimento FRF, Guerra RNM. 2011. Ethnopharmacological use of babassu (*Orbignya phalerata* Mart) in communities of babassu nut breakers in Maranhão, Brazil. Journal of Ethnopharmacology 133: 1-5.
63. Souza PAVR, Palumbo A, Alves LM, et al 2011. Effects of a nanocomposite containing *Orbignya speciosa* lipophilic extract on Benign Prostatic Hyperplasia. Journal of ethnopharmacology 135: 135-146.

CHAPTER 6

USE AND DIVERSITY OF PALM (ARECACEAE) RESOURCES

Renata Corrêa Martins, Tarciso de Sousa Filgueiras, and Ulysses Paulino Albuquerque

INTRODUCTION

Several studies of palm species have highlighted their variety of uses. Palms have often been cited as useful materials for construction, food, handcrafts, rituals, and therapeutics. Quantitative methods have been used in ethnobotanical studies to analyze the knowledge and use of palms by local people, with the work of Bates [1], Byg and Balslev [2], Byg et al. [3], Campos and Ehringhaus [4], Nascimento et al. [5], and Zambrana et al. [6] deserving special mention.

The family Arecaceae offers a high potential value for the Brazilian Cerrado because it has great species diversity (approximately 60 species) [7], exists in all types of habitat (from forest to savanna and grassland) [8], and is part of the culture and family-based economies of many traditional and non-traditional communities of the Cerrado. However, few ethnobotanical studies have been conducted on the use of palms, despite the extensive palm groves and the cultural diversity of the region (see [5, 9, 10]).

The Brazilian Cerrado is also home to a diverse range of Quilombola (Maroon) communities, who represent the descendants of African people who arrived in Brazil as slaves during the colonial period. Ethnobotanical and ethnopharmacological research has been conducted in Quilombola communities with an emphasis on medicinal plants, including studies by Amoroso [11], Rodrigues and Carlini [12], Franco and Barros [13], Rodrigues [14], and Crepaldi and Peixoto [15]. However, because of the wide distribution of these communities across Brazil and their diverse cultures, landscapes, and natural resources, a great deal of ethnobiological knowledge remains unexplored.

This study aims to add to the local botanical knowledge regarding the family Arecaceae (Palmae) in Brazil, especially in the Cerrado, by studying communities closely linked to the Cerrado and its natural resources to understand their relationships with and the use of native palms. This paper aims to answer the following questions.

- What is the diversity of palms known and used by a Quilombola community of the Brazilian Cerrado?
- Do the gender, age, and education of the respondents influence their recognition and use of palm species?
- Which palms are the favorites of the local community, and which parts of those palms are the most important?

MATERIALS AND METHODS

Area of the Study

The field work was conducted in a Quilombola community in Goiás State. The population of Goiás is approximately 1,847,671 (married people over 10 years old) [16], distributed among 246 cities. The 73 Quilombola communities in Goiás are distributed among 24 cities. Cavalcante is home to the greatest number of Quilombola communities (23), followed by Monte Alegre (15) and Teresina de Goiás (6) [17]. People from the Quilombola communities in these three cities are called Kalungas or Calungas.

For the Kalunga people, the name Kalunga refers to a stream that passes near the Contenda farmland. In the past, many people did not like to be referred to as Kalunga because it was synonymous with witchcraft and laziness. However, many became proud of this identity when their presence on the land became legalized [18,19]. The Kalunga territory was established by Africans and African-Brazilians who escaped from slavery and remained in the region after the decline of mining and the abolition of slavery. The freed slave population from other areas also migrated to the region looking for land to cultivate to enable an autonomous and free existence [20].

Citation: Renata Corrêa Martins, Tarciso de Sousa Filgueiras, and Ulysses Paulino Albuquerque, "Use and Diversity of Palm (Arecaceae) Resources," The Scientific World Journal, vol. 2014, Article ID 942043, 14 pages, 2014. DOI:10.1155/2014/942043.

The Kalungas lived for 200 years isolated in the valleys of the tributaries of the Parana river and on the edges of the "Chapada dos Veadeiros" and formed various groups, with four main large groups located in Vão de Almas, Contenda, Vão do Moleque, and Ribeirão dos Bois and other smaller groupings located in Riachão, Sucuri, Saco Grande, Volta do Canto, Olha d'água, Ema, Taboca, Córrego Fundo, Terra Vermelha, Lagoa, Porcos, Brejão, Fazendinha, Vargem Grande, Engenho II, Funil, Capela, and a dozen other places [19].

The Kalungas were scientifically discovered in 1982 by the anthropologist Mari de Nazaré Baiocchi. Prior to this time the Kalungas practiced subsistence living, planting to eat and exchanging products such as cassava flour, leather, and mangaba for fabrics, salt, and other supplies in neighboring towns. The "discovery" of the Kalungas led them to obtain the legal right to own their land, preventing land invaders from displacing them [21].

The area of the Historic Kalunga Site was registered by Law No. 11.406 on January 21, 1991, and regulated by Law No. 19 on January 5, 1996. A large part of this scenic territory consists of mountains, with extensive steep areas, large valleys, and many rivers. Only 30% of this area is arable [19], and accessing many communities is difficult.

The area of the Kalunga territory is a well-preserved Cerrado region, especially in areas with steep topographies featuring hillsides and mountain ranges. The area features extensive palm groves and valleys, and most of the communities are located near rivers, creeks, and streams. Their houses are constructed from adobe, with palm leaf roofs and packed earth floors. These communities practice simple lifestyles, for example, using wood stoves. Some communities (Engenho II and Vão do Moleque) have received benefits from the federal government, and many residents in those communities have acquired new homes built with masonry and with clay roofs.

The social organization of these communities is typically centered around family relationships [18]. The family is the basic unit of society, in which people help each other and the ethical and cultural values of the community are passed on. These communities have cycles of events based on the planting and harvesting times. The Cerrado features two defined seasons (wet summer and dry winter), and these seasons determine the time spent cultivating the land [19, 22].

The Kalunga community lives by subsistence agriculture, including husbandry of cattle, pigs, and poultry and extraction of Cerrado fruits (watermelon, banana), vegetables, and medicinal plants cultivated in home gardens and orchards [22–24]. The main crops are corn (90% of families) followed by beans (80%), cassava (70%), rice (50%), pumpkin (50%), and sugar cane (40%) [25].

Approximately 60% of residents use cassava to produce flour, which is one of the most economically important products for these communities [25]. Kalunga gardens are strategically important due to their growth and maintenance of a wide variety of plants [22].

The Kalunga territory borders Tocantins state, occupying an area of approximately 253.2 thousand hectares at the following geographic coordinates: from 13°20′ to 13°27′ south latitude and from 47°10′ to 47°20′ west latitude from Greenwich. This territory is located in the Chapada dos Veadeiros area, northeast of Goiás state, 600 km from Goiânia and 330 km from Brasilia. The region borders the following cities: Monte Alegre de Goiás, Terezina de Goiás, and Cavalcante (Figure 1).

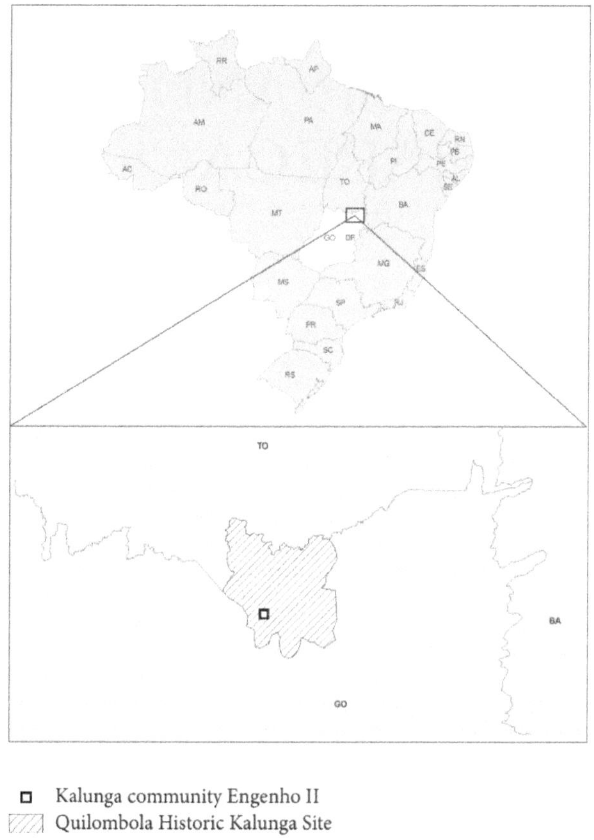

Figure 1: Map of Brazil denoting Goiás State (GO), the Quilombola Historic Kalunga Site, and the Kalunga Community Engenho II in Cavalcante, GO, central western Brazil.

This study examines the Kalunga community Engenho II. This community is located 27 km from the municipality of Cavalcante and is accessed via an unpaved road that reaches the center of the community (13° 34′ 57″ S and 47° 28′ 21″W). The houses of the residents are built of mixed materials, including adobe and masonry, and have roofs made of buriti leaves and clay tiles. At the end of this research, the community consisted of approximately 100 homes and approximately 550 residents.

The residents of the Engenho II practice subsistence agriculture and participate in the regional market, eventually as employees, selling or exchanging products. The cultivated areas are far from the farmhouses of the community. Some residents engage in a small amount of commerce by selling snacks and preparing meals to order. Some Kalungas are public employees and several are tour guides. One of the main income-generating activities in this community is the rural tourism, with attractions including waterfalls, hiking, and homemade food.

Ethnobotanical Data Collection

The process of requesting access to study this traditional community was filed in September 2009 with the request of a signature to provide prior consent. Resolution No. 250 was published in the Official Gazette on April 16, 2010. The process number that authorized this research is 02000.002793/2009-73 and the authorization number is 48/2009.

This work was performed in three phases.

Phase I: Informal Interviews

To develop a preliminary list of palm species known among the residents of the Engenho II community, approximately 20 residents were interviewed, including some tourist guides and local leaders. The interviews took place in a shed used to provide tourist services. We used a diary to record the names of the palms and the locations of the populations, along with information about the use of the species and any other interesting details. In this step, 16 palm species were recognized.

Phase II: Guided Tour and Systematic Botany Collection

A guided tour was conducted to support and validate the names of the plants mentioned in the informal interviews [26]. In this step, a floristic survey of all palm species in the region was conducted. Samples were collected from sites where palms occurred. Areas with forest, savanna, and grassland formations were examined, and all 16 species mentioned during the informal interviews were photographed, collected, and identified.

Jorge Moreira da Silva, Damião Santos Rosa, and João Francisco Maia participated in the botanical collection expeditions. The specimens were deposited in the UB (University of Brasilia) herbarium. Due to difficulties in collecting some of these palm species, some were identified in the field, while others were determined according to the literature [27–29].

Phase III: Sample Selection

This study considered all local residences (n = 100), 12 of which had no residents. We visited the community 12 times, with each visit lasting 2–4 days, beginning in May 2010 and ending in February 2011. The preferred informant was the oldest resident of the house present at time of the visit.

The ethnobotanical data were collected in a semi-structured interview conducted using the checklist/interview method [26, 30]. This technique is considered the best option when there is only one chance to interview the informant [26]. In the first part of the interview, the questions determined the socioeconomic profile of the informant such as gender, age, and education. In the second part, the questions referred directly to the recognition and use of palms, as described below.

Eighty-eight individual interviews were conducted; 56 of the interviewees were women and 32 were men, with ages ranging between 18 and 82 years. The main activity performed by the residents is the cultivation of land (farming) (70; n = 88), and men and women perform other activities, including domestic duties and rural tourism. The respondents also included one health officer, one public employee, two school cooks, two teachers, and three local business owners.

Visual stimuli enable an evaluation of the respondent's knowledge and use of plants (see [30–32]). Considering the size of palm samples, photographs were the best visual stimuli for this study. The systematic use of photographs of plants in ethnobotanical interviews is still poorly represented in the literature [30, 33–35], but this approach has been shown to be more effective than the use of voucher specimens [36].

Sixteen A4-sized photographs of each palm species present in the region were prepared (Figure 2). The photographs show the plant in its habitat and the leaf, flower, and/or fruit shapes. A preliminary compilation of the figures was used in a pilot formulary in the community to determine whether the photographs accurately captured the true plant morphology and fit with local knowledge. Thus, it was possible to ensure that "the informants and the interviewer were talking about the same plant" [33].

Figure 2: A figure used during the interviews in the Kalunga community Engenho II, Cavalcante, GO, central western Brazil. Attalea speciosa (coconut palm).

In each residence, the figures were handed to the informant one at a time. Then, for each species shown, the following questions were asked: Do you know this plant? Have you ever used it? For what purpose did you use this plant? How did you use it? What part did you use? The popular names and descriptions of use were recorded.

A ranking method was used to identify the most culturally significant species, as this method has been shown to be highly useful in studies of local preferences [26]. The informants were asked to rank the species according to their preferences (using the figures).

Data Analysis

The nonparametric Kruskal-Wallis test was used to analyze the influences of gender, age, and education and the number of palms recognized and used by the informants. The analysis of variance was performed using BioEstat 5.0 software [37] and Excel was used for the correlation analysis (Windows 7).

The analysis was performed for the two age groups 18–40 and 41–82 [38–40]. The respondents' levels of education were categorized as illiterate, some elementary education, completed elementary education, some secondary education, completed secondary education and some higher education.

To analyze the rankings of the photos, salience was calculated using the ANTHROPAC software [41]. To analyze the knowledge and use of the palms, the obtained information was analyzed by ethnobotanical indexes according to the consensus among the informants [2, 42, 43] (Table 1). The values of the total species diversity (SDtot) and total species equitability (SEtot) were calculated to characterize the diversity in the use of palms and the uniformity of the degree to which the various species contribute to the daily lives of the informants.

Table 1: Indexes used to analyze the knowledge and use of palms in the Kalunga Community Engenho II, Cavalcante, GO, central western Brazil [2, 42, 43].

Indexes	Calculation	Description
Total species diversity (SDtot)	$SDtot = 1/\sum Ps^2$ (P = total contribution of species s to the total use; number of times species s was cited divided by the total number of citations).	Measures how the species are used and how they contribute to the total palm use. Values range from 0 to n.
Total species equitability (SEtot)	SEtot = SDtot/n (n = number of species used).	Measures how different species contribute to total palm use independent of the number of species used. Values range from 0 to 1.
Importance value (IVs)	IVs = nis/n (number of informants who considered species s the most important). n= total number of informants.	Measures the proportion of informants who cited one species as the most preferred. Values range from 0 to 1.
Use diversity values (UDs)	$UDs = 1/SPc^2$ (Pc = contribution of the use category c to the total utility of species s; the number of times species s was mentioned within each use category divided by the total number of citations of use of species s among all use categories).	Measures how a species is used within a category and the degree to which it contributes to the overall use of this species. The values range from 0 to the number of categories of use for which the plant is used.
Use equitability value (UEs)	UEs = UDs/UDsmax (UDsmax = the maximum use diversity value for a species with a given number of categories).	Measures how different uses contribute to the overall use of a species, independent of the number of categories of use. Values range from 0 to 1.
Informant diversity value (IDs)	$IDs = 1/\sum P_i^2$; P_i = contribution of informant i to the total knowledge of species s(number of uses mentioned by an informant divided by the total number of uses).	Measures how many informants use one species and how its use is distributed among informants. Values range from 0 to the number of informants that use the species.
Informant equitability value (IEs)	IEs = IDs/IDsmax (IDsmax = the maximum diversity value of an informant for species s).	Measures how the use of a species is distributed among informants, regardless of the number of informants who use it. The values range from 0 to 1.
Use consensus value (UCs)	UCs = 2ns/n − 1 (ns = number of people who use species s).	Measures the degree of agreement among informants regarding whether a species is useful. Values range from −1 to +1.
Purpose consensus value (PCs)	$PCs = \sum Pu^2/S$ (Pu = proportional contribution of the use u to the overall utility of the species s; S = total number of different types of use of species s). PCs represents the number of times that use u has been recorded for speciess divided by the total number of recorded uses for species s.	Measures the degree of agreement between informants regarding a given use. Values range from 0 to 1.
Species diversity value (SDi)	$SDi = 1/\sum Ps^2$ (Ps = contribution of a species s to the total use of a species by informanti; the ratio between the number of times that the informant mentions a species divided by the total number of responses of the informant).	Measures how an informant uses many species and how the uses are distributed among species. Values range from 0 to the number of species used by the informant.
Species equitability values (SEi)	SEi = SDi/Sd imax (Sd imax = maximum value of species diversity for informant i)	Measures how an informant uses known plants, regardless of the number of plants used. Values range from 0 to 1.
Part of the plant value (PPV)	The ratio between the total number of reported uses for each plant part ($\sum RU_{plant\ part}$) and the sum of uses reported for that plant ($\sum RU$).	Indicates the difference in the number of uses of the parts of the plant and reveals the most frequently used part of the plant.

RESULTS AND DISCUSSION

The survey generated 1,928 records of use for the 16 palm species. The uses were grouped into 10 categories (Table 2) as follows food/direct use, food/cooking, food/animal feed, handicrafts, construction, medicinal, biofuel, toxic, ritual, and fertilizer.

Table 2: Descriptions of the 10 categories of use identified for palms in the Kalunga community Engenho II, Cavalcante, GO, central western Brazil.

Category of use	Description
Food/direct use	The consumption of fruits or other plant parts occurs in natura.
Food/cooking	The part used (fruit, seed, palm heart) is manipulated for the consumption of juice, cake, dessert, porridge, oil, and meal.
Food/animal feed	Used to feed domestic animals.
Handicrafts	Household items ("tapiti," sieve, "quibano"), rustic furniture (sofa, bookshelf, table), decorative objects, and direct use of any part of the plant (e.g., using the sheath of a buriti palm as a dustpan).
Construction	Construction and/or coverage of ranches, houses, kitchens, hen houses, and pigsties.
Medicinal/cosmetic	Use of any part of the palm, alone or manipulated with other species for the manufacture of homemade medicine and cosmetics (only one cosmetic use was cited).
Ritual	Decoration in religious ceremonies and blessings.
Fertilizer	Use of decomposed plant parts as fertilizer.
Biofuel	Use of dried leaves to light fires and of dried fruits as charcoal.
Toxic	Citations of contraindicated uses.

Diversity of the Known and Used Palms

In the Kalunga community studied here, respondents recognized 16 palm species belonging to nine genera (Table 3). Approximately 51 common names were cited to designate the 16 species recorded (Table 3). All palms in this study are native to the Cerrado, and only Syagrus oleracea (guariroba) was cited as cultivated by one informant. Euterpe edulis (palmiteiro) was the only species identified that has a vulnerable conservation status [44].

Table 3: List of species, common names, parts used, and the uses of palms in the Kalunga community Engenho II, Cavalcante, GO, central western Brazil. In the "Vernacular names" column, the bolded names indicate those most frequently used by the informants; in the "Uses" column, the number indicates the number of different uses for the species.

Scientific name	Vernacular names	Parts used	Uses
Acrocomia aculeata(Jacq.) Lodd. ex Mart.	Macaúba Palm, xodó	Fruit (epicarp, mesocarp, seed), stipe ("shell" and center core), palm heart	(5) Food (direct use, cooking and animal feed), handicrafts, medicinal
Allagoptera campestris(Mart.) Kuntze	Licuri rasteiro do campo, licurizinho, coquinho rasteiro, licurizinho do cerrado, coquinho de licuri	Leaf, fruit (mesocarp, seed), palm heart	(3) Food (direct use, cooking), handicrafts
Allagoptera leucocalyx(Drude) Kuntze	Licuri rasteiro da mata, coco licuri, licurizinho, licuri rasteiro, licurizinho rasteiro, coquinho painha	Mesocarp, seed, palm heart	(6) Food (direct use, cooking), handicrafts, construction, medicinal, toxic
Attalea comptaGlassman	Indaiá, daiá, babaçu	Fruit (mesocarp, endocarp, seed), leaf, palm heart, peduncular bract, infructescence	(6) Food (direct use, cooking), handicrafts, construction, medicinal, fuel
Attalea eichleri(Drude) A.J. Hend.	Pindoba, palhinha, piaçaba, painha, coco-painha, candoba	Fruit (mesocarp, seed), leaf, peduncular bract	(5) Food (direct use, cooking), handicrafts, construction, medicinal
Attalea speciosa Mart. ex Spreng.	Palm, coco-palmeira, babaçu	Leaf, fruit (endocarp, mesocarp, seed), palm heart	(4) Food (direct use, cooking), handicrafts, construction

Butia purpurascens(Mart.) Becc.	Cabeçudo	Leaf, fruit (mesocarp, seed), peduncular bract, inflorescence, palm heart	(7) Food (direct use, cooking), handicrafts, medicinal, ritual, biofuel, toxic
Euterpe edulisMart.	Palmito, palmito-do-brejo, açaí	The whole plant, stipe, palm heart, fruit (mesocarp, seed), leaf, peduncular bract	(5) Food (direct use, cooking), handicrafts, construction, ritual
Geonoma pohlianasubsp.wed-delliana	Palmita, palmito-merim	The whole plant, stipe, mesocarp, seed	(5) Food (direct use, cooking), handicrafts, construction, ritual
Mauritia flexuosa L.f.	Buriti palm	Fruit (epicarp, meso-carp, seed), leaf (the whole leaf, sheath, petiole, new unopened leaf), stipe, root	(7) Food (direct use, cooking, and animal feed), construction, handicrafts, medici-nal, cosmetic
Mauritiella armata(Mart.) Burret	Buritirana, buriti-merim, pati	Stipe, palm heart, leaf (the whole leaf, peti-ole), root, fruit (meso-carp)	(5) Food (direct use, cooking), handicrafts, medicinal, construc-tion
Syagrus comosa(Mart.) Mart.	Garioba catolé, garioba, garioba-do-campo, garioba da serra, gariroba-catolé, coco-catolé, gariobina-do-cerrado, gueiroba, gario-binha, catolezinha, garioba-comum	Palm heart, fruit (meso-carp, endocarp, seed), stipe	(5) Food (direct use, cooking), handicrafts, construction, medici-nal
Syagrus deflexaNoblick and Lorenzi	Licuri da serra, coquinho-do-cerrado, licurizinho-da-serra, coquinho-de-licuri, coquinho-da-serra, paia-de-nicuri	Leaf, fruit (mesocarp, endocarp, seed), palm heart	(5) Food (direct use), handicrafts, construc-tion, biofuel, toxic
Syagrus oleracea(Mart.) Becc.	Garioba verdadeira, gariroba verdadeira, gueroba	Peduncular bract, fruit (mesocarp, endocarp, seed), palm heart	(4) Food (direct use, cooking), handicrafts, biofuel
Syagrus romanzoffiana(Cham.) Glassman	Jarobá	Fruit (mesocarp, seed), palm heart, inflores-cence	(4) Food (direct use, cooking, and animal feed), handicrafts
Syagrus rupicolaNoblick and Lorenzi	Catolé, catolé-rasteiro, catolé-da-serra, catolezinho, licuri-de-raposa, catolez-inho, coquinho-catolé, coquinho-da-serra	peduncular bract, meso-carp, seed, palm heart	(3) Food (direct use, cooking), handicrafts

The number of palms recognized (NPR) varied between 1 and 16 (mean 13) and the number of palms that have some use (NPU) varied between 5 and 16 (mean 13). The total species diversity (SDtot) for all the palms used by the Kalungas was SDtot = 15.47, while the value of total species equitability was SEtot = 0.97. These values indicate that many species were mentioned by many informants and that the palms are homogeneously used throughout the community. In another study about palms, Byg and Balslev [2] suggest a heterogeneous distribution of knowledge about palms, demonstrating that many palm species were used by only a few informants.

The Influence of Gender, Age, and Education on the Recognition and Use of Palms

Among women and men, the average numbers of palms recognized (NPR) were 12.5 (±3.3) and 12.9 (±3.8), respectively, and the average numbers of plants used (NPU) were 12.6 (±3.0) and 13.3 (±2.1), respectively. The NPR (P = 0.69) and NPU (P = 0.57) were not significantly different between men and women (Kruskall-Wallis test). However, the informant diversity index (IDs) shows a difference in knowledge between the genders. The average IDs for women was 34.46 (±9.2), while for men it was 21.69 (±4.8) with $P <$ 0.05. Moreover, the average informant equitability index (IEs) was 0.73 (±0.19) for women and 0.8 (±0.18) for men.

Rural areas often feature stark divisions of labor, and the relationship between gender and plant knowledge seems to be correlated with the genders' different obligations to the community [45]. These authors point to the accumulation of roles for women, who, in addition to the "easier" work (housework, food preparation, gardening, etc.), also participate in the "heavy" work (planting and harvesting in the plantation).

The greater diversity of palm use among women can be explained by the large array of responsibilities that

Kalunga women and other rural women take on to ensure the maintenance and survival of the family, especially the children. Activities such as obtaining firewood or long walks to arrive in the plantation areas are possible explanations for the higher exposure of women to different palm species. The highest value of equitability of use among men was associated with the category of construction, which is primarily performed by men.

Analyzing two age groups (18–40 and 41–82) revealed that the NPR values were 12.8 (±2.8) and 12.4 (±4.4), respectively, while the NPU values were 12.4 (±2.8) and 13.4 (±2.5), respectively. The analysis of variance (Kruskal-Wallis) showed no significant differences in the recognition ($P = 0.52$) and use ($P = 0.07$) of the species between the age groups ($P > 0.05$). However, in the second age group, the NPU is greater than the NPR. This result is explained by the nonrecognition of certain species by some older informants with vision problems. The use of visual stimuli among older informants had limitations. Previous studies using visual stimuli also observed this limitation [46].

Analyses correlating age and knowledge of plants, especially knowledge of medicinal uses, have shown that older age corresponds with greater knowledge about plants [47, 48]. Regarding the use of palms in the Engenho II community, age differences are not very pronounced. The noticeable familial relationship among community members strongly suggests that the family is mainly responsible for the temporal continuity of knowledge. The older people in the community are highly respected, and their homes are visited by younger relatives for several reasons. The young people, in turn, receive training from their elders and learn new ways of using the palms. Thus, the transmission and exchange of knowledge is not interrupted in this community. However, some products derived from palms are being replaced by industrialized products, such as oils, and as some practices are abandoned, knowledge of palms in the Engenho II will likely change over the long term.

In the case of the Engenho II community, education did not have a statistically significant effect on the NPR ($P = 0.2$) or the NPU ($P = 0.75$), the numbers of palms recognized and used by the informants ($P > 0.05$). Some authors assume that formal education decreases traditional knowledge [47, 49]; however, other studies do not observe this effect [50].

Distribution of the Knowledge and Use of Palms

Regarding the number of citations of use ($n = 1928$) by species, Mauritia flexuosa (buriti palm) was the best-represented species, followed by Attalea compta (indaiá) and Acrocomia aculeata (macaúba) (Figure 3 and Table4).

Table 4: Number of citations, species, and parts used for each category of palm use among the Kalunga from the Engenho II community, Cavalcante, GO, central western Brazil.

Category of use	Number of citations ($n = 1928$)	Number of species used ($n = 16$)	Number of parts used ($n = 18$)
Food/cooking	444	15	6
Food/animal feed	12	3	4
Food/direct use	889	16	8
Handicrafts	264	16	12
Biofuel	8	4	2
Construction	240	11	4
Fertilizer	3	1	1
Medicinal/cosmetic	60	8	10
Ritual	19	3	3
Toxic	6	3	2

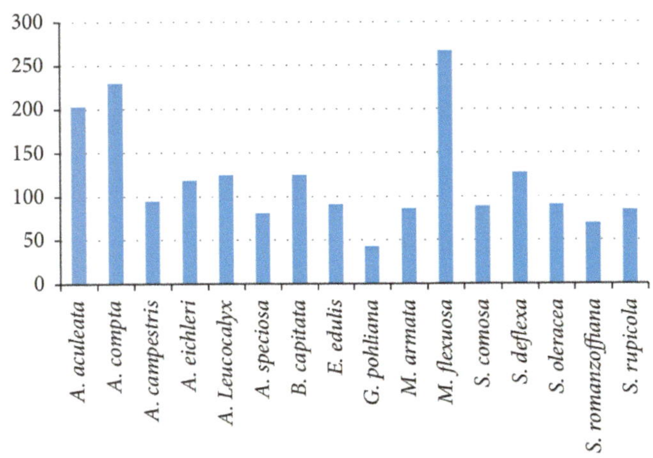

Figure 3: Number of citations of the use of palm species in the Engenho II community, Cavalcante, GO, central western Brazil. (Acrocomia aculeate, Attalea compta, Allagoptera campestris, Attalea eichleri, Allagoptera leucocalyx, Attalea speciosa, Butia capitata, Euterpe edulis, Geonoma pohliana, Mauritiella armata, Mauritia flexuosa, Syagrus comosa, Syagrus deflexa, Syagrus oleracea, Syagrus romanzoffiana, and Syagrus rupicola).

Food Use

Among the Kalungas of the Engenho II community, food use happens either directly (in nature) or in cooking, and all palm types of the area are used in this category. The mesocarp, epicarp, seeds, palm heart, and stipe of various species are used in both ways. In addition to in natura use, the mesocarp is used to prepare juices, desserts, milk, candies, and liquor (immersed with the fruits of Butia purpurascens).

Acrocomia aculeata had the greatest number of parts used for food (6). Many Kalungas extract oil from the seeds of some species for use in cooking and to make milk that is used in other preparations. The palm hearts of 13 species (n = 16) are consumed among the Kalungas of the Engenho II community.

Palms provide a variety of food sources for the Kalunga people. Frequently, palms are cited as an "emergency food," referring to the consumption of fruits and seeds during the planting seasons and during the long walks between the plantation and the housing.

Some food uses are mentioned by the oldest individuals as "uses of another era," from times when there was no food and palms provided an alternative way to end hunger. Using the toasted epicarp to make coffee and extracting starch from the pith of the stipe of A. aculeata are apparently no longer necessary.

The importance of food uses of palms has been noted in studies conducted in different regions [51–54]. Among the Kalungas of the Engenho II community, their survival has been linked to the use of palms as food. However, it is not possible to know whether the survival of palms is threatened by the community's continued use of parts that compromise the palms' survival (palm heart) and reproduction (seeds).

Use in Handicrafts

Among the Kalungas of Engenho II, all palm species of the region (n = 16) were cited as used in handicrafts (Table 4). Different parts of the palms are cited in this category and appear to be useful in a direct way ("in natura") or as raw materials for the manufacture of handicrafts. The use of the peduncular bract of species with underground stems as spoons and the sheaths of buriti palms as dustpans are examples of direct use.

The Buriti palm (Mauritia flexuosa L.f.) contributed the largest number of parts in this category (7; n = 18), and almost all utensils (quibanos, sieves, tapitis) present in residences are manufactured from this species. Details on the use of Buriti palm by the Kalungas were described by Martins et al. [10].

Handicraft production also identifies communities that represent themselves as ethnic or as Quilombo, as this is a source of income and cultural expression [55]. Among the groups of Quilombolas from the Amazon, materials derived from nature are commonly used for handicrafts, and palm species are one common source of these materials [55]. In the same region, the use of palms in handicrafts among indigenous and river communities is described in the works of Balick [56–59].

In the Cerrado of the Krahò indigenous territory (TO), approximately six of the 17 species of palms identified in the region provide raw material for handicrafts, which are sold in villages and neighboring towns [5, 53]. In another community of Tocantins, M. Sousa and A. Sousa [60] highlight the replacement of products manufactured with palms by more modern products, emphasizing the evidence that this substitution is related to the accelerated destruction of native vegetation, particularly in the valleys.

In the present study, while all species were used for handicrafts, most products are made with the Buriti palm petiole, which is collected under the plant after the leaves have fallen. This use is low impact and does not pose any risk to the plant. However, the uses of silk extracted from young leaves of the buriti palm and the stipe of the Euterpe edulis (palm heart) to manufacture beds and doors may have negative impacts on the populations of these species.

Use in Construction

Among the Kalungas, 11 species are used for construction, with Mauritia flexuosa (buriti palm) and Attalea compta (indaiá) being the most frequently cited in this category, followed by Attalea eichleri (Pindoba), Euterpe edulis (palm heart), Attalea speciosa (coco-palmeira), and Mauritiella armata (buritirana). The leaves of the buriti palm and indaiá are used together to cover houses. These leaves are gathered under the waning moon but never during the new moon, because according to local tradition, this is the best time to collect them to avoid infestations of insects and fungi in the leaves and to produce more durable structures. In contrast, Shanley and Rosa [61] observed that among the Caboclos of Pará, it is said that the best time to collect the leaves is actually under the new moon because the leaves will be less damaged by insects at this time.

The use of palms for construction has been observed in ethnobotanical studies in various regions of the country [5, 52, 54, 56, 62], with leaves and stipe as the preferred parts for this category.

Medicinal Use

The Kalungas of Engenho II cited medicinal use for eight species (Table 5). The following parts are used for medicinal purposes: stipe, leaf (petiole, rachis, and whole leaf), fruits (mesocarp, endocarp, liquid endosperm, and seed), roots, and palm hearts.

Table 5: Species cited as medicinal, number of citations, parts used, and therapeutic indications, Engenho II community, Cavalcante, GO, central western Brazil.

Species (number of citations)	Part used	Therapeutic indication
Acrocomia aculeata (17)	Fruit (endocarp oil)	Toothache, ear diseases
	Fruit (seed oil)	Ear diseases
	Palm heart (juice)	Antivenom
Allagoptera leucocalyx (2)	Palm heart (juice)	Ear and digestive system diseases
Attalea compta (8)	Fruit (endocarp oil, mesocarp, and seed)	Toothache
Attalea eichleri (3)	Leaf (rachis juice)	Skin diseases (burning)
	Fruit (liquid endosperm)	Ocular diseases
Butia purpurascens (1)	Fruit (mesocarp)	Skin diseases (healing)
Mauritia flexuosa (26)	Leaf (petiole juice)	Antivenom
	Fruit (mesocarp oil)	Antivenom, cardiovascular, and respiratory diseases
	Fruit (toasted seed)	Reproductive system diseases
	Root	Musculoskeletal and rheumatic diseases
Mauritiella armata (6)	Stipe	Skin diseases (burning)
	Root	Genitourinary system and rheumatic diseases
Syagrus comosa (2)	Palm heart (juice)	Digestive system diseases

Mesocarp oil was the most frequently cited medicinal part, followed by the seed oil, palm heart, and root. The main therapeutic indication of the palms was for the treatment of respiratory diseases such as flu and pneumonia. The second most common use is against snake bites. Two species share these two uses, the Buriti palm (Mauritia flexuosa) and Macaúba (Acrocomia aculeata). Two species were used for tooth pain, Macaúba and Indaiá (Attalea compta).

The importance of palms in medicine and pharmacology is discussed by Sosnowska and Balslev [63]. Based on a review of the literature over the last 25 years, these authors identified 106 species of palms in the Americas with medicinal uses. In this study, the fruit was the most frequently used part for medicinal purposes (56 spp.), followed by oil (19 spp.), mesocarp (16 spp.), and the endosperm (11 spp.). Other parts of the palm were also mentioned, including the root (27 spp.), the leaves (22 spp.), the palm heart (19 spp.), the stipe (17 spp.), and flowers (9 spp.).

Other Uses: Ritual, Fertilizer, Biofuel, and Toxic Uses

Among the Kalungas, of the three species placed in the ritual category, two (Euterpe edulis and Geonoma pohliana) were used to create decorative "lapinhas," large bows used for the closure of a traditional religious festival. In this festival, the plants are placed at the front of the church such that the people pass underneath the plants before entering into the religious space. Butia purpurascens leaves were cited by one informant as used for incense.

In the Amazon region, one species of palm (Socratea exorrhiza) has previously been described as having a ritual use [54]. Among the Krahò Indians, two species are in this category (Mauritia flexuosa, buriti palm, andOenocarpus distichus, bacaba) [5]. The Buriti stipe is used in traditional log races, a practice observed in different indigenous ethnic groups in the central western region of Brazil.

Decomposed Buriti palm pith and stipe were cited as used for fertilizer (manure) by three informants. Use as a biofuel has been cited for four species. The leaves of three species (Butia purpurascens, and Syagrus deflexa, Syagrus oleracea) are used to light fires, and the dried fruit of Attalea compta (indaiá) is used as charcoal.

Three palm species were cited as toxic by five informants. Consumption of mesocarp and seedsofAllagoptera leucocalyx (licuri-rasteiro-da-mata), Butia purpurascens (cabeçudo), and Syagrus deflexa (licuri-da-serra) are contraindicated because they cause stomach pain. The liquid of the mesocarp pulp of the licuri-da-serra is also contraindicated for people with respiratory diseases.

The Importance of Palms

The most prominent plant for the Kalunga people of the Engenho II community is Mauritia flexuosa (buriti palm), followed by Attalea compta (indaiá), Acrocomia aculeata (macaúba), Butia purpurascens (cabeçudo),Attalea eichleri (pindoba), and Syagrus comosa (gariroba-catolé). The buriti palm was the first in 88% of the rankings and was present in all of them (100%) (Table 6).

Table 6: Salience analysis calculated from the rankings of palm species by the Quilombo Kalunga of the Engenho II community, Cavalcante, GO, central western Brazil (in order of salience). IVs: importance values.

Species	Frequency (%)	Mean	Salience index	IVs
Mauritia flexuosa	100.0	1.70	0.882	0.977
Attalea compta	86.0	2.78	0.599	0.841
Acrocomia aculeate	81.4	3.09	0.527	0.795
Butia purpurascens	67.4	4.66	0.257	0.659
Attalea eichleri	50.0	4.05	0.242	0.489
Syagrus comosa	33.7	3.66	0.181	0.329
Euterpe edulis	32.6	4.71	0.124	0.239
Attalea speciosa	29.1	3.88	0.151	0.284
Syagrus oleracea	26.7	4.13	0.126	0.261
Allagoptera leucocalyx	24.4	4.48	0.103	0.239
Syagrus deflexa	18.6	4.69	0.072	0.182
Mauritiella armata	16.3	4.00	0.077	0.170
Syagrus rupicola	8.1	4.00	0.039	0.079
Syagrus romanzoffiana	8.1	4.14	0.039	0.079
Allagoptera campestris	1.2	6.00	0.002	0.011
Geonoma pohliana subsp. weddelliana	1.2	6.00	0.002	0.011

The average importance value in this study was 0.35 (±0.3), with five species showing values above 0.49. Comparing the results of importance values (IVs) with the analysis of salience, the most important species are the same in both analyses (Table 6), and consequently, no significant differences in the results were observed between the methods ($P = 0.1$).

Byg and Balslev [2] found differences in the importance and use of palms between species, indicating that specific characteristics determine how a species is used and appreciated. According to these authors, many ethnobotanical studies assume that the importance of a plant is related to the ways that it is used. To test this premise, we calculated the correlation coefficient between the salience and the number of different uses of the palm species mentioned in the study. The results indicated a moderate to positive correlation ((r = 0.64), demonstrating that the more uses a palm has, the greater its cultural importance to the Kalungas of the Engenho II community is.

Richness and Distribution of Knowledge

Ethnobotanical indexes for the richness and distribution of knowledge among palm species were calculated (Table 7). The average use diversity (UDs) was 2.47 (±1.29), and the use equitability (UEs) was 0.52 (±0.26). The results indicate that the uses of species are distributed almost uniformly among categories.

Table 7: Summary of the quantitative values capturing the uses and importance of palm species among the Kalunga of the Engenho II community, Cavalcante, GO, central western Brazil.

	Mean value (min., max.)	Standard deviation
Per palm species		
Number of citations	1928	
Number of categories	10	
Use diversity (UDs)	2.47 (1.04, 4.8)	1.29
Informant diversity (IDs)	58.85 (32.67, 75.26)	13.76
Informant equitability (IEs)	0.77 (0.42, 1)	0.18
Use consensus (UCs)	0.68 (−0.09, 1)	0.30
Purpose consensus (PCs)	0.12 (0.03, 0.39)	0.08
Per informant		
Number of informants	88	
Number of categories	5 (3, 7)	1.15
Number of species used	12.84 (5, 16)	2.72
Species diversity (SDi)	10.02 (4.26, 13.71)	2.33
Species equitability (SEi)	0.73 (0.31, 1)	0.17

Comparing the results of IVs and salience with those of the use diversity (UDs), three of the six most important species showed the highest values of this index. However, the species with the lowest IVs (Geonoma pohliana) had the fourth highest use diversity index. Considering the number of different uses for each species, the buriti palm (Mauritia flexuosa) and cabeçudo (Butia purpurascens) had the greatest diversity in categories of use (7; n = 10). However, Butia purpurascens showed a low UD value (2.15), as it was almost exclusively used for making brooms.

Regarding the distribution of uses among species, the values of informant diversity indexes (IDs) indicate that the species with the greatest diversities of use were Attalea compta (indaiá) (74.51), Allagoptera leucocalyx(licuri-rasteiro-da-mata) (74.05), and Mauritia flexuosa (buriti) (67.7). Byg and Balslev [2] identified a positive correlation between the informant diversity (Ids) and the importance of the palm, which was also observed in the current study. The relatively high average value of the informant equitability (IEs: 0.75) indicates that almost all people who know a particular species know relatively the same number of uses (Table 7). The average use consensus among informants (UCs) was 0.68 (±0.3). This indicates that expressive agreement exists among informants for most species.

On the other hand, the average purpose consensus (PCs) value, which measures the agreement of the informants regarding a specific purpose, was very low (Table 7). This low value indicates that the informants use the same species for different categories of use. Although it had a high importance value and many use citations, the buriti palm (Mauritia flexuosa) had the lowest PC value, demonstrating a low degree of overlap between informants and that many uses of buriti palm are not very popular, such as cosmetic use, which was mentioned only once. Additionally, statements by the oldest individuals, such as "this was done in difficult times to obtain food to eat," show that some uses have been abandoned.

On average, the informants know 12.65 (±3.5) species and use 12.84 (±2.7) (n = 16). The numbers of informants who recognize and use all species of palms are 19 (n = 88) and 14 (n = 88), respectively. Based on the 10 use categories in the present study, the average number of categories of use of a species was 4.9 (±1.2). The average values of species diversity (SDi) and equitability (SEi) were 10.02 (±2.3) and 0.73 (±0.2), respectively. These values suggest that knowledge about the use of palms is relatively high and is well distributed throughout the community.

Use and Value of Various Palm Parts

The analysis of the palm parts considered all 18 parts that were cited. However, only the results from the five most important parts are presented here (Table 8). The plant part value (PPV) is a value given to a specific part of the plant.

Table 8: Value of use index of the different parts (PPV) of palm species assigned by the Kalunga community Engenho II, Cavalcante, GO, central western Brazil.

Species	Stipe	Leaf	Mesocarp	Seed	Palm heart
Acrocomia aculeate	0.02	—	0.43	0.45	0.05
Allagoptera campestris	—	0.01	0.25	0.72	0.01
Allagoptera leucocalyx	—	0.02	0.31	0.65	0.02
Attalea speciosa	0.01	0.24	0.05	0.67	—
Attalea compta	0.00	0.36	0.04	0.49	0.06
Attalea speciosa	—	0.22	0.05	0.65	0.04
Butia purpurascens	—	0.60	0.16	0.14	0.02
Euterpe edulis	0.51	0.01	0.24	0.07	0.09
Geonoma pohliana	0.30	—	0.30	0.07	0.09
Mauritia flexuosa	0.06	0.29	0.46	0.02	—
Mauritiella armata	0.21	0.49	0.17	0.01	0.01
Syagrus comosa	0.01	0.03	0.02	0.11	0.80
Syagrus deflexa	—	0.04	0.27	0.57	0.07
Syagrus oleracea	0.03	0.01	0.02	0.29	0.59
Syagrus romanzoffiana	—	—	0.56	0.30	0.07
Syagrus rupicola	—	—	0.18	0.75	0.04
Average	0.08	0.16	0.20	0.36	0.13
Standard deviation	0.15	0.20	0.17	0.27	0.24

The fruit was the most frequently cited part in each palm species. The seed and mesocarp were frequently cited for the food category. The fruits of 16 palm species have been described as useful by the local population. All species with a pulpy mesocarp are consumed in natura or used in the manufacture of desserts and in other types of cooking. The following species are highlighted in this group: Acrocomia aculeata (macaúba), Mauritia flexuosa (buriti), Syagrus romanzoffiana (jarobá) and Allagoptera leucocalyx (Licurí-rasteiro-da-mata).

The palm seeds are heavily used by the Kalungas of the Engenho II community. Many species provide seeds that are eaten in natura, such as Acrocomia aculeata (macaúba), Allagoptera campestris (licurí-rasteiro-do-campo), Allagoptera leucocalyx (licurí-rasteiro-da-mata), Syagrus deflexa (licurí-da-serra) and Syagrus rupicola(catolé-da-serra). Another group of species have seeds that are also used to extract oil: Attalea speciosa (coco-palmeira), Attalea compta (indaiá), and Attalea eichleri (pindoba).

Informants also described uses for whole leaves, with Butia purpurascens (cabeçudo) exhibiting the highest value for this part, followed by Mauritiella armata (buritirana), Attalea compta (indaiá), and Mauritia flexuosa(buriti palm) (Table 8). The leaves are used for construction and for the manufacture of handicrafts and household items. The species with the highest extraction is Butia purpurascens. Among the Kalungas and in other regions where it occurs, its leaves are used to make domestic brooms [64]. However, in the Kalunga region, groups of pickers visit the region annually to collect large amounts of leaves. This intense and disordered extraction may be causing serious damage to the populations of this species, which tend to be aggregated in easily accessible locations.

The stipe was used in nine species, including two forest species known to the Kalungas, Euterpe edulis andGeonoma pohliana subsp. weddelliana (Table 8). The stipes of these two species were mentioned by the oldest respondents as a "former" use for building roofs and beds. The fact that Euterpe edulis is included in the endangered species list in Brazil [65] clearly shows the need for an updated and well-reasoned approach to this subject within the community.

The palm heart was a significant part, with 14 species providing useful palm hearts (Table 8). However, theSyagrus oleracea (gariroba-verdadeira) and Syagrus comosa (gariroba-catolé) palm hearts are preferred for consumption, as both have a bitter taste that is highly appreciated by the Kalungas. Unfortunately, these two species have solitary stipes, and extracting this part kills the palm. Throughout this research, few individuals ofS. comosa were observed near the community, and the only two observed individuals of S. oleracea were growing in the backyard of a residence. Some informants were concerned about possible punishment from environmental agencies because of this practice, as demonstrated in the following sentence: "we pick them up, but not too much because if the zimbamba see it, we will have problem." Zimbamba refers to individuals from the Chico Mendes Institute, formerly called the IBAMA (Brazilian Institute of Environment and Renewable Natural Resources).

Final Considerations

- All of the palms present in the region of community Kalunga Engenho II were cited as useful by the informants. Species occur in different vegetation types: rupestrian Cerrado, gallery forest, grassland, shrub savanna, and valleys.
- Palms were most frequently used for food, handicrafts, and construction, demonstrating that the use of palms by the Kalunga people did not differ from the categories recognized in other studies performed in the Cerrado and in the Amazon. All species of palms of the area were cited as useful for food and handicrafts, demonstrating their importance for the survival of the Kalungas over time. The medicinal category featured the largest number of different parts.
- The variables gender, age, and education did not affect the numbers of plants recognized and used by the Kalungas. However, the informant diversity and equitability indexes show that women know more different uses than men, although men's knowledge is distributed more homogeneously.
- The indices calculated to evaluate the distribution of the knowledge among the Kalungas demonstrated that knowledge is well distributed throughout the community. The most important palm for the Kalungas of the Engenho II is Mauritia flexuosa (buriti palm). This species occurs in the vicinity of the community and was recognized by all informants. Despite the high number of use citations for this species, the diversity use index was low, demonstrating that the uses are virtually the same among informants.
- The palm fruits represent an important source of vitamins and protein for the local population, and all species (n = 16) are sources of food. The use of oil seeds was widely cited among the informants, but in many homes, it is now being replaced by commercial oils. Although many residents have knowledge about the practice of extracting oils, all oils identified in the residences come from other, more distant Kalunga communities.
- Use of leaves predominantly occurred in two species, Mauritia flexuosa (buriti palm) and Attalea compta (indaiá). The leaves are extracted due to spontaneous demand, in accordance with indications of traditional knowledge. Based on a study conducted in the Jalapão region (TO-state), Sampaio et al. [9] suggest guidelines for the sustainable extraction of buriti palm leaves. That literature will be incorporated into a "Guide of the palms from the Engenho II," which is in preparation.
- The palm hearts of 14 species (n = 16) were consumed, primarily for family consumption and for festivities in the community. Until now, it was not possible to know whether the extraction of palm hearts was related to any business. This delicacy is special due to its taste and its importance to the welfare of the Kalunga people. The use of the apical meristem, apparently in response to spontaneous demand, is believed to be affecting the abundance of some species of palms near this community. This finding demonstrates the need for studies related to the phenology, phytosociology, and management of species with bitter palm hearts, the variety preferred by this community.

ACKNOWLEDGMENTS

The authors would like to give their thanks to members of the Community Association of Engenho Quilombo Kalunga II and to the community leaders and local guides, especially Mr. Sirilo Santos Rosa, Mr. Jorge Moreira da Silva, Mr. João Santos Rosa, Mrs. Getúlia Moreira da Silva, Ilma Francisco Maia, and Januária Moreira de Sena. The authors thank Ana Paula Gulias for the map. Renata Martins thanks CAPES for the scholarship that allowed her to pursue a doctorate at the Deanship of

Research at the University of Brasilia, the ISPN (Institute for Society, Population, and Nature) for supporting the completion of the field work, the CNPq for post doctorate project process no 5577182009-0 coordinated by Dr. Christopher Fagg, and for the research productivity grant given to UPA.

REFERENCES

1. D. M. Bates, "Utilization pools: a framework for comparing and evaluating the economic importance of palms," Advances in Economic Botany, vol. 6, pp. 56–64, 1988.
2. A. Byg and H. Balslev, "Diversity and use of palms in Zahamena, eastern Madagascar," Biodiversity and Conservation, vol. 10, no. 6, pp. 951–970, 2001.
3. A. Byg, J. Vormisto, and H. Balslev, "Using the useful: characteristics of used palms in south-eastern Ecuador," Environment, Development and Sustainability, vol. 8, no. 4, pp. 495–506, 2006.
4. M. T. Campos and C. Ehringhaus, "Plant virtues are in the eyes of the beholders: a comparison of known palm uses among indigenous and folk communities of southwestern amazonia," Economic Botany, vol. 57, no. 3, pp. 324–344, 2003.
5. A. T. Nascimento, A. A. Santos, R. C. Martins, and T. A. B. Dias, "Comunidade de palmeiras no território indígena krahò, tocantins, brasil: biodiversidade e aspectos etnobotânicos," Interciencia, vol. 34, no. 3, pp. 182–188, 2009.
6. N. Y. P. Zambrana, A. Byg, J.-C. Svenning, M. Moraes, C. Grandez, and H. Balslev, "Diversity of palm uses in the western Amazon," Biodiversity and Conservation, vol. 16, no. 10, pp. 2771–2787, 2007.
7. R. C. Mendonça, J. M. Felfili, B. M. T. Walter et al., "Flora vascular do cerrado," in Cerrado: Ecologia e Flora, Embrapa, Ed., Embrapa-Cpac, Planaltina, Brazil, 2008.
8. J. F. Ribeiro and B. M. T. Walter, "As principais fitofisionomias do bioma cerrado," in Cerrado: Ecologia e Flora, S. M. Sano, S. P. Almeida, and J. F. Ribeiro, Eds., pp. 153–212, Embrapa, Brasília, Brazil, 2008.
9. M. B. Sampaio, I. B. Schmidt, and I. B. Figueiredo, "Harvesting effects and population ecology of the Buriti palm (Mauritia flexuosa L. f., Arecaceae) in the Jalapão Region, Central Brazil," Economic Botany, vol. 62, no. 2, pp. 171–181, 2008.
10. R. C. Martins, T. S. Filgueiras, and U. P. de Albuquerque, "Ethnobotany of mauritia flexuosa (arecaceae) in a maroon community in central brazil," Economic Botany, vol. 66, no. 1, pp. 91–98, 2012.
11. M. C. M. Amoroso, "Uso e diversidade de plantas medicinais em Santo Antônio do Leverger, MT, Brasil," in Acta Botanica Brasilica, vol. 16, pp. 189–203, 2002.
12. E. Rodrigues and E. A. Carlini, "Plants used by a Quilombola group in Brazil with potential central nervous system effects," Phytotherapy Research, vol. 18, no. 9, pp. 748–753, 2004.
13. E. A. P. Franco and R. F. M. Barros, "Uso e diversidade de plantas medicinais no Quilombo Olho D'água dos Pires, Esperantina, Piauí," Revista Brasileira De Plantas Medicinais, vol. 8, no. 3, pp. 78–88, 2006.
14. E. Rodrigues, "Plants of restricted use indicated by three cultures in Brazil (Caboclo-river dweller, Indian and Quilombola)," Journal of Ethnopharmacology, vol. 111, no. 2, pp. 295–302, 2007.
15. M. O. S. Crepaldi and A. L. Peixoto, "Use and knowledge of plants by "Quilombolas" as subsidies for conservation efforts in an area of Atlantic forest in Espírito Santo State, Brazil," Biodiversity and Conservation, vol. 19, no. 1, pp. 37–60, 2009.
16. R. S. A. Anjos, Quilombos Geografia Africana—Cartografia Étnica—Territórios Tradicionais, Mapas Editora & Consultoria, Brasília, Brazil, 2009.
17. A. Silva, Kalunga Identidade étnica de uma comunidade remanescente de quilombos [M.S. thesis], Amsterdam-Holland, Amsterdam, The Netherlands, 1999.
18. A. D. Vellozo, Mapeamento narrativas: uma análise do processo histórico-espacial da comunidade do engenho II—kalunga [M.S. thesis], Universidade de Brasília, 2007.
19. D. Jatobá, A comunidade kalunga e a interpelação do estado da invisibilidade a identidade política [M.S. thesis], Universidade de Brasília, 2002.
20. M. N. Baiocchi, "Calunga—kalumba: universo cultural," Revista Do Instituto Histórico e Geográfico de Goiás, vol. 11, pp. 75–85, 1986.
21. D. B. Ungarelli, A comunidade quilombola kalunga do engenho II: cultura, produção de alimentos e ecologia de sabers [M.S. thesis], Universidade de Brasília, 2009.
22. M. G. Almeida, "Territórios de quilombolas: pelos vãos e serras dos Kalunga de Goiás—patrimônio e biodiversidade de sujeitos do Cerrado," Ateliê Geográfico, vol. 4, no. 9, pp. 36–63, 2010.
23. N. P. Massarotto, Diversidade e uso de plantas medicinais por comunidades quilombolas e urbanas, no nordeste goiano do estado de goiás-go, Brasil [M.S. thesis], Universidade de Brasilia, 2008.
24. A. C. G. R. Neiva, J. R. B. Sereno, S. A. Santos, and M. C. S. Fioravanti, "Caracterização cultural da comunidade quilombola kalunga de cavalcante-goiás, brasil: dados preliminares," in IX Simpósio Nacional Cerrado/ II Simpósio Internacional Savanas Tropicais, 2008.
25. U. P. Albuquerque, "Etnobotânica aplicada para a conservação da biodiversidade," in Métodos e Técnicas Na Pesquisa Etnobotânica, U. P. Albuquerque, R. F. P. Lucena, and L. F. V. Cunha, Eds., pp. 227–240, Comunigraf/NUPEEA, Recife, Brazil, 2008.
26. A. Henderson, G. Galeano, and R. Bernal, Field Guide to the Palms of the Americas, Princeton University, Princeton, NJ, USA, 1995.
27. R. C. Martins and T. S. Filgueiras, "Arecaceae," in Flora Do Distrito Federal, T. A. B. Cavalcanti, Ed., pp. 47–82, 2006.
28. H. Lorenzi, L. Noblick, F. Kahn, and E. Ferreira, Flora Brasileira Arecaceae (Palmeiras), Instituto Plantarum, Nova Odessa, Brazil, 2010.
29. P. M. Medeiros, A. L. S. Almeida, R. F. P. Lucena, F.

J. B. Souto, and U. P. Albuquerque, "The role of visual stimuli in ethnobotanical surveys: an overview," in Current Topics in Ethnobotany, U. P. Albuquerque and M. A. Ramos, Eds., pp. 125–137, Research Signpost, Kerala, India, 2008.

30. P. M. Medeiros, A. L. S. Almeida, R. F. P. Lucena, F. J. B. Souto, and U. P. Albuquerque, "Uso de estímulos visuais na pesquisa etnobiológica," in Métodos e Técnicas Na Pesquisa Etnobotânica, U. P. Albuquerque, R. F. P. Lucena, and L. V. F. C. Cunha, Eds., pp. 109–126, Comunigraf/NUPEEA, Recife, Brazil, 2008.
31. U. P. Albuquerque, P. M. Medeiros, T. A. S. Araújo et al., "The role of ethnobotany and environmental perception in the conservation of atlantic forest fragments in Northeastern Brazil," Bioremediation, Biodiversity and Bioavailability, vol. 2, no. 1, pp. 27–34, 2008.
32. G. J. Martin, Ethnobotany: a Methods Manual, vol. 33, Chapman & Hall, New York, NY, USA, 1994.
33. M. N. Alexiades and J. W. Sheldon, Selected Guidelines for Ethnobotanical Research: a Field Manual, The New York Botanical Garden, New York, NY, USA, 1996.
34. M. L. T. Nguyen, "Comparison of food plant knowledge between urban vietnamese living in vietnam and in Hawai›i," Economic Botany, vol. 57, no. 4, pp. 472–480, 2003.
35. E. Thomas, I. Vandebroek, and P. Van Damme, "What works in the field? A comparison of different interviewing methods in ethnobotany with special reference to the use of photographs," Economic Botany, vol. 61, no. 4, pp. 376–384, 2007.
36. M. Ayres, D. L. Ayres Júnior, and A. A. S. Santos, BioEstat: Aplicações Estatísticas Nas Áreas Das Ciências Biomédicas, Sociedade Civil Mamirauá, Belém, Brazil, 2005.
37. S. C. Rossato, H. D. F. Leitão-Filho, and A. Begossi, "Ethnobotany of caicaras of the atlantic forest coast (Brazil)," Economic Botany, vol. 53, no. 4, pp. 387–395, 1999.
38. N. Hanazaki, J. Y. Tamashiro, H. F. Leitão-Filho, and A. Begossi, "Diversity of plant uses in two caicara communities from the Atlantic forest coast, Brazil," Biodiversity and Conservation, vol. 9, no. 5, pp. 597–615, 2000.
39. T. M. Miranda and N. Hanazaki, "Conhecimento e uso de recursos vegetais de restinga por comunidades das ilhas do Cardoso (SP) e de Santa Catarina (SC), Brasil," Acta Botanica Brasilica, vol. 22, no. 1, pp. 203–215, 2008.
40. S. P. Borgatti and M. A. Natick, Anthropac, vol. 32, Analytic Technologies, 1992.
41. V. A. Silva, U. P. Albuquerque, and V. T. Nascimento, "Técnicas para análise de dados etnobotânicos," inMétodos E Técnicas Na PEsquisa Etnobotânicos, U. P. Albuquerque, R. F. P. Lucena, and L. V. F. C. Cunha, Eds., pp. 127–144, Comunigraf/NUPEEA, Recife, Brazil, 2008.
42. A. Gomez-Beloz, "Plant use knowledge of the Winikina Warao: the case for questionnaires in Ethnobotany," Economic Botany, vol. 56, no. 3, pp. 231–241, 2002.
43. IUCN, "IUCN standards and petitions subcommittee. Guidelines for using the IUCN red list categories and criteria. Version 9. 0.," Standards and Petitions Subcommittee, 2011, http://www.iucnredlist.org/documents/RedListGuidelines.pdf.
44. E. Pastore, N. C. Polese, and L. M. O. Pastore, "O papel da mulher na agricultura diversificada e agroecológica: influências e mudanças nas relações de gênero," 2012, http://www.fazendogenero7.ufsc.br/artigos/P/Pastore-Polese-Pastore_37.pdf.
45. R. J. Case, G. F. Pauli, and D. D. Soejarto, "Factors in maintaining indigenous knowledge among ethnic communities of Manus Island," Economic Botany, vol. 59, no. 4, pp. 356–365, 2005.
46. R. A. Voeks and A. Leony, "Forgetting the forest: assessing medicinal plant erosion in eastern Brazil,"Economic Botany, vol. 58, supplement 1, pp. S294–S306, 2004.
47. A. F. M. Viu, M. A. O. Viu, and L. Z. O. Campos, "Etnobotânica: uma questão de gênero," Revista Brasileira De Agroecologia, vol. 5, no. 1, pp. 138–147, 2010.
48. E. J. Luoga, E. T. F. Witkowski, and K. Balkwill, "Differential utilization and ethnobotany of trees in kitulanghalo forest reserve and surrounding communal lands, Eastern Tanzania," Economic Botany, vol. 54, no. 3, pp. 328–343, 2000.
49. A. Byg and H. Balslev, "Factors affecting local knowledge of palms in Nangaritza valley in south-eastern Ecuador," Journal of Etnhnobiology, vol. 24, pp. 255–278, 2004.
50. G. T. Prance, W. Balée, B. M. Boom, and R. L. Narneiro, "Quantitative ethnobotany and the case for conservation in Amazonia," Conservation Biology, vol. 1, pp. 296–310, 1987.
51. C. R. Clement, E. Lleras Pérez, and J. Van Leeuwen, "O potencial das palmeiras tropicais no Brasil: acertos e fracassos das últimas décadas," Agrociencias, Montevideu, vol. 9, no. 1-2, pp. 67–71, 2005.
52. A. T. Nascimento, "Riqueza e etnobotânica de palmeiras no território indígena Krahô, tocantins, Brasil,"FLORESTA, vol. 40, no. 1, pp. 209–220, 2010.
53. F. R. Araújo, M. A. Lopes, and D. M. Rodrigues, "Caracterização do uso de palmeiras (arecaceae) no mosaico de unidades de conservação (MUC) lago de tucuruí—pará," Cadernos de Agroecologia, vol. 6, no. 2, 2011.
54. A. J. P. Amaral, "Artesanato Quilombola: identidade e etnicidade na amazônia," Cadernos do CEOM, vol. 23, no. 31, 2010.
55. M. J. Balick, "Economic Botany of the Guahibo. I. Palmae," Economic Botany, vol. 33, no. 4, pp. 361–376, 1979.
56. M. J. Balick, "Ethnobotany of palms in the Neotropics," Advances in Economic Botany, vol. 1, pp. 9–23, 1984.
57. M. J. Balick, The Palm—Tree of Life: Biology, Utilization and Conservation, New York Botanical Garden, Bronx, NY, USA, 1986.
58. M. J. Balick, "The use of palms by the apinayé and guajajara Indians of northeastern Brasil," Advances in

Economic Botany, vol. 6, pp. 65–90, 1988.

59. M. R. S. Sousa and A. J. Sousa, "Vestígios artesanais na biodiversidade e plasticidade do Buriti," in II Congresso de Pesquisa e Inovação de Rede Norte e Nordeste de Inovação Tecnológica, 2007.

60. P. Shanley and N. A. Rosa, "Eroding knowledge: an ethnobotanical inventory in eastern Amazonia›s logging frontier," Economic Botany, vol. 58, no. 2, pp. 135–160, 2004.

61. R. C. Martins, T. S. Filgueiras, and S. P. Almeida, "As palmeiras da região do parque nacional grande sertão veredas (PNGSV): uso e sustentabilidade no cerrado," in 54° Congresso Nacional de Botânica, 2003.

62. J. Sosnowska and H. Balslev, "American palm ethnomedicine: a meta-analysis," Journal of Ethnobiology and Ethnomedicine, vol. 5, article 43, 2009.

63. A. C. Marcato, Revisão taxonômica do gênero butia (Becc.) Becc. e filogenia da subtribo butiinae saakov (palmae) [Ph.D. thesis], Universidade de São Paulo, 2004.

64. MMA, Lista Oficial das Espécies da Flora Brasileira Ameaçadas de Extinção, 2008.

CHAPTER 7 THE MEDICAL ETHNOBOTANY OF LESOTHO

A. Moteetee and B-E. van Wyk

Keywords: Ethnobotany, Lesotho, magic, *materia medica*, medicine, traditional healers

INTRODUCTION

Traditional healing has always been practised in Lesotho from time immemorial, but the concepts, practices and *materia medica* are not well known beyond the borders of the country and have not yet been systematically recorded. This paper is therefore aimed at providing a broad review of traditional medicine in Lesotho, focusing on cultural practices, traditional health care practitioners and especially the botanical identity, vernacular names, uses and dosage forms of the plants that are used.

There are three reasons why we are hesitant to call this a review of Sotho traditional medicine. Firstly, it is difficult to delimit the geographical boundaries of the Sotho culture, e.g. should North Sotho (Pedi)—excluded here—be included? Secondly, there are subtle linguistic differences, in terminology and plant names, for example between the Sesotho spoken in Lesotho and that which is spoken in other parts of southern Africa. Thirdly, the plant species used in Lesotho sometimes differ from the species found in other regions. However, even though this review is focused on Lesotho, the information probably has a much wider relevance and application than the boundaries of a single, small country.

Sources of Information

The review is based on the experience and recollections of the first author and on the available literature. These experiences and recollections stem from early life (until 20 years ago) of the first author; many of these plants were used in her home and village in the rural areas of the Qacha's Nek District and later in the Maseru District in Lesotho. Further information regarding plant use was gathered through discussions and interviews with informants using Sesotho names of plants and in some cases with voucher specimens. The most comprehensive reference material is that of Phillips (1917) but the list of plants is incomplete and many of the names are outdated. Jacot Guillarmod (1971) also gives an account of plant use but her main source was Phillips (1917). Other useful sources of information include Watt & BreyerBrandwijk (1962), Motlamelle (1938), Schmitz (1982) and Maliehe (1997), of which the last three are out of print. The lastmentioned is limited to 60 species of useful plants in Lesotho. Some valuable information has also been published in early volumes of *Bantu studies* (Watt & Brandwijk 1927) and in lesser known publications such as Casalis (1861) and Ashton (1967). A comprehensive list of medicinal plants which includes the scientific names, author citations, vernacular names and main uses. Authorities for scientific names are not repeated in the text.

Cultural Practices

Three important aspects, medicine (*moriana, litlhare*), magic (*lithlare*), and sorcery (*boloi*) play an important role in the Basotho culture.

Medicine (Moriana, Litlhare)

Many people in Lesotho, particularly those in rural areas with no clinics or health facilities, are able to diagnose and treat minor common ailments such as colds, coughs, headaches and constipation. This self-medication relies almost exclusively on medicinal plants. Headaches, for example, can simply be cured by inhaling smoke from a burning stick of *monkhoane* (*Heteromorpha aborescens*), stomach aches by chewing roots of a grass known as *hloko* (*Elionurus muticus*) (Motlamelle 1938). If there is no cure, a traditional healer is consulted. Women are traditionally experts in home remedies because they collect vegetables (*moroho*), many of which are also used for medicinal purposes. Historically however, most traditional healers are men. There are two main reasons for this: 1, there is a strong traditional belief that women should perform their

Citation: A. Moteetee and B-E. van Wyk (2011). The medical ethnobotany of Lesotho, Bothalia; Vol 41, No 1, 209-228. DOI: 10.4102/abc.v41i1.52.

duties at home, taking care of their families, including children and the elderly; 2, there are taboos around women handling strong medicines.

If there was an epidemic in a village, the village doctor or chief's doctor (at the chief's expense), usually a diviner (*selaoli*) or herbalist (*ngakachitja*), would prepare a mixture (known as *leliba*) that would have to be taken by all the villagers.

Magic (Lithlare)

Medicines are not only used to cure illnesses but are used for almost any situation, including those that occur when a need for help in controlling natural or social phenomena is perceived. This category includes plants that are used as Category 1, good luck charms, such as *molomomonate* (*Lotononis eriantha* var. *obovata*) which is a very well-known plant used to 'make sure you are heard wherever you go' (e.g. litigants use it to strengthen their case.); Category 2 includes plants that are used as love philtres such as *Cymbopogon marginatus* which may be added to a man's bathing water while he calls out the name of the desired woman; Category 3 includes plants that are used as charms against lightning, thunderstorms and hailstorms such as *Phygelius aequalis* (*metsimatšo*) which is a renowned charm to protect crops from hail damage. The Basotho believe that good health is not only physical but also spiritual; therefore ancestral spirits (*balimo*) form an integral part of their lives. If a person experiences bad luck and misfortune, then this is most likely ascribed to the fact that they have neglected their ancestors and are therefore being punished.

Sorcery (Boloi)

There is a strong belief in sorcery among the people of Lesotho, so much so that misfortune, accident, unusual illness, or even death may be attributed to it. Sorcerers are believed to use special or ordinary medicines maliciously. It was previously believed that they were even capable of and responsible for harming whole communities through disasters and epidemics (Ashton 1967). Therefore, it was (and still is) a socially unacceptable practice. Although killing alleged sorcerers was proclaimed illegal by King Moshoeshoe I in 1855 (Ashton 1967), people suspected of sorcery were often severely punished, sometimes even brutally tortured to death (Casalis 1861). This practice, however, appears to have discontinued in the last 50 years. Sorcery is believed to be driven either by jealousy (mostly) or by revenge. The effects of sorcery range from a minor injury or illness (such as diarrhoea or stomach ache believed to be due to worms, beetles or small snakes introduced into the patient's body, often from eating bewitched food), or more serious conditions such as convulsions (due to strangling by a familiar (a supernatural being), i.e. a *thokolosi*), or death by lightning. When sorcery is suspected, a diviner is consulted, who will then, as the first step, counteract the witchcraft by renewing all protective medicines. Sometimes the patient is taken and hidden away in another village. If nothing works, then people who believe themselves to be victims of sorcery have sometimes been driven to murdering the suspected sorcerer (Ashton 1967).

Traditional Healthcare Practitioners

Several types of traditional healthcare practitioners exist in Lesotho. These include: the diviner or bone thrower (*ngaka, selaoli*), the herbalist (*ngaka chitja*), the seer (*senohe*), the veterinarian (*ngakaealiphoofolo*), the sangoma (*lethuela, mokoma, leqekha*), the apprentice (*lehlahana*) and the charlatan (*ngakanakahetla*). The Sesotho word *ngaka* is a generic term applied to any kind of healer or doctor, but some authors (e.g. Motlamelle 1938) apply it exclusively to the diviner. Although traditional healers do not have any special status in society, the chief formerly had a doctor who occupied a socially highranking position next to him or her. These doctors associated with the chiefs were rainmakers (*seepameru*), diviners (*ngaka, selaoli*) and keepers of war medicine.

When consulting a healer, one is expected to pay an initial consultation fee (usually a sheep or goat) to 'open the bag of medicines'. In the case of a diviner, the fee is known as *pulamolomo,* which literally translated means 'one that opens (*bula*) the mouth (*molomo*)' [*pula* is the noun form of *bula*, not to be confused with *pula* (rain), which is written in the same way but pronounced differently]. *Pulamolomo* therefore has to be paid before the diviner will even speak. It is not customary to say 'thank you' when given medicine because this will weaken its strength. Final payment is done only if the treatment has worked.

Diviner (Ngaka, Selaoli)

The diviner uses bones (*litaola*) to help him make a diagnosis and discern the appropriate treatment. The diagnosis is based on the way in which the various bones land or fall on the ground, having been thrown by the person consulting. Diviners are expected to solve problems of any kind; they prescribe remedies for minor ailments, protection against sorcery, misfortune and accidents, and also for good luck and prosperity. 'There is, in fact, no manner of question which the diviners do not undertake to answer' (Casalis 1861). Traditionally, he is a striking figure and may be recognised by his headdress (*kuoane*), made from monkey skin and decorated with feathers, gall bladders and eagle's claws, a shoulder strap of horns (*lenaka*) from various beasts containing all kinds of powders and ointments (known as *mohlabelo*) and, a bag of divining bones (*khetsi*) consisting of varying pairs of bones from different animals (Ashton 1967).

Herbalist (Ngakachitja)

Traditional healers who rely only on plants are known in Sesotho as *ngakachitja* which means a hornless doctor, because they do not keep medicines in horns. They diagnose and treat minor ailments and, traditionally, if they are unable to cure the disease, they refer patients to a diviner. In recent times they are known also to refer patients to a clinic.

Seer (Senohe)

The seer is a person with the power of second sight, meaning that they divine supernaturally. This power is a skill that cannot be acquired but is believed to be a gift from the ancestors. A seer can diagnose illnesses and foretell the future. A special type of seer can locate (see) people who have drowned or have been swallowed by the water serpent, *khanyapa* or *nohaeametsi* (Ashton 1967). Chief Mohlomi, a very wellknown leader in Lesotho's history, who was an advisor, mentor, and doctor to the founder of the Basotho nation, King Moshoeshoe I, was a seer by profession.

Veterinarian (Ngakaealiphoofolo)

Historically, there were no doctors devoted entirely to animal illnesses; this was the job of diviners, herbalists or seers. However, the existence of veterinarians is now recognised. Animals such as cattle, Basotho ponies, sheep and goats play an important role in the Basotho culture. Cattle are symbolic in Lesotho, being used for *lobola* (*bohali*) and may also be slaughtered during burial rituals to accompany (*ho felehetsa*) deceased persons so they can rest in peace. In addition, cattle represent wealth. Many areas in Lesotho are still only accessible by horses and they therefore serve as a useful mode of transport. Sheep and goats are kept mainly for meat and wool. It is therefore imperative that all these animals are kept healthy. Veterinarians work in a similar manner as herbalists; they are able to diagnose and treat a wide range of animal diseases including gall sickness, 'black leg' and lung sickness.

Sangoma (Lethuela, Mokoma, Leqekha)

The concept of a sangoma is foreign to the Lesotho culture and it was introduced by the Thembus who originated from what was then known as the Cape Colony (Motlamelle 1938; Ashton 1967), perhaps currently the Eastern Cape Province. The language in which sangomas practise their craft points to the foreign origins of this tradition. Furthermore, sangomas are mostly women, whereas in Lesotho, healers were traditionally male only. When consulted, they diagnose problems using the divining method (in a similar manner as diviners, except that their bag of bones also includes sea shells and lately a set of dice) and prescribe medicines.

Apprentice (Lehlahana)

Diviners and herbalists learn their profession by becoming an apprentice (*lehlahana*) to a reputable healer. Among other duties, the *lehlahana* carries the teacher's bag, cooks the food, collects the fees, collects and prepares the medicines and performs techniques such as making incisions (*ho phatsa*) (Motlamelle 1938). Once qualified, the apprentice may establish his or her own practice, or on rare occasions he or she may inherit the master's horn. Other healers inherit prescriptions from their family members who are doctors.

Charlatan (Ngakanakahetla)

While we disagree with Casalis' (1861) assertion that traditional healers are 'a crowd of clever and cunning men, who, under the generic name of *engakas*, or learned men, perform functions of priests, prophets, diviners, and doctors, and enrich themselves at the public expense', there are traditional healers who are charlatans. These are termed *ngakanakahetla* (the verb *hetla* means to look over your shoulder) in Sesotho, a mocking word applied to people who claim to be healers but have not undergone proper training. As a result they have to keep looking over their shoulders to make sure they are not being followed by the patients whom they have cheated.

Dosage Forms and Methods of Administering Medicine

Medicinal plants are used in many different forms. They are taken as decoctions (mostly with water, rarely with milk) or infusions either orally (*ho noa*) or as an enema (*ho tšela*) for various internal ailments. Plants are also incinerated, powdered and used as a snuff (*ho tsuba*), often to induce sneezing to get rid of head and chest colds. Another form is inhalation of either smoke from a burning plant (another form of *ho tsuba*) or vapour from a freshly crushed plant or fresh leaves of *lengana* (*Artemisia afra*) or *koena* (*Mentha longifolia*) stuffed in the nostrils, as a cure for headaches and colds. Smoke can also be introduced into the body through fumigation (*ho futhela*). Boiling decoctions (generally of some *Helichrysum* species, *Artemisia afra* or *Mentha longifolia*) are used for steaming (*ho arubela*) to treat colds. Sometimes plants are also taken as masticatories, although only a few plants are chewed raw (e.g. *Alepidea cordifolia* and *Elionurus muticus*). Other medicines are applied or rubbed into incisions/scarifications (*ho phatsa*) made on the body. These incisions are done for two reasons: 1, to protect people against sorcery or to counteract its effects, in which case several parts of the body (including forehead, cheeks, chin, neck, throat, breasts, and joints) are scarified and the medicine is rubbed in; or 2, as a cure for ailments such as headaches, whereby only the affected part is scarified. Other plants are used externally as lotions, mixed with fat to make ointments or applied as poultices (*ho thoba* or *ho tapisa*).

Materia Medica

A checklist of medicinal plants used in Lesotho. The list is arranged alphabetically according to family and genus. Several plant species used for magical purposes and during the divining process (e.g. for washing the divining bones) have also been included in the list. The botanical nomenclature in older publications has been updated to reflect currently accepted names, following Germishuizen & Meyer (2003).

355 plant species (including 23 exotics), from eight pteridophyte and 76 angiosperm families, are used for various medicinal purposes in Lesotho. In terms of the number of medicinal species, the most important families are Asteraceae (76), Fabaceae (33), Hyacinthaceae (15), Apocynaceae (12), Scrophulariaceae (12), Lamiaceae (11) and Poaceae (11). It is also interesting to note that seven species of ferns are used medicinally—mostly used to treat headaches, either through smoking the leaves or inhaling smoke from smouldering leaves.

Some plants are age and gender specific, others are strictly for livestock, whereas others are used for both humans and animals. Several plants are used to treat children's ailments. The following pattern of exclusive uses emerges: human (287 species), livestock (8 species), women (20 species), men (three species) and children (29 species). The medicinal plant uses can also be grouped on the basis of the ailment treated, such as problems relating to digestion (84 species), respiration (67 species), fertility and reproduction (31 species), circulation (26 species) and skin and wounds (24 species). Another important category is that of magic and sorcery; at least 50 species are used for this purpose. For the sake of brevity, a more detailed analysis is not given here, but the necessary data for more comprehensive analyses.

Interesting examples of specific uses include the following. Plants used exclusively for children include *Brachystelma foetidium* (colds), *Helichrysum ecklonis* (diarrhoea) and *Euphorbia clavarioides* (skin rashes). Numerous plants are used specifically by women for infertility (e.g. *Ajuga ophrydis*, *Equisetum ramosissimum* and *Haplocarpha scaposa*); menstrual pains (e.g. *Gunnera perpensa*, *Scabiosa columbaria*) or for ailments associated with pregnancy such as heartburn, flatulence and colic (e.g. *Anthospermum rigidum*, *Gazania linearis*, and *Convolvulus bidentatus*). Plants used to treat animals only include *Watsonia densiflora* (for diarrhoea in calves) and *Acrotome inflata* (for 'black leg' in cattle).

Based on the frequency of published and unpublished records, it is possible to rank the species and in this way determine which ones are the best known and/or most frequently used. A total of 37 species have at least four citations for medicinal uses—and almost invariably the same number of citations for vernacular name(s). These 37 species may be considered as the most important and best known medicinal plants in Lesotho. The number of citations for vernacular names and for uses were both considered and are given in Table 1.

Table 1.—No. times vernacular names cited and no. times uses cited for taxa

7, 7	6, 6	5, 5	5, 4	4, 4
Artemisia afra	*Gerbera pilloseloides*	*Alepidea cordifolia*	*Bulbine narcissifolia*	*Ajuga ophrydis*
Dicoma anomala	*Helichrysum caespititium*	*Cotyledon orbiculata*	*Dianthus basuticus*	*Albuca setose*
	Mentha longifolium	*Eucomis undulata*		*Aloe striatula*
	Olea europaea	*Gerbera viridiflora*		*Crassula peploides*
	Rhamnus prinoides	*Gunnera perpensa*		*Cussonia paniculata*
		Helichrysum odoratissimum		*Euclea crispa*
		Hermannia depressa		*Gazania krebsiana*
		Heteromorpha arborescens		*Gladiolus crassifolius*
		Pentanisia prunelloides		*Hibiscus microcarpus*
		Trifolium burchellianum		*Hypoxis argentea*
				Mentha aquatic
				Scabiosa columbaria
				Stachys aethiopica
				Urtica urens
				Vernonia hirsuta
				Withania somnifera

It is interesting to note that the two most important medicinal plants in the *materia medica* of Lesotho are used also for the two most common everyday ailments, namely respiratory and digestive problems (*Artemisia afra* and *Dicoma anomala*). *Artemisia afra* (*lengana*) is most commonly used as a nasal plug but the leaves are also used as a decoction or for steaming. The plant is used as a traditional medicine in many parts of Africa, to treat various illnesses including colds, sore throat, headaches and intestinal worms, but should not be used in high doses as it usually contains thujone, which is a central nervous system toxin (Van Wyk & Wink 2004). *Dicoma anomala* (*hloenya*: to grimace, because of the extremely bitter taste) is used for a wide range of illnesses including diarrhoea, toothache, hypertension, diabetes, period pains, pneumonia, backache, wounds and sores, and for gall sickness in cattle. In South Africa and Zimbabwe, it is used for similar digestive and respiratory ailments (Van Wyk & Gericke 2000).

Another plant ranking amongst the top in importance is *Pentanisia prunelloides*, appropriately named *setimamollo* (fire extinguisher) in Sesotho for its use to relieve the burning pain of boils and reducing fevers. *Urtica urens* (*bobatsi*: something that is rough) is interestingly not indigenous to Lesotho but is also amongst the list of commonly used plants. Other important plants include *Gerbera pilloseloides* (*tsebeeapela*: ear of the rockrabbit), *Helichrysum caespititium* (*phateeangaka*: mat of the healer, because of the prostrate habit), *Mentha longifolium* (*koena*: crocodile, because of its association with aquatic environments), *Rhamnus prinoides* (*mofifi*: very dark, because of the dark green foliage), *Alepidea cordifolia* (*lesoko*), *Heteromorpha arborescens* (*monkhoane*), *Felicia muricata* (*'mamileng*: along the road), *Gerbera viridiflora* (*moarubetso:* the one used to fumigate), *Helichrysum odoratissimum* (*phefo*: breeze, allegedly because when burnt during harvesting it brings about a breeze that helps with winnowing), *Cotyledon orbiculata* (*serelile*: slippery), *Trifolium burchellianum* (*'musapelo*: the one who turns the heart around), *Gunnera perpensa* (*qobo*), *Olea europaea* (*mohloare*), *Hermannia depressa* (*seletjane*: a small axe), *Eucomis undulata* (*khapumpu*), *Bulbine narcissifolia* (*khomoeabalisana*: a cow of the shepherds), *Dianthus basuticus* (*hlokoanalatsel*: the grass of the road), *Mentha aquatica* (*koenaenyenyane*: the small crocodile), *Cussonia paniculata* (*motšetse*), *Gazania krebsiana* (*tsikitlane*), *Xysmalobium undulatum* (*pohotšehla*: yellow bull) and *Tulbaghia acutiloba* (*sefothafotha*). The traditional use of *Mentha aquatica* and *M. longifolia* for respiratory problems can be ascribed to the presence of a volatile oil with possible decongestant effects (Van Wyk & Gericke 2000). *Xysmalobium undulatum* is used all over southern Africa for stomach problems and for a similar purpose in modern phytotherapy—it contains uzarin which has proven antidiarrhoeal properties (Van Wyk & Wink 2004).

Although the meaning of some Sesotho plant names is not obvious, many of them are descriptive and indicate either the habit or the usage of the plant (Moteetee & Van Wyk 2006). For example, *Stapelia grandiflora* (*bohatsu*: numbness) is used to cure numbness of the limbs, *Asclepias aurea* (*mohlatsisa*; emetic) is used as an emetic to treat constipation, *Senecio erubescens* (*letapiso*; poultice) is used as a hot poultice, *Scabiosa columbaria (selomi*: one that causes pain or bites) is used to relieve period pains and *Gomphocarpus fruticosus* (*moithimolo*: one who causes sneezing) is used to cure headaches and colds by sneezing. Several plants known as '*musapelo* are used as sedatives and adaptogens, for dropsy of the heart or heart palpitations. This concept has been dealt with in detail elsewhere (Moteetee & Van Wyk 2007).

Animal products are also used for medicinal purposes. Examples include cow dung (used to cure chickenpox), dog fur (placed on a dogbite wound to speed up the healing process), a mixture of powdered mare's afterbirth and crocodile skin (to assist difficult childbirth; fumigation with smoke from crocodile skin is an alternative). Snake flesh mixed with *Linum thunbergii* (*bohloko*) is used to treat snakebite.

CONCLUSIONS

The most important plants that are used in Basotho traditional medicine and the main health care practitioners and their practices have been described and conceptualized for the first time. A total of 355 plant species are used, representing eight pteridophyte and 76 angiosperm families of which the Asteraceae, Fabaceae, Hyacinthaceae, Apocynaceae, Scrophulariaceae, Lamiaceae and Poaceae are the most important. The traditional system appears to be dynamic and adaptive, because at least 25 alien species have been incorporated into the *materia medica*.

Plants are used for a wide range of ailments, the most frequent indications being digestive ailments and respiratory conditions. *Dicoma anomala* (used mainly for digestive ailments) and *Artemisia afra* (used mainly for respiratory ailments) appear to be the best known and most widely used medicinal plants. Fiftytwo species are variously employed for magic and sorcery. Other important medicinal species include *Gerbera pilloseloides, Helichrysum caespititium, Mentha longifolium, Rhamnus prinoides, Alepidea cordifolia, Heteromorpha arborescens, Felicia muricata, Gerbera viridiflora, Helichrysum odoratissimum, Cotyledon orbiculata, Trifolium burchellianum, Gunnera perpensa, Olea europaea, Pentanisia prunelloides, Hermannia depressa, Eucomis autumnalis, Bulbine narcissifolia* and *Dianthus basuticus.*

Although hitherto scattered in the literature, the medicinal plants of Lesotho seem to be relatively well recorded, as this review did not reveal any new species records. Detailed comparisons with other African healing cultures are now possible. The data suggest that the uses of the plants were not comprehensively recorded, as 36 new uses are reported. Field surveys in remote parts of Lesotho may reveal additional species and use records, but we believe that we have presented a reasonably comprehensive insight into the diversity of Lesotho medicinal plants and practice.

ACKNOWLEDGEMENTS

The University of Johannesburg is thanked for financial support. The first author acknowledges all interviewees and family members who provided some information on traditional plant use.

REFERENCES

1. Ashton, H. 1967. *The Basuto, a social study of traditional and modern Lesotho*: 283–318. Oxford University Press, London.
2. Brendler, T. & van Wyk, B-E. 2008. A historical, scientific and commercial perspective on the medicinal use of *Pelargonium sidoides* (Geraniaceae). *Journal of Ethnopharmacology* 119: 420–433.
3. Casalis, E. 1861. *The Basutos; twenty-three years in South Africa*: 1–355. Nosbert, London.
4. Germishuizen, G. & Meyer, N.L. 2003. Plants of southern Africa: an annotated checklist. *Strelitzia* 14. National Botanical Institute, Pretoria.
5. Germishuizen, G., Meyer, N.L., Steenkamp, Y. & Keith, M. 2006. *A checklist of South African plants.* Southern African Botanical Diversity Network Report No. 41. SABONET, Pretoria.
6. Hutchings, A. & van Staden, J. 1994. Plants used for stress-related ailments in traditional Zulu, Xhosa and Sotho medicine. Part 1: plants used for headaches. *Journal of Ethnopharmacology* 43: 89–124.
7. Jacot Guillarmod, A. 1971. *Flora of Lesotho.* Cramer, Lehre. MALIEHE, E.B. 1997. *Medicinal plants and herbs of Lesotho.* Mafeteng Development Project, Lesotho.
8. Moteetee, A. & van Wyk, B-E. 2006. Sesotho names for exotic and indigenous edible plants in southern Africa. *Bothalia* 36: 25–32.
9. Moteetee, A. & van Wyk, B-E. 2007. The concept of 'musapelo and the medicinal use of shrubby legumes (Fabaceae) in Leso-tho. *Bothalia* 37: 75–77.
10. Motlamelle, M.P. 1938. *Ngaka ea Mosotho.* Morija Printing Works, Lesotho.

11. Phillips, E.P. 1917. A contribution to the flora of the Leribe Plateau and environs. *Annals of the South African Museum* 16: 1–379.

12. Schmitz, M.O. 1982. *Wild flowers of Lesotho*. ESSA, Roma, Lesotho. Shale, T.L., Stirk, W.A. & van Staden, J. 1999. Screen-ing of medicinal plants used in Lesotho for anti-bacterial and anti-inflammatory activity. *Journal of Ethnopharmacology* 67: 347–354.

13. van Wyk, B-E. & Gericke, N. 2000. *People's plants: a guide to useful plants of southern Africa*. Briza Publications, Pretoria.

14. van Wyk, B-E. & Wink, M. 2004. *Medicinal plants of the world*. Briza Publications, Pretoria.

15. Watt, J.M. & Brandwijk, M.G. 1927. Suto (Basuto) medicines. *Bantu Studies* 3: 73–100, 154–178, 296–319.

16. Watt, J.M. & Breyer-Brandwijk, M.G. 1962. *The medicinal and poisonous plants of southern and eastern Africa*. Living-stone, Edinburgh.

CHAPTER 8

ETHNOBOTANY OF GINGEE HILLS, VILLUPURAM DISTRICT, TAMIL NADU, INDIA

S. Muruganandam, E. Kamaraj, V. Kadirvelmurugan, and S. Ravikumar

Keywords: Medicinal plants, Irula tribes, Ethno medicines.

INTRODUCTION

There exists a wide variety of life forms in the world and every one interacts with other in the biosphere. Ethnobotany deals with, how the different ethnic groups of humans are dependent on plants and their knowledge on medicinal property of plants which has been using from the time unknown to cure various diseases with-out any side effects which is common in synthetic drugs of today. The plants are useful in many ways to the mankind as it provides food, drugs, commercial products, numerous industrial benefits etc[1-7]. The traditional knowledge of people on medicinal property of plants is still being used by the tribes and urban people. This traditional knowledge reveals how the economically backward local and tribal people were getting treatment for various diseases[8-9]. According to World Health Organization (WHO), more than 80% population of the world, are using traditional medicine, for primary health care needs[10]. Traditional systems of medicine have been popularized over past few years, due to the side effects of modern medicine systems caused by some chemical drugs. The ancient Indian literature, Rigveda and Materia Medica have also highlighted with the documentation of numerous medicinal plants. But the ethnomedicinal knowledge of tribes has been passing from generation to generation through oral communication and there is no documentation in any form. The state, Tamil Nadu has 1.05% tribal population who belongs to 36 sub groups and they are well adapted to survive in the forests and adjoining areas[11]. Dr. Janaki Ammal (1956) has initiated the work on importance of ethnobotanical studies, followed by Jain (1963) extended using novel methods in these studies [12].

MATERIALS AND METHODS

Gingee hills is one of the archeologically important places and also the natural bioresource of north-east of Tamil Nadu, India. The study area lies in a complicated region located between Lat. 12° 10'N and Long. 79°.17'E in the Villupuram District of Tamil Nadu, India (Figure 1). The study area is found to be with rockier outer surface and receives maximum rain fall during North-east monsoon. It is located at a distance of 170 Kms South-west of Chennai city; 90 Kms North-west of Pondichery (Union-Territory) and 35 Kms east of Thiruvannamalai town. Frequent field trips were made to 17 villages namely Sethavarai (Citravarai), Naraayanapuram, Nallanpillaipetral, Rathaapuram, Pakkam, J ag an ad h ap u r am, P o d h u p aly am, S o k u p p am, D h v ad h an amp atai, Kanakkankupam, Gangavaram, Udhoo-kuttai, Malaiyarankupam, Pazhavalam, Pothuvaai, Ramarajanpatai, and Thadagam. The Sethavarai (Citra-varai), village is an Archeologicaly important place as it has the paintings drawn on the rocks of natural cavern of Ayyanar hill which is 3500 year old. These ancient paintings depict the pictures like Fire, Fish, Deer, and buffalo etc., especially a Deer and a Fish are seen in big size (Figure 2)[13-14].

Citation: S. Muruganandam, E. Kamaraj, V. Kadirvelmurugan, and S. Ravikumar. Ethnobotany of Gingee Hills, Villupuram District, Tamil Nadu, India, International Educational Scientific Research Journal [IESRJ], August 2016; Volume:2, Issue: 8. E-ISSN No : 2455-295X.

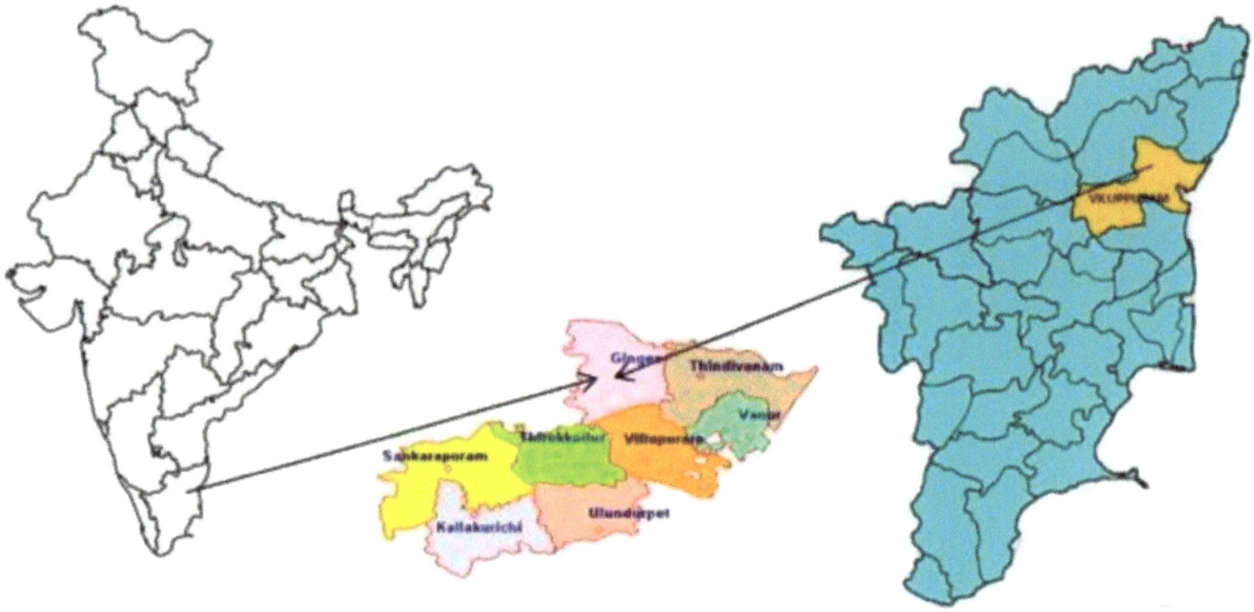

Figure 1. Location of study area of Gingee hills, Villupuram District, Tamil Nadu.

Figure 2. Archeological rock painting of Sethavarai (Citra-varai), village

Table 1: Ethnomedicinal plants, Family, Vernacular name, Method of administration and uses in Gingee hills in Villupuram District of Tamil Nadu

Table 1: Ethnomedicinal plants, Family, Vernacular name, Method of administration and uses in Gingee hills in Villupuram District of Tamil Nadu							
Botanical name	**Family**	**Vernacular name**	**H**	**Utilized parts**	**Formation**	**Method of administration**	**Disease**
Abutilon indicum (Link) Sweet.	Malvaceae	Periyathuthi	Sh	Flower	Paste	Flowers are ground with water and taken in empty stomach for one week.	Menstrual problems
Acacia arabica Wild.	Mimociaceae	Karuvelam	Tr	Gum	Paste	The Gum is mixed with hot water and taken orally three times a day.	Dysentery
Acacia chundra (Roxb.exRottl) Willd.	Mimosaceae	Karungali	Tr	Wood	Powder	Dried wood powder mixed with honey is taken in empty stomach monthly four times.	Body strength
Acalypha fruticosa Forssk.	Euphorbiaceae	Soothanthazhi	Sh	Leaves	Paste	Fresh leaves are made into paste with water and taken in empty stomach weekly once.	Skin diseases

Achyranthes aspera L.	Amaranthaceae	Naaiyuruvie	Hr	Stem	Paste	Fresh stem is made into paste and applied on the teeth surface.	Toothache
Aegle marmelos (L.) Correa	Rutaceae	Vilvam	Tr	Fruit	Paste	Fruit paste is mixed with ghee and taken in empty stomach.	Stomach ache
*Aerva lanata (L.)*Juss.ExSchult.	Amaranthaceae	Siru Peelai	Hr	Leaves	Powder	Shade dried leaves powder is mixed with goat milk and taken orally monthly once.	Kidney stone
Albizia amara (Roxb.) Boiv.	Mimosaceae	Thruingee	Tr	Leaves	Powder	Shaded dried leaves powder mixed with clay and water and applied on head bath.	Hair fall
Anisomeles malabarica (L.) R.Br.exims	Lamiaceae	Nettaperamettai	Sh	Leaves	Decoction	Fresh leaves are boiled in water and the vapour is inhaled.	Headache
Aristolochia indica L.	Aristolochiaceae	Perumarundhu	Cl	Root	Powder	Shade dried root powder mixed water is taken for 48 days.	Skin diseases
*Azadirachta indica*A.Juss.	Mealiaceae	Veapan	Tr	Flower	Decoction	Shade dried flowers are boiled and the decoction is ta`ken monthly twice.	Stomach ache
Barleria prionitis L.	Acanthaceae	Kaatu-kanagambaram	Hr	Flower	Paste	Fresh flower paste is applied on fore head.	Headache
*Calotropis gigantia(L.)*W.T.Aiton	Aasclepiadaceae	Erukku	Sh	Latex	Paste	Latex mixed with clay is applied on sprains.	Sprains
Canthium parviflorum Lam.	Rubiaceae	Kala chedi	Sh	Flower	Extract	Two drops of filtered clear flower extract is applied on the eyes.	Eye problems
Capparis horrida Linn.	Capparaceae	Adhandan	Li	Unripe fruit	Paste	The unripe fruit is ground with butter milk and taken in empty stomach.	Sinus
Cardiospermum helicacabum L.	Sapindaceae	Mudakathan	Cl	Leaves	Decoction	Fresh leaves decoction is taken weekly once.	Join pains
Chloroxylon swietenia (Roxb.) Dc.	Rutaceaea	Porusumaram	Tr	Bark	Paste	Fresh bark paste is applied on skin.	Skin diseases
Cissus quadrangularis L.	Vitaceae	Perandai	Cl	Stem	Pickle	Tender stems mixed with spices are pickled and taken with food.	Piles
Clausena dentata (Willd.) M.Roem.	Rutaceae	Konjee	Sh	Leaves	Paste	Half teaspoon leaves paste is taken in empty stomach.	Mucus
Commiphora caudata (Wight &Arn.) Engl.	Burseraceae	Kiluvaimaram	Tr	Bark	Extract	Fresh bark extract is taken orally.	Poisons bite
Corallocarpus epigaeus (Rottler) C.B.Clarke	Cucurbitaceae	Karudankizhing u	Cl	Tuber	Powder	Shade dried tuber powder is mixed with honey and taken orally.	Poisons bite
Crataeve magna (Lour.)Dc	Capparaceae	Mavalingam	Tr	Bark	Paste	The bark paste is consumed with water.	Skin diseases
Cynodon dactylon (L.) Pers.	Poaceae	Arugampullu	Hr	Leaves	Decoction	Leaves decoction is taken in empty stomach.	Diabetes
Delonix elata (L.)Gamble	Caesalpiniaceae	Vathanaaryanan	Tr	Leaves	Decoction	Leaves decoction mixed with spices is taken orally after food.	Join pains
Desmostchya bipinnata(L.)Stapf	Poaceae	Dharpai pull	Hr	Leaves	Decoction	Leaves decoction is taken orally.	Body cooling
Dichrostachys cinerea Wight et Arn.	Mimosaceae	Vedutharam	Tr	Gum	Paste	The Gum paste mixed with honey is taken orally.	Dysentery
Diospyros ferrea (Willd.) Bakh.	Ebenaceae	Erubuli	Tr	Leaves	Powder	Dried leaves powder mixed with water is applied on insect bite area and taken orally.	Insect bite.
Enicostemma littorala Blume.	Gentianaceae	Vellarugu	Hr	Leaves	Paste	Leaves paste with water is taken orally.	Snake bite
Evolvulus alsinoides (L.)L.	Convolvulaceae	Vishnukarinthai	Hr	Whole plant	Decoction	Plant decoction is consumed twice a day.	Fever
Ficus recemosa L.	Moraceae	Aththi	Tr	Unripe fruit	Paste	The unripe fruit paste is taken with curd.	Diarrhoea
Ficus virens L.	Moraceae	Kallarasu	Tr	Latex	Paste	Stem latex with clay is applied on the sprains.	Sprains
Glinu slotoides L.	Mollugiaceae	Serucerupadi	Hr	Leaves	Paste	Leaves paste mixed with cow milk is taken orally and applied on wounds.	Syphilis
Gmelina asiatica L.	Verbanaceae	Kumlankaai	Sh	Unripe fruit	Paste	Unripe fruit paste is applied on head bath.	Hair fall
Grewia orientalis L.	Sterculiaceae	Valachal	Sh	Fruit	Paste	Fruit paste mixed with curd is taken orally.	Body strength
Hildegardia populifolia (Roxb.) Sehott & Endl. Conservation	Malvaceae	Mala poovarasu	Tr	Bark	Decoction	The bark decoction is consumed for two days.	Fever
*Indgofera aspalathoides*Vahl.	Fabacee	Sivanar vembu	Hr	Whole plant	Paste	The Plant paste half teaspoon mixed with water is taken orally weekly once.	Cancer
Jasminum sambac (L.)Aiton	Oleaceae	Malli poo	Hr	Flower	Paste	Fresh flower paste is applied on face.	Sinus
Justicia adhatoda L.	Acanthaceae	Adathodai	Sh	Leaves	Decoction	50ml of Leaves decoction is consumed twice a day for 3 days.	Cough
Kedrositis foetidissima (Jacq.)Cogn.	Cucurbitaceae	Appakovi	Cl	Leaves	Extract	Fresh leaves mixed with *Mukia maderaspatna leaves* are oil fried and 5 ml filtered oil is given to baby orally twice a day.	Mucus
Leucas aspera (Willd.) Link.	Lamiaceae	Thumbai	Hr	Leaves	Juice	Plant juice is taken orally.	

Limonia acidissima L.	Rutaceae	Velaamaram	Tr	Fruit	Paste	Fruit paste with butter milk is taken in empty stomach weekly once.	Body strength
Madhuca longifolia (J.Koing) J.F.Macbr.	Sapotaceae	Iluppai	Tr	Leaves	Paste	Leaves paste is applied on skin allergies.	Allergies
Marsilea quadrifolia L.	Marsileaceae	Araakeeri	Hr	Leaves	Juice	Leaves juice with ghee is taken orally.	Body strength
Melia azedarach L.	Meliaceae	Malaivembu	Tr	Leaves	Juice	Leaves juice is consumed orally after Menstruation.	Menstrual problems
Memecylon umbellatum Burm.f.	Melastomaceae	Kachan	Tr	Leaves	Paste	Fresh leaves paste with water is taken orally.	Body pain
Mimosa pudica L.	Mimosaceae	Thotta surungi	Hr	Leaves	Juice	100 ml Leaves juice with hot water is consumed twice a day.	Allergies
Mimusops elengi L.	Sapotaceae	Magezha maram	Tr	Flower	Powder	Shade dried flower powder mixed with hair oil is applied on head.	Hair fall
Moringa oleifera Lam.	Moringaeceae	Murungai	Tr	Seed	Powder	Dried seeds powder mixed with ghee is taken in empty stomach once a day for a month.	Nerves disorder
Mucuna pruiens (L.)Dc.	Fabaceae	Punaikalai	Cl	Seed	Powder	Seeds powder mixed with honey is taken orally for 48 days.	Infertility
Mukia maderaspatna (L.) M.Roem.	Cucurbitaceae	Musumusukai	Cl	Leaves	Extract	Fresh leaves mixed with the leaves of *Kedrositis foetidissima* are oil fried. 5 ml of the filtered oil is given to baby twice a day.	Mucus
Mussaenda tomentosa Wight ex Hook.f.	Rubiaceae	Sanjeevi	Sh	Leaves	Juice	Leaves juice mixed with ghee is taken orally.	Body strength
Oldenlandia umbllata L.	Rubiaceae	Inburan	Hr	Whole plant	Paste	Plant paste is taken with hot water.	Dry Cough
Ormocarpum cochinchinens (Lour.) Merr.	Fabaceae	Elumbotti	Sh	Leaves	Powder	Shade dried leaves powder with cow milk is taken for 15 days.	Bone fracture
Pavetta indica L.	Rubiaceae	Pavattan	Sh	Bark	Powder	Barks powder mixed with cow milk is taken for ten days.	Body pain reduce
Phoenix loureiroi Kunth.	Arecaceae	Eecha maram	Tr	Young shoot	Decoction	Young shoot decoction is consumed in empty stomach.	Diarrhoea
Physalis angulata L.	Solanaceae	Panbuthkkalai	Hr	Fruit	Juice	10 fruits with 15 leaves of *Solanum nigrum* were made into juice and consumed in empty stomach for 3 days.	Ulcer
Plecospermum spinosum Trec.	Moraceae	Achankrunai	Li	Wood	Powder	Dried wood powder mixed with coconut oil is applied on wound.	Wound
Polyalthia longifolia (Sonn.) L.	Annonaceae	Nettilingam	Tr	Bark	Decoction	200 ml of Bark decoction is consumed in empty stomach before menstruation.	Infertility
Premna tomentosa Willd.	Verbenaceae	Pudangainari	Sh	Leaves	Decoction	Fresh leaves decoction mixed with cow urine is consumed for 10 days.	Obesity
Prosopis spicigera L.	Mimosaceae	Vannimaram	Tr	Bark	Decoction	Bark decoction is consumed once a day.	Asthma
Psidium guajava L.	Myrtaceae	Koiyah	Tr	Young shoot	Juice	Young shoot Juice with cow milk is consumed in empty stomach.	Piles
Pterolobium hexapetalum (Roth) Santapau&Wagh	Caesalpiniaceae	Pill thanakkan mull	Sh	Flower	Paste	Flower paste is applied on fore head.	Headache
Sansevieria roxburghiana Schult.f.	Asparagaceae	· Marul	Hr	Leaves	Juice	1ml fresh leaves extract is dropped in ear.	Earache
Sapindus emarginata Vahl.	Sapindaceae	Kottapungan	Tr	Fruits	Paste	Seed removed fruits paste is applied on hair before head bath.	Hair fall
Senna auriculata (L.) Roxb.	Caesalpiniaceae	Aavarai	Sh	Flower	Powder	Shade dried flowers powder with goat milk is taken twice a day.	Diabetes
Sesbania grandiflora (L.) Poiret	Fabaceae	Agathi	Tr	Leaves	Decoction	Leaves decoction is consumed for 3 days.	Ulcer
Sida acuta Burm.f.	Malvaceae	Aruvamanai poondu	Hr	Leaves	Paste	Leaves paste is applied on wounds.	Wound
Sida humilis Cav.	Malvaceae	Pazhampasi	Hr	Leaves	Powder	Shade dried leaves powder mixed with water is taken orally.	Body strength
Solanum nigrum L.	Solanaceae	Manathakkali	Hr	Leaves	Paste	15 fresh leaves and fruits are made into paste and taken in empty stomach.	Ulcer
Syzygium cumini (L.) Skeels.	Myrtaceae	Naval	Tr	Bark	Decoction	500 ml bark decoction is consumed orally.	Over bleeding
Tinospora cordifolia (Thunb.) Miers.	Menispermaceae	Seendil	Cl	Leaves	Paste	Fresh leaves paste is taken orally for 2 days.	Mucus
Toddalia asiatica (L.) Lam.	Rutaceae	Milagaranai	Sh	Leaves	Powder	Dried leaves powder with water is taken once a day for a week.	Gas problem
Tridax procumbens L.	Asteraceae	Mookuthi poondu	Hr	Leaves	Paste	Fresh leaves paste is applied on wound.	Wound
Vitex negundo L.	Lamiaceae	Kattu notchi	Sh	Leaves	Decoction	Fresh leaves are boiled in water and vapour is inhaled.	Headache
Ziziphusjujuba Mill.	Rhamnaceae	Ilanthai	Tr	Leaves	Paste	Leaves paste with water is taken for 15 days in empty stomach.	Lever problems

***Abbreviation: H**-Habit; **Hr**-Herb; **Sh**-Shrub; **Ci**-Climber; **Tr**-Tree.*

Ethnobotanical Survey

Ethnobotanical investigations are conducted randomly around the study area, in which, totally 30 informants of Irula tribes and non tribal people between the ages of 22 to 81 have been consulted. 'Irula' tribes have been recognized as a major group of tribes by the Government of Tamil Nadu and they speak tamil dialect. They are sound skilled in handling snakes, catching rat, and in agriculture, honey gathering, etc. Early in the 20th century, anthropological literature classified the *Irular* under the Negrito ethnic group, and few *irula* were also found in Western Ghats[15-16]. They migrated to various parts of the state for economic purposes. As they are frequently changing their places of living for various reasons their children were largely illiterates. Few *Irulas* are trained on folk and traditional medicine and are treating others. The medicinal knowledge of them has been recorded in field note and cross checked with the vernacular names said by local people. The collected plants specimens were enumerated in alphabetical order by their botanical names followed by families, vernacular names, habit, parts used, etc. The collected plant specimens were identified using *The Flora of Presidency*[17] and *The Flora of Tamil Nadu Carnatic*[18]. The plant specimens are preserved as Herbarium and deposited in the PG and Research Department of Plant Biology and Plant Biotechnology, Presidency College, Chennai, Tamil Nadu, India.

RESULTS & DISCUSSION

Totally 75 plant species belonging to 38 families, were identified and these species were grouped into the following three groups viz. Dicot, Monocot and Pteridophyte. These plants were used to cure 40 type of problems like allergies, asthma, body pain, body cooling, body pain reduce, body strength, bone fracture, cancer, cough, diabetes, diarrhea, dry cough, dysentery, earache, eye problems, fever, gas problem, hair fall, headache, infertility, insect bite, join pains, kidney stone, lever problems, menstrual problems, mucus, nerves disorder, obesity, over bleeding, piles, poisons bite, sinus, skin diseases, snake bite, sprains, stomach ache, syphilis, toothache, ulcer and wound. Mostly Trees species are utilized and Leaves are the most useful part for medicine. Some disease are cured by using drugs with different combination of plants like *Kedrositis foetidissima* and *Mukia maderaspatna*, used for Mucus, and *Physalis angulata* and *Solanum nigrum* used for Ulcer. The leaves of *Mussaenda tomentosa* has been recommended as a substitute for water loss in human. The plant products were given along with butter milk, cow milk, cow urine, curd, ghee, goat milk, oil, honey, spices and water. The plants species are arranged alphabetically with their family name, vernacular name (Tamil), parts used, plant habit, Method of administration, and remedy for disease are tabulated in the Table 1.

CONCLUSION

The main objective of this study is to document the ethno medicinal diversity and ethno medical knowledge from the resource persons. Some plants were exploited by human activity for their economical prosperous. The awareness to conserve these plant species is the need of the hour, otherwise we may not find many valuable plants hereafter which are significant in life saving. The ethno knowledge of Irula tribes and non-tribal people may have high potential for research for the discovery of new drugs and new source of food with novel uses. This study indicates the importance of herbal plants and how the urban people have been depending on forest medicinal sources, rather than modern chemical medicines.

ACKNOWLEDGEMENT

The authors are grateful to the resource persons of the study area, for revealing their traditional and folk knowledge. The corresponding author gratefully acknowledges Dr. R. Dhamotharan Head, Department of Plant Biology and Plant Biotechnology, Presidency College, Chennai for his help in identification of plant species and suggestions on crafting the article.

REFERENCES

1. Ayyanar M and Ignachimuthu S. Plants used for non-medicinal purposes by the tribal people in Kalakad Mundanthrai Tiger Reserve, Southern India. Indian J. of T. knowl. 2010, 9: 515-518.
2. Rasingam L, Ethnobotanical studies on the wild edible plants of Irula tribes of Pillur Valley, Coimbatore District, Tamil Nadu, India. Asian P. J. of T. Biomed., 2012, 1493-1497.
3. VermaRachna, ParkashVipin, Kumar Dinesh Ethno-Medicinal uses of some plants of Kanag Hill in Shimla, International J. of Res. Article. 2012, 3(2): 319-322.
4. Samydurai P, Jagatheshkumar S, Aravinthan V, Thangapandian V. Survey of wild aromatic ethnomedicinal plants of Vellingiri Hills in the Southern Western Ghats of Tamil Nadu, India. Int. J. Med. Arom. Plants. 2012; 2: 229-234.
5. Karuppusamy S. Medicinal plants used by Paliyan tribes of Sirumalai hills of Southern India. Natural P. Radiance, 2007; 6 (5): 436-442.
6. Kadirvelmurugan V, Raju K, Arumugam T, Karthik V and Ravikumar S. Ethnobotany of medi-flora of Kolli Hills, Tamil Nadu, Archives of A. Science Research, 2014, 6 (1):159-164.
7. Kadirvelmurugan Venkatachalam, Dhamotharan Ramasamy and Ravikumar Sundaramurthy. Ethnomedicinal study on tribal area of Kathiri Hills in Erode district of Tamilnadu, India. Central Euro. J. of Experimental Biology, 2015, 4 (1): 20-26.
8. ChellaiahMuthu, MuniappanAyyanar, Nagappan Raja, and SavarimuthuIgnacimuthu, Medicinal plants used by traditional healers in Kancheepuram District of Tamil Nadu, India. J. of Ethnobio. and Ethnome., 2006, 2: 43.
9. Sankaranarayanan S, Bama P, Ramachandran J, KalaichelvanP. T, Deccaraman M, Vijayalakshimi M, Dhamotharan R, Dananjeyan B, and SathyaBama S.Ethnobotanical study of medicinal plants used by traditional users in Villupuram district of Tamil Nadu, India J. of Medicinal Plants Research. 2010. 4(12): 1089-1101.
10. WHO general guidelines for methodologies on research and evaluation of traditional medicinal, Geneva, Switzerland. 2000: 1-80.

11. Ministry of Tribal affairs, Annual Report, 2005-2006.
12. JanakiAmmal, E.K. Introduction to the subsistence economy of India. In: Mans role in changing face of the earth (Ed. William LT Jr) University of Chicago press, Chicago.1956: 324-335.
13. Theyagarajan E. History of Vettavalam (Tamil edition), 2014: 258 – 259.
14. http://www.google.co.in/url?sa=t&rct=j&q=&esrc=s&source=web&cd=1&ved=0ahUKEwicOZ9tHKAhUVCI4KHTtmAd4QFggbMAA&url=http%3A%2F%2Fwww.tnarch.gov.in%2Fcons%2Fpaint%2Fpaint2.htm&usg=AFQjCNE2xmJMpNnv44P2KIGAQr7iqICXWA&sig2=1bcbF2d0BK1wj-GVJNsHQg&bvm=bv.113034660,d.c2E&cad=rjt. Archeological of rock painting of Sethavarai, Gingee Hills, Villupuram District.
15. Sinu E. Living Conditions of IrulaTribes in GingeeTaluk, Villupuram District of Tamil Nadu (India). http://www.theinternationaljournal.org/ojs/index2.htm RJSSM, 2013. 03 (2): 141-149.
16. Thurston E, Castes and Tribes of Southern India, e-Book. Version Aug 18, 2013: 209-219.
17. Gamble, J.S., The Flora of Presidency of Madras Adlard & son, Ltd, London, 1935.
18. Matthew KM. The Flora of Tamil Nadu Carnatic, Vol I-III, The Rapinat Herbarium, St. Josep's College, Tiruchirapalli, India. 1983

CHAPTER 9

TWO DECADES OF ETHNOBOTANICAL RESEARCH

Rainer W. Bussmann and Douglas Sharon

Keywords: William L. Brown Center, Traditional knowledge, ethnobotany, ethnopharmacology, Nagoya Protocol, repatriation of knowledge

INTRODUCTION

William L. Brown Center (WLBC) at Missouri Botanical Garden and its role in Traditional Medicine

WLBC

Plants provide humankind with our most basic resources – food, medicines, fiber, and a whole array of other useful products. Relatives of wild crops and traditional varieties – the repository of genetic diversity within and among food plants – have been the foundation of crop domestication, plant breeding, and indeed the modern agriculture that feeds the earth's 6 billion people. Plants provide the molecular basis of many pharmaceuticals, as direct compounds or molecular blueprints. Modern science begins to confirm that the distinction between nutrition and medicine is blurred. With economic development empowering a greater percentage of the world's people, urban areas continuing to expand, and human populations projected to double in the next 50 years, it seems certain that natural resources will face increasing threat. Habitat loss, unsustainable extraction of plants, spread of invasive species, climate change, and other human activities will have tremendous impact. Plant species will be lost, genetic diversity of surviving species will be diminished, and traditional knowledge associated with plant use will be eroded. Perhaps never before in human history has there been a more pressing need to discover, understand, conserve and sustainably use the plant resources that are essential for the benefit of humanity.

The William L. Brown Center is uniquely positioned to respond to these issues and to play a leading role in addressing the problems outlined above. The Center is located in one of the largest herbaria in the world, making a wealth of plant data available from collections. Access to advanced scientific methodologies allows more rapid characterization of useful species, chemicals or genes that lead to new nutritional and pharmaceutical products. The Center has access to improved information technologies that facilitate the rapid communication of data, and allow repatriation of data to the countries where it is needed to make intelligent decisions about the use of natural resources. Appropriate partnerships between the Center and collaborators in developing countries enable capacity building to ensure that countries have the infrastructure to make sound development and conservation plans. Finally, partnerships between the Center and both national institutions and local communities permit the implementation of integrated conservation and sustainable development programs.

WLBC focuses to a large extent on mountainous ecosystems. Mountain systems are highly bio-diverse, they are the most important sources of crop species and medicinal plants, and are heavily affected by climate-change. The programs of the WLBC concentrate on medicinal and food species in mountain regions, especially the Andes, the

Citation: Rainer W. Bussmann and Douglas Sharon. Two decades of ethnobotanical research, Ethnobiology and Conservation, 2014; 3:3. DOI:10.15451/ec2014-6-3.2-1-50. ISSN 2238-4782.

Himalayas and the African Highlands. Numerous projects utilize information from local people to better understand how communities rely on locally available natural resources and to design solutions so that they can be used sustainably. The WLBC ethnobotany programs are internationally recognized as leading scientific efforts to understand the relationship between plants, people, the environment, and associated traditional knowledge. WLBC is a leader in the compilation of databases on traditional medicinal and food species. The center plays a leading role in the preservation of traditional knowledge for indigenous and local communities, and in the production of useful plant volumes for ongoing floristic projects. The Center plays a leading role organizing symposia, meetings, and workshops to facilitate scientific exchange. The Garden is an international leader in developing the legal and ethical frameworks for conducting international programs in natural products discovery, medicinal plant research, and ethnobotany. These frameworks ensure compliance with the Convention on Biological Diversity and equitable distribution of benefits that arise from such research, including efforts to value, acknowledge and equitably compensate traditional knowledge. Projects conducted by the WLBC include training programs ranging from botanical field techniques and ethnobotanical research to collections management, market development, sustainable production and public health. Projects have significant conservation components and these activities include programs aimed at medicinal and other useful plants as well as those that incorporate traditional knowledge in the conservation of natural resources.

Traditional Medicine

Traditional Medicine is used globally and has a rapidly growing economic importance. In developing countries, Traditional Medicine is often the only accessible and affordable treatment available. In Uganda, for instance, the ratio of traditional practitioners to the population is between 1:200 and 1:400, while the availability of Western doctors is typically 1:20,000 or less. Moreover, doctors are mostly located in cities and other urban areas, and are therefore inaccessible to rural populations. In Africa, up to 80% of the population uses Traditional Medicine as the primary healthcare system. In Latin America, the WHO Regional Office for the Americas (AMRO/ PAHO) reports that 71% of the population in Chile and 40% of the population in Colombia have used Traditional Medicine. In many Asian countries Traditional Medicine is widely used, even though Western medicine is often readily available. In Japan, 60–70% of allopathic doctors prescribe traditional medicines for their patients. In China, Traditional Medicine accounts for about 40% of all healthcare, and is used to treat roughly 200 million patients annually. The number of visits to providers of Complementary-Alternative Medicine (CAM) now exceeds by far the number of visits to all primary care physicians in the US (WHO 1999a,b; 2002).

Complementary-Alternative Medicine is becoming more and more popular in many developed countries. Forty-eight percent of the population in Australia, 70% in Canada, 42% in the US, 38% in Belgium and 75% in France, have used Complementary-Alternative Medicine at least once (WHO 1998; Fisher et al. 1971; Health Canada 2001). A survey of 610 Swiss doctors showed that 46% had used some form of CAM, mainly homeopathy and acupuncture (Domenighetti et al. 2000). In the United Kingdom, almost 40% of all general allopathic practitioners offer some form of CAM referral or access (Zollmann et al. 2000). In the US, a national survey reported the use of at least 1 of 16 alternative therapies increased from 34% in 1990 to 42% in 1997 (Eisenberg et al. 1998; UNCTD 2000).

The expenses for the use of Traditional and Complementary-Alternative Medicine are exponentially growing in many parts of the world. In Malaysia, an estimated US $ 500 million is spent annually on Traditional Medicine, compared to about US $ 300 million on allopathic medicine. The 1997 out-of-pocket Complementary-Alternative Medicine expenditure was estimated at US $ 2,700 million in the USA. In Australia, Canada, and the United Kingdom, annual Complementary-Alternative Medicine expenditure is estimated at US $ 80 million, US $ 2,400 million and US $ 2,300 million, respectively. The world market for herbal medicines based on traditional knowledge was estimated at US $ 60,000 million in the late 1990s (Brevoort 1998). A decade later it was around US $ 60 billion (Tilbert et al. 2008) with estimates for 2015 at US $ 90 billion (GIA 2012). The sales of Herbs and herbal nutritional supplements in the US increased 101% between May 1996 and May 1998. The most popular herbal products included Ginseng (*Ginkgo biloba*), Garlic (*Allium sativum*), *Echinacea* spp. and St. John's Wort (*Hypericum perforatum*) (Brevoort 1998).

Traditional and Complementary-Alternative Medicine are gaining more and more respect by national governments and health providers. Peru's National Program in Complementary Medicine and the Pan American Health Organization recently compared Complementary Medicine to Allopathic Medicine in clinics and hospitals operating within the Peruvian Social Security System. A total of 339 patients - 170 being treated with Complementary-Alternative Medicine and 169 with allopathic medicine - were followed for one year. Treatments for osteoarthritis; back pain; neurosis; asthma; peptic acid disease; tension and migraine headache; and obesity were analyzed. The results, with 95% significance, showed that the cost of using Complementary-Alternative Medicine was less than the cost of Western therapy. In addition, for each of the criteria evaluated - clinical efficacy, user satisfaction, and future risk reduction - Complementary-Alternative Medicine's efficacy was higher than that of conventional treatments, including fewer side effects, higher perception of efficacy by both the patients and the clinics, and a 53–63% higher cost efficiency of Complementary-Alternative Medicine over that of conventional treatments for the selected conditions (EsSalud 2000).

According to WHO (2002), the most important challenges for Traditional Medicine / Complementary-Alternative Medicine for the next years are:

- Research into safe and effective Traditional Medicine and Complementary Alternative Medicine treatments for diseases that represent the greatest burden, particularly among poorer populations.
- Recognition of the role of Traditional Medicine practitioners in providing healthcare in developing countries.
- Optimized and upgraded skills of Traditional Medicine practitioners in developing countries.

- Protection and preservation of the knowledge of Indigenous Traditional Medicine.
- Sustainable cultivation of medicinal plants.
- Reliable information for consumers on the proper use of Traditional Medicine and Complementary-Alternative Medicine therapies and products.

Northern Peru represents the "health axis" of the old Central Andean cultural area stretching from Ecuador to Bolivia. The traditional use of medicinal plants in this region, which encompasses in particular the Departments of Piura, Lambayeque, La Libertad, Cajamarca, Amazonas, and San Martin possibly dates as far back as the first millennium B.C. (north coastal Cupisnique Culture) or at least to the Moche period (A.D. 100-800), with healing scenes and healers frequently depicted in ceramics.

Precedents for this study have been established by early colonial period chroniclers (Monardes 1574; Acosta 1590; Cobo 1653/1956; the plant collections (293 plants in crates 11 and 12 of 24) of Bishop Baltasar Jaime Martínez Compañón sent to the Palacio Real de Madrid along with cultural materials in 1789 under the title *Trujillo del Perú* in 9 illustrated volumes (Martinez Compañon 1789; Schjellerup 2009; Sharon et al. 2006); the travel journals of H. Ruiz from 1777-88 (Ruiz 1777-1788); the work of Italian naturalist Antonio Raimondi (1857); ethnoarchaeological analysis of the psychedelic San Pedro cactus (Sharon 2000); *curandera* depictions in Moche ceramics (Glass-Coffin et al. 2004), and research on the medicinal plants of Southern Ecuador (Bejar et al. 1997, 2001; Bussmann 2006; Bussmann & Sharon 2007).

ANTECEDENTS - MEDICINAL PLANT RESEARCH AND TRADITIONAL MEDICINE IN PERU

Containing 78 of the 107 eco-regions of the world, in 1993, it was estimated that Peru had 17,143 taxa of spermatophytes in 2,485 genera and 224 families (Brako et al. 1993). It is thought that only 60% of the Peruvian flora has been studied, with 1,400 species described as medicinal (Brack-Egg 2004).

The importance of biodiversity for the Peruvian economy is enormous since 25% of all exports are living species, the uses of which are essential to local populations in terms of firewood, meat, lumber, medicinal plants, and many other products. Of particular importance are vegetal species, with 5,000 plants applied in 49 different uses of. Of the 5,000 plants in use some 4,400 are native; only 600 are introduced. The majority of useful native species are not cultivated; only 222 can be considered to be domesticated or semidomesticated (Brack-Egg 1999).

Transculturation is resulting in an enormous loss of traditional knowledge of great value to the science and technology of Peru. The flora of the country represents 10% of the world's total, of which 30% is endemic. Peru is the fifth country in the world in number of plant species with known properties utilized by the population (4,400 species); it is the first in domesticated native species (182) (Brack-Egg 1999).

In all Peruvian ethnic groups, plant knowledge is invaluable because it reinforces national identity and values, which are being lost in the complementary processes of modernization and globalization. In the current situation the emerging recognition and incipient application of these resources and associated knowledge emphatically underscores the critical need for ethnobotanical research.

Over the last 15 years considerable progress has been made in the overall taxonomic treatment of the flora of Peru. However, while the Amazon rainforests have received a great deal of scientific attention, the mountain forests and remote highland areas are still relatively unexplored. Until the late 1990s little work had been done on vegetation structure, ecology, and ethnobotany in the mountain forests and coastal areas of the North. In spite of the fact that this region is the core of what Peruvian anthropologist Lupe Camino (1992/1999) calls the "health axis" of Central Andean ethnomedicine, little ethnobotanical and ethnomedical research has been published on the rich flora found here.

ISSUES IN THE GLOBALIZATION OF TRADITIONAL MEDICINE

Moran, King and Carlson (2001) trace the emergence of biodiversity prospecting. On 5 June 1992, in order to alleviate the loss of earth's flora and fauna the Convention on Biological Diversity (CBD) was inaugurated at the UN Earth Summit in Rio de Janeiro, Brazil. CBD objectives are: 1) conservation of biodiversity, 2) sustainable use of components of biodiversity, and 3) equitable sharing of benefits derived from commercial use of genetic resources.

For biodiversity-rich developing countries the most critical element in the CBD is sovereignty over bioresources by nation states, since the treaty recognizes their right to regulate and charge outsiders for access to their biodiversity. The sovereignty component is meant to replace the "common heritage" paradigm, which provides unrestricted access to biological resources. Ideally this paradigm shift is supposed to balance the way in which all involved interest groups can gain from biodiversity use by recognizing the economic, sociocultural, and environmental values of bioresources and the cost of their preservation.

In the time since the CBD was initiated, few of the 178 signatory nations have introduced legislation requiring benefit sharing for outside commercial access to their national bioresources, although some suggestions for implementation of the CBD have been brought forward (Iwu 1996; Buitron 1999). Despite the lukewarm response to the CBD by nation states, the global shift in awareness concerning tropical deforestation provided an opportunity for ethnobotanists to assert that everyone has an interest in preserving rainforests because they might contain compounds that could cure cancer, HIV-AIDS, and other diseases (Schultes et al. 1990; Elisabetsky et al. 1990); Cox et al. 1994; Schultes 1994; Brown 2003). In addition, income derived from the marketing of traditional medicinal knowledge was seen as an instrument to alleviate poverty and to finance conservation efforts (Reis 1993; Mooney 1993; Baker et al. 1995). Within a few years, however, for its critics, ethnobotany--initially seen as instrument that could help to salvage declining traditional knowledge and biodiversity-had simply become an instrument of theft and "biopiracy."

In his book *Who Owns Native Culture*? anthropologist Michael Brown (2003) has a chapter entitled "The Ethnobotany

Blues" which documents high-profile projects launched in Africa and Latin America in the early 1990s. They were organized under the U.S. initiative known as the International Cooperative Biodiversity Groups (ICBG), administered by the Fogarty International Center for Advanced Study in Health Sciences, part of the National Institutes of Health (NIH), with additional funding from the National Science Foundation (NSF) and the U.S. Agency for International Development (USAID). Projects involved partnerships between American and host-country scientists as well as major drug companies, including Monsanto, Bristol-Myers Squibb, and American Cynamid. Brown (2003) describes ICBG-Peru's troubled relationship between the Aguaruna of the Peruvian Amazon and Washington University (St. Louis), criticizing "paternalistic interventions that leave native peoples on the margins of decision-making and profit-taking" (p. 114). In Mexico, he documents how ICBG-Maya was shut down by an indigenous healers' organization and their activist allies on the grounds that it was an effort to steal native knowledge and resources. And he traces the failure of Shaman Pharmaceuticals, a California company which folded in 1999, in trying to adapt ethnobotanical bioprospecting to the "magic-bullet" paradigm of the pharmaceutical industry.

In the late 1990s, anthropologist Cori Hayden (2003) conducted an ethnography of an ICBG bioprospecting agreement inaugurated in 1993 between the University of Arizona and its pharmaceutical partners (whose contribution was a discount on the use of their equipment!) and a team of plant researchers at Mexico's National Autonomous University (UNAM) headed by ethnobotanist Robert Bye. Under the agreement, UNAM researchers sent extracts of Mexican medicinal plants to the US in exchange for research funds and promises of a percentage of royalties 10 to 20 years in the future—should a drug result from the collaboration. The project was also designed to collect ethnobotanical knowledge and to direct some royalties back to source-communities. It concluded in 2003 when UNAM opted out of a second renewal.

Hayden elucidates the complex issues that emerged during the project, in particular the paradoxical effects of NIH's advocacy of benefit-sharing according to the neo-liberal paradigm of bioprospecting. For NIH, this meant that field researchers were supposed to sign contracts with each individual supplier of plants. Suppliers—and, by implication, their communities—were presumed to be "authors" and "stewards" of resources as well as future benefit-recipients. For UNAM ethnobotanists, drawing on a well-established research methodology, this meant collecting initial plant species from urban marketplaces and rural roadsides, a major disruption of a fundamental biopropecting assumption that plants and knowledge "come with" clearly identified local stewards, authors, and claimants.

In stark contrast with the ICBG approach, there is the Mexican Institute for Social Security (IMSS) model put into practice at its Southern Center for Biomedical Research (CIBIS) in Cuernavaca and focused on the production of herbal medicines. On 20 February 1997, Hayden (2003) interviewed Miguel Antinori, a prominent CIBIS official who denigrated bioprospecting agreements for using Mexico's chemists as "cheap labor" and for sending extracts abroad for "more sophisticated" work. Further, he added, "It's hard to see an assertion of [Mexican] national identity in these contexts—up north, they just see Mexico as a source of raw material and certainly not as research partners or collaborators. Why don't they locate more of the development part here? Because they don't trust Mexican science."

Shaman Pharmaceuticals scientists Moran, King and Carlson (2001) discuss the irony in the situations described above, indicating that the majority of the biotech industry is not involved in bioprospecting, since most companies favor the use of cheaper and faster synthetic technologies over the complex process involved in exploring for natural products. Nonetheless, biotechnology spawns ethical, social, and legal debates at the margins of pharmaceutical bioprospecting, including the collaboration between big business and big science, the ethics of genetic engineering, and the patentability of life forms as well as ideas about genetics and racism, culture and ethnicity. However, it is significant to note that, since the inauguration of the CBD, no pharmaceutical bioprospecting product developed by using traditional knowledge has generated an economic profit. (But this does not mean that pharmaceutical companies do not try to impede or co-opt efforts to get natural plant products to market.) Also, only a small number of bioprospecting research expeditions begin by using ethnobotany as a discovery methodology, with the work soon evolving into economic botany as the laboratory focus shifts to the plant's chemistry, biological activity, and pharmacology/toxicology. During drug discovery, active chemical components are isolated, often modified, and patented. Patented information then becomes a commodity in itself.

Peruvian pharmaceutical researcher Angulo (2009) discusses new approaches to research on medicinal plants contrasting Western and Eastern methodologies. For example, whereas the West does not value popular wisdom and usage developed over centuries by local cultures, the East uses this knowledge as a paradigmatic base for its model of science. Whereas the West has exclusively followed the Cartesian model of scientific skepticism, Eastern pragmatism, building on tradition, has formalized usage and then applied the methods of modern science. Whereas the West has ignored traditional knowledge in designing artificial studies that isolate chemical components and evaluate their toxicity and bioactivity to later take finished products into clinical settings, the East has followed an inverse strategy, i.e., valuing traditional knowledge by applying original remedies and therapies in the medical clinic and then subjecting those that work to biochemical research and development. Whereas the West followed a basic research paradigm of random screening, component analysis and synthesis, the East recognized the holistic action of herbal medicines in seeking ways to industrialize them. As a result of the foregoing factors, Western science has developed economic botany, which uses a methodology of chemical taxonomy based on the assumption that only by knowing the chemistry of plants can we discover their active principles and bioactivity. This has led to the current emphasis on synthetic chemistry for the development of modern medicines.

Angulo (2009) points out that, by uncritically following the Western model for biochemical research promoted by large European and American pharmaceutical corporations, Peru has acquiesced to the notion that countries like Peru and Mexico lack the technical and economic resources necessary to compete with foreign consortiums. As a result, these countries, for the most part, have denigrated their own indigenous knowledge and neglected the development of viable national research programs in ethnobotany and ethnopharmacology. Joining Elisabetski et al. (1990), Angulo suggests that:

"Traditional medicine should be the basis for the development of drugs, given that it includes the knowledge of the therapeutic value of local flora. Thus, knowledge of the practices of Traditional Medicine plays a crucial role in the selection of species to subsequently be considered as potential sources of universally applicable drugs. Elizabetsky concludes that the interaction between anthropology and ethnopharmacology is the basis on which should be developed the holistic investigation of medicinal plants in particular and healthcare in general."

We would only add that applied research on natural plant remedies should also be on the national agendas of Peru and neighboring republics.

By focusing on indigenous knowledge as it relates to the environment, the Convention on Biological Diversity managed to sidestep some of the more politically charged aspects of the intellectual property rights (IPR) issue. The greatest impact on concerns over indigenous and local-community rights can be traced to the mercurial rise of biotechnology on the international trade front and the 1995 version of the World Trade Organization (WTO) Agreement on Trade Related Aspects of Intellectual Property Rights (TRIPS). These two factors have created a large potential market for indigenous and local knowledge and resources, while at the same time raising concerns about the risk that these resources will be misappropriated. Thus this knowledge is receiving increasing international attention in terms of its relationship to human rights as well as its relevance to modern science. The situation has created opposing pressures calling for the rights of local and indigenous peoples on the one hand and further exploitation of their knowledge on the other. Moran, King and Carlson (2001), and Greaves (1995), indicate that the biggest problem with the orthodox intellectual property system is its focus on material aspects of knowledge at the expense of the cultural. They advocate recognition of alternative worldviews in the formulation of new indigenous knowledge rights that are localized, relevant, pertinent, and effective.

In their article in *Cultural Survival Quarterly* Bannister and Barrett (2001) contend that bioprospecting is a form of economic botany that can run contrary to the ethnobotanical objectives of protecting biological and cultural diversity. The economic focus of this activity highlights issues concerning indigenous rights, cultural knowledge, and traditional resources - areas in which current intellectual property protection regimes are inadequate and inappropriate. However, indigenous communities are increasingly forced to employ intellectual property rights to protect these resources. Protection issues ought to be addressed well before the point at which employing intellectual property mechanisms seems to be the only alternative. Significant control lies at the point of decision about publication and dissemination of knowledge to the wider community, which raises important questions about facilitating the appropriation of cultural knowledge. The authors advocate a more "precautionary" approach to ethnobotanical inquiry in assisting indigenous communities in protecting their cultural heritage and intellectual property rights.

Probably the major concern in many traditional communities is that their spiritual legacies will be profaned by a secularized and consumer-driven outside world. Often, however, legitimate economic considerations also play a role in the defensive reactions of these societies to the well-intended but naïve desire of the academic world to place its findings in the public domain. Greaves (1995) and Bannister and Barrett (2001) have warned that the downside in this approach is that a "colonializing archive" can become easily "mined" for clues in the search for new drugs without the inconvenience of fieldwork or inclusion of source communities in the benefits derived from products resulting from research.

However, although acknowledging genuine concerns about neocolonialism and biopiracy, we would submit that each situation has to be considered on its own merits, especially with regard to its specific cultural context. A first step in the evaluation process should involve the important distinction between "indigenous peoples" and "local communities" (King et al. 2001). The latter for the most part are farmers who speak the national language, practice the majority religion, and identify with the nation-state, especially with regard to their socioeconomic aspirations, whereas the former tend to be tribal and/or ethnic minorities, who seek collective rights and self-determination for their biological and cultural resources. Although it is often the case that in both communities' traditional knowledge and resources are undocumented and in danger of disappearing, this danger tends to be more pressing in local communities as their members continue to adapt to privatization and globalization. In cases such as these successful ethnobotanical intervention requires a methodology that combines "salvage ethnography" with "rapid assessment". This is the methodology that we initially applied in Peru, motivated by our prior experience in Southern Ecuador where traditional knowledge of medicinal plants similar to those found in Northern Peru is diminishing at an alarming rate. However, with our database firmly established as a research vehicle, we can now turn our attention to facilitating proactive issues of education, conservation, and sustainable development of natural plant products.

India provides a positive example of the proactive application of this approach. Taking advantage of the "novelty" criterion in international patent law, with regard to the documentation of Ayurvedic and other traditional medicine, millennial Sanskrit texts as well as modern publications are included in a traditional knowledge database, which is subsequently provided to patent agencies. The expectation is that, by placing the knowledge about long-term cultural precedents for traditional uses in the public domain, this research will prove that contemporary patent applications derived from local medicinal knowledge lack originality, i.e., that they are not "novel" enough to qualify as inventions warranting protection under international patent law, and are thus not patentable.

Fortunately, in 2002, with the support of the International Phyto-Genetic Resource Institute (Rome, Italy), Peru promulgated Law 27811 for the protection of the collective knowledge of indigenous peoples related to biological resources. Article 17 of the law establishes a National Public Register to include collective knowledge that is in the public domain. This register is administered by INDECOPI (National

Institute of Competitive Defense and Intellectual Property), with the obligation to send the information recorded to principal patent offices around the world, a protective defense mechanism intended to prevent the granting of patents which do not meet the criteria of novelty and degree of inventiveness (Venero 2005; Ferro et al. 2005).

As noted earlier, Peru has also activated the Peruvian National Commission Against Biopiracy. In the Congressional Forum of 2004 which led to the formation of the Commission a number of important issues were addressed, including intellectual property , the high protein cereal Quinua and biopiracy, passage of the law for the protection of Peruvian biodiversity and the collective knowledge of indigenous peoples, and efforts to annul the US patent for the virility stimulant Maca as well as suggestions for combating biopiracy(Ferro et al. 2005). Briefly noted was the issue of genetically modified foods, anticipated as a concern that was likely to emerge with approval of a free trade agreement with the]. When the Commission was legally mandated, later in 2004, 19 plant claims were slated for review. By 2010, claims for 69 plants were being researched, 17 cases of biopiracy had been identified, and seven (from France, Japan, and South Korea) had been successfully blocked. One hopes that in all these deliberations the following remarks by forum panelist Jorge Agurto (2005) will be borne in mind:

> *"The problem underlying biopiracy is the open recognition of the rights of the indigenous peoples and communities. Many times they have been excluded and marginalized from the politics of Government. Even today we encounter members of Congress who are either unaware of the existence of indigenous peoples or who do not recognize their rights. It is impossible to speak of biopiracy if we do not defend the holders of many genetic resources, those who have achieved the domestication, knowledge, and technology to utilize biodiversity in a sustainable fashion. They are also the holders of the right to prior informed consent, a fundamental right to know the objectives of the exploration and exploitation of their resources and traditional knowledge and the consequences or potential benefits that can come with industrial, commercial or scientific uses. "*

Spanish anthropologist Luisa Abad (2003) concludes in her book *Ethnocide and Resistance in the Peruvian Amazon* that foreign and domestic development policies contribute to the marginalization of indigenous peoples:

> *"Underdeveloped, developing, Third World, North-South..., perhaps the language has been changing in these times and the terminology has been adapting itself to partially new habits, but the unequal, hierarchical reality remains the same, given that those who exercise power continue to be the same. International assistance also keeps promoting unequal development between peoples."*

BIODIVERSITY CONSERVATION AND T RADITIONAL MEDICINE

A policy report, *Biodiversity, Traditional Knowledge and Community Health: Strengthening Linkages,* published by the United Nations University--Institute of Advanced Studies in Yokohama, Japan addresses many of the issues discussed above (Unnikrishnan et al. 2012).

Building on the WHO Alma Ata Declaration of 1978 relating to Traditional Medicine and primary health care, the UN Convention on Biological Diversity of 1992, and the UN's Middle Development Goals (MDGs) of 2011, this document shows that links between Traditional Medicine and biodiversity are strengthened by three processes: 1) a medical approach involving national efforts to integrate Traditional Medicine into institutional healthcare delivery which includes challenges related to safety, quality, efficacy, access, and regulation; 2) a market-oriented approach focused on drug development or tourism promotion focused on biomedical products and services as marketable commodities; and 3) a community-focused approach activated by civil society organizations focused on conservation implemented through a grassroots mobilization process involving health professionals, botanists, conservationists, and community activists.

The community-based approach shows allegiance to the Alma Ata primary health care model. Examples include the barefoot doctors strategy in China and the social health activist programs in India. Given the centrality of biodiversity in human lives, there still is a need to develop sustainable strategies for health maintenance combined with conservation of biological resources and linked to local knowledge and practices. This is relevant even in developed countries where there is an increasing demand for alternative and complementary medicine.

TWO DECADES OF ETHNOBOTANICAL RESEARCH IN SOUTHERN ECUADOR AND NORTHERN PERU

Work up to 2012 - besides developing a database of 510 medicinal plants (Bussmann et al. 2006 a,b: 2007a) and 974 remedies of mixtures (Bussmann et al. 2010a) has demonstrated that herbal commerce in Peru is a major economic resource (Bussmann et al. 2007b), which, although used alongside modern pharmaceutical products, is showing signs of diminished popular knowledge of applications (Bussmann et al. 2007 b,c, 2009 a). Laboratory research on most of the database has ranged from minimum inhibition concentrations (Bussmann et al. 2010b) to toxicity screening (Bussmann et al. 2011) as well as bioassays to determine antibacterial activity (Bussmann et al. 2008 a,b,c; 2009 b) and phytochemical analysis (Bussmann et al. 2009c; Perez et al. 2012) with more focused analyses of herbal treatments for acne (Bussmann et al. 2008a) and malaria (Bussmann et al. 2010c). Other studies have sought to identify Ullucho, a ceremonial plant of the pre-Hispanic Moche culture (Bussmann et al. 2009d) as well as surveying colonial sources of medicinal plants in Northern Peru and Southern Ecuador (Bussmann et al. 2006b). An ethnography of peasant herbalists which documented aspects of the market supply chain showed that suppliers are not adequately remunerated and revealed threats posed by lack of conservation measures and overharvesting (Revene et al. 2008; Carillo 2012) criticized the scientific reductionism of laboratory research in attempting to appropriately verify traditional remedies. Anthropological studies of traditional *curanderos* and their curing altars (*mesas)* include articles by Sharon et al. (2009), Sharon & Gálvez (2009), Sharon (2009); and Glass-Coffin et al. (2004).

It is worth noting that, during the decade that we have been working in the field and the laboratory, there has been a sea change in attitudes and perceptions of Traditional Medicine. In Trujillo, Lima and Arequipa a pilot program prescribing medicinal plants, scientifically validated by WHO/PAHO, has been initiated by EsSalud's National Program for Complementary Medicine, an initiative begun in 1999 with three centers which has grown to 26 to date (Fernandez 2009). In Trujillo, the Missouri Botanical Garden´s Sacred Seeds program has started an herbal garden and educational outreach program at the site museum of the pre-Hispanic Chimú city of Chan Chan. In Huamachuco, a program of ethnobotany and conservation manifest in community gardens and seed banks of medicinal and food plants is slowly emerging through collaboration between three local peasant communities, the Beneficencia Publica and regional hospital, MBG's Sacred Seeds program, MHIRT, and the Peace Corps. Future work will involve developing a supply chain between Huamachuco and CCM-Trujillo with scientific validation by MBG, UB(SUNY), the Biotransformation & Natural Products Laboratory at UNT, and the Interdisciplinary Research Group at UPAO as coordinated by MHIRT and MBG.

Plant Nomenclature in Northern Peru

The naming of plant species follows three general patterns. Plant names already used by original indigenous populations are often maintained, although slightly modified. Plants similar to species already known, or with similar habitus, often receive the same name (transposition). In other cases, completely new names are created (neology) (Van den Eynden et al.2004).

The vernacular names of the plants used in Northern Peru reflect the historical development of plant use in the region. Introduced species (e.g. *Apium graveolens* - Apio, *Foeniculum vulgare* - Hinojo), native species similar to species found in Spain (e.g. *Adiantum concinnum* - Culantrillo, *Matricaria frigidum* - Manzanilla), as well as species growing mostly in the coastal regions of the area (e.g. *Alternanthera porrigens* - Sanguinaria), are often addressed with names derived from Spanish roots. Plants from the mountain forests and especially the Andean highlands or the Amazon are often known by their Quechua names (e.g. *Pellaea ternifolia* - Cuti Cuti, *Amaranthus caudatus* - Quihuicha, *Banisteriopsis caapi* - Ayahuasca), and a few plant names can be traced back to Mochica (the original indigenous language spoken at the coast of Northern Peru) roots (e.g. *Nectandra* spp. - Espingo) (Bussmann et al. 2009e). Van den Eynden et al. (2004) observed similar patterns in Southern Ecuador, although her study focused only on edible species. Nine hundred thirty-eight vernacular names were recorded for 510 plant species. About one third if all names represented Quechua names or had Mochica roots, while 66.5% of all names were of Spanish origin or at least had Spanish components. In comparison, 41% of the vernacular names of edible plants in Southern Ecuador were found to be of Spanish origin. More than half of the indigenous species carried only one vernacular name, with the remaining species carrying a variety of indigenous names, often derived from the same root. In comparison, almost 75% of the introductions were known by one name only. The slight differences in plant names indicate that the species have been used in the region for a long time, and that their names reflect small variations in the local dialects.

Two Decades of Ethnobotany in Northern Peru and Southern Ecuador

Since 1995 ethnobotanical data were collected in Northern Peru and Southern Ecuador (Fig. 1), from plant sellers while purchasing plant materials in local markets, by accompanying local healers (*curanderos*) to the markets when they purchased plants for curing sessions, and into the field when they were harvesting. In addition, plants were collected by the project members in the field, and--together with the material purchased in the markets--taken to the homes of *curanderos* to discuss the plants' healing properties, applications, harvesting methodology, and origins. The project region represents a floristic hotspot in Peru, with striking gradients from coastal deserts and dry foreats, to some of the wettest Paramos known, to the Amazon lowlands (Fig. 2). At the *curanderos'* homes the authors also observed the preparation of remedies and participated in healing rituals. Plant uses were discussed in detail with informants, after seeking prior informed consent from each respondent. Following a semi-structured interview technique, respondents were asked to provide detailed information about the vernacular plant name in Spanish or Quechua; plant properties (hot/cold); harvesting region; ailments for which a plant was used; best harvesting time and season; plant parts used as well as mode of preparation and application; and specific instructions for the preparation of remedies, including the addition of other plant species. All interviews were carried out in Spanish, with at least one of the authors present. Both authors are fluent in Spanish, and no interpreter was needed to conduct the interviews. Data on plant species, families, vernacular names, plant parts used, traditional uses, and modalities of use were recorded.

Figure 1. Study area: Peruvian Departments of Amazonas, Piura, Lambayeque, La Libertad, Cajamarca, San Martin, and the Ecuadorian Province of Loja. Dry forest (above) and Cloudforest (below) indicating the striking moisture gradient.

Figure 2. A 50km transect in Southern Ecuador from dry forest (250mm precipitation, left) to Podocarpus National Park Paramo (10000mm precipitation, middle) to Amazon lowlands (right).

Many of the species reported from northern Peru are widely known by *curanderos* and herb vendors as well as the general population of the region, and are employed for a large number of medical conditions. One hundred fifty to two hundred plant species, including most of the introductions, are commonly sold in the local markets (Bussmann et al. 2006b). Rare indigenous species were either collected by the healers themselves, or are ordered from special collectors or herb vendors. The same plants were frequently used by a variety of healers for the same purposes, with only slight variations in recipes. However, different healers might give preference to different species for the treatment of the same medical condition. All species found were well known to the healers and herb vendors involved in the study, even if they themselves did not use or sell the species in question. Many species were often easily recognized by their vernacular names by other members of the population. This indicates that these remedies have been in use for a long time by many people. The use of some species, most prominently San Pedro (*Echinopsis pachanoi*), Maichil (*Thevetia peruviana*) and Ishpingo (various species of *Nectandra*), can be traced back to the Moche culture (AD 100-800). Representations of these plants are frequently found on Moche ceramics, and the remains of some were found in a variety of burials of high-ranking individuals of the Moche elite, e.g., the tomb of the Lord of Sipán (Bussmann et al. 2009e).

Medicinal Uses

Five hundred-ten plants with medicinal properties were registered in Northern Peru. The same species was often used for various medical conditions and applied in different ways for the same condition. For example, nervous disorders might be treated using different parts of a plant in different applications, e.g., topical (as a Poultice or Bath), oral (ingestion of plant extracts), and by supplying the patient with a *seguro*, a bottle filled with herbs and perfumes, which serves as a protective charm. Two thousand four hundred ninety-nine different uses were registered for the 510 species encountered. In the following, the total number of uses/applications and the number of species used are given, rather than only the number of plant species used to treat a condition, in order to emphasize the importance of the treatment of specific conditions.

The highest number of species (207, 40.4%) is used for the treatment of "magical" ailments, with 682 (27.3%) of all conditions. Respiratory problems (95 species, 18.5%) were mentioned as 233 (9.3%) of all uses; 98 species (19.1%) are used to treat psychosomatic and nervous system problems, with 176 applications (7%). Kidney and Urinary tract disorders are treated with 85 species (16.6%), for 111 conditions (4.4%). Rheumatic and arthritic symptoms are mentioned in 103 uses (4.1%), with 45 species (8.8%) used for treatment. Infections of female organs are treated with 66 species (12.9%), and comprised 100 (4.4%) of all conditions.

Treatments are most often performed in the homes of the individual healers, who normally have their *mesas* (healing altars) set up in their backyards (Fig. 3). Healers also treat patients at altars and consultation chambers (*consultorios*) in their homes, at sacred sites in the countryside, or at sacred lagoons high in the mountains. Healing altars (*mesas*) bearing a large number of power objects are often employed (Fig. 4). A curing ceremony normally involves purification of the patient by orally spraying blessed and enchanted herbal extracts on the whole body to fend off evil spirits and by nasal ingestion of tobacco juice and perfumes. Plant supplies come mostly from the local markets (Fig. 5).

Figure 3. Contrremporary healing altar (mesa) in Southern Ecuador.

Figure 4. Contemporary healing altar (mesa) in Northern Peru.

Figure 5. Plant preparations and raw plants for sale in Mercado Mayorista, Trujillo.

Two hundred seventy-eight different medical conditions were recorded. Most plants were used for the treatment of multiple ailments. The large variety of conditions is grouped into 72 main categories.

Magical Uses

Mental, neurological and psychosomatic disorders are highly prevalent on a global scale. The burden of mental health problems has been seriously underestimated. Although neurological problems are only responsible for about 1% of global deaths, they contribute to over 11% of the global disease burden. It is estimated that this share will rise to 15% by 2020 (WHO 2009). Western medicine often offers little help for patients afflicted by these disorders.

Healing altars (*mesas*) in Northern Peru often follow the old tradition by including a large variety of "power objects," frequently with a "pagan" background. Objects such as seashells, pre-Columbian ceramics, staffs, stones, etc. are very common on Peruvian *mesas* and are blended with Christian symbols such as crosses and images of saints. Treatments are most often performed in the homes of the individual healers, who normally have their *mesas* set up in their backyards. Healers also treat patients at altars and consultation chambers (*consultorios*) in their homes, at sacred sites in the countryside, or at sacred lagoons high up in the mountains. A curing ceremony normally involves purification of the patient by orally spraying blessed and enchanted herbal extracts on the whole body to fend off evil spirits and by "Spiritual Flowerings" (*baños de florecimiento*). In most cases, the cleansing of the patients involves drinking boiled *San Pedro* juice and the nasal ingestion of tobacco juice and perfumes. Sometimes extracts of Jimson weed (*Datura ferox*), *Brugmansia* spp., and tobacco are also used to purify the patients. While the incantations used by healers during their curing sessions include Christian components (e.g., the invocation of Christ, the Virgin Mary, and any number of saints), references to Andean cosmology (e.g., to the *apus* or the spirits of the mountains) are very common. The use of guinea pigs as diagnostic instruments is standard in Northern Peru (Sharon 1978, 1980, 1994, 2000; Joralemon et al. 1993).

Traditional Medicine is also gaining more attention by national governments and health providers. Peru's National Program in Complementary Medicine and the Pan American Health Organization recently compared Complementary Medicine to allopathic medicine in clinics and hospitals operating within the Peruvian Social Security System (EsSalud 2000).

Mal Aire (Bad Air), *Mal Viento* (Bad Wnd), *Susto* or *Espanto* (Fright), *Mal Ojo* (Evil Eye) and *Daño* or *Brujería* (Sorcery) are seen as very common illnesses in Andean society. Causes include sudden changes in body temperature (*Mal Aire, Mal Viento*), any kind of shock (*Susto, Espanto*), "humors" or spells cast by other people (*Mal Ojo*), poisoned food, curses, etc. (*Daño, Brujeria*). Medical problems caused by outside influences were reported in a wide variety of studies (Girault 1987; Oblitas 1992). The Western concept of "psychosomatic disorders" comes closest to characterizing these illnesses.

These illness categories are deeply rooted in Andean society, and Western medicine does not offer efficient alternatives to traditional treatment. This might explain why this category has still such outstanding importance. Treatment in many cases involved the participation of the patient in a cleansing ceremony or *limpia*. This could either be a relatively simple spraying with perfumes and holy water, or an allnight ceremony involving the healer's curing altar (*mesa*). In the days after an all-night ceremony, patients are normally treated with a *baño de florecimiento* (flowering bath) in order to relieve them of any remaining adversary symptoms or spirits. In addition, patients frequently receive *seguros* (herbal amulets) for protection against further evil influences and for good luck. *Seguros* are flasks filled with powerful herbs, as well as perfumes, pictures of saints, and the hair and fingernails of the patient.

The enormous number of plant species used for the treatment of psychosomatic disorders indicates that the *curanderos* of Northern Peru are valued specialists who are consulted mainly for these conditions. This is all the more interesting since Western medicine has still not found efficient treatments for psychosomatic disorders. The plant species used for "magical or ritual" disorders come mostly from the high Andes, especially from the vicinity of sacred lakes, since plants from those regions are regarded as especially powerful. This links the present day curing practices directly to ancient Andean cosmology. The use of purgatives and laxatives, to literally "expel" evil spirits is also very common.

A total of 222 plant species belonging to 172 genera and 78 families were documented and identified as herbal remedies used to treat nervous system problems in Northern Peru. Most species used were Asteraceae (36 species, 16.21%), followed by Solanaceae (15 species, 6.76%) and Lamiaceae (14 species, 6.31%). The most important nervous system families are somewhat over-represented in comparison to the overall medicinal flora, while some other medicinally important families (e.g. Poaceae, Cucurbitaceae, Euphorbiaceae) are completely missing or under-represented from the nervous disorder portfolio (Bussmann et al. 2006b).

The majority of herbal preparations were prepared from the whole plant (31.56%), while the leaves (24.48%), stems (21.24%), and flowers (8.55%) were used less frequently. Whole plants and stems were more often used than characteristic for the overall medicinal preparations found in the region (Bussmann et al. 2006b). This indicates that the local healers count on a very well developed knowledge about the properties of different plant parts. In over 60% of the cases fresh plant material was used to prepare remedies, which differs slightly from the average herbal preparation mode in Northern Peru. Interestingly, only about 36% of the remedies were applied orally, while the majority was applied topically (46.65%), often as bath, and the remaining ones were used as spiritual safeguard (*seguro*). This is different from the regional average of application, and underlines the importance of spiritually oriented treatments. Over 79% of all remedies were prepared as mixtures with multiple ingredients by boiling plant material either in water or in sugarcane spirit.

Respiratory System

The WHO reports that respiratory illnesses are of high importance as a cause of death and morbidity at a global scale. WHO elaborated a Strategy for Prevention and Control of Chronic Respiratory Diseases (CRDs), (WHO 2002), and respiratory problems are a major cause for infant deaths in Peru (Weil 1978).

A total of 91 plant species belonging to 82 genera and 48 families were documented and identified as respiratory system herbal remedies in Northern Peru. Most species used were Asteraceae (15 species, 16.67%), followed by Lamiaceae and Fabaceae (8.89% and 5.56%%). Most other families contributed only one species each to the pharmacopoeia. The most important families are clearly similarly well represented in comparison to the overall medicinal flora, although some other medicinally important families (e.g. Euphorbiaceae, Lycopodiaceae, Cucurbitaceae) are completely missing from the respiratory portfolio (Bussmann et al. 2006a).

The majority of respiratory disorder herbal preparations were prepared from the leaves of plants (27.69%), while the whole plant (18.46%), flowers (13.85%) and stems (17.69%) were used less frequently [125]. This indicates that the local healers count on a very well developed knowledge about the properties of different plant parts. In almost 55% of the cases fresh plant material was used to prepare remedies, which differs little from the average herbal preparation mode in Northern Peru. About 86% of the remedies were applied orally, while the remaining ones were applied topically. Over half of all remedies were prepared as mixtures of multiple ingredients by boiling plant material either in water or in sugarcane spirit.

Respiratory disorders are so common globally, and over-the counter remedies, both allopathic and complementary, so frequently sold, that much effort has been put into the verification of traditional remedies. Almost 50% of the plants found in the respiratory pharmacopoeia of Northern Peru, or their congeners have been studied for their medicinal properties. The original hypothesis that many species employed for respiratory illnesses would be non-native, introduced to treat diseases that were originally also introduced by colonialists, did not hold however. Quite contrarily, many remedies for respiratory ailments are native to the study area. From this perspective it is surprising to see how many species have actually been studied at least preliminarily.

Urinary System (Kidneys, Bladder)

The recent WHO report on urinary tract infections (UTI) indicates that UTIs are one of the most common bacterial infections seen, in particular in children. It has been estimated that UTI are diagnosed in 1% of boys and 3-8% of girls. In the first year of life UTI is more prevalent in boys with rates of 2.7% compared with 0.7% in girls. The reported rate of recurrent UTI is around 12-30% with risk greater in infants < 6 months, severe vesico-ureteric reflux and abnormal nuclear renal scans at time of first infection (WHO 2005). Studies have shown a higher UTI prevalence of 8-35% in malnourished children, with the risk of bacteriuria increasing significantly with the severity of malnutrition (WHO 2005).

A total of 69 plant species belonging to 61 genera and 43 families were documented and identified as herbal remedies for kidney- and urinary tract problems in Northern Peru. Most species used were Asteraceae (8species, 11.43%), followed by Fabaceae and Poaceae (both 5 species, 7.14%). All other families mostly contributed only one species each to the pharmacopoeia. The most important families are represented similarly as in the overall medicinal flora, while some other medicinally important families (e.g. Lycopodiaceae, Cucurbitaceae) are completely missing from the kidney portfolio (Bussmann et al. 2006b)..

The majority of kidney herbal preparations were prepared from the whole plant (27.78%), while the leaves of plants (25.56%), flowers (12.22%) and stems (16.67%) were used less frequently (Bussmann et al. 2006b). This indicates that the local healers count on a very well developed knowledge about the properties of different plant parts. In almost 64% of the cases fresh plant material was used to prepare remedies, which differs little from the average herbal preparation mode in Northern Peru. About 88% of the remedies were applied orally, while the remaining ones were applied topically. Over half of all remedies were prepared as mixtures of multiple ingredients by boiling plant material either in water or in sugarcane spirit.

Kidney and urinary system problems are very common globally, but allopathic treatments, in particular with regard to renal calculi, are mostly focused on dilation of the ureter, and pain management. Although a large number of plants are used in traditional medicine to treat this problem, less than 35% of the plants found in Peru or their congeners have been studied at all for their medicinal properties. Kidney and urinary tract diseases are a major health challenge worldwide. Many plant species are traditionally used for kidney disease treatment, and some have been investigated for their efficacy with positive results. An often-limiting factor to these investigations is lack of comprehensive ethnobotanical data to help choose plant candidates for potency/efficacy tests. Since the plant parts utilized in preparation of kidney remedies are reported in this survey, it serves as an indication of species that may need further ecological assessment on their regeneration status.

Rheumatic Problems

The National Institutes of Health (NIH) reports that an estimated 23.5 million Americans suffer from autoimmune diseases and that this number is expected to grow. Medical research has currently identified 80-100 autoimmune diseases, and 40 additional diseases are suspected to have an autoimmune basis. Autoimmune diseases collectively rank in the top ten leading causes of death for women aged from adolescents up to age 64. In Western medicine, the most common treatments are immuno-suppressants, which are known to have devastating long-term side effects (Gillett et al. 2000).

The housing conditions already described, as well as difficult working conditions, lead to a wide spectrum of muscular-skeletal disorders, including rheumatism, arthritis, and bone- and muscle-pain. A total of 55 plant species belonging to 53 genera and 43 families were documented and identified as autoimmune herbal remedies in Northern Peru. Most species used were Fabaceae (4 species, 7.27%), followed by Rosaceae and Myrtaceae (both 3 species, 5.45%). All other families contributed only one or two species each to the. The most important families are clearly over-represented in comparison to the overall medicinal flora, while some other medicinally important families (e.g. Asteraceae, Lamiaceae, Euphorbiaceae, Apiaceae, Lycopodiaceae, Cucurbitaceae) are less commonly used for the treatment of autoimmune disorders and pain or are completely missing from the portfolio (Bussmann et al. 2006b).

The majority of the herbal preparations were prepared from the leaves of plants (35%), while the whole plant (21.25%), and stems (17.5%) were used less frequently [126]. This indicates that the local healers count on a very well developed knowledge about the properties of different plant parts. In 60% of the cases fresh plant material was used to prepare remedies, which differs little from the average herbal preparation mode in Northern Peru. Only about 55% of the remedies were applied orally, while the remaining ones were applied topically. This is little different from the regional average of application. Over half of all remedies were prepared as mixtures of multiple ingredients by boiling plant material either in water or in sugarcane spirit. Very little western scientific evidence exists to prove the efficacy of the species employed as remedies in Northern Peru to treat autoimmune problems. Less than a pitiful 22% of the plants found or their congeners have been studied at all for their medicinal properties.

Internal Organs (Liver, Gallbladder)

Disorders of internal organs fall far behind the most commonly treated medical conditions. (Bussmann et al. 2006b). This is an indication that *curanderos* in Northern Peru are to a large extent specializing in the treatment of psychosomatic disorders, and that "bodily" illnesses are treated more as a sideline. However, a large number of plant species was used by local healers to treat liver and gallbladder ailments.

A total of 51 plant species belonging to 43 genera and 31 families were documented and identified as liver and gallbladder herbal remedies in Northern Peru. Most species used were Asteraceae (9 species, 17.66%), followed by Euphorbiaceae (4 species, 7.85%) and Gentianaceae (3 species, 5.89%). All other families contributed only one or two species each to the pharmacopoeia. Asteraceae are clearly over-represented in comparison to the overall medicinal flora, while some other medicinally important families (e.g. Solanaceae, Lycopodiaceae, Cucurbitaceae, Rosaceae) are completely missing from the liver ailment portfolio (Bussmann et al. 2006b).

The majority of herbal preparations employed for liver ailments were prepared from the whole plants (35.38%), while the leaves (24.61%), flowers (9.23%) and stems (12.32%) were used less frequently. Whole plants were more often used than characteristic for the overall medicinal preparations found in the region, while stems of plants were employed much less frequently (Bussmann et al. 2006b). This indicates that the local healers have a less well-developed knowledge about the constituents of individual plant parts in the case of liver and gallbladder treatments than for other applications. In almost 65% of the cases fresh plant material was used to prepare remedies, which differs little from the average herbal preparation mode in Northern Peru. Most of the remedies were applied orally (over 90%), while the remaining ones were applied topically. This is highly different from the regional average of application. Over 71% of all remedies were prepared as mixtures with multiple ingredients by boiling plant material either in water or in sugarcane spirit. This indicates that the local healers have a very profound knowledge about the synergistic effects of plants in multi-ingredient preparations.

Almost no scientific evidence exists to date to prove the efficacy of the species employed as liver and gallbladder remedies in Northern Peru. Only 8% of the plants found or related species in the same genus have been studied at all.

Diarrhea, Stomach Problems and Other Intestinal Ailments

Foodborne diseases are a serious public health problem worldwide. Some foodborne diseases are well recognized, but have recently become more common. Outbreaks of salmonellosis have been reported for decades, but within the past 25 years the disease has increased in incidence on many continents. While cholera has devastated much of Asia and Africa for years, its introduction for the first time in almost a century on the South American continent in 1991 makes it another example of an infectious disease that is both well-recognized and emerging. While cholera is often waterborne, many foods also transmit infection. Infection with *Escherichia coli* serotype O157:H7 (*E. coli*) was first described in 1982. Subsequently, it has emerged rapidly as a major cause of bloody diarrhea and acute renal failure. Outbreaks of infection, generally associated with beef, have been reported in Australia, Canada, Japan, United States, in various European countries, and in southern Africa (WHO 2002).

A total of 75 plant species belonging to 62 genera and 39 families were documented and identified as herbal remedies for intestinal ailments in Northern Peru. Most species used were Lamiaceae (13.33%), followed by Asteraceae and Rutaceae (both 5 species, 6.67%). Most other families contributed only one species each to the pharmacopoeia. The most important anti-infectious families are clearly over-represented in comparison to the overall medicinal flora, while some other medicinally important families (e.g. Asteraceae), are much less important (Bussmann et al. 2006b).

The majority of anti-infectious herbal preparations were prepared from the leaves of plants (29.25%), the whole plant (22.64%), and stems (16.04%). This indicates that the local healers count on a very well developed knowledge about the properties of different plant parts. In almost 60% of the cases fresh plant material was used to prepare remedies, which differs little from the average herbal preparation mode in Northern Peru. Interestingly, only about 83% of the remedies were applied orally, while the remaining ones were applied topically. Over half of all remedies were prepared as mixtures of multiple ingredients by boiling plant material either in water or in sugarcane spirit.

Large parts of the species used for intestinal disorders in Northern Peru are introductions from other parts of the world, especially Europe. Many of these are well known, and almost 50% of the plants found in this study have shown efficacy in scientific studies.

Reproductive Problems and Female Health

According to 1999 WHO estimates reproductive problems, including, 340 million new cases of curable Sexually Transmitted Diseases (STIs; syphilis, gonorrhoea, chlamydia and trichomoniasis) occur annually throughout the world in adults aged 15-49 years. In developing countries, STIs and their complications rank in the top five disease categories for which adults seek health care. Infection with STIs can lead to acute symptoms, chronic infection and serious delayed consequences such as infertility, ectopic pregnancy, cervical cancer and the untimely death of infants and adults (WHO 2007).

A total of 105 plant species belonging to 91 genera and 62 families were documented and identified as herbal remedies for reproductive problems in Northern Peru. Most species used were Asteraceae (9.52%), followed by Lamiaceae and Fabaceae (8.57% and 6.67%). Other families were less important, and 44 contributed only one species each to the pharmacopoeia. The most important families are clearly represented very similarly to their overall importance in the local pharmacopoeia (Bussmann et al. 2006b).

The majority of herbal preparations for reproductive issues were prepared from the leaves of plants (22.72%), the whole plant (21.97%), and stems (21.21%), while other plant parts were used much less frequently. This indicates that the local healers count on a very well developed knowledge about the properties of different plant parts. In almost 62% of the cases fresh plant material was used to prepare remedies, which differs little from the average herbal preparation mode in Northern Peru. Over 70% of the remedies were applied orally, while the remaining ones were applied topically. Many remedies were prepared as mixtures of multiple ingredients by boiling plant material either in water or in sugarcane spirit.

Little scientific evidence exists to prove the efficacy of the species employed as reproductive disorder remedies in Northern Peru. Only 34% of the plants found or their congeners have been studied at all for their medicinal properties.

Heart and Circulatory System

Cardiovascular diseases are collectively the number one cause of death on the globe, accounting for over 30% of all deaths worldwide, 80% of which occur in lower income countries with often little western healthcare available. Lower income groups have generally a higher prevalence of risk factors (WHO 2009). Traditional Medicine is used globally and has rapidly growing economic importance. In developing countries, Traditional healers are frequently consulted to treat heart problems and disorders of the circulatory system. The healers encountered used a wide variety of terms relating to heart problems, that in part generalized the condition (e.g. "heart disease"), included references to conditions as underlying cause of heart problems (e.g. "cholesterol"), or simply used terms to indicated treatment options (e.g. "blood irrigation" as term referring to "thin" a patients blood, "blood purification," or "refreshing the heart" as terms indicating a process cleansing the blood from suspected toxins, or "blood circulation," indicating a treatment that would improve circulation). The use of western style bio-medicinal terms is not surprising, given that all informants were of Mestizo origin, and lived in an urban environment.

Most treatments of the circulatory system involved the purification of the blood in order to improve the general condition of the patient. In addition, the fashionable concept of "weight management" and conditions related to obesity has entered into the domain of Peruvian healers. All healers readily acknowledge the negative influence of high cholesterol levels, and plant remedies were used specifically to lower cholesterol as well as weight loss therapies, while plants used for weight gain were insignificant.

A total of 60 plant species belonging to 52 genera and 33 families were documented and identified as heart herbal remedies in Northern Peru. Most species used were Asteraceae (7 species, 11.67%), followed by Lamiaceae (6 species, 10%), and Solanaceae (4 species, 6.67%). Fabaceae, Amaranthaceae and Cucurbitaceae each contributed 3 species (5%) to the heart pharmacopoeia. All other twenty-seven families contributed only one or two species each to the pharmacopoeia. Asteraceae are in general under-represented as heart remedies in comparison to the medicinal flora used in Northern Peru; Lamiaceae and Euphorbiaceae are clearly over-represented in comparison to the overall medicinal flora, while some other medicinally important families (e.g. Poaceae, Lycopodiaceae, Rosaceae) are completely missing from the heart portfolio (Bussmann et al. 2006b).

The majority of heart remedies were prepared from whole plants (37.18%), while the leaves (24.36%), stems (15.38%), and flowers (7.69%) were used less frequently. Whole plants were more often used than characteristic for the overall medicinal preparations found in the region (Bussmann et al. 2006b). In almost 70% of the cases fresh plant material was used to prepare remedies, which differs little from the main herbal preparation mode in Northern Peru. Over 90% of the remedies were applied orally, while the remaining ones were applied topically. This is very different from the regional average of application. Over 65% of all remedies were prepared as mixtures with multiple ingredients by boiling plant material either in water or in sugarcane spirit. This indicates that the local healers have a very profound knowledge about the synergistic effects of plants in multi-ingredient preparations.

Little scientific evidence exists to date to prove the efficacy of the species employed as heart remedies in Northern Peru. Only 33% of the plants found or related species in the same genus have been studied at all.

Inflammation and Bacterial Infections

Bacterial infections and inflammation are among the ailments responsible for a large number of deaths worldwide, and are often treated by traditional healers (Bussmann et al. 2006a; WHO 2009).

A total of 96 plant species belonging to 84 genera and 46 families were documented and identified as anti-infective herbal remedies in Northern Peru. Twenty percent of the species were introductions, while 80% belonged to the native flora of Peru. Most species used belong to Asteraceae (18.95%), followed by Fabaceae and Euphorbiaceae (7.37% and 5.26%). Most other families contributed only one species each to the pharmacopoeia. The most important anti-infectious families were over-represented in comparison to the overall medicinal flora, while some other medicinally important families (e.g. Lycopodiaceae, Cucurbitaceae) are completely missing from the anti-infective portfolio.

The majority of herbal preparations were prepared from the leaves of plants (31.34%), while the whole plant (18.66%), flowers (12.69%) and stems (17.16%) were used less frequently. In almost 67% of the cases fresh plant material was used to prepare remedies. Only about 55% of the remedies were applied orally, while the remaining ones were applied topically. Over half of all remedies were prepared as mixtures of multiple ingredients by boiling plant material either in water or in sugarcane spirit. Infections, in particular by strains of *Staphylococcus aureus* are very common, and increasingly difficult to treat, due to widespread formation of drug resistance. Fungal infections, due to the structure of the organisms involved, have always been a hard task to treat. Given the high importance of infections, it is not surprising that anti-infective agents are high on the list for drug development, and a large number of species used traditionally, have undergone screening. Almost 43% of the plants used in Northern Peru to treat infections, or their congeners have been studied for their medicinal properties.

Malaria and Fever

Malaria is still a major global public health problem in most tropical countries. It is thought that malaria is by far the most serious tropical disease causing one to two million deaths per year, and it plays a major role in the high mortality seen in infants and children (El Kamali et al. 1997; Milliken 1997). It is also responsible for miscarriages, premature deliveries, growth retardation, low birth weight and anemia (Connally 1996; Gbile 1984; Minakawa et al. 2002; Hay et al. 2003).

The World Health Organization (WHO) has estimated that about 2 billion people in over 100 countries are exposed to malaria, with 247 million cases in 2006 alone, and half of the world's population is potentially exposed to the disease (WHO 2009). The worsening global economic situation makes it difficult to expand modern health services, hence effective low-cost delivery medical system is urgently needed (El Kamali et al. 1997).

This is even more pressing because the use and misuse of over the counter anti-malaria remedies like chloroquine to prevent and treat *falciparium* malaria has led to widespread appearance of resistant parasites (Milliken 1997). This is complicated by the fact that global warming may lead to expansion of areas in which the ambient temperature and climatic conditions are suitable for *Plasmodium* transmission. Climatic variability has been associated with some of the recent epidemics (Minakawa et al. 2002).

A total of 17 plant species belonging to 17 genera and 13 families were documented and identified as anti-malarial herbal remedies in Northern Peru. Most species used were Asteraceae (3 species, 17.66%), followed by Fabaceae and Solanaceae (both 2 species, 11.77%). All other families contributed only one species each to the pharmacopoeia. The most important anti-malarial families are clearly over-represented in comparison to the overall medicinal flora, while some other medicinally important families (e.g. Lamiaceae, Euphorbiaceae, Poaceae, Apiaceae) are completely missing from the anti-malarial portfolio (Bussmann et al. 2006b). In the context of the questionnaires healers and venders often referred to "Fever" when talking about malaria. Fever however included a variety of conditions, from fevers accompanying flu to fever as a result of malaria. Malaria was recognized as a parasitic infection, and treated accordingly, while other plant species were used to treat fever as a symptom, mainly focusing on lowering body temperature.

The majority of anti-malarial herbal preparations were prepared from the leaves of plants (38.46%), while the whole plant (26.92%), flowers (15.38%) and stems (11.54%) were used less frequently. Leaves and stems were used more often for malaria treatments than would have been expected in comparison to the overall medicinal preparations found in the region, while seeds of plants were employed much less frequently and other plant parts not at all (Bussmann et al. 2006b). This indicates that the local healers count on a very well developed knowledge about the properties of different plant parts. In almost 70% of the cases fresh plant material was used to prepare remedies, which differs little from the average herbal preparation mode in Northern Peru. Interestingly, only about 55% of the remedies were applied orally, while the remaining ones were applied topically. This is little different from the regional average of application. Over half of all remedies were prepared as mixtures of multiple ingredients by boiling plant material either in water or in sugarcane spirit.

The very limited number of plants employed at the Peruvian coast to treat malaria and fevers might on a first glance surprise, if compared to studies from other regions of the country (Kvist et al. 2006; Roumy et al. 2007). However, malaria has always been of relatively minor importance in the coastal desert areas, and thus it is not surprising that few remedies are employed. There are indications that health practices are in the process of changing, and traditional healers start to treat a patient with prepared western remedies (e.g. Aspirin, Primaquin, Malaraquin or Lariam) although plant preparations are still important (Bussmann et al. 2006a; 2007c; 2009a).

Little scientific evidence exists to prove the efficacy of the species employed as malaria remedies in Northern Peru. Only 41% of the plants found or their congeners have been studied at all for their medicinal properties.

Cancer and Tumors

Fourty-seven plant species belonging to 42 genera and 30 families were used by curanderos in Northern Peru to treat cancerous conditions and diabetes symptoms. Most species used were Asteraceae (9 species, 19.15%), followed by Gentianaceae (3 species, 6.37%), and 7 familes with 2 species each (4.25%). All other families contributed only one species each to the pharmacopoeia. Asteraceae as the most important anti-cancer and anti-diabetic family is clearly over-represented in comparison to the overall medicinal flora, while most other medicinally important families are either under-represented or completely missing from the portfolio (Bussmann et al. 2006b).

The majority of anti-cancer and anti-diabetic herbal preparations were prepared from the leaves of plants (30.77%), while the whole plant (20%), stems (20%) and flowers (6.15%) were used less frequently. Leaves and stems were more often used than characteristic for the overall medicinal preparations found in the region, while whole plants were employed less frequently (Bussmann et al. 2006b). This indicates that the local healers count on a very well developed knowledge about the properties of different plant parts. In almost 60% of the cases fresh plant material was used to prepare remedies, which differs little from the average herbal preparation mode in Northern Peru. Over 90% of the remedies were applied orally, while the remaining ones were applied topically. This is significantly different from the regional average of application. More than 50% of the remedies included multiple plants.

Little scientific evidence exists to date to prove the efficacy of the species employed as anti-cancer and anti-diabetic remedies in Northern Peru. Only 38.71% of the plants found as diabetes treatments and 17.65% employed as anti-cancer remedies or related species in the same genus have been studied at all.

Parts of Medicinal Plants Used and Mode of Application

Northern Peruvian *curanderos* prefer to use either the leaves (in 25% of all uses) or the whole plant (24%) for the preparation of their remedies. In 19% of the cases the stems of the plants were used, most commonly together with the leaves. Flowers (10%), seeds (7%), fruits and roots (4% each), bark (3%), fruit peel (2%), and latex and wood (1% each) were only used for a small number of preparations.

Almost two-thirds (64%) of the remedies employed in Northern Peru are prepared using fresh plant material. Many of the introduced species are cultivated in fields and gardens, but the majority of the indigenous species are collected wild. This indicates that a widespread system of plant collectors is needed to supply the fresh plant material needed in Traditional Medicine. Most healers agreed, however, that in most cases dried material could be used if fresh plants were not available. In 36% of all cases the remedies were prepared using specifically dried plant material. Fresh material was not used in these situations (Bussmann et al. 2006b).

Healers in Northern Peru often employ very sophisticated mixtures of a variety of plants in their treatments. The use of single species for treatments was rare. Most commonly, plant material was boiled in water, or in some cases in sugarcane alcohol (*aguardiente*) to extract the active compounds. In some cases, plant material was macerated in cane alcohol or wine for longer periods of time before use.

The *curanderos* all had strikingly exact recipes for treatment, with very specific quantities of plant material used to prepare remedies. These quantities did not differ greatly from one healer to another. Also, the amount of a specific remedy that was given to a patient was very similar among the different *curanderos*.

The most frequent way to administer remedies was to prepare a decoction and ingest it orally (52% of all uses) followed by application as a poultice (38%, plant crushed and/or boiled and applied). Seven percent of all plant uses entailed the preparation of a *seguro*, a bottle or small flask filled with plant material along with various perfumes. This amulet has to be carried by the patient at all times, or it is placed in the house and used for periodic blessings. *Seguros* contained anything from a handful to more than three-dozen different ingredients. In two percent of the plant uses the material was employed to fabricate charms, and in one percent of all applications the plant material was burned as incense, with the smoke inhaled for treatment.

Many traditional healers rely on herbal preparations, often consisting of complex ingredients and with very specific preparations, to treat their patients' illnesses, rather than just employing single plant extracts. However, studies documenting these preparations and analyzing the composition of the mixtures are almost non-existent. Most ethnobotanical studies to date document the "use" of single species, without asking the important question if the plants in question are really employed alone, or if they are in fact part of a more complex preparation. Cano et al. (2004) were amongst the first authors to respond to this challenge, and reported on plant mixtures employed in Cuba and the Middle East, and Vandebroek et al. (2010) demonstrated the great complexity of plant preparations in the Dominican Republic. No information however was available for the very species rich Andean pharmacopoeia.

The present publication attempts to give a detailed overview on the herbal mixtures employed by traditional practitioners in Northern Peru and the specific applications they are used for, in order to provide a baseline for more in-depth studies on efficacy and safety of these preparations, as well as the possible applications in the public health system.

The investigation of plant mixtures used in traditional medicine in Northern Peru yielded a total of 974 herbal preparations used to treat 164 different afflictions (Bussmann et al. 2010a). The classification of diseases followed the curandero's terminology. To allow a better overview the different disease concepts were grouped in more inclusive disease categories, according to their similarity. Psychosomatic disorders were the most outstanding afflictions treated with traditional herbal mixtures, with almost 30% of all recipes applied, followed by respiratory illnesses, female issues, kidney problems and heart problems. *Susto* (fright), problems of the nervous system, general systemic inflammation and bronchitis together accounted for almost 25% of all remedies used. In many cases, healers used only one or two common mixtures to treat an illness. This degree of consensus between different healers shows great sophistication in the diagnosis and treatment of specific disorders. On the contrary, when it came to the treatment of unspecific disease categories like "inflammation" or "bronchitis", every healer seemed to use her/his own specific mixture to treat the problem. This was particularly obvious in the treatment of neurological and psychosomatic problems, for which the majority of plants and mixtures was employed. Up to 49 different preparations were used to treat the same disease. This seems to indicate a high degree of experimentation that is still ongoing in order to find a working cure for unspecific symptoms, and that there is very little consent amongst the individual healers as which cure to employ. This low consensus, especially where spiritual and nervous system / psychosomatic aspects are involved, might also indicate that the individual healers are reluctant to exchange knowledge about their dedicated, specific and guarded treatment methodology in these areas, while the knowledge about "simple" treatments is much more widespread..

Altogether 330 plant species, representing almost 65% of the medicinal flora used in the region (Bussmann et al. 2010a) were applied in mixtures. Of these, 64 species (19.39%) were introductions, which falls within the range of introduced species as percentage of the whole medicinally used flora. Amongst the plants employed, Asteraceae expectedly stood out, and the number of species of this family used was comparable to the percentage of Asteraceae in the medicinal flora of the region (Bussmann & Sharon 2006). The overwhelming number of plant mixtures contained 2-7 different plant species, although in the most extreme case 27 distinct species were included. A large number of species appeared in various mixtures. The plant species for each mixture are listed in the order given by the *curanderos* in order to express the importance of the individual species, rather than providing an alphabetical listing. For a detailed overview on quantities and parts of each plant use see (Bussmann et al. 2006b).

The cluster analysis confirmed that mixtures used for applications like inflammations, infections and blood purification, as well as cough, cold, bronchitis or other respiratory disorders, or urinary infection and kidney problems had similar floristic compositions. However, a few interesting clusters stood out: Mixtures used for nervous system disorders, anxiety and heart problems often had a similar composition for example, as did mixtures for prostate and bladder problems; kidney problems, gallbladder disorders, diabetes and cholesterol were treated with the same preparations; as were rheumatic illnesses and asthma. Our research suggests that this indicates that the local healers have a very detailed understanding of disease concepts, and are choosing their remedies very carefully, based on what underlying cause they diagnose, i.e. heart problems get treated differently if they are caused by stress, versus a physical agent, kidney infections are treated differently from kidney problems linked to diabetes and/or obesity.

The floristic composition as well as the complex phytochemistry of traditional herbal mixtures remain woefully understudied. This is the more surprising as traditional one-plant one single-compound based drug discovery efforts have yielded very little results in the last decades, and might in fact be an explanation as to why so many plant species that have been documented for a certain use, are "inefficient" or "toxic" when introduced to clinical trials.

Our research indicates that a large number of plants used in traditional healing in Northern Peru are employed in often-sophisticated mixtures, rather than as individual plants. Peruvian *curanderos* appear to employ very specific guidelines in the preparation of these cocktails, and seem to have a clear understanding of disease concepts when they diagnose a patient, which in turn leads them to often apply specific mixtures for specific conditions. There seems to be a widespread exchange of knowledge about mixtures for treatment of bodily diseases, while mixtures for spiritual, nervous system and psychosomatic disorders appear to be more closely guarded by the individual healers.

Traditional herbal mixtures, with their wealth of compound fragments and new compounds originating in the preparation process, could well yield new clues to the treatment of a wide variety of disease. The present paper provides detailed baseline information on composition and use of traditional mixtures in Northern Peru, and further studies to compare the compound composition of these preparations versus single plant extracts, as well as investigations comparing efficacy and toxicity of herbal preparations versus their single plant ingredients are in progress.

Does Traditional Medicine Work? A Look at Antibacterials Used in Northern Peru

Plants with potential medicinal activity have recently come to the attention of Western scientists, and studies have reported that some are bioactive (Perumal Samy et al. 2000). Potentially active compounds have been isolated from a few of the plants tested (D'Agostino et al. 1995 a,b; Okuyama et al. 1994, Rodriguez et al. 1994).

In order to evaluate the antibacterial activity of species used in TM in Northern Peru, 525 plant samples of at least 405 species were tested in simple agar-bioassays for antibacterial activity against *Staphylococcus aureus, Escherichia coli, Salmonella enterica Typhi* and *Pseudomonas aeruginosa.* A much larger number of ethanolic plant extracts showed any antibacterial activity compared to water extracts for all antibacterial activity. One-hundred-ninety-three ethanolic extracts and 31 water extracts were active against *S. aureus.* In twenty-one cases only the water extract showed activity (for all bacterial species) compared to ethanol only. None of the aqueous extracts were active against the other three bacteria, with the activity of the ethanolic extracts also much reduced, as only 36 showed any activity against *E. coli*, and 3 each against *S. enterica Typhi* and *P. aeruginosa.*

Eighteen ethanol extracts were effective against both *E. coli* and *S. aureus*, while in two cases the ethanol extract showed activity against *E. coli* and the water extract against *S.*

aureus. The ethanol extract of *Dioscorea trifida* was effective against *E. coli, S. aureus and P. aeruginosa. Caesalpinia spinosa* was the only species that showed high activity against all bacteria, including *Salmonella enterica Typhi* and *Pseudomonas aeruginosa*, when extracted in ethanol (Bussmann et al. 2010b).

Two hundred twenty-five extracts came from plant species that are traditionally employed against bacterial infections. One hundred sixty-six (73.8%) of these were active against at least one bacterium. Of the three hundred extracts from plants without traditional antibacterial use, only 96 (32%) showed any activity. This shows clearly that plants traditionally used as antibacterial had a much higher likelihood to be antibacterially active than plants without traditional anti-bacterial use. However, the efficacy of plants used traditionally for antibacterial related applications did vary, which underlines the need for studies aiming to clearly understand traditional disease concepts. Plants used for respiratory disorders, inflammation/ infection, wounds, diarrhea, and to prevent post-partum infections were efficacious in 70-88% of the tests. Plants used for "kidney inflammation" had a much lower efficacy against bacteria, and fell within the range of species that are traditionally used to treat other bodily disorders. Only species used for spiritual/ritual treatments scored worse. Of these only 22% showed some antibacterial; activity. However, amongst the "spiritual" plants 38% of the species used for cleansing baths did in fact show activity, while only 15% of the plants often used in protective amulets (mostly species with the families of Lycopodiaceae and Valerianaceae) showed limited antibacterial activity.

A variety of species showed higher efficacy than the control antibiotics employed, e.g. *Ambrosia peruviana, Iresine herbstii, Niphogeton dissecta, Opuntia ficus-indica, Smilax kunthii* were particular effective against *Escherichia coli. Berberis buceronis*, *Caesalpinia paipai*, *Caesalpinia spinosa*, *Cestrum strigilatum*, *Cydista aequinoctialis*, *Dioscorea trifida*, *Escallonia pendula*, *Escobedia grandiflora*, *Eucalyptus citriodora*, *Eucalyptus globulus*, *Eugenia obtusifolia*, *Eustephia coccinea*, *Gallesia integrifolia*, *Geranium sessiliflorum*, *Hedyosmum racemosum*, *Iresine herbstii*, *Lycopersicon hirsutum*, *Mauria heterophylla*, *Phyllanthus niuriri*, *Porophyllum ruderale*, *Salvia cuspidata*, *Senecio chionogeton*, *Smilax kunthii*, *Tagetes erecta* and *Taraxacum officinale* showed high activity against *Staphylococcus aureus*. The same holds true for *Ephedra americana, Gentianella bicolor*, *Mandevilla* cf. *trianae*. However, extracts of these three species were highly inconsistent in their efficacy.

The comparison of closely related species traditionally employed for different purposes (e.g. different *Alternanthera* spp., *Passiflora* spp., *Senecio* spp. and *Salvia* spp. for spiritual purposes and against bacterial infections) showed that the "spiritual" species normally were not effective against bacteria, while the species used as antibacterials had increased effectiveness. The example of *Plantago sericea* var. *sericea* (used in seguros, no efficacy) and *Plantago sericea* var. *lanuginosa* (used for vaginal infections, high efficacy against *S. aureus*) is a particularly compelling case that indicates the sophistication of traditional knowledge. However, we did find examples like *Chuquiragua* spp., where closely related species were used as antibacterials, but only one of them did in fact show efficacy, clearly indicating that in this case traditional knowledge did not produce reliable results.

On the other hand, extracts of the same species traditionally used to treat infections often produced vastly diverging results when collected from different localities. Good examples are *Iresine herbstii*, *Schinus molle*, *Eustephia coccinea*, *Oreopanax eriocephalus*, *Myroxylum balsamum*, *Spartium junceum*, or *Gentianella dianthoides*. Most of these species did not produce particularly high inhibition rates in any case, and were not the first choice of healers when trying to find remedies for bacterial infections. Many traditional remedies for concepts like "kidney inflammation" did not produce any antibacterial results, which underlines that research into efficacy does need to closely take traditional disease concepts into account.

Many remedies used for spiritual healing and other non-infection purposes did show antibacterial efficacy in vitro, but were not listed as such by the local healers. This might be explained by the fact that they either are very inconsistent in their activity (e.g. *Mandevilla trianae*, *Loricaria* spp., *Lonicera japonica*, *Hypericum laricifolium*, *Hyptis sidifolia*, *Mentha piperita*, *Brachyotum naudinii*, *Cydonia oblonga*), or are so closely related that identification, especially when dried, can be a problem, e.g. in the case of *Baccaris* spp., *Gentianella* spp., and *Valeriana* spp., or are prone to toxic side effects like *Ephedra americana* and *Brugmansia* spp.

Almost all remedies are traditionally prepared as water extracts, although ethanol (in the form of sugarcane spirit) is readily available. This might at a first glance seem astonishing, given the low efficacy of water extraction found in this study. However, initial results from Brine-Shrimp toxicity assays indicate that the ethanolic extracts are by far more toxic than water extracts of many species, and thus ethanolic extraction might in many cases not e suitable for application in patients. This again indicates the considerable sophistication and care with which traditional healers in northern Peru chose their remedies for a specific purpose.

If the botanical documentation of Peruvian medicinal plants has been neglected, investigations of the phytochemical composition of useful plants is lagging even further behind. Most studies on the phytochemistry of Peruvian plants concentrate on a few "fashionable" species that have been marketed heavily on a global scale, especially Maca (*Lepidium meyenii*), Sangre del Drago or del Grado (*Croton lechleri*), and Uña de Gato (*Uncaria tomentosa* and *Uncaria guianensis*). The number of other Peruvian plants for which at least some phytochemical studies exist is still miniscule, and most efforts are fuelled by the fads and fashions of the international herbal supplement market. Studies involving multiple species were initiated as late as the 1990s.

Minimum inhibitory concentrations foud for Peruvian plant extracts ranged from 0.008 to 256mg/ml. The very high values in many species indicate only a very limited antibacterial efficacy. The ethanolic extracts exhibited stronger activity and a much broader spectrum of action than the water extracts. The most interesting activity on *E. coli* was obtained from ethanolic extracts of *Baccaris* sp., *Ochroma pyramidale*, *Croton lechleri*, *Banisteriopsis caapii*, *Miconia salicifolia*, and *Eugenia obtusifolia*. Only the latter species also showed strong activity in the aqueous extract. A much wider range of species, including most species active against *E. coli* showed inhibition of *S. aureus*. *Poropohyllum ruderale, Senecio* sp., *Corynaeae crassa*, *Dioscorea trifida*, *Senna monilifera*, *Spartium junceum*, *Pelargonium odoratissimum*, *Satureja pulchella*, *Cuphea* sp., *Malva parviflora*, *Brosmium rufescens*, *Syzygium aromaticum*, *Sanguisorba minor*, *Citrus limetta*, *Verbesine* sp. and 2 unidentified species all showed MIC values between 1-4mg/ml. Most of them however did not portray any efficacy in aqueous extract. *Hypericum laricifolium*, *Hura crepitans*, *Caesalpinia paipai*, *Cassia fistula*, *Hyptis sidifolia*, *Salvia* sp., *Banisteriopsis caapi*, *Miconia salicifolia* and *Polygonum hydropiperoides* showed the lowest MIC values and would be interesting candidates for future research (Bussmann et al. 2010b).

Most species effective against *S. aureus* are traditionally used to treat wound infection, throat infections, serious inflammations, or are post partum infections. Interestingly many species used in cleansing baths also showed high activity against this bacterium. Many of these species are either employed topically, or in synergistic mixtures, so that possible toxicity seems not to be an issue. The species effective against *E. coli* were mostly employed in indications that traditional healers identified as "inflammation".

Most of the plants used by the healers have antibacterial activity, but only 8 of the 141 plants (5.6%) examined in this study show any MIC values of 200 or less mg/ml of extract. Of these 8 plants 5 are used to treat diseases believed to be in bacterial origin by TM, one is a disease not believed to be caused by bacteria and one is used for undefined treatment purposes.

Nine out of 141 plants (6.3%) tested that were not used for diseases believed to be bacterial in origin by TM, 5 showed high antibacterial activity with MIC values below 16 mg/ml. Four of these were among the most potent plants tested with MIC values of 2 or less mg/ml including the hallucinogen and extracts used to treat diabetes and epilepsy. Diseases such as diabetes often compromise the health of the individual and antibacterial treatments can be warranted for secondary complications of the disease.

Toxicity in Traditional Medicine

Crude medicinal activities have been investigated for a wide variety of plants, but while toxicity assays are available for a wide variety of countries no data exists on the potential toxicity of Peruvian medicinal species.

Brine shrimp *(Artemia)* are frequently used as agent in laboratory assays to determine toxicity values by estimating LC_{50} values (median lethal concentration). The Brine shrimp lethality activity of 501 aqueous and ethanolic extracts of 341 plant species belonging to 218 genera of 91 families used in Peruvian traditional medicine was tested (Bussmann et al. 2011). The aqueous extracts of 55 species showed high toxicity values (LC_{50} below 249 mg/ml), 18 species showed median toxicity (LC_{50} 250-499 mg/ml) and 18 low toxicity (LC_{50} 500-1000 mg/ml). The alcoholic extracts proved to be much more toxic: 220 species showed high toxicity values (LC_{50} below 249 mg/ml, with 37 species having toxicity levels of >1mg/ml), 43 species showed median toxicity (LC_{50} 250-499 mg/ml) and 23 species low toxicity (LC_{50} 500-1000 mg/ml). Over 24% of the aqueous extracts and 76% of the alcoholic extracts showed elevated toxicity levels to brine-shrimp. Traditional preparation methods are taking this into account – most remedies are prepared as simple water extracts, thus avoiding potential toxic effects. Excellent examples where the water extracts are non-toxic, while the ethanolic extracts show high toxicity are *Ocimum basilicum* L., *Salvia* sp., or *Laccopetalum giganteum* (Wedd.) Ulbrich. In contrast, *Cinchona officinalis* L. ethanolic extracts were non-toxic, and are traditionally used, while the highly toxic water extract has no traditional use.

Species which showed higher levels of toxicity were *Bejaria aestuans* L., *Erodium cicutarium* (L.) L'Her., *Brachyotum naudinii* Triana, *Miconia salicifolia* (Bonp. ex Naud.) Naud., *Cuscuta foetida* Kunth, *Caesalpinia spinosa* (Molina) Kuntze, and *Phyllactis rigida* (Humb. and Bonpl.) Pers. *Achillea millefolium* L., *Artemisia absinhitum* L., and *Eucalyptus globulus* Labill all frequently used as medicinal teas also fall in this group, as do *Lupinus mutabilis* Sweet, and *Ilicium verum* Hook. f. Solanaceae (e.g. *Nicotiana tabacum* L. and *Solanum americanum* Mill.) proved highly toxic, while other species, known to be highly toxic when ingested (e.g. *Datura* sp. and *Brugmansia* spp.) did not show toxicity in Brine Shrimp.

Multiple extracts from different collections of the same species showed in most cases very similar toxicity values. However, in some cases the toxicity of extracts from different collections of the same species varied from non-toxic to highly toxic. Examples for such variation in toxicity were found for *Chersodoma deltoidea* M.O. Dillon and Sagast., *Satureja sericea* (C. Presl. and Benth.) Briq., *Eugenia obtusifolia* Cambess., *Epidendrum* sp., *Capparis crotonoides* Kunth, *Sambucus peruviana* Kunth, and *Malva* sp. In case of these frequently used species, harvest time, collection locality or use of specific plant parts might be important for a reduction of toxicity.

Markets and Sustainability

The pharmacopoeae of Southern Ecuador and Northern Peru - Colonial regimes and their influence on plant use The differences in medicinal plant use between Southern Ecuador and Northern Peru are striking. Both regions share the same cultural background, and have a very similar flora, with a comparable number of plant species that to a large extent overlap. Many plants mentuioned in Martinez Compañon (1789) are still found in local markets today (Fig. 6). However, the medicinal flora of Southern Ecuador includes only 40% of the species used in Northern Peru. The differences in traditional medicinal use can be explained by comparing the development of the pharmacopoeiae of both areas from the start of the colonial period until today. Colonial chroniclers often included detailed descriptions of useful plants in their reports. The most comprehensive early accounts of the economically interesting flora of Northern Peru and Southern Ecuador were provided by Monardes (1574), Acosta (1590), and Cobo (1653, 1965). Later treatments were included in Alcedo (1776). Martínez Compañon, Archbishop of Trujillo, had a complete inventory of his dioceses prepared (1789). Finally, Ruiz and Pavón provided the first real botanical inventory of the region (1777-1788). The account of Martínez Compañon (1789) provides the best baseline for a comparison of the colonial and modern medicinal flora of the region. The work includes detailed paintings for every species, which allows a close comparison with the modern medicinal flora, indicating that the vernacular names of useful plants have not changed significantly since colonial times. It contains 526 useful plant species. A preliminary review of this work seems to indicate that the number of plants used has not changed significantly since the late 1700's, with over 500 plant species still found in modern Peruvian markets. A closer comparison shows, however, that only 41% of the species mentioned by Martinez Compañon (1789) are still sold nowadays in Peru. An additional 32% are still used in the Amazon basin, but do not reach the coastal markets anymore. Twenty-seven percent have completely disappeared from modern day use. This means that 58% of the species sold in Peruvian markets and 41% of the species used in Ecuador were added to the pharmacopoeia within the last 200 years (Fig. 7).

Figure 6. Flor de San Juan from drawing in Martinez Compañon (1789) and recent market purchase.

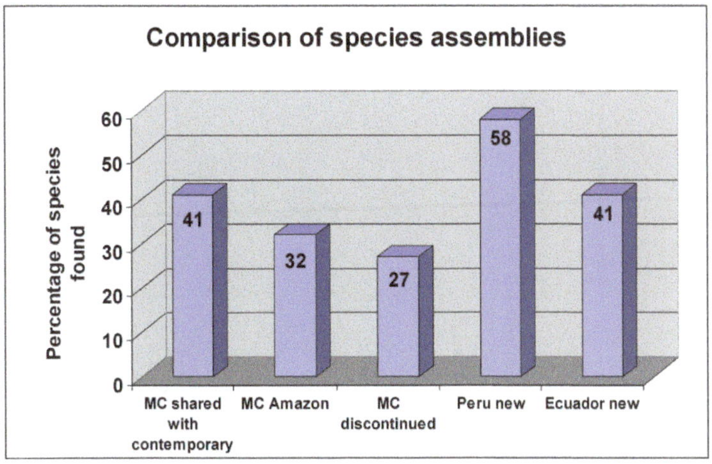

Figure 7. Differences in plant usage between Peru and Ecuador.

A cluster analysis of the colonial and modern plant inventories showed a striking explanation for the use differences between Ecuador and Peru, and helps to explain why the plant inventories changed so significantly in the 18th century: The current pharmacopoeia of useful flora in Ecuador was most similar to the early colonial flora mentioned in Monardes (1574), Acosta (1590), Cobo (1653, 1956) and Alcedo (1776). This indicates that the Ecuadorian medicinal flora did not develop much between early and late colonial times. In contrast, the modern Peruvian healing flora was much more similar to later collections (Fig. 8). An explanation for this lies in the different treatment of traditional practices in Ecuador and Peru: In Ecuador, traditional medicinal practitioners were immediately persecuted once the colonial administration took hold, while the Peruvian administration was much more tolerant. This also reflects in the establishment of a National Institute for Traditional Medicine in Peru in the 1980s, while traditional medicine was illegal in Ecuador, until a constitutional change in 1998. This meant that Ecuadorian healers had no opportunity to experiment with new species to cure diseases introduced by Europeans, while Peruvian healers were able to explore the rich flora of the region in order to find new remedies. This experimentation also extended to "magical" disease concepts like *Mal Aire, Mal Ojo, Susto,* and *Envidia* that were introduced from Spain during the colonial regime. Peruvian healers developed a vast array of medicinals to treat these conditions, which, to a large extent explains the shift in the medicinal flora between the late 1700s and modern times. Experimentation in Ecuador remained restricted to the treatment of common diseases, while spiritual treatments were outlawed until a constitutional revision in 1998 recognized the right of the population to use traditional medicinal practices (Fig. 9) (Bussmann et al. 20009e).

Figure 8. Cladogram of linkage of historic sources and current pharmacopoeiae.

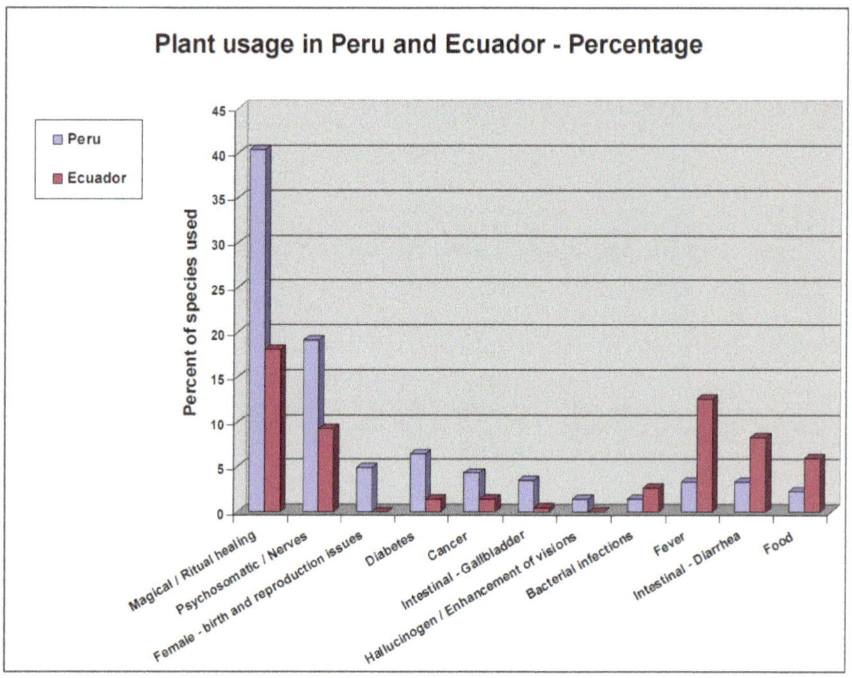

Figure 9. Plant use differences in Ecuador and Peru.

THE CASE OF ULLUCHU

Ulluchu is the common name assigned to a plant frequently depicted in the art of the Moche culture, which thrived on the north coast of Peru from A.D. 100 to 800. It is a grooved, comma-shaped fruit with an enlarged calyx found mainly in fine-line scenes painted on Moche ceramics (Fig. 10). The term first appeared without linguistic explanation in the work of pioneer Moche scholar Rafael Larco Hoyle [1939: Fig. 58; 1940: 98, Figs. 166 & 167]. In his 1939 publication, he reported that the peoples of the sierras and the coastal region (Viru and Moche valleys) believed that the fruit had to be picked silently to prevent it from turning bitter. He wondered if the plant symbolized the silence and discretion of richly attired Moche messengers, some of whom wear belts adorned with *Ulluchus*. In his 1938 publication, he labeled a Moche fine-line drawing of *Ulluchu* as *Phaseolus* sp. (a bean). Larco clearly recognized that *Ulluchu* had nothing whatsoever in common with "*ulluco*" (*Ullucus tuberosus*), and Andean tuber still widely cultivated and consumed in Peru nowadays.

Figure 10. *Ulluchu* in Archaeology. **A.** *Ulluchu* fruit from cache at Sipán. After photograph by Christopher B. Donnan. B. *Ulluchu* fruit from Dos Cabezas burial;, after photograph from Donald McClelland. C. Bone beads in form of *Ulluchus* fron Huaca de la Luna. After photograph of Donald McClelland. D. Golden *Ullucho* bead, after photograph by Donald McClelland. E. *Spondylus* shell bead in form of *Ulluchu.* After photograph by Donald McClelland. F. Supernatural figure seated holding a gourd, possibly containing ground *Ulluchu* seeds. *Ulluchus* painted on headdress. Private collection. After photograph by Christopher B. Donnan, from McClelland (McClelland 2008). G. Anthropomorphized hawk runners. *Ulluchu* on belts, *Ulluchu* seeds floating above, runner carrying snuff tube. The Art Institute of Chicago. After drawing by Donna McClelland (McClelland 2008). H. *Ulluchu* harvest. Note tree with opposite leaves, and extracted seeds on bottom right. After McClelland (2008), Private collection.

The symbolic importance of *Ulluchu* in Moche iconography was firmly established by Moche scholar Donnan McClelland (1977). Based upon a meticulous review of the UCLA Moche Archive, she showed that its distribution was non-random and that its varied usage displayed definite patterns with the greatest variability among background elements and the most frequent representation found on the belts of warriors and runners. She demonstrated that "the leaves of the *Phaseolus* do not resemble the *ulluchu* leaf depictions" (McClelland 1977*:* 43) *Pepino (Solanum muricatum)* and *aji (Capsicum annum),* which are clearly depicted in Moche art and do not resemble *Ulluchu* were also eliminated "since the *ulluchi* [*sic*] fruit is suspended from the plant by its smaller pointed end, whereas these two are suspended by the large end" (McClelland 1977*:* 437). She also indicated that the plant had not been botanically identified, pointing out that, if it turned out to be a mythical plant, no identification would be possible.

A decade after McClelland's seminal article, S. Henry Wassen (1987) of the Gothenburg Ethnographical Museum, eliminated *Persea americana* Miller var. *americana* (a wild relative of *avocado*) as a candidate, and concluded that *Ulluchu* was *Carica candicans* A. Gray (a species of wild *papaya*). He also co-authored an article describing the enzyme papain, which can be extracted from unripe papaya, for use as a blood anti-coagulant (Hulten et al. 1987). In the latter article, the authors proposed that papain was used in the Moche sacrifice ceremony to prevent the coagulation of blood drawn from sacrificed warriors for later consumption by priests.

In a paper presented at the Sibley Conference at the University of Texas at Austin, in November 2003, McClelland (2008), in addition to updating her 1977 paper in the light of a vastly expanded Moche Archive and archaeological discoveries of real *Ulluchu*, refuted the papaya hypothesis. She also discussed the presence in the art of yellow oleander seeds (*Thevetia peruviana*) as legging rattles as well as *espingo* seeds (*Nectandra* sp.) which (wassen 1976) had earlier suggested might have been added to corn beer for medicinal and psychotropic purposes. McClelland concluded that the largest remaining challenge was an identification of *Ulluchu.*

Issues Surrounding the Name Ulluchu

The name *Ulluchu* seems to have been coined by Larco (1939). According to his description, the name originated in the Virú River valley, and is supposedly of Mochica origin. However, there is no linguistic evidence that such a term indeed existed in the Mochica or Yunga language. The most comprehensive Mochica-Spanish dictionary available, compiled from the writings of Moche scholar E. Brüning (2004), has no such term. In addition, the local population, as well as market vendors, plant collectors, and *curanderos* interviewed, had no knowledge of *Ulluchu* whatsoever, other than what they derived from Larco. Since this first publication the term has been copied by all subsequent authors (McClelland 1977, Wassen 1987), without any regard to its validity. It is unlikely that Brüning (2004) would have missed the name when doing his research early in the 20th century, if it indeed was still being. Brüning lists quite a few Mochica plant names, some of which are still used for the same plants today, e.g. "*faik*" = *Acacia macracantha* (*faique, espino*), from "*fáçek, fáçke*" = spine.

The only language that has a somewhat similar word from which *Ulluchu* could be derived is Quechua: "*uchu*" translates to "chili, pepper," while "*ullu*" translates as "penis." The term "*ullu uchu*" is sometimes used as a name for *Columellia ovata* R. & P. (Columelliaceae), a small high-Andean plant, described as "a very thick tree; its wood is suitable for various purposes, and its leaves have febrifugal properties and are very bitter" (Ruiz 1777). However, this plant has no likeness whatsoever to the Moche *Ulluchu*. Thus we must conclude that the term *Ulluchu* was most likely coined by Larco (1940: 98) based on a Quechua term for a species with somewhat similar fruits that has no relation to the species used by the Moche.

Botanical Identification: Why is Ulluchu not Carica Candicans?

Carica candicans is a wild relative of papaya (*Carica papaya* L.). Although the fruits are not marketed, they are occasionally consumed by the local population, and some market vendors and healers interviewed did know the plant under its vernacular name, "*mito.*" Larco (1939; 1940) never mentions the plant in relation to *Ulluchu*. Assuming that he indeed encountered a plant with that name, it cannot have been *C. candicans*, because this species would have been named "*mito.*"

McClelland (2004) argues that *Ulluchu* "cannot be a papaya, which belongs to a group of plants called 'cauliflory' [i.e., stem flowery]. . . . The flowers and fruit of a cauliflory grow on the trunk of the tree and not on the limbs . . . . *Ulluchus* depicted in Moche art, however, hang from limbs. Papaya leaves do not resemble *ullucho* leaves, which are triangular, ovoid, or boomerang shapes hanging from limbs. Each large palmate papaya leaf grows on a stem from the top of the tree.*"* However, further complicating this matter, it turns out that *C. candicans* happens to be one of the few papayas that are not cauliflorous, that have triangular leaves with entire margins, and the fruits do hang from branches. Thus, judging from the iconography alone, *C. candicans* actually could be *Ulluchu*.

Based on recent archaeological evidence however (Fig. 11), (McClelland 2008; Alva et al. 1993; Alva 1994; Donnan et al. 1994), it has become clear that the actual fruits found in burials do not resemble *C. candicans*. In addition, the explanation that papain, might have been extracted by the Moche from unripe papaya for use as a blood anti-coagulant (Hultin ert al. 1987), albeit reasonable, does not make much sense from a phytochemical perspective. Cultivated papaya (*C. papaya*) is often depicted in Moche pottery, and the species contains large amounts of papain. Why would the Moche have resorted to a rare wild species, when they could have used a cultivar with the same properties that grew on their very doorstep? Also, *C. candicans* fruits are often 10-15 cm. long, and, while this would relate to the size of some of the *Ulluchus* in the iconography, it is vastly larger than many of the fruits depicted and much larger than the fruits found in burials. Finally, the anatomy of *C. candicans* simply does not correlate with the fruits encountered in burials.

Figure 11. Front view of crushed *Ulluchu*. Photo Don McClelland.

What is Ulluchu, and what was it really used for?

Moche fine-line drawings of *Ulluchu* normally depict seed pods or seeds floating in the air in sacrificial scenes (McClelland 2008: Fig. 3.14), associated with runners and messengers e.g., McClelland 2008: Fig. 3.1) or intoxicated priests (e.g., McClelland 2008: Fig. 3.6). The *Ulluchu* fruits vary greatly in size, ranging from about 1-15 cm. They are normally comma-shaped, often with an "exaggerated round calyx" [6: 43] with lines on the body of the fruit (e.g. McClelland 2008: Fig. 3.4). Some illustrations show *Ulluchu* harvested by monkeys, and in such cases the fruit is mostly shown growing along the axes of the plant's leaves (e.g., McClelland 2008: Figs. 3.27 & 3.28).

Starting from this basis in 2002, we built on the work of Donna McClelland and the archaeological excavations at Sipán in the Lambayeque Valley (McClelland 2008; Alva 1994; Alva et al. 1994) and at Dos Cabezas in the Jequetepeque Valley in the 1990s (McClelland 2008; Donnan 1994). Botanically, all these depictions resemble capsules or drupe-like fruits. It became apparent that in a biodiversity hotspot like Peru, with a flora of more than 18.000 species, a large number of plant families have fruits that vaguely resemble Moche fineline drawings of *Ulluchu*, and many of these families contain more than one genus with similar fruits. Examples include: (Apocynaceae: *Ambelania*; Caricaceae: *Carica*; Celastraceae: *Maytenus*; Chrysobalanaceae: *Chrysobalanus*, *Hirtella*, *Licania*; Convolvulaceae: *Dicranostyles*; Fabaceae: *Aldina*, *Alexa*, *Andira*, *Dipteryx*, *Dussia*, *Ormosia*; Guttiferae: *Tovomita*; Hippocrateaceae: *Cheiloclinum*, *Salacia*; Icacinaceae: *Calatola*; Meliaceae: *Guarea;* Menispermaceae: *Abuta*, *Curarea*, *Elephantomene*, *Telitoxicum*; Myristicaceae: *Virola*; Olacaceae: *Cathedra*; Quiinaceae: *Lacunaria*; Sabiaceae: *Meliosma*; Sapindaceae: *Cupania*, *Paullinia*; and Sapotaceae: *Pouteria*. Some of these are still highly important in traditional societies. For example, *Ambelania* fruit is often consumed; *Ormosia* contains potent poisonous compounds, but is now mostly used in crafts; *Curarea* is one of the ingredients of "*curare,*" the famous Amazonian arrow poison; *Virola* species are still used as potent snuffs in the Amazon, and *Paullinia* is the source of "*yopo,*" an important stimulant. However, none of these carry the vernacular name *Ulluchu*.

Fortunately, at this tine the archaeological evidence provides good clues for identification. The *Ulluchu* fruits found in burials in the 1990's are clearly capsules or drupes, slightly comma-shaped, between 1.5 and 5 cm. long, and slightly grooved. They closely resemble bone, gold, and *Spondylus* beads found *in situ* in size, form, and texture. In the iconography, the fruits are often depicted as located on both sides of branches, e.g. on headdresses. It is important to note that this figure has widely extended nostrils, as is often seen in people inhaling hallucinogenic snuffs, and is holding a gourd and pestle. McClelland (2008) interpreted this as lime gourd used for chewing coca. We suggest that this might also be a gourd used to grind the seeds of *Ulluchu* for inhalation. Further iconographic evidence supports this hypothesis. Runners and messengers associated with *Ulluchu* are often winged – they literally fly, i.e., the inhalation of *Ulluchu* gives them wings. The personages have *Ulluchu* on their belts, *Ulluchu* seeds floating above their heads, and they hold instruments that closely resemble a typical double snuff tube that would serve to inhale powdered hallucinogenic substances. Thus, it seems possible that one of the uses of *Ulluchu* may have been as a mind-altering snuff. Another reason for identifying the seeds in the iconography as hallucinogenic *Ulluchus* is underscored from Moche paintings, e.g a famous Moche scene (McClelland 2008: Fig. 3.34) where monkeys are picking fruits from an *Ulluchu* tree. It is important to note that the tree depicted has opposite leaves and that seeds are extracted from the fruit, possibly for roasting in a typical oven on the bottom right. The roasted seeds could then be ground to powder and inhaled. The fruits themselves seem to be 5-valved. The function of *Ulluchu* as a hallucinogen is reinforced by other imagery (e.g. McClelland 2008: Fig. 3.6), where personages, surrounded by *Ulluchu* fruits lie on the ground in what clearly appears to be an intoxicated state. In addition, prisoners in sacrificial scenes (e.g. Hocquenghem 2008, Figs. 2.2, 2.3, 2.12 & 2.24), especially the well known "lines of prisoners" at Huaca El Brujo and Huaca de la Luna all show clearly visible erections (Fig. 12), which may be seen as another indication of the ingestion of some substance causing this effect. From this perspective Larco's term *Ulluchu* (Larco 1940), if derived from Quechua "*ullu-uchu*" – "penis pepper," would in fact make sense in describing the possible effects of the plant in question. Also, the association with sexual arousal is reinforced by mythical scenes where an *Ulluchu* tree grows out of the back of an erotic couple.

Figure 12. Line of prisoners at Huace El Brujo. Photo: Rainer W. Bussmann.

In light of the above, *Ulluchu* is a tree with opposite leaves and fruits that are drupes between 1-15 cm. long possibly containing active ingredients that would elevate the blood pressure and cause erections, and psycho-active substances. The only plant family from the list above having representatives that meet all these criteria is Meliaceae, and the genus *Guarea* is the one that most closely fits the description. It includes mostly trees with pinnate leaves (which is unusual for Meliaceae), and fruits that are 3-5 valved capsules, with large, pseudo arillate seeds. The genus *Guarea* is found throughout Peru, but is mostly restricted to tropical lowland forests, with some species reaching could forest habitat. No species is found along the dry coast of Peru, which indicates that the material must have been widely traded in Moche times. A typical representative is *Guarea grandifolia* DC. (Fig. 13). The species has clearly pinnate leaves, and the fruits very clearly resemble the archaeological samples. In addition, *Guarea* contains a large number of species with varying fruit sizes (from 1-15 cm.), calyx swelling, and grooving on the body of the fruit (Fig. 14), which all correlates with the varied *Ulluchu* imagery in Moche fineline drawings. The seeds of *Guarea species, with a distinct white navel, very much resemble the seeds depicted in Moche fineline paintings.*

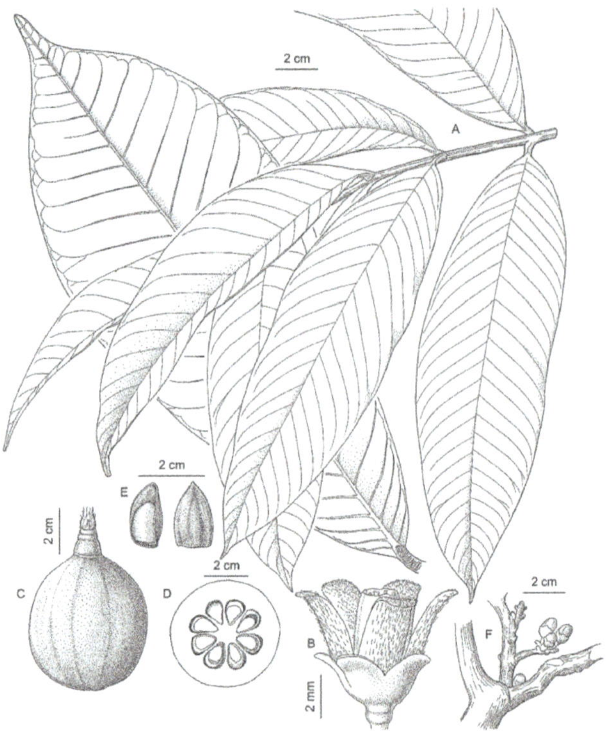

Figure 13. *Guarea grandifolia.* **A.** Mature branch; B. Flower; C. Mature fruit; D. Fruit cross section; E. Seeds; F. Branching pattern.

Figure 14. Sustainable plant use – from plant documentation with local healers to production of correctly identified bulk herbal packages and teas for local markets.

Many species of *Guarea* are known by a wide variety of vernacular names, e.g., *Guarea* spec.: requia, kushímsakish; *G. glabra:* yecheñor, yechemor; *G. grandifolia:* bola requia; *G. guidonia:* atapio, latapi, latapi caspi, requia colorada, requia latapi, xoro; *G. kunthiana:* requia, paujil ruro; *G. purusana:* latapi, requia). The wood of many species is used as timber for construction. The wood, bark, and leaves contain compounds that act as abortive, emetic, purgative, and antiamoebiac agents, and the bark is often sold as *Coccilliana* in expectorant preparations (Kraemer 1915; Rätsch 1998). The fruits and seeds contain a variety of alkaloids that are very rarely used due to their high toxicity (Kraemer 1915). Some of the alkaloids found, e.g. rusbyine, have a structure and effects like emetine, an alkaloid found in *Psychotria ipecacuanha* (Brot.) Stokes, which has been widely used as an emetic and expectorant. Other species of *Psychotria* are well known as components in *ayahuasca* preparations due to their high content of N,N-DMT [30]. In large dosages, *ipecac* preparations cause high blood pressure, arrhythmia, spasms, and extension of the blood vessels. While the existing literature on *Guarea* seed compounds is rather fragmentary, it seems clear that a concentrated dosage of *Ulluchu* seeds, if ingested, would increase a prisoner's heartbeat, elevate the blood pressure, and widen blood vessels thus causing erection. All of this would make it much easier to extract sacrificial blood. Also, when inhaled by priests, the active compounds could have a mind-altering effect, which would not necessarily lead to high levels of toxicity, and could induce very rapid, short-term hallucinations.

We conclude that *Ulluchu* can be identified as a group of species of the genus *Guarea* (Meliaceae) based on morphological characteristics. In addition, the chemical composition of the plant's compounds supports the theses that it was used in a sacrificial context to improve the extraction of blood from sacrificial victims. We also suggest that a ground preparation of *Guarea* seeds, when inhaled, may have been used as a hallucinogen. However, more detailed phytochemical research is needed to corroborate the latter hypothesis.

CHANGING MARKETS

Exotics played an important role amongst all pants sold in Norhern Peruvian markets. Fifty-nine species (15%) found in all markets were exotics. However, amongst the species most commonly encountered in the inventories, 40-50% were exotics. *Matricaria recutita* (chamomile) was found in the inventory of approximately 70% of vendors. The next most popular species sold in these markets included *Equisetum giganteum*, *Phyllanthus urinaria, Phyllanthus stipulatus*, *Phyllanthus niruri* (chanca piedra – stone breaker), *Eucalyptus globulus (eucalyptus), Piper aduncum, Uncaria tomentosa* (cat's claw), *Rosmarinus officinalis* (rosemary), *Peumus boldus, Bixa orellana* (achiote) and *Buddleja utilis*. However, when taking sales volume into account, *Croton lechleri* (dragon's blood), *Uncaria tomentosa,* and *Eucalyptus globulus* were clearly the most important species (Bussmann et al. 2009f).

While it was very easy for all vendors to name their most important and frequently sold species, it proofed impossible to get detailed information about species that vendors observed as "rare" or "disappearing". In most cases, vendors mentioned species as rare because they themselves did not sell them, in many cases these plants were very common outside the market (e.g. *Plantago major* or common plantain), or because demand was so low, that it would not have made sense to carry them in their inventories.Very small vendors had inventories that represented the most common medicinal plants available, and excluded most species in the large "witchcraft" segment of the pharmacopoeia. On the other hand, well-established large stands specialized in supplies for healers (including "magical" plants).

All four markets had inventories containing more than 50% of all inventoried plant species, but lacked many of the "generalist" plants sold by other vendors. The portfolio of these stands focused almost entirely on "magical" species that are needed to cure illnesses like *"susto"* (fright), *"mal aire"* (evil wind), *"daño"* (damage), *"envidia"* (envy) and other "magical" or psychosomatic ailments. At the same time, all four vendors catered also to the esoteric tourism crowd that tends to frequent the large markets, and carried a variety of plants that were not used by *curanderos*, but instead were sold to meet tourist demand.

A Look on Sustainability - How Much Plant and for Which Price?

More than two thirds of all species sold in Northern Peruvian were claimed to originate from the highlands (*sierra*), above the timberline, which represents areas often heavily used for agriculture and livestock grazing. The overall value of medicinal plants in these markets reaches a staggering 1.2 million US$/year. This figure only represents the share of market vendors, and does not include the amount local healers charge for their cure. Thus, medicinal plants contribute significantly to the local economy. Such an immense market raises questions of the sustainability of this trade, especially because the market analysis does not take into account any informal sales.

Most striking was the fact that 7 indigenous and 3 exotic species, i.e. 2.5% of all species traded, accounted for more than 40% of the total sales volume (with 30 and 12% respectively). Moreover, 31 native species accounted for 50% of all sales, while only 16 introduced plants contributed to more than a quarter of all material sold. This means that little over 11% of all plants in the market accounted for about three fourths of all sales. About one third of this sales volume includes all exotic species traded. None of these are rare or endangered. However, the rising market demand might lead to increased production of these exotics, which in turn could have negative effects on the local flora (Bussmann et al. 2007b).

A look at the indigenous species traded highlights important conservation threats. *Croton lechleri* (dragon's blood), and *Uncaria tomentosa* (cat's claw) are immensely popular at a local level and each contributes to about 7% to the overall market value. Both species are also widely traded internationally. The latex of *Croton* is harvested by cutting or debarking the whole tree. *Uncaria* is mostly traded as bark, and again the whole plant is normally debarked. *Croton* is a pioneer species, and apart from *C. lechleri* a few other species of the genus have found their way in the market. Sustainable production of this genus seems possible, but the process has to be closely monitored, and the current practice does not appear sustainable because most *Croton* is wild harvested. The cat's claw trade is so immense, that in fact years ago collectors of this primary forest liana started complaining about a lack of resources (Cabieses Molina 2000) and during the years of this study other *Uncaria* species, or even *Acacia* species have appeared in the market as "cat's claw" (own observation). As such, the *Uncaria* trade is clearly not sustainable.

Some of the other "most important" species are either common weeds (e.g. *Desmodium molliculum*), or have large populations (e.g. *Equisetum giganteum*). However, a number of species are very vulnerable. *Tillandsia cacticola* grows in small areas of the coast as epiphyte (Downer 2006). The habitat, coastal dry forest and shrub, is heavily impacted by urbanization and mechanized agriculture, the impact of the latter worsened by the current bio-fuel boom.

Gentianella alborosea, G. bicolor, G. graminea, Geranium ayavacense and *Laccopetalum giganteum* are all high altitude species with very limited distribution. Their large-scale collection is clearly unsustainable, and in case of *Laccopetalum* collectors indicate that supply is harder and harder to find. The fate of a number of species with similar habitat requirements raises comparable concern. The only species under cultivation at this point are exotics, and a few common indigenous species.

When looking at the reasons why people chose medicinal plants or pharmaceuticals for greater consumption, it seemed as though the major reasons were fairly obvious. Many people preferred using plants more often because they are natural and safe. Pharmaceutical products have too many synthetic chemicals and foreign substances that can affect the body. Using plants that have been in use for centuries seems to be a safer and healthier alternative. Many people said that pharmaceuticals were used for particular illnesses, but often had side effects that result in negative impacts elsewhere in the body. Respondents agreed however, that pharmaceuticals products were more effective than medicinal plants. Even though they still used plants, they would not completely depend on them, knowing that there a limit to their use. A lot of agreement was registered for use of doctor's prescriptions. Many people have faith in their doctor, and if he recommends using a certain medicine, they will. This faith is based on the confidence people have in science and medicine with a great deal of research available, which has gained the public's trust. Because of this, people feel safer relying on modern medicine. Along with the research, people know that medicine has noticable effects that can be more easily obtained than those from plants. Plant remedies take longer and are more subtle in their effects. These are reasons why pharmaceuticals are used more often. Although the number was minimal, there were respondents who did say that they used the two kinds of medicine in the same amounts. What was interesting was that people said that they used both together. For example, often people said that they would drink a cup of herbal tea while taking pills. Although people felt that each type of medicine has a role, most agreed that pharmaceuticals provide the best route taken for fighting certain sicknesses.

Traditional Medicine is experiencing increasing demand, especially from a Peruvian perspective, as indicated by the fact that the number of herb vendors, in particular in the markets of Trujillo, has increased in recent years. Also, a wide variety of medicinal plants from Northern Peru can be found in the global market. While this trend might help to maintain traditional practices and to give traditional knowledge the respect it deserves, it poses a serious threat, as signs of over-harvesting of important species are becoming increasingly apparent.

Today the most serious threat to this millennial tradition is the destruction of medicinal plant habitats. Urban sprawl and the sugar industry have already greatly altered the coastal plains around Trujillo and Chiclayo. Climatic change and deforestation are threatening the mountain forest systems that are the source of many medicinal species. Most importantly, the high Andean ecosystems and sacred lagoons where many medicinally active species are found are in danger of being destroyed by large-scale mining activities (Zamora Perez 2007). In order to counteract such influences, close work with local healers, in order to document local knowledge, and use of the results to develop sustainably sourced, and scientifically identified plant material for local marketsis being developed (Fig. 14). The knowledge also gets repatriated in bilingual form (Fig. 15).

Figure 15. Dissemination of traditional knowledge in bilingual books.

It is apparent that the respondents used medicinal herbs more often than pharmaceutical medicines, but only to a small degree. People generally assumed that plants are healthier and better to use because they are natural and are thought to not have any side-effects. It is difficult to determine if the knowledge of the use of medicinal plants is growing or decreasing, but the indications are that the last generation knows more than the present. However, most of the present generation does teach their children about the use of medicinal plants. The present study also showed what medicinal plants the respondents used for which purposes. It would be interesting to evaluate the properties of the species used in bioassays.Similarly, the plant knowledge of patients at both facilities was largely identical, with an essentially overlapping selection of common, mostly introduced, species, and basically the same number of medicinal plants mentioned overall. This indicates that traditional medicinal knowledge is a major part of a people's culture that is being maintained while patients are also embracing the benefits of western medicine.

This attitude does however lead to profound challenges when it comes to the safety of the plants employed, in particular for applications that require long-term use. Bussmann et al. (2013) found that various species were often sold under the same common names. Some of the different fresh species were readily identifiable botanically, but neither the collectors nor the vendors do made a direct distinction between species (Fig. 16). However, often material was sold in finely powdered form, which makes the morphological identification of the species in the market impossible, and greatly increases the risk for the buyer. The best way to ensure correct identification would be DNA bar-coding. The necessary technical infrastructure is however not available locally. The use of DNA bar-coding as quality control tool to verify species composition of samples on a large scale would require to carefully sample every batch of plant material sold in the market. The volatility of the markets make this is an impossible logistical task. Often the same or closely related species mentioned in literature sell under wide variety of common names. Worse, one species might be sold e.g. as "Hercampuri" in one location or market stand, while selling under a different name at a neighboring stand. As expected there is no consistency in the dosage of plants used, nor do vendors agree on possible side effects.

Figure 16. Medicinal plant presentation in Mercado Aviación Lima: A. packaged ground plat material; B. traditional presentation; C-E. *Geranium sessiliflorum*; D. unidentifiable *Geranium* fragment; F. *Cheilanthes bonariensis*; G. *Argyrochosma nivea*; H. *Gentianella thyrsoidea*; I. *Gentianella nitida*.

Studies indicate that the plant use in Northern Peru, although footing on a millennial tradition, has changed considerably even during the last decades. Even even in case of plant species used for very clearly circumscribed applications, patients run a considerable risk when purchasing their remedies in the local markets, and the possible side effects can be serious. Much more control, and a much more stringent identification of the material sold in public markets, and entering the global supply chain via Internet sales, would be needed.

SACRED SEEDS, THE NAGOYA PROTOCOL AND REPATRIATION OF TRADITIONAL KNOWLEDGE

Sacred Seeds is one of the programs of the William L. Brown Center at the Missouri Botanical Garden. Creating a global model of sustainable development through botanical research programs, in Sacred Seeds the WLBC works closely with communities in host countries and has long-term success in finding practical solutions to conservation issues (Fig. 17). The program aims to establish in-situ collections of locally important plant species used in traditional medicine, as food, and for other purposes. Participating sanctuaries receive direction, feedback, guidance and advice from the WLBC team, linking experiences in botany, anthropology, traditional healing and conservation with the collective experience of all Sacred Seed sanctuaries. Sacred Seeds strives to facilitate seed storage training, with the possibility of building a relationship with large seed storage facilities, training in plant taxonomy, and publishing assistance and templates for creating educational, scientifically valid literature, and creating a

"Sacred Seeds Book" for every garden, in local language, but also for worldwide distribution (Fig. 18). The Sacred Seeds program has currently almost 40 participating foundational gardens, with a network of close to 1500 participating garden units.

Figure 17. Sacred Seeds Ethnobotanical Garden Network website.

Figure 18. Sacred Seeds – garden guide example.

The Nagoya Protocol and Repatriation of Traditional Knowledge

The final implementation of the "Nagoya Protocol on Access to Genetic Resources and the Fair and Equitable Sharing of Benefits Arising from their Utilization to the Convention on Biological Diversity" last month, has brought a great boost for the rights of indigenous and local communities.

The main objective of the protocol is *"the fair and equitable sharing of the benefits arising from the utilization of genetic resources, including by appropriate access to genetic resources and by appropriate transfer of relevant technologies, taking into account all rights over those resources and to technologies, and by appropriate funding, thereby contributing to the conservation of biological diversity and the sustainable use of its components,"* including that *"traditional knowledge associated with genetic resources that is held by indigenous and local communities is accessed with the prior and informed consent or approval and involvement of these indigenous and local communities, and that mutually agreed terms have been established."*

This is naturally of great importance. The establishment of prior informed consent has, fortunately, been widely practiced already, although there are still projects that place only limited emphasis on permits and consent, because the process is often long and tedious. Under the Nagoya Protocol prior informed consent and providing benefits for knowledge holders is no longer only good ethics, it is also international law. It is to hope that all major granting agencies, whether private or governmental, will soon make proof of prior informed consent a requirement for funding.

The concept, that *"benefits"* might result from the documentation of traditional knowledge, is somewhat new to many colleagues. In globalized science, where data is easily circulated, it is only just to make sure as much as possible, that the knowledge our counterparts share with us, is not simply appropriated by parties not involved in the original study, whether for scientific or commercial purposes. In practice, this means that the establishment of prior informed consent valid under the Nagoya Protocol, needs to include an explicit statement along the lines that *"any work conducted in a community iscarried out under the Nagoya Protocol on Access to Genetic Resources and the Fair and Equitable Sharing of Benefits Arising from their Utilization, and that the right to use and authorship of any traditional knowledge all informants is maintained, as well as that any use of the information obtained, other than for the scientific publication does require additional prior consent of the traditional owners, as well as a consensus on access to benefits resulting from subsequent use."*

From our perspective at William L. Brown Center at MBG, the Nagoya Protocol is to be applied with immediate effect, and does not only cover future research, but any previous data gathered by our institution, unless a different agreement with the original owners exists. Benefit Sharing in this context needs to also not only include the repatriation of the new data gathered, in a language and form accessible to the traditional owners, but also the translation and repatriation of the results of previous studies conducted in the same indigenous or local community, if not already done by the original researchers. In addition, informants, should they so desire, must be allowed full participation as authors in all publications of a study, rather than simply being mentioned as a sideline in the acknowledgments.

Along these lines WLBC has edited a series of books, authored by the members of local communities, that repatriate knowledge on plant use, in local language, and in an appropriate format, as requested by the communities, to the original knowledge holders. These are given to all participating community members, as well as local schools. The first reaction in most communities, when presented with the results of such a collaborative investigation, is astonishment that researchers actually came back, and in fact brought some useful material in local language. This is normally followed by great satisfaction of the participants to be actual authors of the material (Fig. 19). All publications acknowledge the local communities intellectual property under the Nagoya Protocol (Fig. 20), and the authorship of the local collaborators (Fig. 21). In the case of publications on palm use, WLBC developed together with the communities easy to understand symbols for usage categories and plant parts used (Fig. 21). The books include graphic descriptions of plant parts (Fig. 23), with color examples for different growth-, leaf- and fruit-forms (Figs. 24 and 25), followed by in depth description of the individual species (Fig. 26), and detailed, well illustrated desriptions of usage (Fig. 27). In addition to these efforts, WLBC strives to translate previously published studies on traditional knowledge, to make them available in local knowledge, to the benefit of the communities that were involved in the original research. A good example is Bussmann and Paniagua Zambrana (2011). "La etnobotanica de los Chacobo: Traduccion de "Boom, B. – The Ethnobotany of the Chacobo Indians, Beni, Bolivia" (Fig. 28).

Figure 19. Repatriation of knowledge – Presentation of palm books to local communities.

"KAMPANAK SE USA PARA EL TECHO PERO YA NO HAY"
USO Y CONSERVACIÓN DE PALMERAS ENTRE LOS AWAJUN, AMAZONAS, PERÚ

Este libro forma parte del material generado por el Proyecto Impacto de la cosecha de palmeras en los bosques tropicales (PALMS), financiado por el 7º Programa Marco de la Unión Europea, contrato nº 212631 (http://www.fp7-palms.org) y las instituciones colaborantes.

Herbario Nacional de Bolivia
Instituto de Ecología – Universidad Mayor de San Andrés (UMSA)
Campus Universitario, Calle 27 Cota Cota s/n
Casilla 10077 correo central
La Paz, Bolivia

Departamento de Biología - Área de Botánica
Universidad Autónoma de Madrid (UAM)
Edificio de Ciencias Biológicas

Impreso en GRAFICART SRL:
San Martín 375 - Trujillo, Perú
ISBN-10: 0984841555
ISBN-13: 978-0-9848415-5-4
Hecho el Depósito Legal en la Biblioteca Nacional del Perú N° 2012-07672
Primera edición, Trujillo, Perú, julio del 2012

Todos los trabajos se ejecutaron bajo el reglamento del "Protocolo de Nagoya sobre acceso a los recursos genéticos y participación justa y equitativa en los beneficios que se deriven de su utilización al Convenio sobre la Diversidad Biológica" y se mantiene el derecho de autoría de sus conocimientos tradicionales de todos los informantes. Cualquier uso comercial de la información presentada en esta publicación requiere del consenso previo con los informantes y comunidades, y un acuerdo sobre la distribución de beneficios.

IMPRESO EN PERÚ - PRINTED IN PERU

Todos los trabajos se ejecutaron bajo el reglamento del "Protocolo de Nagoya sobre acceso a los recursos genéticos y participación justa y equitativa en los beneficios que se deriven de su utilización al Convenio sobre la Diversidad Biológica" y se mantiene el derecho de autoría de sus conocimientos tradicionales de todos los informantes. Cualquier uso comercial de la información presentada en esta publicación requiere del consenso previo con los informantes y comunidades, y un acuerdo sobre la distribución de beneficios.

IMPRESO EN PERÚ - PRINTED IN PERU

Figure 20. Book page: Acknowledging local intellectual property and the Nagoya Protocol.

AUTORES POR COMUNIDAD

PALMA REAL

Adela Shanocua	Florentino Montesinos	María Biaeja
Aidé Chetu	German Biaeja	María Isabel Saavedra
Alejandro Mohahe	German Biaeja	María Naji
Alfonsa Sehua	Gloria Saavedra	Maricruz Shanoco
Alfredo Biaeja	Graciela Huohehua	Mateo Biaeja
Alicia Saavedra	Gregorio Mejía	Melchor Vauhehua
Alina Shoe	Héctor Ioshi	Mercedes Shanocua
Ana Elinei	Henry Posho	Mercedes Shehua
Asunta Zapata	Inés Meshi	Midio Huane
Benito Huane	Asunta Biaeja	Miguel Domínguez
Betti Biaeja	Jesús Meshi	Nina Saavedra
Carlitos Tirina	Jesusa Hudomela	Pastora Huaquibeline
Carlitos Tirina	Jordán Kioshi	Pilar Valdivia
Carmen Machuquia	Jorge Shanoqua	Pio Meshi
Cecilio Yoyaje	José Gongora	Raúl Meshi
Celestina Hejayha	José Liaqui	Ricardo Yoyaje
Zenón Yohahe	Geovana Chaeta	Roger Mahio
Clementino Meshi	Juan Shanoqua	Rosa Paketanaji
Dali Tipuna	Juanita Ekinei	Rosario Soria
Dorilda Posho	Julia Huane	Rosy Viasteno
Doris Elinei	Lidia Meshi	Rudi Sehua
Elda Hudenda	Lidia Shehua	Ruth Tilihua
Elisa Huajohuaho	Lorenzo Shoe	Sara Saavedra
Eliseo Meshi	Lucio Yohahe	Segundo Saavedra
Eliser Biaeja	Luis Biaeja	Teófila Saavedra
Elvira Saavedra	Luzmila Shanokua	Verónica Beshu
Esperanza Heyahijia	Marcela Biaeja	Victoria Meshi
Exilda Shanoqua	Marcial Soria	Walter Ayca
Felipe Vauhehua	Mardilina Biaeja	Wilma Huahojehua
Félix Huali	Margarita Meshi	Yasira Shehua

Figure 21. Community authors.

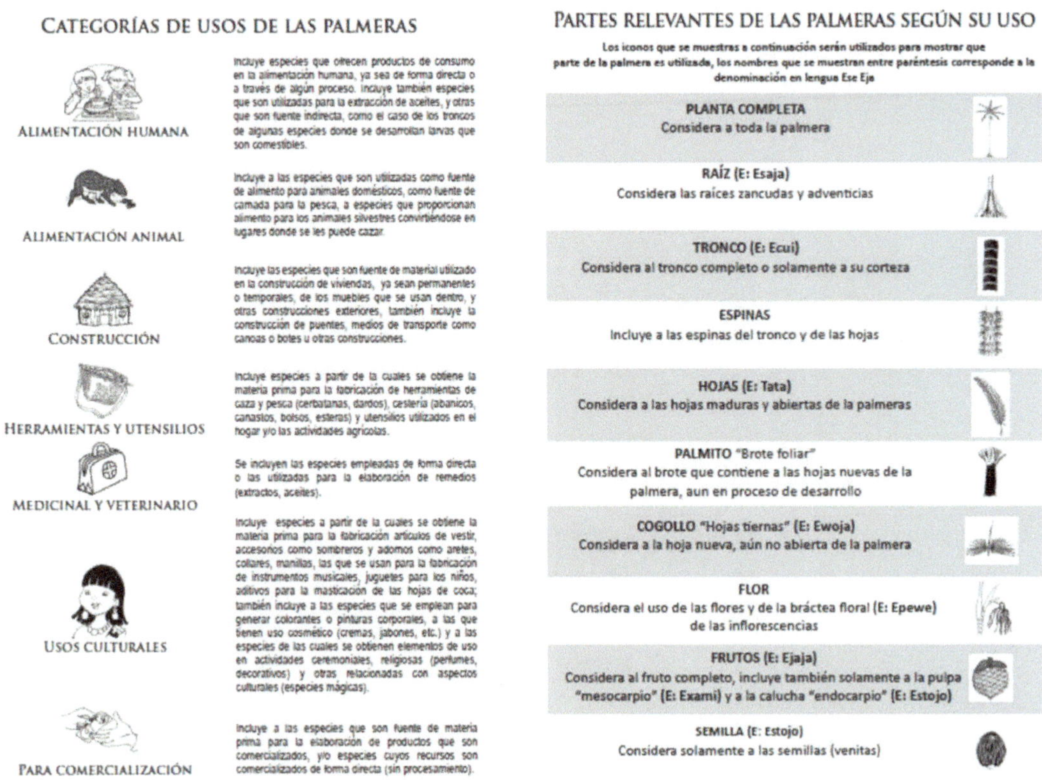

Figure 22. Symbols to explain plant uses and plant parts, developed with local communities.

Conceptos previos
La estructura de las palmeras

Para entender la diversidad que existe entre las palmeras que conocen las comunidades indígenas Ese Eja en el Sur de Perú, es necesario familiarizarse con la estructura de la planta y los nombres correspondientes. Para ello presentamos en esta sección una vista rápida de la estructura de las palmeras.

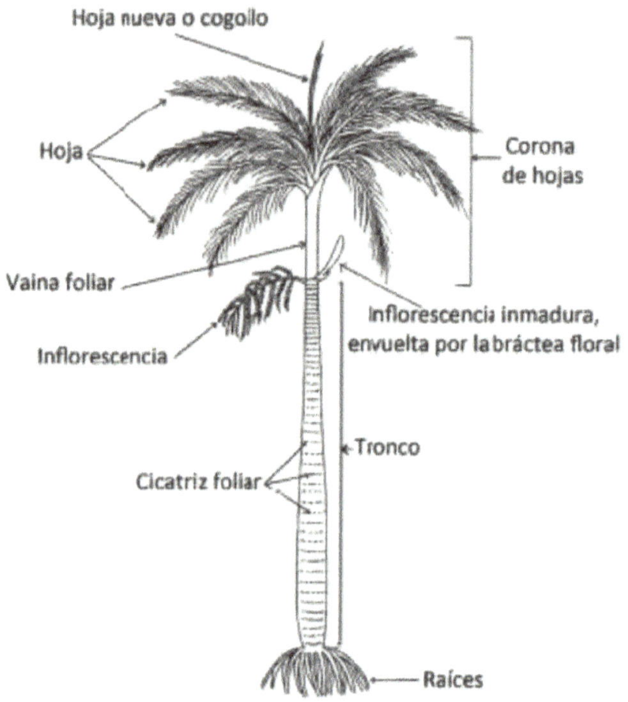

Figure 23. Palm profile – plant parts in local language.

Figure 24. Examples for different palm leaf and growth forms.

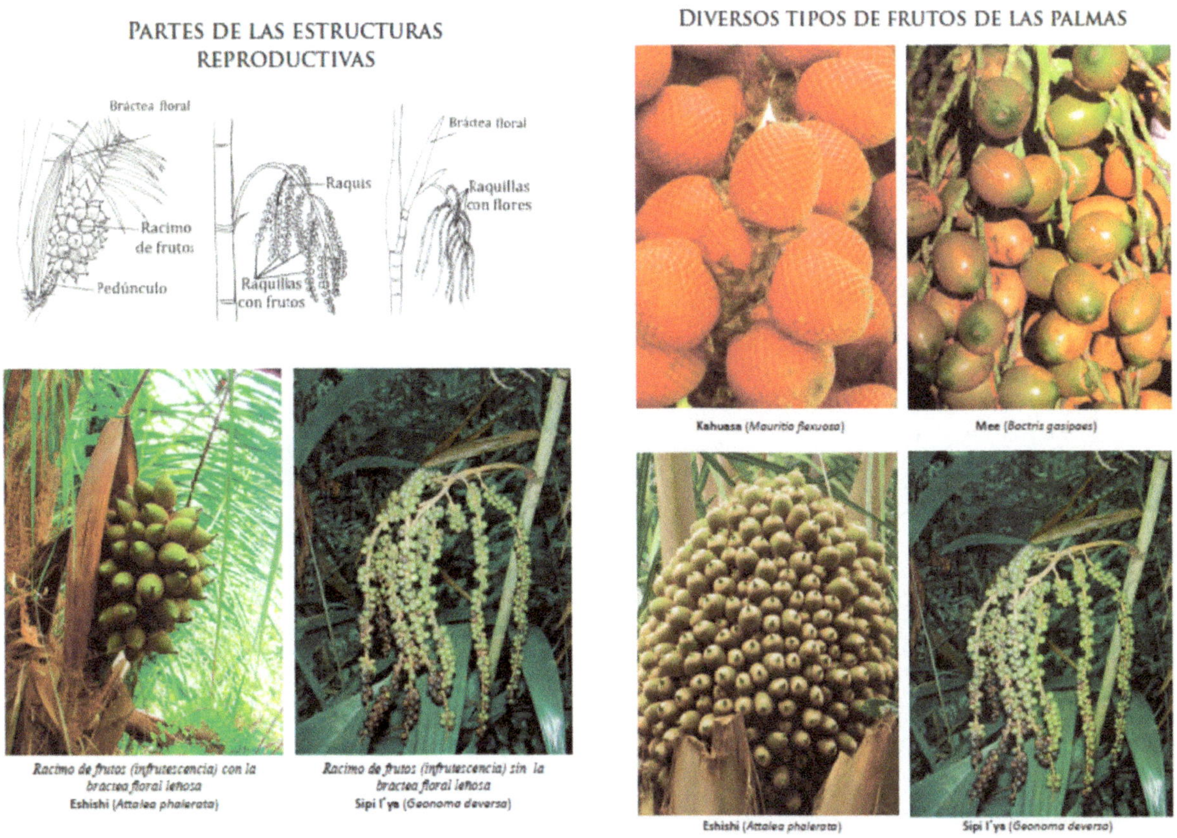

Figure 25. Examples for palm fruits.

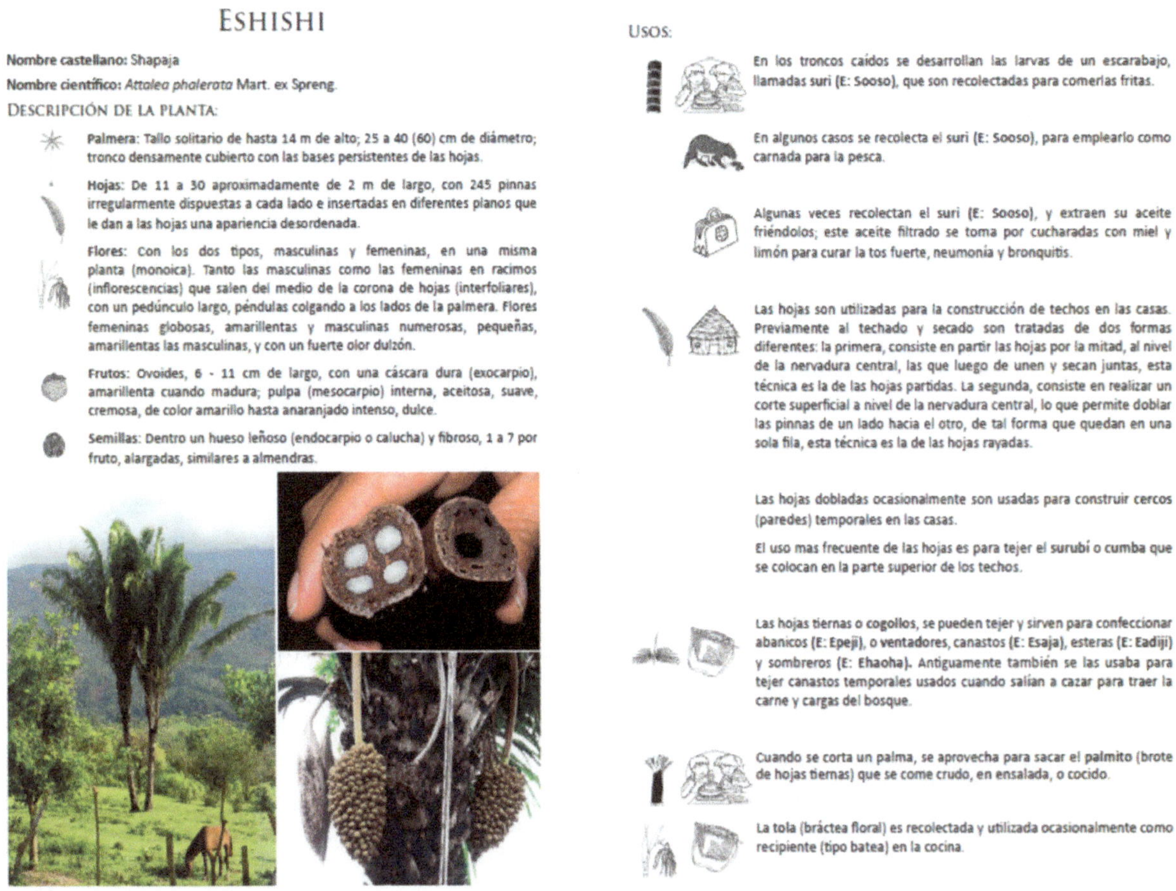

ESHISHI

Nombre castellano: Shapaja

Nombre científico: *Attalea phalerata* Mart. ex Spreng.

DESCRIPCIÓN DE LA PLANTA:

Palmera: Tallo solitario de hasta 14 m de alto; 25 a 40 (60) cm de diámetro; tronco densamente cubierto con las bases persistentes de las hojas.

Hojas: De 11 a 30 aproximadamente de 2 m de largo, con 245 pinnas irregularmente dispuestas a cada lado e insertadas en diferentes planos que le dan a las hojas una apariencia desordenada.

Flores: Con los dos tipos, masculinas y femeninas, en una misma planta (monoica). Tanto las masculinas como las femeninas en racimos (inflorescencias) que salen del medio de la corona de hojas (interfoliares), con un pedúnculo largo, péndulas colgando a los lados de la palmera. Flores femeninas globosas, amarillentas y masculinas numerosas, pequeñas, amarillentas las masculinas, y con un fuerte olor dulzón.

Frutos: Ovoides, 6 - 11 cm de largo, con una cáscara dura (exocarpio), amarillenta cuando madura; pulpa (mesocarpio) interna, aceitosa, suave, cremosa, de color amarillo hasta anaranjado intenso, dulce.

Semillas: Dentro un hueso leñoso (endocarpio o calucha) y fibroso, 1 a 7 por fruto, alargadas, similares a almendras.

USOS:

En los troncos caídos se desarrollan las larvas de un escarabajo, llamadas suri (E: Sooso), que son recolectadas para comerlas fritas.

En algunos casos se recolecta el suri (E: Sooso), para emplearlo como carnada para la pesca.

Algunas veces recolectan el suri (E: Sooso), y extraen su aceite friéndolos; este aceite filtrado se toma por cucharadas con miel y limón para curar la tos fuerte, neumonía y bronquitis.

Las hojas son utilizadas para la construcción de techos en las casas. Previamente al techado y secado son tratadas de dos formas diferentes: la primera, consiste en partir las hojas por la mitad, al nivel de la nervadura central, las que luego de unen y secan juntas, esta técnica es la de las hojas partidas. La segunda, consiste en realizar un corte superficial a nivel de la nervadura central, lo que permite doblar las pinnas de un lado hacia el otro, de tal forma que quedan en una sola fila, esta técnica es la de las hojas rayadas.

Las hojas dobladas ocasionalmente son usadas para construir cercos (paredes) temporales en las casas.

El uso mas frecuente de las hojas es para tejer el surubí o cumba que se colocan en la parte superior de los techos.

Las hojas tiernas o cogollos, se pueden tejer y sirven para confeccionar abanicos (E: Epeji), o ventadores, canastos (E: Esaja), esteras (E: Eadiji) y sombreros (E: Ehaoha). Antiguamente también se las usaba para tejer canastos temporales usados cuando salían a cazar para traer la carne y cargas del bosque.

Cuando se corta un palma, se aprovecha para sacar el palmito (brote de hojas tiernas) que se come crudo, en ensalada, o cocido.

La tola (bráctea floral) es recolectada y utilizada ocasionalmente como recipiente (tipo batea) en la cocina.

Figure 26. Illustrated plant description.

Figure 27. Richly illustrated description of plant use, with local language terms in addition to Spanish.

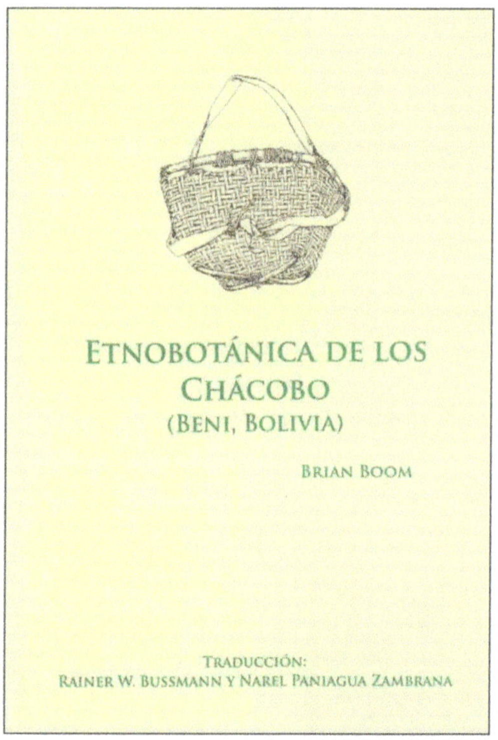

Figure 28. Repatriation of local knowledge in local language: Translation of Boom "Ethnobotany of the Chacobo Indians".

A Global Program on Conservation of Useful Plants and Traditional Knowledge: A Call to Action

On 1–2 May 2013 WLBC brought international experts on plants used by humankind together in St Louis Missouri to consider the ways in which a global crisis now underway - the loss of tens of thousands of plant species - can be addressed. These threatened plants include species vital to the lives of people throughout the world, including plants used for food and nutrition, medicine, cultural and spiritual purposes, and the maintenance of livelihoods; they are needed to redress poverty, provide food security, and ensure sustainable development in many nations. Plants and their associated biocultural knowledge play an essential role in the ecosystem services that support all life on Earth (Barve at a. 2013).

The resulting statement is not only an appeal to the international community to address the tragic loss of plant diversity but a call for the development of a concerted effort worldwide to address the loss of essential knowledge about plants and their uses, especially at the level of local communities.

The meeting specifically focused on the objectives of the Global Strategy for Plant Conservation (GSPC), an initiative adopted by the U.N. Convention on Biological Diversity in 2002, and subsequently updated in 2010, as well as the GSPC targets pertaining to the maintenance and preservation of useful and culturally significant plants. The participants concluded that there is also a great urgency to address the vital importance of traditional knowledge about plants, their utility, management, and conservation. This unique, often ancient, and detailed knowledge is typically held and maintained by local and indigenous communities.

The workshop contributors urged the development of a global program on the conservation of useful plants and associated knowledge, taking into account the need to:

- Call on the international community and governments to recognize the importance of wild and cultivated plant diversity, as well as the associated knowledge of their usefulness as a vital present-day and future resource. This should be accomplished through the successful implementation of the GSPC objectives and targets by 2020.
- Highlight the need for a concerted international effort to compile a widely accessible global catalogue of useful plants of importance for human kind, while respecting intellectual property rights, local ownership of knowledge and appropriate benefit sharing.
- Assist local peoples in the preservation of their traditional knowledge in a culturally appropriate manner.
- Stress the need for cross cultural and multilevel partnerships in the effort to build on and share experiences on conservation of culturally significant plants, their sustainable use, and associated knowledge.
- Develop an international research platform to address gaps in scientific knowledge of useful plants.
- Facilitate capacity building and training opportunities in ethnobotany, particularly in countries and regions with significant gaps in such resources.
- Support and encourage biocultural knowledge transmission and custodianship.
- Develop the appropriate facilities, methodologies, and techniques to support culturally sensitive curation of biocultural collections (artifacts, herbarium vouchers, produces, living collections, etc.) and associated traditional knowledge.
- Elaborate and disseminate educational materials and resources in appropriate languages that support and promote the study and use of traditional knowledge, and insure their inclusion in educational curricula.
- Develop a toolbox of methodologies, case studies, manuals, and good practices in order to support the conservation of useful plant and associated knowledge.
- Highlight the need for measurable indicators that monitor progress in the conservation of useful plants and associated knowledge.
- Follow the framework provided by the Nagoya Protocol on Access to Genetic Resources and the Fair and Equitable Sharing of Benefits Arising from their Utilization (ABS) to the Convention on Biological Diversity and FAO's International Treaty on Plant Genetic Resources for Food and Agriculture, in order to manage and achieve ethical standards for access, fair and equitable benefit sharing, traditional resource and farmers' rights, and the protection of intellectual property.

CONCLUSIONS

Current research indicates that the composition of the local pharmacopoeia in Northen Peru and Southern Ecuador has changed since colonial times (Martinez Compañon 1789; Ruiz 1777-1788; Bussmann et al. 2009e). However, in Northern Peru the overall number of medicinal plants employed seems to have remained at a comparable level, while plant use in Southern Ecuador has decreased. This indicates that the Northern Peruvian health tradition is still going strong, and that the healers and public are constantly experimenting with new remedies. One example of this is the sudden appearance of Noni (*Morinda citrifolia)* fruits and products in large quantities in plant pharmacies and markets in the region since 2005. This plant was not available before, but is heavily marketed worldwide. Peruvian sellers are clearly reacting to a global market trend and are trying to introduce this new species to their customers. This indicates that local herbalists and herb merchants are carefully watching international health trends to include promising species in their own repertoire. In Southern Ecuador, healers were not able to experiment with new remedies due to persecution and legal restrictions. As a result, the pharmacopoeia in this region remained on an early colonial level, with loss of significant knowledge.

The use of hallucinogens, in particular the *San Pedro* cactus (*Echinopsis pachanoi*) is still a vital component of Andean healing practices, and has been around for millennia. *San Pedro* can often be found in Cupisnique and Moche iconography. Five hundred years of suppression of traditional healing practices by Western medicine have not managed to destroy this tradition in Peru. The use of *San Pedro*, together with additives like Angel's Trumpet (*Brugmansia* spp.), Jimsonweed (*Datura ferox*), and tobacco, is still a central part of curing ceremonies in Northern Peru. Healers are in fact experimenting with new hallucinogens, and some northern *curanderos* have started to include decoctions of Ayahuasca (*Banisteriopsis caapi*) in their rituals.

Although not formally acknowledged, Southern Ecuador falls into the Northern Peruvian cultural area. It appears to represent a region where traditional plant knowledge, though important, has declined considerably. Southern Ecuadorian *curanderos* and *parteras* (midwives) having almost entirely abandoned indigenous rituals. In fact, *San Pedro* usage was not mentioned as a mind-altering plant by any healer or midwife interviewed, and was not used in curing ceremonies. Centuries of prohibition have led to a pronounced abandonment of traditional knowledge. This is also reflected in the current study. Many plants used for "magical" purposes in Peru have disappeared from traditional use in Ecuador. The fear of prosecution is still very deeply rooted in the healer community, and most healers interviewed stated that they did not wish to be cited by name. Most healing altars or *mesas* in Southern Ecuador are almost entirely devoid of any "pagan" objects such as seashells, pre-Columbian ceramics, etc. Patients are cleansed, by spraying them with holy water and perfumes. In rare cases tobacco juice and extracts of Jimson weed (*Datura ferox*) are used to purify the patients. Southern Ecuadorian *mesas* are also much less elaborate than the *mesas* of Peruvian *curanderos*. The incantations used by healers during their curing session's center on Christian symbolism. References to Andean cosmology are almost entirely absent, and the use of guinea pigs as diagnostic instruments has all but disappeared from the tool kit of these healers.

Interestingly, Peruvian *curanderos* have started to fill this spiritual void in Southern Ecuador. Healers from the Northern Peruvian Mountains and coastal plains frequently cross over to Ecuador to offer their services to patients – including increasing numbers of foreigners with a "New Age" orientation -- who are not satisfied with the more Westernized approach of Ecuadorian healers. These Peruvian colleagues have much more elaborate plant knowledge, and their *mesas* as well as their incantations follow a more traditional pattern.

REFERENCES

1. Acosta J de (1590) Historia natural y moral de las Indias. Sevilla.
2. Agurto J (2005) ¿Como prevenir la biopirateria en el Peru? Reflexiones y Propuestas pp. 70-73. Lerma Gomez E.I.R.L., Lima-Miraflores.
3. Alcedo A de (1786-89) Diccionario geografico histórico de las indias occidentales o América. Madrid.
4. Alva W (1994) Sipán: descubrimiento y investigación. Cultura y artes del antiguo Perú. Backus & Johnston, Lima.
5. Alva W, Donnan CB (1993) Royal Tombs of Sipán. Fowler Museum of Cultural History, University of California, Los Angeles.
6. Alva W, Donnan CB (1994) Tales from a Peruvian Crypt. Natural History 103(5):26-36.
7. Angulo P (2009) Nuevos enfoques en la investigación de plantas medicinales. In Garrafa R (ed) Medicina Tradicional Andina, Planteamientos y aproximaciones. CBC/CMA, Cuzco, pp. 351-384.
8. Baker J, Borris R, Carté B, Cordell G, Soejarto D, Cragg G, Gupta M, Iwo M, Madulid D, Tyler V (1995) Natural Product Discovery and Development, New Perspectives on International Collaboration. Journal of Natural Products 58(9):1325-1357.
9. Bannister K, Barrett K (2001) Challenging the status quo in ethnobotany, A new paradigm for publication may protect cultural knowledge and traditional resources. Cultural Survival Quarterly 10-12.
10. Barve V, Bhatti R, Bussmann RW, Bye R, Chen J, Dulloo E, Giovannini P, Linares E, Magill R, Roguet D, Salick J, On TV, Vandebroek I, Wightman G, Wyse Jackson P (2013) A Global Program on Conservation of Useful Plants and Traditional Knowledge - A Call to Action. https://www.researchgate.net/ publication/236633499_A_Global_Program_on_Conservation_ of_Useful_Plants_and_Traditional_Knowledge_A_Call_to_ Action?ev=prf_pub
11. Béjar E, Bussmann RW, Roa C, Sharon D (1997) Pharmacological search for active ingredients in medicinal plants of Latin America. In: Shuman T, Garrett M, Wozniak L (eds), International Symposium on Herbal Medicine, A Holistic Approach. SDSU International Institute for Human Resources Development, San Diego, pp. 63-81.
12. Béjar E, Bussmann RW, Roa C, Sharon D (2001) Herbs of Southern Ecuador – Hierbas del Sur Ecuatoriano. Latin Herbal Press, San Diego.
13. Brack Egg A (1999) Diccionario enciclopédico de plantas útiles del Perú. PNUD – CBC, Cuscso.
14. Brack Egg A (2004) Biodiversidad, pobreza y bionegocios. Lima.
15. Brako L, Zarucchi JL (eds) (1993) Catalogue of the Flowering Plants and Gymnosperms of Peru. Missouri Botanical Garden, Saint Louis, MO.
16. Breevort P (1998) The Booming U. S. Botanical Market, A New Overview. *HerbalGram* 44.
17. Brown M (2003) *Who owns native culture?* Harvard University Press, Cambridge.
18. Brüning HH (2004) Diccionario Mochica. Universidad de San Martin de Porres, Lima.
19. Buitron X (1999) Ecuador, uso y comercio de plantasmedicinales, situacion actual y aspectosimportantesparasuconservación. TRAFFIC International, Cambridge, UK.
20. Bussmann RW (2006) Manteniendo el balance de naturaleza y hombre, La diversidad florística andina y su importancia por la diversidad cultural – ejemplos del Norte de.Perú y Sur de Ecuador. Arnaldoa 13(2):382-397.
21. Bussmann RW, Glenn A, Meyer K, Townesmith A, Rothrock A, Sharon D, Castro M, Cardenas R, Regalado S, Toro R, Chait G, Malca G, Perez F (2009a) Antibacterial Activity of Medicinal Plants of Northern Peru-Part II. Arnaldoa 16(1):93-103.
22. Bussmann RW, Glenn A, Meyer K, Rothrock A, Townesmith A, Sharon D, Castro M, Cardenas R, Regalado S, Toro R, Chait G, Malca G, Perez F (2009b) Phyto-Chemical Analysis of Peruvian Medicinal Plants. Arnaldoa 16(1):105-110.
23. Bussmann RW, Glenn A, Meyer K, Kuhlman A, Townesmith A (2010a) Herbal mixtures in traditional medicine in Northern Peru. Journal of Ethnobiology and Ethnomedicine 6(10).
24. Bussmann RW, Malca G, Glenn A, Sharon D, Chait G, Diaz D, Pourmand K, Jonat B, Somogny S, Guardado G, Aguirre C, Chan R, Meyer K, Kuhlman A, Townesmith A, Effio J, Frias F, Benito M (2010b) Minimum inhibitory concentrations of medicinal plants used in Northern Peru as antibacterial remedies. Journal of Ethnopharmacology

132:101-108.

25. Bussmann RW, Glenn A (2010c) Cooling the Heat – Traditional remedies for malaria and fever in Northern Peru. Ethobotany Research and Applications (8):125-134.
26. Bussmann RW, Glenn A, Sharon D, Chait G, Diaz D, Pourmand K, Jonat S, Somogy S, Guardado G, Aguirre C, Chan R, Meyer A, Townesmith A (2010d) Proving that Traditional Knowledge Works, The antibacterial activity of Northern Peruvian medicinal plants. Ethnobotany Research and Applications 9:67-96.
27. Bussmann RW, Malca G, Glenn A, Sharon D, Nilsen B, Parris B, Dubose D, Ruiz D, Saleda J, Martinez M, Carillo L, Kuhlman A, Townesmith A (2011) Toxicity of medicinal plants used in traditional medicine in Northern Peru. Journal of Ethnopharmacology 137:121-140.
28. Bussmann RW, Paniagua Zambrana N, Rivas Chamorro M, Molina Moreira N, Cuadros Negri ML, Olivera J (2013) Peril in the market–classification and dosage of species used as anti-diabetics in Lima, Peru. Journal of Ethnobiology and Ethnomedicine 9(37).
29. Bussmann RW, Sharon D (2006a) Traditional plant use in Southern Ecuador. Journal of Ethnobiology and Ethnomedicine 2(44).
30. Bussmann RW, Sharon D (2006b) Traditional plant use in Northern Peru, Tracking two thousand years of health culture. Journal of Ethnobiology and Ethnomedicine 2(47).
31. Bussmann RW, Sharon D (2007a) Plants of the four winds - The magic and medicinal flora of Peru. Plantas de los cuatro vientos - La flora mágica y medicinal del Perú. Graficart, Trujillo.
32. Bussmann RW, Sharon D (2007b) Plants of longevity - The medicinal flora of Vilcabamba. Plantas de la longevidad – La flora medicinal de Vilcabamba. Graficart, Trujillo.
33. Bussmann RW, Sharon D (2009d) Naming a phantom—the quest to find the identity of Ulluchu, an unidentified ceremonial plant of the Moche culture in Northern Peru. Journal of Ethnobiology and Ethnomedicine 5(8).
34. Bussmann RW, Sharon D (2009e) Shadows of the colonial past-diverging plant use in Northern Peru and Southern Ecuador. Journal of Ethnobiology and Ethnomedicine 5(4).
35. Bussmann RW, Sharon D (2009f) From collection to market and cure-Traditional medicinal use in Northern Peru. In: Albuquerque U (ed) Recent Development and Case Studies in Ethnobotany. Recife, pp.184-207.
36. Bussmann RW, Sharon D, Diaz D, Barocio Y (2008a) Peruvian plants canchalagua (Schkuhria pinnata (Lam.) Kuntze), hercampuri (Gentianella alborosea (Gilg.) Fabris), and corpus way (Gentianella bicolor (Wedd.) J. Pringle) prove to be effective in the treatment of acné. Arnaldoa 15(1):149-152.
37. Bussmann RW, Sharon D, Garcia M (2009a) From Chamomile to Aspirin? Medicinal Plant use among clients at Laboratorios Beal in Trujillo, Peru. Ethnobotany Research and Applications 7:399-407.
38. Bussmann RW, Sharon D, Lopez A (2007c) Blending Traditional and Western Medicine, Medicinal plant use among patients at Clinic Anticona in El Porvenir, Peru. Ethnobotany Research and Applications 5:185-199.
39. Bussmann RW, Sharon D, Perez F, Díaz D, Ford T, Rasheed T, Silva R (2008b) Antibacterial activity of Northern-Peruvian Medicinal Plants - a low cost laboratory.approach to assess biological activity. Arnaldoa 15(1):127-148.
40. Bussmann RW, Sharon D, Vandebroek I, Jones AA, Revene Z (2007b) Health for sale, the medicinal plant markets in Trujillo and Chiclayo, Northern Peru. Journal of Ethnobiology and Ethnomedicine 3(37).
41. Cabieses Molina F (2000) La Uña de Gato u su entorno. De la Selva a la farmacia. Universidad de San Martin De Porres, Lima.
42. Camino L (1992/1999) Cerros, plantas y lagunas ponderosas - la medicina al norte del Perú. Lluvia Editores, Lima.
43. Cano JH, Volpato G (2004) Herbal mixtures in the traditional medicine of Eastern Cuba. Journal of Ethnopharmacology 90:293-316.
44. Carrillo L (2012) Scientific Validation? How Bioprospecting Laboratory Practices Contribute to the Devaluation of Traditional Medicinal Knowledge. The Berkeley McNair Research Journal, vol. 19, Spring.
45. Cobo B (1653) Historia del Nuevo Mundo, 2 tomos. Sevilla.
46. Cobo B (1956) In: Mateos F de (ed), Historia del Nuevo Mundo. Ediciones Atlas, Madrid.
47. Connally MPE, Fabiano E, Patel IH, Kinyanjui SM, Mberu EK, Watkins WM (1996) Antimalarial activity in crude extracts of Malawian medicinal plants. Annals of Tropical Medicine and Parasitology 90:597-602.
48. Cox P, Balick M (1994) The Ethnobotanical Approach to Drug Discovery. Scientific American 270(6):82-87.
49. D'Agostino M, De Simone F, Tommasi N, Pizza C (1995a) Constituents of Culcitium canescens. Fitoterapia 66:550-551.
50. D'Agostino M, Pizza C, De Simone F (1995b) Flavone and flavonol glycosides from Desmodium mollicum. Fitoterapia 66:384-385.
51. Domenighetti G, Grilli R, Gutzwiller F, Quaglia J (2000) Usage personnel de pratiques relevant des médecines douce sou alternatives parmi les médecins suisses. Médecine & Hygiène 58: 22-91.
52. Donnan CB, Castillo LJ (1994) Excavaciones de tumbas de sacerdotes Moche en San José de Moro, Jequetepeque. In: Uceda S, Mujica E (eds) Moche: Propuestas y Perspectivas, 5. Travaux de l'Institut Français d'Etudes Andines 79:415-442.
53. Downer CC (2006) Insights, Mining Peru's Andean Forest Puts Unique Species, Ecosystem at Risk. Environmental News Service, Feb 6.
54. Eisenberg DM, Davis RB, Ettner SL, Appel S, Wilkey S, Rompay M van, Kessler RC (1998) Trends in alternative medicine use in the United States, 1990–1997, results of a follow-up national survey. Journal of the American

Medical Association 280(18):1569-1575.

55. El Kamali H, El Kijalifa KE (1997) Treatment of malaria through herbal drug in the central Sudan. Fitoterapia 6:527-528.
56. Elisabetsky E, Castilhos C (1990) Plants used as analgesics by Amazonian caboclos as a basis for selecting plants for investigation. International Journal of Crude Drug Research 28:309-320.
57. EsSalud/Organización Panamericana de Salud (2000) Estudio Costo-Efectividad, Programa Nacional de Medicina Complementaria. Seguro Social de EsSalud (Study of Cost Effectiveness, National Program in Complementary Medicine. Social Security of EsSalud). Lima, EsSalud/Organización Panamericana de Salud.
58. Fernández M (2005) La OMS y los sistemas médicos tradicionales. In Garrafa R, Garrafa R (eds) Medicina Tradicional, Planteamientos y aproximaciones. CBC/ CMA, Cuzco.
59. Ferro M, Ruiz P (eds) (2005) ¿Cómo prevenir la Biopiratería en el Perú? Reflexiones y Propuestas. Lerma Gómez E.I.R.L., Lima.
60. Ferro M, Ruiz P (eds) (2005) Apuntes sobre Agrobiodiversidad, Conservación, Biotecnología y Conocimientos Tradicionales. Lerma Gómez E.I.R.L., Lima.
61. Fisher P, Ward A (1971) Medicine in Europe, complementary medicine in Europe. British Medical Journal 309:107-111.
62. Gbile ZO (1984) Vernacular Names of Nigerian Plants (Yoruba). FRIN, Ibadan.
63. Gillett NA, Chan C (2000) Applications of immunohistochemistry in the evaluation of immunosuppressive agents. Human & Experimental Toxicology 19(4):251-254.
64. Girault L (1987) Kallawaya, Curanderos itinerantes de los Andes. UNICEF-OPS-OMS, La Paz.
65. Glass-Coffin B, Sharon D, Uceda S (2004) Curanderos a la sombra de la Huaca de la luna. Bulletin Instute francais d'Etudes Andines 33(1):81-95.
66. Global Industry Analysts Inc (2012) Herbal Supplements and Remedies, A Global Strategic Business Report. Global Industry Analysts, San Jose.
67. Greaves T (1995) Cultural Rights and Ethnography. General Anthropology 1(1): 3-6.
68. Hay SI, Were EC, Renshaw M, Noor AM, Ochola SA, Olusanmi I, Alipui N, Snow RW (2003) Forecasting, warning, and detection of malaria epidemics, a case study. Lancet 361(9370):1705-1706.
69. Hayden C (2003) When Nature Goes Public, The Making and Unmaking of Bioprospecting in Mexico. Princeton University Press, Princeton and Oxford.
70. Health Canada (2001) Perspectives on Complementary and Alternative Health Care. A Collection of Papers Prepared for Health Canada. Ottawa, Health Canada.
71. Hocquenghem AM (2008) Sacrifices and Ceremonial Calendars in Societies of the Central Andes: A Reconsideration. In: Bourget S, Jones KL (eds): The Art and Archaeology o fthe Moche. University of Texas Press, Austin, pp. 23-42.
72. Hultin E, Wassén H, Bondeson W (1987) Papain in Moche Blood ceremonies. Journal of Ethnopharmacology 19(2):227-228.
73. Iwu M (1996) Implementing the Biodiversity Treaty, how to make international cooperative agreements work. Trends in Biotechnology 3-4(146):67-107.
74. Joralemon D, Sharon D (1993) Sorcery and Shamanism, Curanderos and Clients in Northern Peru. University of Utah Press, Salt Lake City.
75. Kraemer H (1915) Scientific and Applied Pharmacognosy. Philadelphia.
76. Larco Hoyle R (1938) Los mochicas I. Casa editora "La Crónica" y "Variedades", S. A. Ltda. Lima.
77. Larco Hoyle R (1939) Los Mochicas II. Casa editora "La Crónica" y "Variedades", S. A. Ltda. Lima.
78. Marínez Compañon DB (1789) Razón de las especies de la naturaleza y del arte del obispado de Trujillo del Perú. Tomos III-V, Sevilla.
79. McClelland D (1977) *The Ullchu: A Moche Symbolic Fruit*. In: Cordy-Collins A, Stern J (eds) Pre-Columbian Art History. Peek Publications, Palo Alto, pp. 435-452.
80. McClelland D (2008) Ulluchu - An elusive fruit. In: Bourget S, Jones KL (eds) The Art and Archaeology of the Moche. University of Texas Press, Austin, pp. 43-65.
81. Milliken W (1997) Traditional anti-malarial medicine in Roraim, Brazil. Economic Botany 51(3):212-237.
82. Minakawa N, Sonye G, Mogi M, Githeko A, Yan GY (2002) The effects of climatic factors on the distribution and abundance of malaria vectors in Kenya. Journal of Medical Entomology 39(6):833-841.
83. Monardes N (1574) Primera y segunda y tercera partes de la história medicinal de las cosasque se traen de nuestras Indias Occidentales, que sirven en medicina; Tratado de la piedra bezaar, y de la yerva escuerçonera; Diálogo de las grandezas del hierro, y de sus virtudes medicinales; Tratado de la nieve, y del beuer frio. Alonso Escrivano, Seville.
84. Mooney P (1993) Aprovechando la Diversidad, Una Nota Sobre la Diversidad Biológica y el Conocimiento Indígena. América Indígena 3:41-55.
85. Moran K, King SR, Carlson T (2001) Biodiversity prospecting, lessons and prospects. Annual Review of Anthropology 30:505-526.
86. Oblitas E (1992) Plantas medicinales de Bolivia. Editorial Los Amigos del Libro, La Paz.
87. Okuyama E, Umeyama K, Ohmori S, Yamazaki M, Satake M (1994) Pharmacologically active components from a Peruvian medicinal plant, Huira-Huira (Culcitium canescens H. & B.) Chemical and Pharmaceutical Bulletin 42:2183-2186.
88. Perez F, Rodríguez F, León G, Sharon D, Bussmann RW, Willsky G, Guerrero G, Willner K, Castro I (2012) Estudio fitoquímico y antibacteriana de mezclas de plantas medicinales. En búsqueda de nuevos componentes. Pueblo continente 23(2):339-343.

89. Perumal Samy R, Ignacimuthu S (2000) Antibacterial activity of some medicinal plants used by tribals in Western Ghats, India. Journal of Ethnopharmacology 69:63-71.

90. Raimondi A (1857) Elementos de Botánica aplicada a la medicina y la industria. Lima.

91. Rätsch C (1998) Enzyklopädie der psychoaktiven Pflanzen. AT Verlag, Aarau.

92. Reid W (1993) The Economic Realities of Biodiversity. Issues in Science and Technology 10(2):48-55.

93. Revene Z, Bussmann RW, Sharon D (2008) From Sierra to Coast, Tracing the supply of medicinal plants in northern Peru - A plant collector's tale. Ethnobotany Research & Applications 6:15-22.

94. Rodriguez J, Pacheco P, Razmilic I, Loyola JI, Schmeda-Hirschmann G, Theoduloz C (1994) Hypotensive and diuretic effect of Equisetum bogotense and Fuchsia magellanica and micropropagation of E. bogotense. Phytotherapy Research 8:157-160.

95. Ruiz H (1777-1788) Relación del viaje hecho a los reynos del Perú y Chile. Translated by Schultes ER, Nemry von Thenen de Jaramillo-Arango MJ as "The Journals of Hipólito Ruiz, Timber Press, Portland, 1998.

96. Schjellerup I (2009) Razon de las Especies de la Naturaleza y del Arte del Obispado de Trujillo del Peru del Obispo D. Balazar Martinez Compagñon In: Vergara E, Vásquez R (eds) Medicina Tradicional, Conocimento Milenario. Serie Antropología 1:128-152.

97. Schultes RE (1994) Amazonian ethnobotany and the search for new drugs. In Ciba Foundation Symposium, vol. 185, Wiley, Chichester, pp. 106-115.

98. Schultes RE, Raffauf R (1990) The Healing Forest. Dioscorides Press, Portland, OR.

99. Sharon D (1978) Wizard of the Four Winds, A Shaman's Story. Free Press, New York.

100. Sharon D (1980) El Chamán de los Cuatro Vientos. Siglo veintiuno editores, México DF.

101. Sharon D (1994) Tuno y sus colegas, notas comparativas. In: Millones L, Lemlij M (eds) En el Nombre del Señor, Shamanes, demonios y curanderos del norte del Perú. Australis S.A., Lima, pp. 128-147.

102. Sharon D (2000) Shamanismo y el Cacto Sagrado – Shamanism and the Sacred Cactus. San Diego Museum Papers 37.

103. Sharon D (2009) Tuno y sus colegas: Notas comparativas. In Galvez C. (Ed) Medicina Tradicional Conocimiento Milenario. Serie Antropología, no 1. Museo de Arquelogía, Antropología e Historia, Facultad de Ciencias Sociales, Universidad Nacional de Trujillo, pp. 255-267.

104. Sharon D, Bussmann RW (2006) Plantas Medicinales en la Obra del Obispo Don Baltasar Jaime Martínez Compagñon (Siglo XVIII). In: Millones L, Kato T (eds) Desde el exterior, El Perú y sus estudios. Tercer Congreso Internacional de Peruanistas, Nagoya, 2005, UNMSM, Lima, pp. 147-165.

105. Sharon D, Galvez C (2009) La mesa de Leoncio Carrión. In: Galvez C (ed) Medicina Tradicional Conocimiento Milenario. Serie Antropología, no 1. Museo de Arquelogía, Antropología e Historia, Facultad de Ciencias Sociales, Universidad Nacional de Trujillo, pp. 236-244.

106. Sharon D, Glass-Coffin B, Bussmann RW (2009) La mesa de Julia Calderón de Ávila. In Galvez C (ed) Medicina Tradicional Conocimiento Milenario. Serie Antropología, no 1. Museo de Arquelogía, Antropología e Historia, Facultad de Ciencias Sociales, Universidad Nacional de Trujillo, pp. 245-254.

107. Tilbert JC, Kaptchuk TJ (2008) Herbal medicine research and global health, an ethical analysis. Bulletin of the World Health Organization 86:594-599.

108. United Nations Conference on Trade and Development (2000) Systems and National Experiences for Protecting Traditional Knowledge, Innovations and Practices. Background Note by the UNCTAD Secretariat Geneva, United Nations Conference on Trade and Development, (document reference TD/B/COM.1/ EM.13/2).

109. Van den Eynden V, Cueva C, Cabrera O (2004) Of "Climbing Peanuts" and "Dog's Testicles". Mestizo and Shuar plant nomenclature in Ecuador. Journal of Ethnobiology 24(2):279-306.

110. Vandebroek I, Balick MJ, Ososki A, Kronenberg F, Yukes J, Wade C, Jiménez F, Peguero B, Castilloin D (2010) The importance of botellas and other plant mixtures in Dominican traditional medicine. Journal of Ethnopharmacology 128:20-41.

111. Venero B (2005) Componentes de la Diversidad Biológica Peruana Patentados en el Extranjero, La Experiencia de Maca. ¿Cómo combatir la biopiratería? In: Ferro P, Ruiz M (eds) ¿Cómo prevenir la Biopiratería en el Perú? Reflexiones y Propuestas. Lerma Gómez E.I.R.L., Lima, pp. 50-55 and 74-78.

112. Wassen H (1976) Was Espingo (Ispincu) of Psychotropic and Intoxicating Importance for Shamans in Peru? In: Agehananda Bharati (ed) The Realm of the Extra-Human Agents and Audiences. Mouton Publishers, The Hague-Paris. Distributed in the United States and Canada by Aldine, Chicago.

113. Wassen H (1987) "Ulluchu" in Moche Iconography and Blood Ceremonies: The Search for Identification. Göteborg Etnografiska Museum, Annals 1985/86.

114. Weil AT (1978) Coca leaf as therapeutic agent. American Journal of Drug and Alcohol Abuse 5(1):75-86.

115. World Health Organization (1998) Technical Briefing on Traditional Medicine. Forty-ninth Regional Committee Meeting, Manila, Philippines, 18 September 1998, Manila, WHO Regional Office for the Western Pacific.

116. World Health Organization (1999a) Consultation Meeting on TM and Modern Medicine, Harmonizing the Two Approaches. Geneva, World Health Organization, (document reference. (WP)TM/ICP/TM/001/RB/98–RS/99/GE/32(CHN)), World Health Organization, Geneva.

117. World Health Organization (1999b) Traditional, Complementary and Alternative Medicines and Therapies. Washington DC, WHO Regional Office for the Americas/Pan American Health Organization

(Working group OPS/OMS).

118. World Health Organization (2002a) Implementation of the WHO Strategy for Prevention and Control of Chronic Respiratory Diseases. WHO/MNC/CRA/O2.2, World Health Organization, Geneva.

119. World Health Organization (2002b) WHO Traditional Medicine Strategy 2002–2005. World Health Organization, Geneva.

120. World Health Organization (2002c) Foodborne disease. World Health Organization, Geneva.

121. World Health Organization (2005) Urinary Tract Infections in infants and children in developing countries in the context of IMCI. World Health Organization, Geneva.

122. World Health Organization (2007) Sexually transmitted infections fact sheet. World Health Organization, Geneva.

123. World Health Organization (2009a) Declaración de Alma Ata. World Health Organization, Geneva.

124. World Health Organization (2009b) World health fact sheet. World Health Organization, Geneva.

125. Zamora Pérez DI (2007) Creación de un Órgano Administrativo Especializado en imponer sanciones a los concesionarios mineros en caso de incumplimiento de sus obligaciones ambientales. Tesis de Abogado Universidad Privada Antenor Orrego, Trujillo, Perú.

CHAPTER 10 ETHNOBOTANY IN THE NEPAL HIMALAYA

Ripu M Kunwar and Rainer W Bussmann

BACKGROUND

The term ethnobotany was coined by John W. Harsberger in 1896 [1] and was considered as the art of collection of useful plants by a group of people and the description of the uses of plants. Ford developed the science of ethnobotanical study [2] and included the understanding of knowledge systems through the use of anthropological methods [3]. Over the last century, ethnobotany has evolved into a scientific discipline that focuses on the people-plant relationship in a multidisciplinary manner, incorporating not only collection and documentation of indigenous uses but also ecology, economy, pharmacology, public health, and other disciplines [4]. Presently, ethnobotany has become increasingly valuable in the development of health care and conservation programs in different parts of the world [5]. Ethnobotanical studies that explore and help to preserve knowledge are therefore urgently needed before traditional folklores are lost ever [6].

Ethnomedicine, a branch of ethnobotany, is a set of empirical local practices embedded in the indigenous knowledge of a social group often transmitted orally from generation to generation [7] with intent to understand social, cultural, and economic factors [8] influencing health problems and to overcome such problems. It is a suitable source of information regarding useful medicinal plants that can be targeted for sustainable domestication and management [9]. Ethnobotany and ethnomedicine as sciences addressing indigenous knowledge and practices therefore are now very important for establishing management programs [10].

Justification

The dialectical relationship between indigenous knowledge and practices shape, the ecosystem and affects the constituent plant populations [11]. By incorporating indigenous knowledge and use in the process of scientific research, new hypotheses for the sustainable conservation of the resources can be developed [12]. Indigenous knowledge and use have to be analyzed to develop appropriate management measures that build on both scientific and local knowledge [13]. Due to changing perception of the local people, and the ever increasing influence of global commercialization and socio-economic transformation, indigenous knowledge on plant resource use is constantly diminishing [14, 15]. Due to the lack of organized sustainable and scientifically monitored cultivation and harvesting, proper management techniques, and lack of awareness of social factors, the number of useful plant resources is decreasing at an alarming rate [16]. Furthermore, the indigenous knowledge on the use of lesser-known plants is also rapidly declining [17]. The present study therefore reviews the indigenous knowledge and use of plant resources of Nepal Himalayas along the altitudinal and longitudinal gradient, an area so far unstudied in Nepal.

Context

The first scientific study of Nepalese useful and medicinal plants was conducted by Francis Buchanan, who collected plants from 1802–1803. He was followed by Nathaniel Wallich 1820–1821 [18]. Don [19] and Wallich [20] also collected plants in the Nepal Himalayas and recorded their uses. The earliest published ethnomedicinal-botanical study of Nepal was a paper on medicinal and food plants from eastern Nepal by Banerji [21]. His study was followed by many other researches [e.g. [22, 23, 24, 25, 26, 27, 28, 29, 30]]. Since the 1980s, extensive ethnobotanical studies were conducted. However, none of them has analyzed patterns of the usage of plant resources along the altitudinal and longitudinal gradient. All previous studies enumerated only the plant diversity and usage of plant resources of selected small sites and selected tribal groups, without attempting to synthesize the plant use information in Nepal.

Citation: Ripu M Kunwar and Rainer W Bussmann. Ethnobotany in the Nepal Himalaya, Journal of Ethnobiology and Ethnomedicine, 2008; 4:24. DOI: 10.1186/1746-4269-4-24.

Review and Field Visits

A total of 264 studies, with emphasis on ethnobotany, ethnomedicine, and diversity of medicinal and aromatic plants, conducted between 1979 to 2006, were consulted for the present analysis. Of these papers only 76 were including data on the diversity and distribution of medicinal and aromatic plants, ethnobotany, as well as aspects of ethnomedicine and conservation. Four field visits were carried out in May and December 2006, January–February 2007 and April–May 2008, focusing on the Baglung, Baitadi, Dadeldhura, Darchula, Doti, Gulmi, and Kanchanpur districts of west Nepal, stretching from 80°5' to 83°15' E longitude 28°27' 30" to 30°15' N latitude and 390 m to 4570 m altitude. Field visits were meant for data verification and cross checking.

Nepal extends along the Himalayan range between the latitudes 26°22' – 30°27' and longitudes 80°04' – 88°12'. Altitudes vary from less than 60 m in the lowland of Terai in the South to the crest of the Himalaya reaching 8848 m in the North. In the present study, for easier interpretation and analysis along the longitudinal gradient, the country's longitude was divided into three bio-geographical regions: West Nepal (80°E to 83°E), Central Nepal (83°E to 86°E) and East Nepal (86°E to 88°E), [31, 32]. An analysis of the altitudinal gradient was made in 1000 m intervals as 60 – 1000 m, 1000 – 2000 m, 2000 – 3000 m, and 3000 – 4000 m.

RESULTS AND DISCUSSION

History and Plant Usage

Plants have been one of the most important sources of food and medicine since the dawn of human civilization. Archaeological evidence of 60,000 year-old Neanderthal burial grounds in Shanidar, Iraq, points to the use of plants like Marshmallow, Yarrow and Groundsel, which are still used in contemporary folk medicine [33]. Evidence for the medicinal use of *Papaver somniferum*, the opium poppy, dates back 8000 years [34, 35]. Concomitantly, the earliest written record of plants used as medicine originating from the Himalayas are found in the 6,500 year old texts of the Rigveda [36], followed by Atharveveda (2000-1000 BC) and Auryveda (600-100 BC) [37, 38]. The oldest account recording the uses of 278 Nepalese medicinal plants is 'Saushrut Nighantu,' written in 878 AD (935 BS). Later 'Nepali Nighantu,' an elaborated encyclopedia with information on the traditional knowledge of 750 plant species was published by the Royal Nepal Academy in 1969 [39].

Until the middle of the 19th century, plants were the main therapeutic agents used by humans, and even today almost 80% of the world population rely to some extent on medicinal plants for their primary healthcare needs. The use of nearly 3000 plant species as food during the course of human civilization has been documented, but only about 150 species have been cultivated [40] and less that 10 plant species are meeting over 90% of the world food demand [41]. Human survival still cannot be imagined without plants. The importance of plants is substantial and reflected in the large variety of products such as food, fiber, fodder, vegetables, medicinal plants, and aromatic plants. The contribution of medicinal plants in Nepal is important [42]. A total of 1012 useful plants are documented for Nepal [43], of which 554 (54.74%) have ethnomedicinal in properties. An review of research conducted in the Nepal Himalayas indicated that nearly 40% of the studies were related to medicinal plants and ethnomedicine [43], underlining that ethnomedicine is utmost important in the Nepal Himalayas.

Medicinal Plants and Their Distribution

About 60% of the world population and 60–90% of the population of developing countries (80% in Nepal [44], 70% in India, 80% in Pakistan, 65% in Sri Lanka, 90% in Bangladesh, 85% in Burma, and 60% in Indonesia [45] rely on traditional medicine [46], and about 85% of the traditional remedies for primary health care are derived from plants [47].

In the world, 10–18% of all plant species are used medicinally [48] while in Nepal medicinal plants account about 20–28% of the local flora [49, 50, 51, 52, 53, 54]. The data for India and China is significantly higher i.e. 44% and 29–41% respectively [55, 56] (Table 1). The present analysis found that an average of 56% of higher plants were ethnobotanically important, and 54% were used as ethnomedicine in the Nepal Himalayas (Table2). This indicates that there are more ethnobotanically and ethnomedicinally important plant species in the Nepal Himalayas than previously estimated. Similar accounts of a much higher incidence of plants used in ethnomedicine (more than the estimated 10–18% [57] was recorded all over the world; e.g. about 24% in Costa Rica [58], 25–42% in Kenya [59], 27% in Israel [60], 31–51% in Brazil [61], and 46% in Guyana [62], 64% in Jammu Kashmir, India [63], 54% in Similipahar, Orissa, India [64], and 48% in Beni, Bolivia [65].

Table 1. Ethnobotanical and ethnomedicinal plants in the world

SN	Location and total no. of higher plants	Ethnobotanical plants as % of total flora	Ethnomedicinal plants	Source
1	Nepal, 7000	28%	21% of total flowering plants	[49, 51]
2	Nepal, 7000	56.35% *	53.59% of total plants in use	Present study
3	Hindu Kush Himalaya, 25000	33–40%		[120]
4	Indian Himalaya, 8000		21% of total flowering plants	[55]
5	India, 17000	14%	44% of total flowering plants	[122]

6	Pakistan, 6000		12% of total flowering plants	[103]
7	China, 29700		41% of total flowering plants	[121]
8	World, 297000 <		10–18% of total plants	[57]

* Percentage of plants used in terms of species richness of sites

Table 2. Useful plants of Nepal Himalaya

Ecological zones {m}	Distribution of medicinal plants {%}			Average {%}	% of ethnobotanically important plants			Average {%}	% of ethnomedicinally important plants			Average {%}
	W	C	E		W	C	E		W	C	E	
60–1000	32.06	44.28	42.85	39.73 49.20 [36]	40.72 (3)	-	68.20 (1)	54.46	56.76 (8)	38.55 (7)	51.57 (6)	48.96
1000–2000	40.0	50.95	47.46	46.13 53.96[36]	70.78 (1)	65.29 (4)	-	68.03	70.33 (2)	43.57 (17)	48.71 (2)	54.20
2000–3000	29.68	33.49	30.0	31.05 35.70[36]	38.38 (1)	45.46 (1)	59.25 (1)	47.69	76.24 (6)	60.98 (5)	48.51 (3)	61.91
3000–4000	16.03	17.30	15.07	16.13 18.09[36]	-	-	-	-	60.42 (6)	71.56 (1)	18.46 (2)	50.14
Average	**29.44**	**36.50**	**33.92**	**39.23**	**49.96**	**55.37**	**63.72**	**56.72**	**65.93**	**53.66**	**41.81**	**53.80**

The figure in parenthesis represents the number of sample publications reviewed

Asia represents one of the most important centers of knowledge with regard to the use of plant species for treatment of various diseases. Examples are the Ayurveda, Amchi (traditional healing system of Tibet and mountain areas of Nepal), Siddha, Unani, and Chinese systems of medical care [66, 67]. Folklore medicinal systems (traditional healing and faith healing) are also important in Nepal. In this context, it is interesting to note that the Himalayan medicinal plants are the major contributors to the aforementioned systems. The topographical characteristics of the Himalayas have resulted in a variety of ecological niches that host diverse medicinal plants [68]. It has been estimated that the Himalayan region harbors over 10,000 species of medicinal and aromatic plants, supporting the livelihoods of about 600 million people living in the area [69]. The Nepal Himalayas include about 2,000 species with medicinal and aromatic values, and more than 1,400 of these are known to be used locally [49, 50, 51, 53] particularly as medicines [52].

Medicinal and aromatic plant species richness increases along increasing altitude up to 2000 m [70] and then continues to decline. The temperate and alpine zones harbor highly valued medicinal plants [71]. The diversity of medicinal plants in a mid-elevation peak model, was comparable to other studies [72, 73, 74]. Bhattarai and Vetaas [75] highlighted an insignificant relationship between herbaceous species richness and elevation and climatic factors in the Himalayas. The distribution of medicinal herbs at high altitudes is mainly controlled by both environmental and ecophysiological factors [74, 76]. The maximum species richness at 1000–2000 m could be associated with optimum energy and rainfall [75]. The hard boundary effect might also influence the species richness at high elevation [77]. The trans-Himalaya ecosystem is a fragile biome, characterized by a reduced growing season [76], low productivity [78], high intensity of solar radiation, and high degree of resource seasonality [79], resulting in the control of distribution, richness, physiological stress, and metabolic compounds of medicinal plants. The high altitude medicinal plants contain secondary metabolic compounds as an adaptive strategy to reduce the damaging effects [80, 81].

Ethnobotany and Ethnomedicine

Many people use plants as remedies as an alternative or in addition to visiting western health care practitioners. The extent of plant use differs with location. The use of plants for subsistence, medicines, and source for supplementary income is highly variable. About 85% of the rural population of Nepal are said to use herbal remedies [82]. The use of plant resources as herbal remedies is especially important at higher altitudes. We found that the percentage of plants used as medicine increased along increasing altitude but there was no significant trend in increased usage of plants for other purposes. The overall plant use was proportionate to the number of plant species used in ethnomedicine (Fig. 1).

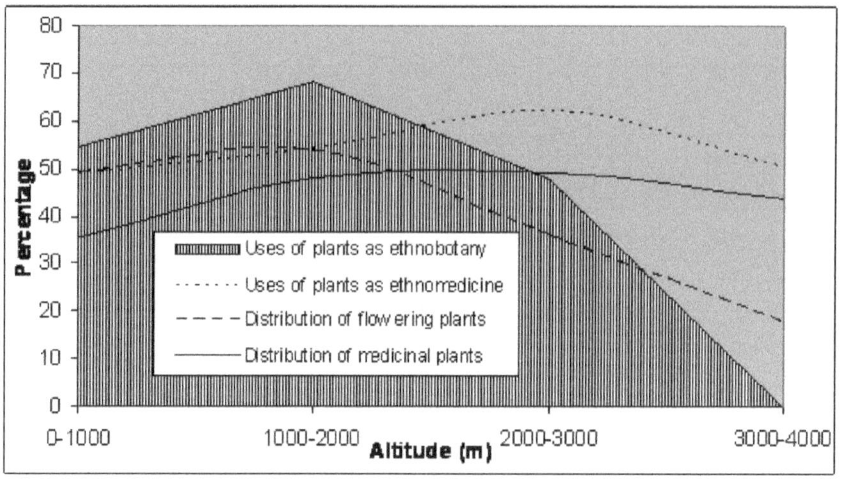

Figure 1. Species richness and usage of plants along altitude in Nepal Himalaya.

A positive correlation between species richness and number of ethnobotanically useful plants was found, but the relationship was not significant. In contrast, a significant positive correlation was found by Salick et al [83,84]. The ethnobotanical use as percentage of plant species richness (number of enumerated plant species) decreased from east to west Nepal, whereas the ethnomedicinal usage increased from east to west Nepal (Tab. 2). Though there was a record of the lowest number of medicinal plants in west Nepal [36], we found the highest usage of plants as ethnomedicine from the same region. Similar relationships were found by other studies [85, 86]. The reason for this could be the excessive utilization of local plant resources as remedies. Lack of modern facilities and services and less development in west Nepal compel the population to use existing plant resources for their immediate needs. Eastern Nepal is comparatively moister than western Nepal [87] and the moisture gradient accounts for differences in plant species richness [70].

High altitude medicinal plants provide quality products [80, 81], and this is the reason why they are often the first choice of local users as immediate therapy and by pharmaceutical companies as precious ingredients. Our analysis shows that the proportion of usage of plants as ethnomedicine steadily increased with increasing altitude (Tab. 2). Pohle [88] found, however, an insignificant relationship between the diversity of the flora and the number of wild plants used by local people in mid hills. Higher uses of plants as ethnomedicine at higher altitude (Fig. 1) was attributed to the absence of modern medical facilities [89], and intensive uses of plants as remedies in mountain areas. In remote and high altitude areas, medicinal herbs are the main ingredients of local medicines and the traditional health care system is considered as the main lifeline [90] and frequently the first choice [91]. There is greater dependence among the population upon natural resources [88]. The optimum uses of the limited plant resources revealed a higher percentage of uses in high altitude areas. However, this trend could not be observed in lowland Terai. The usage of plants as ethnomedicine was less in Terai, with the richness of flowering plants and medicinal plant was higher. The low number (48.96%) of ethnomedicinal plants found in Terai was consistent with the findings of Bishokarma et al [[92], (35.15%)] from the same site. Lesser usage of local plants for therapy was attributed to the availability and accessibility of infrastructures, communication, transport, development, and as such less dependant on the availability of plant resources; but a higher preferences towards allopathic medicines. It was also associated with limited knowledge and limited skilled individuals. However, some marginalized communities and tribal groups of lowland Terai, with poor access to the modern medical facilities, were still utilizing local plant resources for local therapy. Subsequently, most communities still use folk herbal remedies as a readily and cheaply available alternative. A decline of the traditional knowledge has been underway for hundreds of years [93], and the knowledge on medicinal plant use is clearly disappearing fast in low lands of Nepal [94].

The higher percentage of ethnomedicinal uses of plants in high altitude areas was due to the preferences given by local people to the traditional herbal remedies, and a situation of having no alternative choices [95,96], as well as poverty and belief in the effectiveness of folklore herbal remedies [46]. The greater usage of plants as herbal remedies in mountain areas is also due to the prevalence of traditional health systems such as folklore medicinal systems and the Amchi system. The traditional healing systems are both culturally acceptable [97, 98] and induce locales to use plant resource most. Because of the high level of cohesion and strong cultural links with nature in mountain areas [11, 99], local people have a close affinity with locally available resources and easily internalize the resources for their livelihood and needs because these are also readily available.

Home herbal treatment is not merely a medical system but a part of culture [100], and the extensive use of locally available plant resources was observed in mountain areas. Most of the people of mountain areas, particularly elderly people, farmers, and mid-wives, are well acquainted to medicinal herbs and their usage [101], and plants are prescribed to the patients whenever required [102]. The availability of traditional healers is also higher in rural and mountain areas. The use of herbal remedies in mountain areas to alleviate suffering is perhaps as old as the origin of man itself [103], and there is a general trend of increase in the use of medicinal plants [104].

Besides health care and subsistence uses, medicinal plants have a high potential as alternative income-generating sources of the rural hilly populace of the region [105]. Therefore, sustainable use of the resources and conservation of indigenous knowledge of medicinal plants may compliment the income of local people. Medicinal plants are mainly harvested in the wild, traded, and eventually consumed as processed form in lowland cities [106]. Up to 50% of the Nepal's rural household's income is derived from commercial collection of medicinal and aromatic plants [107, 108]. The collection of medicinal plants in a sustainable manner is an integrated process with potential for development and conservation [109].

The indigenous knowledge and practice of usage of medicinal plants in rural areas of Nepal is passed down through oral tradition and personal experiences [110]. The knowledge clearly decreased with age. People of ages between 40–60 possessed greater knowledge on identification and uses of medicinal and aromatic plants in Nepal Himalaya [111], which was consistent to the observation of Philips and Gentry [112]. The young generations tend to leave ancestral practices behind, refocusing their interests on treatments offered by western medicine [113]. Due to changing lifestyles, perception as well as social transformation, the plant resource and indigenous knowledge of utilization are being severely degraded [114]. This impact is inevitable to the Nepal Himalayas and plant resources are in great peril. Indigenous knowledge systems are not only of value for the cultures from which they evolve, but also for scientists and planners striving to improve the living conditions in rural societies [115]. Lambert et al [116] pointed out that preserving and enhancing the indigenous plant knowledge and use was equivalent to 'rescuing a global heritage'. Indigenous plant based traditional knowledge and use has become a recognized tool in search for new sources of drugs and pharmaceuticals [117]. In countries like Kenya [118] and Nepal [119], where the indigenous knowledge is predominantly used for utilization of plant resource for various purposes, and high priority needs to be given to the documentation of indigenous knowledge and use of plant resources to help their conservation.

CONCLUSION

The richness of medicinal plants was decreasing with increasing altitude, but the percentage of plants used as medicine steadily increased with increasing altitude. This was due to preferences given to herbal remedies in high altitude areas and a situation of having no alternative choices, poverty and belief on effectiveness of folklore herbal remedies. Enhancing the sustainable use and conservation of indigenous knowledge of useful and medicinal plants may benefit and improve the living standard of poor people.

ACKNOWLEDGEMENTS

We thank Water Resource Consult, Kathmandu; the Canadian Center for International Studies and Co-operation, Kathmandu; the Department of National Park and Wildlife Conservation, Kathmandu; and the Resource and Environmental Conservation Society Nepal, Kathmandu for providing facilities for field studies and data analysis.

REFERENCES

1. Davis, EW, Schultes RE, Reis SV, null. Ethnobotany: an old practice, a new discipline. Ethnobotany: Evolution of Discipline. 1995: 40-51
2. Ford, RL, Ford RL, null. The nature and status of ethnobotany. Anthropological Papers. 1978
3. Shengji, P, Thomas YA, Karki M, Gurung K, Parajuli D, null. Ethnobotany and modernization of traditional Chinese medicine. Proceeding of Himalayan Medicinal and Aromatic Plants, Balancing Use and Conservation, December 15–20, 2002. 2002: 70-78
4. Gomez-Beloz, A. Plant use knowledge of the Winikina Warao: the case for questionnaires in ethnobotany. Economic Botany. 2002; 56: 231-241
5. Balick, MJ. Annals of the Missouri botanical garden. 1996; 4: 57-65
6. Chaudhary, RP. Biodiversity in Nepal: Status and Conservation. 1998
7. Bussmann, RW, Sharon, D. Traditional medicinal plant use in Northern Peru: tracking two thousand years of healing culture. Journal of Ethnobiology and Ethnomedicine. 2006; 2: 47
8. Bhattarai, NK. Folk herbal remedies of Sindhupalchok district, central Nepal. Fitoterapia. 1992; 63: 145-155
9. Njoroge, GN, Bussmann, RW, Gemmill, B, Newton, LE, Ngumi, VW. Utilization of weed species as source of traditional medicines in central Kenya. Lyonia. 2004; 7: 272-287
10. Cohen, JI, Alcorn, JB, Poster, CS. Utilization and conservation of generic resources: international projects for sustainable agriculture. Economic Botany. 1991; 45: 190-199
11. Ghimire, SK, McKey, D, Aumeeruddy-Thomas, Y. Heterogeneity in ethnoecological knowledge and management of medicinal plants in the Himalayas of Nepal: implications for conservation. Ecology and Society. 2005; 9: 6
12. Henfrey, TB. Ethnoecology, Resource Use, Conservation and Development in a Wapishana Community in the South Rupununi Guyana [PhD dissertation]. 2002
13. Ticktin, T, Johns, T. Chinanteco management of Aechmea magdalenae: implications for the use of traditional ecological knowledge and traditional resource management in management plans. Economic Botany. 2002; 56: 117-191
14. Gadgil, M, Birkes, F, Folkes, C. Indigenous knowledge of biodiversity conservation. Ambio. 1993; 22: 151-160
15. Kunwar, RM, Adhikari, N. Ethnomedicine of Dolpa district, Nepal: the plants, their vernacular names and uses. Lyonia. 2005; 8: 43-49
16. Kunwar, RM, Duwadee, NPS. Ethnobotanical notes on flora of Khaptad national park, far western Nepal. Himalayan Journal of Sciences. 2003; 1: 25-30
17. Kala, CP. Current status of medicinal plants used by traditional Vaidyas in Uttaranchal, India. Ethnobotany Research and Applications. 2005; 3: 267-278
18. Rajbhandari, KR. Ethnobotany of Nepal. 2001
19. Don, D. Prodromus Florae Nepalensis London. 1825
20. Wallich, N. Tentamen Flora Nepalensis. 1826
21. Banerji, ML. Some edible and medicinal plants from east Nepal. Journal of Bombay Natural History Society.

1955; 53: 153-156

22. Pandey, BD. The Wealth of Medicinal Plants of Nepal. 1964: 133-140
23. Singh, SC. Some wild plants of food value in Nepal. Tribhuvan University Journal. 1968; 4: 50-56
24. Bhatt, DD. Natural History and Economic Botany of Nepal. 1970
25. Medicinal Plants of Nepal. 1970
26. Jest, C. Plants Sauvages Utilizes Comme Ailements a Dolpo Haulte Vallee Himalay Ethmedu Nepal. 1972: 325-332
27. Rajbhandari, KR. Some plants of economic value in Nepal. The Science Magazine. 1974; 2: 24-32
28. Dobremej, JF. Exploitation and Prospects of Medicinal Plants in Eastern Nepal. 1976: 97-107
29. Adhikari, PM, Shakya, TP. Pharmacological screening of some medicinal plants of Nepal. Journal of Nepal Pharmacological Association. 1977; 5: 41-56
30. Sacherer, J. The high altitude ethnobotany of the Rolwaling Sherpas. Contribution to Nepalese Studies. 1979; 4: 45-64
31. Banerji, ML. Outline of Nepal phytogeography. Vegetatio. 1963; XI: 5-6
32. Hara, H, Stearn, WT, Williams, LHJ. An Enumeration of the Flowering Plants of Nepal. 1978; 1
33. Lietava, J. Medicinal plants in middle Palaeolithic grave Shanidar IV. Journal of Ethnopharmacology. 1992; 35: 263-266
34. Stockwell, C. Nature's Pharmacy: A History of Plants and Healing. 1989
35. Lewington, A. Plant for People. 1990
36. Malla, SB, Shakya, PR, Majupuria TC, null. Medicinal plants of Nepal. Nepal – Natures' Paradise. 1984: 261-297
37. Nambier, VPK, Bhattarai N, Karki M, null. Improved harvesting, processing and storage of medicinal plants: their role in conservation and quality of plant based drugs. Proceeding of Sharing Local and National Experience in Conservation of Medicinal and Aromatic Plants in South Asia, January 21–23, 2001. 2002: 42-45
38. Kunwar, RM, Nepal, BK, Kshetri, HB, Rai, SK, Bussmann, RW. Ethnomedicine in Himalaya: a case study from Dolpa, Humla, Jumla and Mustang districts of Nepal. Journal of Ethnobiology and Ethnomedicine. 2006; 2: 27
39. National Register of Medicinal and Aromatic Plants. 2004
40. Ecological Aspect of the Development in the Humid Tropics. 1982
41. Wilkes, HG. New or potential crop or what to anticipate for the future Paper presented at the annual meeting of the American Association for the Advancement of Science, January 1981, Toronto, Canada. 1981
42. Kunwar, RM. Non-timber Forest Products of Nepal: A Sustainable Management Approach. 2006
43. Shrestha, KK, Rajbhandary, S, Tiwari, NN, Poudel, RC, Uprety, Y. Ethnobotany in Nepal: Review and Perspectives [Report]. 2004
44. Manandhar, NP. Ethnobotanical notes on unexploited wild food plants of Nepal. Ethnobotany. 1995; 7: 95-101
45. Shinwari, ZK, Watanabe, T, Yousaf, Z. Medicinal plants of Pakistan: an overview. Proceeding of Nepal-Japan Joint Symposium on Conservation and Utilization of Himalayan Medicinal Resources, November 6–11, 2000, Kathmandu, Nepal. 2000: 298-304
46. Shrestha, PM, Dhillion, SS. Medicinal plant diversity and use in the highlands of Dolakha district, Nepal. J Ethnopharmacol. 2003; 86: 81-96
47. Farnsworth, JD, Wilson EO, null. Screening plants for new medicines. Biodiversity. 1988: 83-97
48. Farnsworth, NR, Soejarto, DD, Akerele O, Heywood V, Synge H, null. Global importance of medicinal plants. The Conservation of Medicinal Plants. 1991: 25-51
49. Shrestha, GL, Shrestha, B, Shrestha R, Shrestha B, null. An overview of wild relatives of cultivated plants in Nepal. Wild Relatives of Cultivated Plants in Nepal Proceedings of National Conference on Wild Relatives of Cultivated Plants in Nepal, June 2–4, 1999. 1999: 19-23
50. Tiwari, NN. Wild Relatives of Cultivated Medicinal and Aromatic Plants (MAPs) in Nepal. Proceedings of National Conference on Wild Relatives of Cultivated Plants in Nepal, June 2–4, 1999. 1999: 141-148
51. Shrestha, KK, Tiwari, NN, Ghimire, SK. Medicinal and aromatic plants database of Nepal. Proceeding of Nepal-Japan Joint Symposium on Conservation and Utilization of Himalayan Medicinal Plant Resources, Nov 6–11, 2000, Kathmandu, Nepal. 2000: 53-74
52. Manandhar, NP. Plants and People of Nepal. 2002
53. Baral, SR, Kurmi, PP. Compendium of Medicinal Plants in Nepal. 2006
54. Ghimire, SK, Jha PK, Karmacharya SB, Chettri MK, Thapa CB, Shrestha BB, null. Medicinal plants in the Nepal Himalaya: current issues, sustainable harvesting, knowledge gaps and research priorities. Medicinal Plants in Nepal: an anthology of contemporary research. 2008: 187-193
55. Samant, SS, Dhar, U, Palni, LMS. Medicinal Plants of Indian Himalaya: Diversity, Distribution and Potential Values. 1998
56. Shengji, P, Huying, H, Lixin, Y. Important plant areas for medicinal plants in Chinese Himalaya: national report Regional workshop on identification and conservation of important plant areas for medicinal plants in the Himalayas 19–22, September, 2006, Kathmandu, Nepal. 2006
57. Schippmann, U, Leaman, DJ, Cunningham, AB. Impact of Cultivation and Gathering of Medicinal Plants on Biodiversity: Global Trends and Issues. 2002
58. Kappelle, M, Avertin, G, Juarez, ME, Zamora, N. Useful plants within a Campesino community in a Costa Rican Montane cloud forest. Mountain Research and Development. 2000; 20: 162-171
59. Bussmann, RW, Gilbreath, GG, John, S, Lutura, M,

Lutuluo, R, Kunguru, K, Wood, N, Mathenge, S. Plant use of the Maasai of Sekenani valley, Maasai Mara, Kenya. Journal of Ethnobiology and Ethnomedicine. 2006; 2: 22

60. Ali Shtayeh, MS, Abu Ghdeib, SI. Antimycotic activity of twenty-two plants used in folkloric medicine in the Palestinian area for the treatment of skin diseases suggestive of dermatophyte infection. Mycoses. 1999; 42: 665-672

61. Hanazaki, N, Tamashiro, JY, Leitão-Filho, HF, Begossi, A. Diversity of plant uses in two Caicara communities from the Atlantic forest coast, Brazil. Biodiversity and Conservation. 2000; 9: 597-615

62. Johnston, M. Tree population studies in low diversity forests, Guyana. II. Assessment on the distribution and abundance of non-timber forest products. Biodiversity and Conservation. 1998; 7: 73-86

63. Kapur, SK, Shahi, AK, Sarin, YK, Moerman, DE. The medicinal flora of Majouri-kirchi forests (Jammu and Kashmir), India. Journal of Ethnopharmacology. 1992; 36: 87-90

64. Girach, RD, Shaik, AA, Singh, SS, Ahmad, M. The medicinal flora of Similiphar forests, Orissa, India. Journal of Ethnopharmacology. 1999; 65: 165-172

65. Boom, BM. Ethnobotany of the Chacobo Indians, Beni, Bolivia. Advances in Economic Botany. 1987; 4: 1-68

66. Karki, M, Thomas YA, Karki M, Gurung K, Parajuli D, null. The organic production of medicinal and aromatic plants: a strategy for improved value addition and marketing of products from the Himalayas. Proceeding of Himalayan Medicinal and Aromatic Plants, Balancing Use and Conservation, December 15–20, 2002. 2002: 56-69

67. Kala, CP, Farooquee, NA, Dhar, U. Prioritization of medicinal plants on the basis of available knowledge, existing practices and use value status in Uttaranchal, India. Biodiversity and Conservation. 2004; 13: 453

68. Singh, JS, Singh, SP. Forests of Himalaya: Structure, Functioning and Impact of Man. 1992

69. Shengji, P. Ethnobotanical approaches of traditional medicine studies: some experiences from Asia. Pharmacoceutical Biology. 2001; 39: 74-79

70. Bhattarai, KR, Ghimire, MD. Commercially important medicinal and aromatic plants of Nepal and their distribution pattern and conservation measure along the elevation gradient of the Himalayas. Banko Janakari. 2006; 16: 3-13

71. Ghimire, SK, McKey, D, Aumeeruddy-Thomas, Y. Himalayan medicinal plant diversity in an ecologically complex high altitude anthropogenic landscape, Dolpo, Nepal. Environmental Conservation. 2006; 33: 128-140

72. Dobremez, JF, Jha PK, Ghimire GPS, Karmacharya SB, Baral SR, Lacoul P, null. Foreword. Environment and Biodiversity: in the context of South Asia. 1996: i-iii

73. Kunwar, RM, Chaudhary, RP. Status, vegetation composition and biomass of forests of Arun valley, East Nepal. Banko Janakari. 2004; 14: 13-18

74. Kala, CP. Status and conservation of rare and endangered medicinal plant in the Indian trans-Himalaya. Biological Conservation. 2000; 93: 371-379

75. Bhattarai, KR, Vetaas, OR. Variation in plant species richness of different life forms along a subtropical elevation gradient in the Himalayas. Global Ecology and Biogeography. 2003; 12: 327-340

76. Körner, C. The alpine life zone under global change. Gayana Botanica. 2000; 57: 1-8

77. Colwell, RK, Lees, DC. The mid-domain effect: geometric constraints on the geography of species richness. Trends in Ecology and Evolution. 2000; 15: 70-76

78. Kunwar, RM. Some threatened medicinal and aromatic plants: Status, trade and management practice in Dolpa district, mid-west, Nepal. Journal of Natural History Museum. 2002; 21: 173-186

79. Kala, CP, Mathur, VB. Patterns of plant species distribution in the trans-Himalayan region of Ladakh, India. Journal of Vegetation Science. 2002; 13: 751-754

80. Mikage, M, Mouri, C. Altitudinal variations of Berberine content in the bank of Berberis plants from West Nepal. Newsletter of Himalayan Botany. 2000; 26: 16-19

81. Iwashina, T, Omori, Y, Kitajima, J, Akiyama, S, Suzuki, T, Ohba, H. Flavonoids in translucent bracts of the HImalayan Rheum nobile (Polygonaceae) as ultraviolet shields. Journal of Plant Research. 2004; 117: 101-107

82. Dani, DD, Joshi SC, null. Population and society in Nepal: an overview. Nepal Himalaya: Geoecological Perspective. 1986: 163-173

83. Salick, J, Biun, A, Martin, G, Apin, L, Beaman, R. Whence useful plants? A direct relationship between biodiversity and useful plants among the Dusun of Mt Kinabalu. Biodiversity and Conservation. 1999; 8: 797-818

84. Salick, J, Anderson, D, Woo, J, Sherman, R, Cili, N, Dorje, S. Bridging Scales and Epistemologies: Linking Local Knowledge and Global Science in Multi-scale Assessments. 2004: 1-12

85. Burlakoti, C, Kunwar, RM, Jha PK, Karmacharya SB, Chettri MK, Thapa CB, Shrestha BB, null. Folk herbal medicines of Mahakali Watershed Area, Farwest Nepal. Medicinal Plants in Nepal: an anthology of contemporary research. 2008: 187-193

86. Kunwar, RM, Uprety, Y, Burlakoti, C, Chowdhary, CL, Bussmann, RW. Indigenous use and ethnopharmacology of medicinal plants in far west Nepal. Journal of Ethnobotany and Research Application. 2008

87. Singha, IL, Majupuria TC, null. Rainfall distribution. Nepal – Natures' Paradise. 1999: 56-58

88. Pohle, P. Useful plants of Manang district: a contribution to the ethnobotany of the Nepal-Himalayas. Nepal Research Center, Publication no 16. 1990

89. Uniyal, SK, Kumar, A, Lal, B, Singh, RD. Quantitative assessment and traditional uses of high value medicinal plants in Chhota Bhangal area of Himanchal Pradesh, Western Himalaya. Current Science. 2006; 91: 1238-1242

90. Dhar, U, Manjkhola, S, Joshi, M, Bhatta, A, Bisht, AK, Joshi, M. Current status and future strategy for development of medicinal plant sector in Uttranchal, India. Current Science. 2002; 83: 956-964

91. Bhattarai, NK. Traditional herbal medicines used to treat wounds and injuries in Nepal. Tropical Doctor. 1997; 27: 43-47

92. Bishokarma, BK, Kinsey, CK, Dangol, DR, Chaudhary, P. Folk use of plant resource at Madi valley of Chitwan, Nepal. Banko Janakari. 2005; 15: 43-48

93. Hamilton, AC. Medicinal plants, conservation and livelihoods. Biodiversity and Conservation. 2004; 13: 1477-1517

94. Mandar, LN, Chaudhary, RP. Medicinal plants and their traditional use by tribal people of Saptari district, Nepal. Proceeding of 1st National Botanical Conference Botanical Society of Nepal, Kathmandu, April, 1988. 1992: 33-41

95. Bhattarai, NK. Folk anthelmintic drugs of central Nepal. International Journal of Pharmacognosy. 1992; 30: 145

96. Bhattarai, NK. Medical ethnobotany in the Karnali zone, Nepal. Economic Botany. 1992; 46: 257-261

97. Bhatta, LR. Ethnobotanical study in a village at Rukum district, Nepal. Banko Janakari. 1999; 9: 40-43

98. Ghimire, SK, McKey, D, Aumeeruddy-Thomas, Y. Conservation of Himalayan medicinal plants: harvesting patterns and ecology of two threatened species. Nardostachys grandiflora and Neopicrorhiza scrophulariiflora Biological Conservation. 2005; 124: 463-475

99. Kunwar, RM, Parajuli, RR. Good governance in natural resource management: a case study from Dolpa district, west Nepal. Banko Janakari. 2007; 17: 17-24

100. Bhattarai, NK. Folk herbal remedies for diarrhea and dysentery in central Nepal. Fitoterapia. 1993; 64: 243-249

101. Niraula, K. Vegetation Analysis and Ecology of the Medicinal Plants in and around Tinjure Hill (Terhathum and Sankhuwasabha Districts), Eastern Nepal [MSc Thesis]. 2001

102. Storrs, A. Jhankris: the faith healers of Nepal. Nepal Traveler. 1997; 11: 34-39

103. Shinwari, ZK, Gilani, SS, Kohjoma, M, Nakaike, T. Status of medicinal plants in Pakistani Hindu Kush Himalayas. Proceeding of Nepal-Japan Joint Symposium on Conservation and Utilization of Himalayan Medicinal Resources, November 6–11, 2000, Kathmandu, Nepal. 2000: 257-264

104. Weckerle, CS, Huber, FK, Yongping, Y, Weibang, S. Plant knowledge of the Shuhi in the Hengduan mountains, Southwest China. Economic Botany. 2006; 60: 3-23

105. Bhattarai, NK. Ethnobotanical studies in central Nepal: the ceremonial plant foods. Contributions to Nepalese Studies. 1989; 16: 35-41

106. Olsen, CS. Valuation of commercial central Himalayan medicinal plants. Ambio. 2005; 34: 607-610

107. Edwards, DM. Non Timber Forest Products from Nepal: Aspects of Trade in Medicinal and Aromatic Plants Forest Research and Survey Centre monograph 1/96. 1996

108. Olsen, CS, Helles, F. Medicinal plants, markets and margins in the Nepal Himalaya: trouble in paradise. Mountain Research and Development. 1997; 17: 363-374

109. Hall, P, Bawa, KS. Methods to assess the impact of extraction of non-timber forest products on plant populations. Economic Botany. 1993; 47: 234-247

110. Bhattarai, NK, Gautam PL, Raina R, Srivastava V, Raychaudhuri SP, Singh BB, null. Traditional medicines: role of medicinal plants in present and future health cure. Prospects of Medicinal Plants. 1998: 96-104

111. Sherpa, S. The High Altitude Ethnobotany of the Walung People of Walangchungola, Kanchenjunga Conservation Area, East Nepal [MSc thesis]. 2001

112. Philips, O, Gentry, AH. The useful plants of Tambopata, Peru: II. Additional hypothesis testing in quantitative ethnobotany. Economic Botany. 1993; 47: 33-43

113. Schultes, RE, Von Reis, S. Ethnobotany: Evolution of Discipline. 1995

114. Silori, CS, Rana, AR. Indigenous knowledge on medicinal plants and their use in Narayan Sarovar Sanctuary, Kachchh. Ethnobotany. 2000; 12: 1-7

115. Warren, DM. Indigenous knowledge and development Background paper for seminar session on sociology Natural Resource Management and Agriculture Development December 3, 1990. 1990

116. Lambert, J, Srivastava, J, Vietmeyer, N. Medicinal Plants: Rescuing a Global Heritage. 1997

117. Sharma, PP, Mujundar, AM. Traditional knowledge on plants from Toranmal Plateau of Maharastra, India. Indian Journal of Traditional Knowledge. 2003; 2: 292-296

118. Heine, B, Heine, I, Koenig, C. Plant use and concepts – an ethnobotanical survey of the semi-arid and arid lands of east Africa. Part V. Plants of the Samburu (Kenya). Koelner Beitraege zur Entwicklungsforschung. 1988; 10: 1-286

119. Kala, CP, Dhyani, PP, Sajwan, BK. Developing the medicinal plant sector in northern India: challenges and opportunities. Journal of Ethnobiology and Ethnomedicine. 2006; 2: 32

120. Shengji, P, Tewari DW, null. Mountain culture and forest resource management of Himalayas. Himalayan Ecosystems. 1992

121. Shengji, P, Shinwari ZK, Hamilton A, Khan AA, null. Brief review of ethnobotany and its curriculum development in China. Proceeding of Workshop on Curriculum Development in Applied Ethnobotany, May 2–4, 2002, Lahore. 2002: 21-33

122. Shiva, MP. Inventory of Forestry Resources for Sustainable Management and Biodiversity Conservation. 1996

CHAPTER 11

ETHNOBOTANY AND ANTIBACTERIAL POTENTIALITY OF BLUMEA LACERA L.

Fakhruddin Ali Ahmed, Anisur Rahman, Sanzida Mubassara, and Gazi Mosharof Hossain

Keywords: Ethnobotany, antibacterial activity, Blumea lacera, Sundarban mangrove forest

INTRODUCTION

Primary tropical forest is generally considered to be the most likely habitat to discover new pharmaceuticals due to high biodiversity and endemism. However, many indigenous groups rely on forested and non-forested environments for medicinal plants. An ethnobotanical approach to drug discovery suggests that extreme environments may also harbor potentially useful medicinal compounds (Gentry, 1993; Macilwain, 1998). During the last ten years pace of development of new antimicrobial drugs has slowed down while the prevalence of resistance has increased astronomically (Akinpelu *et al.*, 2009). According to WHO, medicinal plants would be the best source to obtain a variety of drugs. Systematic screening of folk medicines and plants may result in the discovery of novel effective compounds (Tomoko *et al.*, 2002).

Blumea lacera L. belonging to the Asteraceae family is a common plain land weed of Bangladesh. This plant species also grows in Sundarban area and known as Kukursunga in that locality (Uddin *et al.*, 2011). *B. lacera* has an enormous medicinal value and been widely used in the traditional medicinal system of Bangladesh for a long time. It is used as antipyretic, stimulant, astringent, anthelmintic, febrifuge, diuretic and in hemorrhoids (Pandit *et al.*, 1996) bronchitis (Jha & Verma, 1996) as antimicrobial for healing cuts (Joshi, 1980) in urinary troubles (Singh, 1988) in irritation, as haemostatic (Kakrani & Saluja, 1994) and anti-inflammatory (Pal, 1972). There are scanty reports on ethnobotany and antimicrobial activity of *B. lacera* that are found in mesophytic habitat (Tiwari *et al.*, 2012; Singh *et al.*, 2010) but yet now there is no aforesaid information of this plant grows in mangrove or halophytic habitat. Considering the above fact, present study was aimed to document the indigenous knowledge on the utilization of *B. lacera* by the ethnic communities living in the vicinity of Sundarban mangrove forest of Bangladesh followed by comparative analysis of antibacterial activity of the aerial parts of the plant collected from Sundarban as well as from Jahangirnagar University campus.

MATERIALS AND METHODS

Ethnobotanical Study

The survey was conducted among the interim fisherman, forest guard, and honey collector of Kalagachia, Dublar Char and Kotka sites of Sundarban mangrove forest of Bangladesh. Data were collected in the field note book on the plant parts used, preparation of the drug, its dosage and administration etc. as given by the informants. Interviews were conducted with the help of a semi-structured questionnaire and the guided field-walk method of Martin (1995) and Maundu (1995). Interviews were conducted in the Bengali language. Notably, the Kavirajes practiced among the mainstream Bengali-speaking population.

Collection and Identification

Plant samples were collected from two different habitats namely Jahangirnagar University campus and Sundarban and were identified and authenticated by the Bangladesh National Herbarium (DACB), Mirpur, Dhaka, Bangladesh. The voucher specimens (accession no. of JU sample: DACB 38210 and Sundarbans sample: DACB 32624) have been deposited in DACB for further reference.

Preparation of Crude Extract and Fraction

The collected aerial parts of both the samples were sun-dried followed by drying in a hot air oven (Gallenkamp) at reduced temperature (<50°C). About 200 g powder was digested with 1000 ml of ethanol for three days accompanying with

Citation: Fakhruddin Ali Ahmed, Anisur Rahman, Sanzida Mubassara, and Gazi Mosharof Hossain. Ethnobotany and antibacterial potentiality of Blumea lacera L., Jahangirnagar University Journal of Biological Sciences, 2014; vol. 3, no. 2, p. 17-24. http://dx.doi.org/10.3329/jujbs.v3i2.28282.

occasional shaking and stirring. The whole mixtures was filtrated by a piece of clean, white cotton material. The extract was concentrated at 45°C under reduced pressure using a rotary evaporator.

Partition of the Crude Methanolic Extract

Solvent-solvent partitioning was done by using the protocol designed by Kupchan & Tsou (1973) and modified version of Wagenen *et al.* (1993). The crude ethanolic extract of the sample was first partitioned with *n*-hexane and methanol at ratio 1:1 using separating funnel and these two fractions were evaporated separately to dryness by rotary evaporator at 40°C. Then the methanol fraction was further partitioned to ethyl acetate and methanol fractions and evaporated to dryness. Fraction samples were then marked and stored in dark, cold and dry place.

Determination of Antibacterial Activity

Antibacterial potentiality of the fractions (*n*-hexane, ethyl acetate and methanol fractions) were determined by disc diffusion method following Saad *et al.* (2012). As per availability, two Gram positive bacteria *Bacillus cereus* and *B. subtilis* and eight Gram negative bacteria *Erwinia* sp., *Proteus mirabilis*, *Pseudomonas* sp., *Salmonella* sp., *Salmonella typhi*, *Serratia* sp., *Shigella flexneri* and *Vibrio cholera* were used to determine antibacterial activities. The sterile filter paper disc (8 mm diameter) containing three different concentration of the dry extracts (40 µg/ml, 80 µg/ml and 120 µg/ml), standard antibiotic disc (5 µ g/ml Streptomycine) and negative control (blank disc soaked with methanol solvent only) were used in the experiment and the results recoded as the mean values of three replications. The bacterial strains used for the experiment were collected as pure cultures from Microbiology laboratory, Department of Botany, Jahangirnagar University, Savar, Dhaka.

RESULTS AND DISCUSSION

Traditional medicinal practitioners tend to keep the treatment methods within the family, where the knowledge is passed on from generation to generation. As a result over the centuries, traditional medicine practitioners have developed a familiarity with plant species found in their region or habitat and the applicability of those plant species in treatment of various types of diseases. In the present study, interim peoples of the three sites namely Dublar char, Kalagachia and Kotka were surveyed to collect the information on medicinal use of *B. lacera*. About 25 people were asked to collect the information on *B. lacera* growing in Sundarbans. The plant was found non familiar to the local people as most of them could not recognize it. Restricted use and application was another sign of its less popularity in traditional medicinal system of Sundarban area. However, five of the respondent, all of them were illiterate, including kavirajes, fishermen and forest guard took the interviewers in areas from where they collected the plant, pointed out the plant, and mentioned their uses. They learned this knowledge from their forefathers. So it was difficult to gather much information in shortened period field works.

Table 1. Ethnobotanical information on *B. lacera* at different sites of Sundarban mangrove forest

Site	GPS coordinates and forest type*	Name and age of the informant	Profession	Uses
Dublar char	N21°43'47" E89°36'15.2" (Gewa-Kewra type forest)	Aziz (37)	Local practitioner	1. Warm leaf paste is applied to lower abdomen as diuretic. 2. Leaves juice (approx. 1 tea spoonful) with honey orally taken to treat most kinds of colds.
		Mrinmoy Nandi (63)	Kaviraj	1. Root and stem decoction is taken to treat blood dysentery. 2. Leaf paste (2½ tea spoonful) with black pepper powder (1 tea spoon) is applied in dog bite. 3. Leaf juice (1 cup) mixed with a tea spoonful of mustard seed is orally taken to relieve piles.
Kalagachia	N22°13.124' E89°14.329' (Gewa-Goran-Poshur type forest)	Shafiuddin (51)	Fisherman	1. Leaf paste is applied on forehead for the treatment of fever and to the joints for gout. 2. Pellet of leaf and seed powder is taken in leucorrhoea.
		Malek (45)	Forest Guard	Native people use the herb in the treatment of bleeding piles.
Kotka	N21°51.263" E89°47.257' (Sundari-Gewa-Kewra type forest)	Idris miah (39)	Forest Guard	Warm decoction of the plant is applied to hasten the expulsion of placenta after childbirth.

* Source: Hossain, 2003.

Kumar & Bhagat (2012) reported that roots and leaves of *B. lacera* are used as anti-pyretic and root decoction of *Blumea hieracifolia* for gastric trouble in Nepal. Leaf juice used as an astringent, febrifuge, diuretic and anti-helminthic whilst leaf paste along with a pinch of common salt is applied externally to cure eczema and ring worm (Satapathy *et al.*, 2012).

Table 2. Antibacterial activities of different fractions of *B. lacera* collected from JU campus and Sundarban mangrove forest

Fractions	Sample	Dose (µl)	*Bacillus subtilis*	*Bacillus cereus*	*Vibrio cholerae*	*Pseudomonas* sp.	*Serratia* sp.	*Erwinia* sp.	*Protius mirabilis*	*Shigella flexneri*	*Salmonella* sp.	*Salmonella typhi*
n-hexane	JU	40	8a	8a	8a	8a	8a	8a	8a	8a	8a	8a
		80	8a	8a	8a	8a	8a	8a	8a	8a	8a	8a
		120	8a	8a	8a	8a	8a	8a	8a	8a	8a	8a
	Sundarban	40	8a	8a	8a	8a	8a	8a	8a	8a	8a	8a
		80	8a	8a	8a	8a	8a	8a	8a	8a	8a	8a
		120	8a	8a	8a	8a	8a	8a	8a	8a	8a	8a
Ethyl acetate	JU	40	8a	9.33a	8b	8a	8a	8a	8a	8a	8a	8a
		80	8a	11a	9.17b	8a	8a	8a	8a	8a	8a	8a
		120	8a	14a	10.33b	8a	9.50b	8a	8a	8a	8a	8a
	Sundarban	40	8a	10.67a	8a	8a	8a	8a	8a	8a	8a	8a
		80	8a	13.83a	8a	8a	8a	8a	8a	8a	8a	8a
		120	8a	14.50a	8a	8a	8a	8a	8a	8a	8a	8a
Methanol	JU	40	8a	8a	8a	8a	8a	8a	8a	8a	8a	8a
		80	8a	8a	8a	8a	8a	8a	8a	8a	8a	8a
		120	8a	8a	8a	8a	8a	8a	8a	8a	8a	8a
	Sundarban	40	8.83cde	8a	9.50c	8a	12.50a	8.50de	11.33b	9cd	8a	9.50c
		80	9.33cd	8a	10.17c	8a	13.17a	8.83de	12.17b	9.83cd	8a	10.33c
		120	10.83b	10.17b	11.33b	10.33b	14a	10.33b	13a	11.17b	8a	10.67b

NB: In a column means followed by common letters are not significantly different at 5% level by DMRT.
*Disc diameter = 8 mm

The use of plant extracts or phytochemicals with known antimicrobial properties can be of great significance for therapeutic treatments (Jaimini *et al.*, 2011). Plants have been found *in vitro* to have antimicrobial property as they are rich in a wide variety of secondary metabolites, such as tannins, terpenoids, alkaloids and flavonoids (Cowan, 1999). In this study, the antibacterial property of methanolic extracts of the aerial parts have been tested against ten bacterial strains to evaluate the broad-spectrum antimicrobial properties of *B. lacera* (Table 2).

Two gram positive bacteria namely *Bacillus subtilis* and *B. cereus* and eight gram negative bacteria namely *Vibrio cholerae*, *Pseudomonas* sp., *Serratia* sp., *Salmonella typhi*, *Salmonella* sp., *Proteus mirabilis*, *Erwinia* sp., and *Shigella flexneri* were used in the present experiment. *n*-hexane fractions of JU and Sundarban samples showed no antibacterial activity whereas ethyl acetate fraction of both the samples showed inhibitory affect against all the tested bacteria under investigation (Fig. 1). Maximum zone of inhibition produced by ethyl acetate fraction was 14 mm for 120 µg/ml against *Bacillus cereus* followed by 13.83 mm for 80 µg /ml and 10.67 mm fo r 40 µg/ml respectively. Thus the finding from the present experiment suggest that the ethyl acetate fraction of *B. lacera,* irrespective of habitat choice was active against the gram positive bacteria *Bacillus cereus* following dose dependency (Fig. 2).

The methanol fraction of JU sample did not produce any inhibition zone whereas Sundarban sample showed broad spectrum inhibition against both gram positive and gram negative bacteria except *Salmonella* sp. The methanol fraction of Sundarban sample showed significant dose dependent increase in antibacterial activity against the gram negative bacteria namely, *Serratia* sp., *Proteus mirabilis*, *Salmonella typhi* and *Vibrio cholerae*. The highest inhibition zone of this fraction recorded as 14 mm and 13 mm against *Serratia* sp. and *Proteus mirabilis*, respectively whereas *Bacillus cereus* showed the lowest inhibition zone (10.17 mm) (Table 2). These observations were more likely to be the fact that an outer membrane in gram negative bacteria, which acts as a barrier to many environmental substances including antibiotic (Panda *et al.*, 2012).

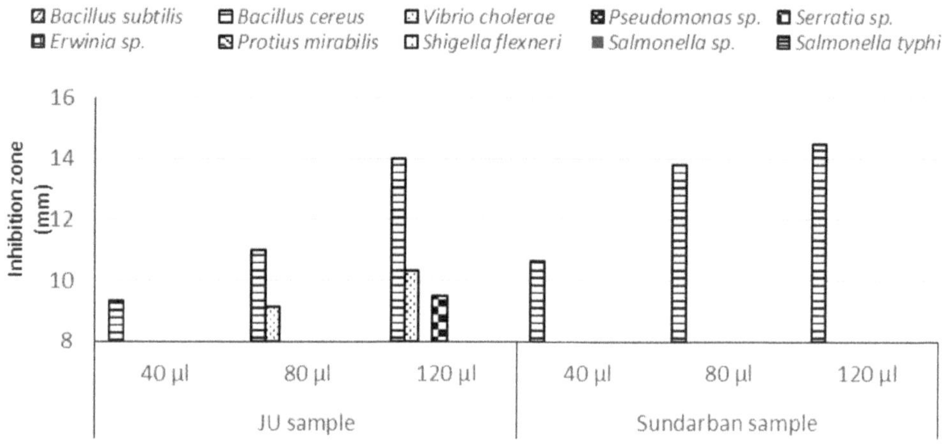

Figure 1. Antibacterial activity of ethyl acetate fractions of *B. lacera* collected from JU campus and Sundarban mangrove forest.

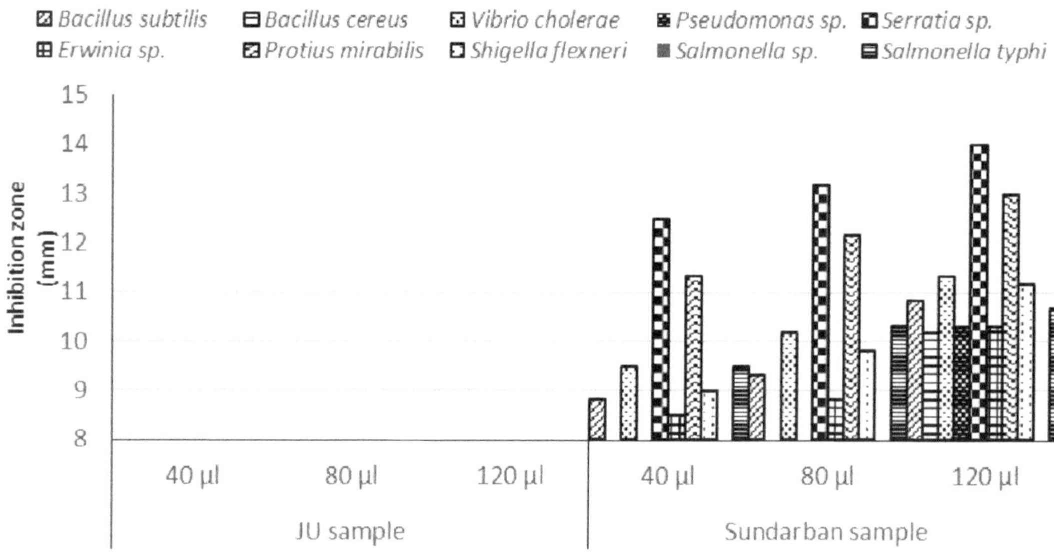

Figure 2. Antibacterial activity of methanol fractions of *B. lacera* collected from JU campus and Sundarban mangrove forest.

A number of review and research article provide the information of the antibacterial activities in *B. lacera* and the members of Asteraceae family. Singh *et al.* (2010) reported that the methanol extract showed good inhibitory activity against *Staphylococcus aureus, Bacillus subtilis, Serratia marcescens, Staphylococcus aureus and Candida albicans* which corroborates with findings of Sundarbans sample of the present study. In contrast, Tiwari *et al.* (2012) reported 13mm (100µg/ml) inhibition zone a gainst *Bacillus subtilis* and *Escherichia coli* respectively with the chloroform fraction of *B. lacera*. However, Islam *et al.* (2008) reported that *B. lacera* showed most promising antibacterial properties indicating the potential for discovery of new antibacterial drugs. The present study suggests that the methanol fraction of the aerial parts of *B. lacera* collected from Sundarban could be relatively potential source of antibacterial agent than the other fractions. Thus further extensive work on *B. lacera* growing in Sundarban may lead to the isolation of interesting therapeutic compound and future work in this regard is needed.

ACKNOWLEDGEMENT

The authors would like to express their gratitude to University Grants Commission (UGC) of Bangladesh for providing the research fund during 2011-2012 financial year.

REFERENCES

1. Akinpelu, D.A., Aiyegoro, A.O. and Okoh, A.I. 2009. Studies on the biocidal and cell membrane disruption potentials of stem bark extracts of *Afzelia africana* Smith. *Biol Res.* 42: 339-349.
2. Cowan, M.M. 1999. Plant product as antimicrobial agents. *Clin. Microbiol. Rev.* 12: 564-582. Gentry, A.H. 1993. Tropical forest biodiversity and the potential for new medicinal plants. *In:* Human Medicinal Agents from Plants. (Eds. by Kinghorn, A.D., Balandrin, M.F.), American Chemical Society, Washington, DC. pp.13-24.
3. Hossain, A.B.M.E. 2003. Final report on the undergrowth species of Sundarban mangrove ecosystem (Bangladesh). Sundarban biodiversity conservation project. pp. 10-11.

4. Islam, M. J., Barua, S., Das, S., Khan, M. S. and Ahmed, A. 2008. Antibacterial Activity of Some Indigenous Medicinal Plants. *J .Soil .Nature*. 2(3): 26-28.

5. Islam, M.T., Das, P.R., Kabir, M.H., Akter, S., Khatun, Z., Haque, M.M., Roney, M.S.I., Jahan, R. and Rahmatullah, M. 2012. Acanthaceae and Asteraceae family plants used by folk medicinal practitioners for treatment of malaria in Chittagong and Sylhet divisions of Bangladesh. *American-Eurasian J. Sustainable Agric*., 6(3): 146-152.

6. Jaimini, M.; Mahida, Y. and Mohan, J.S.S. 2011. Screening of Indian plant extracts for antibacterial activity. *Pharma. Biol.* 44(8): 627-631.

7. Jha, R.R. and Verma, S.K. 1996. Ethnobotony of Sauria Paharias of Santhal Pargana, Bihar: Indian Medicinal plants. *Ethnobotany*, 8: 31-35.

8. Joshi, M.C. 1980. Some folk medicine of Danga, Gujarat state. *Bull Med. Ethnobot Res*. 1: 8-24. Kakrani, H.K.N. and Saluja, A.K. 1994. Traditional through herbs in Kutch district Gujarat state,

9. India. Analgesic, anti-inflammatory, antirheumatic, antiarthritic plants, *Fitoterapia*, 65: 427-430.

10. Kumar, R. and Bhagat, N. 2012. Ethnomedicinal plants of district Kathua (J&K). *Int. J. Med. Arom. Plants.* 2(4): 603-611.

11. Kupchan, S.M. and Tsou, G. 1973. Tumor inhibitors-A new antileukemic simaroubolide from *Brucea antidysenterica*. *J. Org. Chem*. 38: 178-179.

12. Macilwain, C. 1998. Plants Used Against Cancer, Quarterman, Lawrence, Massachusetts. *Ann. Clin. Psychiatry*, 392: 535–540.

13. Martin, G.J. 1995. Ethnobotany: a 'People and Plants' Conservation Manual, Chapman and Hall, London, pp. 268.

14. Maundu, P. 1995. Methodology for collecting and sharing indigenous knowledge: a case study. Indigenous Knowledge and Development Monitor, 3: 3-5.

15. Mollik M.A.H., Hossan, M.S., Paul, A.K., Rahman, M.T., Jahan, R. and Rahmatullah, M. 2010. A comparative analysis of medicinal plants used by folk medicinal healers in three districts of Bangladesh and inquiry as to mode of selection of medicinal plants. *Ethnobotany Res. Applications*, 8:195-218.

16. Pal, R. 1972. Campesterol from *Blumea lacera*. *Phytochemistry*. 11: 18-55.

17. Panda, S.K., Pati, D., Mishra. S.K., Sahu, S., Tripathy, B. and Nayak, L. 2012. Phytochemical investigation & antimicrobial activity of methanolic extract of *Sonneratia apetala* Buch-Ham. Areal parts. *Int. J. Pharma. Biol. Arch.* 3(1): 79-83.

18. Pandit, B.R., Kotiwar O.S., Oza, R.A. and Kumar, R.M. 1996. Ethnomedicinal plant lore from Gir forest Gujarat. *Adv. Plant Sci*. 9 (1): 81-84.

19. Rahman, A.H.M.M. 2013. An ethnobotanical investigation on Asteraceae family at Rajshahi, Bangladesh. *J. Business Admin. Management Sci. Res.* 2(5): 133-141.

20. Saad, S., Taher, M., Susanti, D., Qaralleh, H. and Awang, A.F.I.B. 2012. *In vitro* antimicrobial activity of mangrove plant *Sonneratia alba*. *Asian Pac. J. Trop. Med.* 1:427-429.

21. Satapathy, K.B., Binod, B.S. and Jena, G.S. 2012. Crop weeds diversity and their ethnomedicinal uses in the treatment of common ailments in Jaipur district of Odisha, India. *Int. J. Med. Arom. Plants*. 2(1): 80-90.

22. Singh, B., Singh, S., Singh, S., Ali, L. and Mahajan, R. 2010. Ethnobotanical significance and antimicrobial activity of *Blumea lacera* (Roxb.) D.C. *Intl. J. Pharma. Biol. Arch*., 1(3): 314 – 316.

23. Singh, H. 1988. Ethnobotonical treatment of piles by Bhoxas of Utterpradesh. *Ancient Sci Life*. 8: 167-170.

24. Singh, V.K., Siddiqui, M.K. and Aminuddin. 2007. Folk medicinal plants used for the treatment of bronchial asthma in India. Proc. IS on Med. and Nutraceutical Plants. (Ed. by Yadav A.K.), *Acta Hort*. pp. 63-66.

25. Tiwari, P., Saluja, G., Pandey, A.S. and Sharma, N. 2012. Isolation and biological evaluation of some novel phytoconstituents from *Blumea lacera* (Burn F.) D.C. *Int. J. Pharm. Pharm. Sci*. 4(4):148-150.

26. Tomoko, N., Takashi, A., Hiromu, T., Yuka, I., Hiroko, M., Munekazu, I., Totshiyuki, T., Tetsuro, I., Fugio, A., Iriya, I., Tsutomu, N. and Kazuttito, W. 2002. Antibacterial activity of extracts prepared from tropical and sub-tropical plants on methicillin resistant *Staphylococcus aureus*. *J. Health. Sci*. 48: 273-289.

27. Uddin, S.J., Grice, I.D. and Tiralongo, E. 2011. Cytotoxic effects of Bangladeshi medicinal plant extracts. *Evidence-Based Compl. Alter. Med.* 2011:1-7.

28. Wagenen, B.C.V., Larsen, R., Cardellina, J.H., RanDazzo, D., Lidert, Z.C. and Swithenbank, C. 1993. Ulosantoin, a potent insecticide from the sponge *Ulosa ruetzleri*. *J. Org. Chem*. 58: 335-337.

CHAPTER 12

GAS EXCHANGE AND LEAF ANATOMY OF A C_3–CAM HYBRID, YUCCA GLORIOSA (ASPARAGACEAE)

Karolina Heyduk, Nia Burrell, Falak Lalani, and Jim Leebens-Mack

Keywords: Crassulacean acid metabolism, gas exchange, hybrid, leaf anatomy, photosynthesis, physiology

INTRODUCTION

The daily fixation of atmospheric carbon dioxide is a defining trait of green plants, and is arguably the basis for the major part of terrestrial biodiversity. Plants experience a number of stresses that make the photosynthetic machinery less than optimal, including drought, shade, and high temperatures. As they cannot relocate to avoid the stress, plants have instead modified their photosynthetic pathways in ways to circumvent limitations caused by abiotic stress. One such modification, Crassulacean acid metabolism (CAM), has evolved a number of times independently across angiosperms in response to carbon limitation due to water stress. Most plants that use the C_3 carbon metabolism pathway fix CO_2 during the day via the carboxylation of ribulose-1,5-bisphosphate by the enzyme Rubisco. However, Rubsico has both carboxylase and oxygenase activity, the latter of which is favoured in high temperatures or under low CO_2 levels, resulting in costly photorespiration. For C_3 plants during times of water stress, stomata close to prevent water loss, which leads to a depletion of internal CO_2 and initiation of photorespiration. To simultaneously increase carbon uptake while minimizing water loss, CAM plants instead open their stomata at night, when transpirational rates are lowest. Atmospheric CO_2 is converted to a four-carbon intermediate in the form of malic acid and stored in the vacuole until day time, when stomata close and the malic acid is decarboxylated (Osmond, 1978; Winter and Smith, 1996). The resulting internal concentrations of CO_2, which is fixed by Rubisco behind closed stomata, are as high as 2% (Cockburn *et al.*, 1979), compared with 0.38–0.4% outside the leaf. CAM species have estimated water use efficiency (moles of CO_2 fixed per mole H_2O transpired) of $(6–30)\times10^{-3}$, compared with $(0.6–1.3)\times10^{-3}$ for C_3 plants (Lüttge, 2004).

CAM is found in about 6% of flowering plants and is distributed across 35 plant families (Cushman, 2001; Silvera *et al.*, 2010). CAM species share a suite of physiological characteristics, including the ability to fix carbon at night, the accumulation of malic acid in the vacuoles over the dark period, increased leaf succulence due to enlarged vacuoles for the storage of malic acid, and daily carbohydrate turnover for the regeneration of phospho*enol*pyruvate (PEP), the molecule carboxylated for the initial fixation of carbon by the enzyme PEP carboxylase (PEPC). Leaf anatomy has also been implicated in the evolution of CAM, with previous research indicating tight cell packing (or low intercellular airspace; IAS) and large mesophyll cells as requirements for the optimal CAM function (Nelson and Sage, 2008; Zambrano *et al.*, 2014). Changes in these two traits may be linked responses to selection for increased vacuolar storage of malic acid in CAM species, but this hypothesis needs to be tested. Vein density has been shown to be a critical trait for the evolution of C_4 photosynthesis and is central to Kranz anatomy (Hattersley, 1984; Ueno *et al.*, 2006; McKown and Dengler, 2007; Christin *et al.*, 2013), but venation has been largely overlooked in anatomical studies of CAM plants. Interveinal distance has been found to correlate positively with succulence (Ogburn and Edwards, 2013). In addition, stomatal densities in *Clusia* were found to be lower in plants with higher night-time carbon uptake (Zambrano *et al.*, 2014). In general, it is thought that there are tradeoffs between CAM-promoting traits and efficient C_3 photosynthesis. The optimal anatomy for CAM plants includes large cells and a decrease in the amount of IAS; these same traits would limit efficient conductance of gas throughout the leaf (Nelson *et al.*, 2005; Nelson and Sage, 2008; Zambrano *et al.*, 2014).

Despite the extremes in phenotypes, CAM is often described as a continuum or spectrum, with full CAM at one end, C_3 at the other, and various intermediate forms between (Winter *et al.*, 2015). Although CAM plants are defined by

Citation: Karolina Heyduk, Nia Burrell, Falak Lalani, and Jim Leebens-Mack. Gas exchange and leaf anatomy of a C_3–CAM hybrid, Yucca gloriosa (Asparagaceae), Journal of Experimental Botany, 2016; Vol. 67, No. 5 pp. 1369–1379. DOI:10.1093/jxb/erv536.

night-time carbon uptake, this dogma ignores the high degree of plasticity found in CAM plants (Dodd *et al.*, 2002). Some CAM lineages have 'weak' CAM plants, predominantly C_3 plants with very low levels of CAM expression (as measured by leaf titratable acidity) (Silvera *et al.*, 2005). In addition, some species, including those in *Clusia*and *Mesembryanthemum*, are known to be facultatively CAM, whereby they use the C_3 pathway under non-stressed conditions but can upregulate the CAM cycle in response to a variety of abiotic stresses. Detailed comparative studies between C_3, CAM, and intermediate forms may help bridge the evolutionary gap between the ends of the spectrum (Silvera *et al.*, 2010; Winter and Holtum, 2014; Garcia *et al.*, 2014). Correlations between the ability of intermediate species to use the CAM pathway and their physiology and leaf anatomy could pinpoint traits that are vital—or not—to night-time carbon fixation.

To assess how physiology and anatomy correlate to a species' ability to use CAM, a natural hybrid system in the genus *Yucca* L. (Asparagaceae) was explored: *Yucca aloifolia* (CAM) and *Yucca filamentosa* (C_3), which are sympatric in the southeastern USA and have hybridized to form *Yucca gloriosa* (Rentsch and Leebens-Mack, 2012). *Yucca gloriosa* is unlikely to be a recent F1 hybrid (Rentsch and Leebens-Mack, 2012), and this species may be segregating for parental phenotypes as well. The parental species and the hybrid overlap in habitats along the dunes of the southeastern coastline, but grow in different parts of the dune system and are morphologically distinct. *Yucca filamentosa* inhabits the scrub-pine forests behind the dunes, except in the northern parts of its range, where it lives on the dunes under the protection of nurse plants. *Yucca aloifolia*and *Y. gloriosa* are both foredune species, with *Y. gloriosa* typically found on the ocean-side of the foredune. *Yucca gloriosa* was previously shown *in situ* to accumulate significant amounts of malic acid during the night, but showed no concurrent night-time $^{14}CO_2$ uptake (Martin *et al.*, 1982). The lack of night-time carbon uptake could be an artifact of sampling methodology used, especially if gas exchange rates are low. Genotypes from each of the three species were collected from across the species' ranges and assessed for photosynthetic pathway. Plants were assayed for carbon uptake in growth chambers under well-watered conditions, as well as while drought stressed, as CAM has been shown to be up-regulated under drought in various species (Ting, 1985; Lee and Griffiths, 1987; Winter and Ziegler, 1992; Winter *et al.*, 2011). These same *Yucca*individuals were phenotyped for titratable acidity and leaf anatomical characteristics. To better understand what suites of traits are important for CAM function, and how they impact the ability of *Y. gloriosa* to use either photosynthetic pathway, physiological traits were assessed for correlations.

The extent to which the hybrid species *Y. gloriosa* exhibits characteristics of its C_3 and CAM parents was also evaluated. Phenotypic assessment shows parental species are true to type, displaying gas exchange patterns and leaf anatomical traits that are predicted by their respective photosynthetic pathways. The hybrid species *Y. gloriosa* uses both C_3 and CAM pathways to assimilate carbon, and converts to fully CAM when drought stressed. Moreover, it is argued that *Y. gloriosa* can serve as a new study system for investigating the genetic architecture and evolution of CAM within the Agavoideae, a group that includes some of the most iconic CAM species.

MATERIALS AND METHODS

Plant Acquisition and Maintenance

All three *Yucca* species are clonal and readily generate ramets from the base of larger maternal plants. In summer 2013, ramets between 10 and 15cm tall were collected from all three species across the southeastern US seaboard, from the Outer Banks of North Carolina to the barrier islands of Georgia. Two clonal ramets were collected from each maternal plant and transplanted to the University of Georgia greenhouses in a 50:50 by volume mix of sand:pine bark (with vermiculite and limestone added). Plants were watered and fertilized as needed for at least 9 months prior to experimentation (described below). Diseased plants or those where size dimorphism between clones was very large were excluded.

Genotyping of Individuals

To assess a degree of genetic variation among individuals of all three species, microsatellite markers were used from Flatz *et al.* (2011) orRentsch and Leebens-Mack (2012), or were developed *de novo* for this study. *De novo* development of primers took advantage of unpublished sequence capture data in *Y. aloifolia* (see Heyduk *et al.*, 2015 for an overview of methods). An assembled set of contigs from the sequence capture data was first analysed for a single *Y. aloifolia* individual (Y45) through BatchPrimer3 for microsatellite discovery. These putative loci were then screened in the sequence capture assemblies from an additional five genotypes of *Y. aloifolia*. Polymorphism across all six genotypes was assessed *in silico,* and loci that varied in repeat length across the individuals of *Y. aloifolia* were kept for screening via PCR. A final screening left seven loci from each of the three resources (Flatz *et al.* 2011; Rentsch and Leebens-Mack, 2012; this study) that amplified successfully in all three species.

DNA from individuals used in phenotypic analysis was isolated using a modified cetyltrimethylammonium bromide (CTAB) method (Doyle, 1987; Štorchová *et al.*, 2000). DNA was amplified for the microsatellite loci using a three-primer system, where one primer (M13) is fluorescently tagged with either a FAM or HEX fluorophore, and the forward primer has an additional sequence that is complementary to the M13 primer sequence (Schuelke, 2000). Loci were amplified with the following PCR mix: 3.6 μl of PCR buffer (100mM Tris-HCl pH 8.0, 500mM KCl), 0.9 μl of 25mM $MgCl_2$, 0.6 μl of 10mM dNTPs, 0.4 μl of reverse primer (10 μM), 0.4 μl of M13 (10 μM), 0.375 μl forward primer (2 μM), 6.725 μl of H_2O, 1 μl of *Taq* polymerase, and 1 μl of template DNA diluted to ~10ng μl^{-1} for a total of 15 μl per PCR reaction. Amplification used a touchdown programme as follows: initial denaturation at 95°C for 2min; 10 cycles of 95°C for 15s, 64°C for 15s with a 1degree drop per cycle, and 72°C for 30s; 25 cycles of 95°C for 15s, 54°C for 15s, and 72°C for 30s; a final extension at 72°C for 1min. PCR products were diluted 1:15, pooled when appropriate, then 3 μl was mixed with 10 μl of a formamide–ROX dye-labelled size standard (1ml formamide, 100 μl ROX ladder). Fragment analysis was conducted on an Applied Biosystems 3730xl DNA Analyzer. Alleles were called using Geneious 8.1.6 (Kearse *et al.*, 2012), exported, and analysed for hybrid index scores using the introgress package (Gompert and Buerkle, 2010) in R 3.2.2 (R Core Team, 2013).

Gas Exchange

A total of 16 genotypes were phenotyped: three from *Y. aloifolia,* four from *Y. filamentosa*, and nine from *Y. gloriosa*. In order to assess gas exchange patterns on a large number of plants, genotypes from all three species were split between three independent experiments conducted in July and October 2014 and February 2015. For a given experimental period, five to six genotypes each with two clones were measured. Genotypes were randomly assigned to one of three time blocks, and individual clones of each genotype were randomly assigned to the water or drought treatment. Plants were moved into the growth chamber 1 week prior to the onset of any experimental treatment and all plants regardless of assigned treatment were watered daily. Growth chamber conditions were set to day/night temperatures of 32/17 °C, with a relative humidity of 30% and day length of 12h. Photosynthetically active radiation (PAR) at plant level was between 400 and 500 μmol m^{-2} s^{-1}.

Leaf gas exchange measurements were conducted with a LiCOR-6400XT portable photosynthesis system (LiCOR, Lincoln, NE, USA). The first set of measurements was taken while soil for all plants was still at field capacity (day 1). Gas exchange measurements were taken every 4h for 24h, beginning 1h after the onset of light in the growth chamber. After the initial 24-h interval of gas exchange measurements under well-watered conditions was complete, water was withheld from the drought-treatment plants to initiate a dry-down, while well-watered treatment plants continued to be watered daily. Soil moisture probe measurements taken every other day were used to ensure a relatively even dry-down. Well-watered and drought-treated plants were then measured for gas exchange rates every other day for three 24-h intervals (days 3, 5, and 7). For each measurement, plants were measured in the same order and the same leaf was used for the gas exchange measurements, unless the leaf showed signs of damage from the LiCor chamber. After the fourth 24-h interval (day 7) of gas exchange measurements were complete, all plants were watered, and a final day of gas exchange measurements was conducted 1 day later (day 9).

Leaf Titratable Acidity

Discs of leaf tissue were taken from plants 2h before lights turned off in the evening (p.m.) and two hours before lights turned back on in the morning (a.m.). Samples were taken on the initial day while all plants were well watered (day 1), on two of the three drought-treatment days (days 5 and 7), and on the final re-watered day (day 9). Because of limited tissue, two to three disc samples per plant were taken. Tissue was immediately frozen in N_2 and stored at –80 °C until leaf titrations commenced. Titrations were done on each sample independently by first measuring frozen weight, then boiling for 20min in 20ml of H_2O. Boiled samples were allowed to cool to room temperature, then titrated with 100mM NaOH to the initial pH of the water used, which varied slightly in pH from 7.0 to 8.0. ΔH^+ was calculated as the difference in microequivalents of H^+measured in the morning and the evening, with values above 0 indicating CAM activity. Student's *t*-test was performed to check for ΔH^+ values that were significantly different from zero, as well as to check for significant differences between watered and drought stressed samples on any given day.

Leaf Anatomy

Leaf thickness was averaged from five replicate leaves per plant, on one to two plants per genotype; the midrib was avoided, as were young or very old leaves. To measure succulence, the most recent, fully mature leaf per plant was collected, cutting just above the lighter-coloured petiole. Fresh leaves were immediately weighed, scanned, and dried in an oven at 60 °C. The final dry weight was recorded when the mass changed less than 0.01 g d^{-1}. Leaf area was calculated in ImageJ (Rasband). Succulence was calculated as grams of water per centimetre of leaf area: (FW (g) – DW (g))/area (cm^2), where FW and DW are fresh and dry weight, respectively. Genotype succulence values were calculated by averaging across replicate plants, and as residuals were not normally distributed, a Wilcoxon rank sum test was used to test for significant differences.

Leaf cross sections from all species were prepared as follows: tissue was harvested from leaves, fixed in formalin, and embedded in paraffin. Samples were then sliced with a microtome, mounted, and stained with Toluidine blue by the University of Georgia Veterinary Histology Laboratory. Slides were imaged by a Zeiss microscope using ZEN software with one to three images taken per slide. Leaf anatomical characteristics were measured in ImageJ. All traits were measured independently on each image, then averaged across all images for a given genotype. Cell size was measured separately on adaxial and abaxial portions of the leaf and also averaged together for a cumulative average cell size. In many CAM plants, mesophyll cells fail to differentiate into palisade or spongy forms (Gibson, 1982), so these subdivisions were not taken into account when measuring cell sizes for either *Y. aloifolia* or *Y. gloriosa. Yucca filamentosa* does have clear mesophyll differentiation, and palisade and spongy mesophyll cells were used for adaxial and abaxial measurements, respectively. Leaf IAS was calculated by measuring the airspace in a given demarcated area of each cross section and dividing by the total area of mesophyll tissue in that area.

To address whether vein density is related to a plant's use of CAM, vein spacing and the degree of 3D venation in the three species were measured. Average distances between major and minor veins were measured horizontally only when veins were clearly in the same plane and when xylem and phloem were visibly developed. The degree of 3D venation was assessed by counting the number of independent planes of veins (major or minor) in each cross section. Residuals of traits measured from cross sections were not normally distributed and the Wilcoxon rank sum test was used to assess species trait differences.

Stomatal density was measured using fresh tissue for the same genotypes used in gas exchange experiments many months after any treatment was imposed to avoid any effects of drought stress on stomatal density. In addition, leaves that were fully developed before treatment in the growth chamber were used. Leaves were painted with varnish on both adaxial and abaxial sides. The varnish was allowed to dry, then removed by adhering tape and gently removing the imprinted varnish. Images of stomatal peels were captured using the same Zeiss microscope and ZEN software as was used for cross sections. The stomatal density was calculated as number of stomata per area of epidermal tissue. An ANOVA was used to test for effects of species, side of leaf (upper or lower), and an interaction of the two. Neither the side of leaf nor the interaction of side with species was significant (F=1.174, P=0.286 for side of leaf; F=0.284, P=0.754 for the interaction of side×species), so these were not considered in further tests of species differences, which were evaluated with the Wilcoxon rank sum tests.

To assess the phenotypic space described by the traits measured, anatomical traits (with the exception of the number of planes of veins in the leaf, which was invariable) were combined with titratable acidity ΔH^+values and gas exchange data to perform a principle coordinates analysis (PCA). ΔH^+ values were averaged for both clones of a genotype measured from day 1, as all plants were well watered. ΔH^+ values were included from drought-stressed clones from day 7 as a measure of the ability of a plant to accumulate malic acid under drought stress. As a proxy for the ability to use the CAM pathway, the carbon uptake values that occurred at night during a given day–night cycle were summed, and divided by the total carbon uptake value (the sum of all CO_2 uptake values measured for that plant across a 24-h period). While it excludes the carbon uptake that occurred in the 4h between measurements, those values are not likely to be largely different from the ones recorded, and the overall proportion calculated is an accurate estimate of relative contribution of night-time carbon uptake.

On the same matrix of phenotype data, a Spearman rank correlation matrix with Holm–Bonferroni adjusted P-values was calculated. Each trait was also tested for bimodality using Hartigan's dip test, as correlation results may be influenced by strongly bimodal distributions.

RESULTS

Genetic Diversity

Hybrid index scores for the hybrid ranged from ~0.35 to ~0.65, with two individuals having an index of 0.5 (Fig. 1A). These two individuals had loci that were monomorphic for parental alleles, indicating they are not F1 hybrid genotypes. Hybrid indices indicate *Y. gloriosa* samples used in this study were largely later generational hybrids segregating for alleles from each parent. The first two principal components of the genetic distance PCA explain 81.1% of the variation among the three species (Fig. 1B). The species cluster into three distinct groups; there is no evidence of ongoing back-crossing of *Y. gloriosa* to either parental species in the samples used in this study. Rather, the marker data suggest that *Y. gloriosa* is distinct from both *Y. aloifolia* and *Y. filamentosa,* and is likely to be on an independent evolutionary trajectory from either parental species.

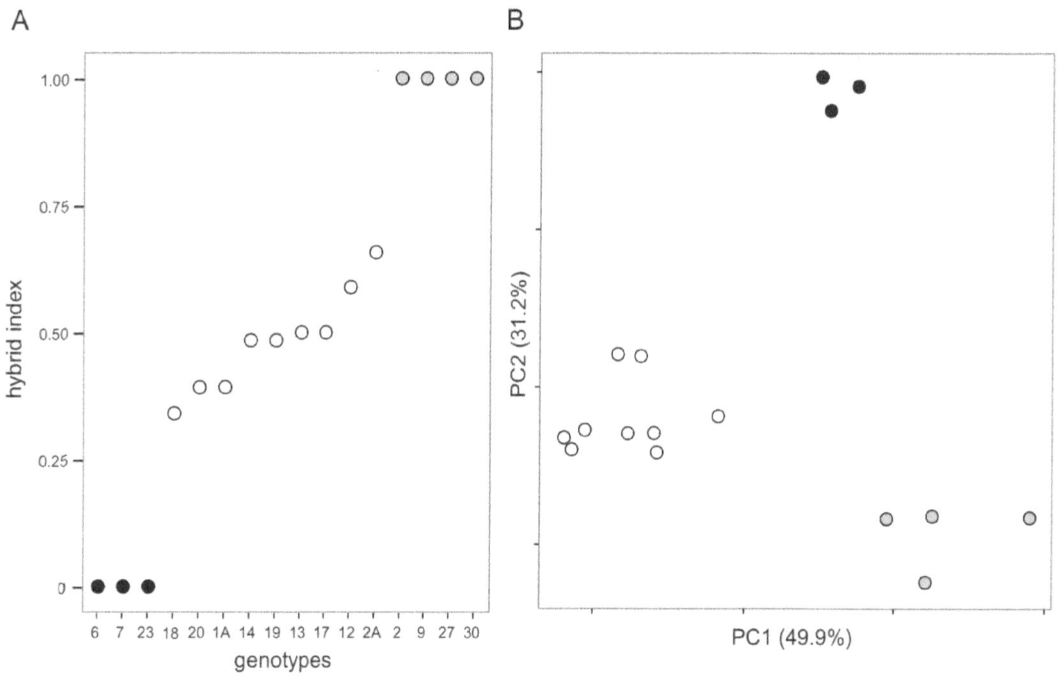

Figure 1. Microsatellite variation among genotypes represented by hybrid index scores (A), with parental scores set to 0 or 1 and the index value for the hybrid species representing proportion of ancestry from each parent, and PCA of distance matrix of multi-locus genotypes (B). *Yucca aloifolia* is represented by black filled circles, *Y. gloriosa* by open unfilled circles, *Y. filamentosa* by grey circles.

Gas Exchange Pattern

Both parental species behaved as their photosynthetic types would predict under well-watered and drought conditions. *Yucca aloifolia* showed predominantly night-time CO_2 uptake along with late afternoon uptake (Fig. 2). This pattern remained under well-watered conditions for all five days. Under drought-stressed conditions, carbon still entered the leaves of *Y. aloifolia* at night but photosynthetic rates were reduced, and daytime uptake became negligible. For *Y. filamentosa*, well-watered plants showed no ability to take up carbon at night (Fig. 2). Drought-stressed *Y. filamentosa* plants likewise showed no transition to night-time uptake, and total photosynthetic rates were reduced to nearly zero by day 5 across all time points measured. *Yucca gloriosa* showed high levels of daytime carbon uptake, but night-time carbon gain happened under well-watered conditions as well (Fig. 2). As drought stress was induced in the hybrid, daytime uptake of carbon dropped to zero and net carbon gain occurred entirely in the dark, although absolute values were never as high in the hybrid as in *Y. aloifolia*. Upon re-watering, the drought stressed clones of each species began returning to their original, well-watered phenotype.

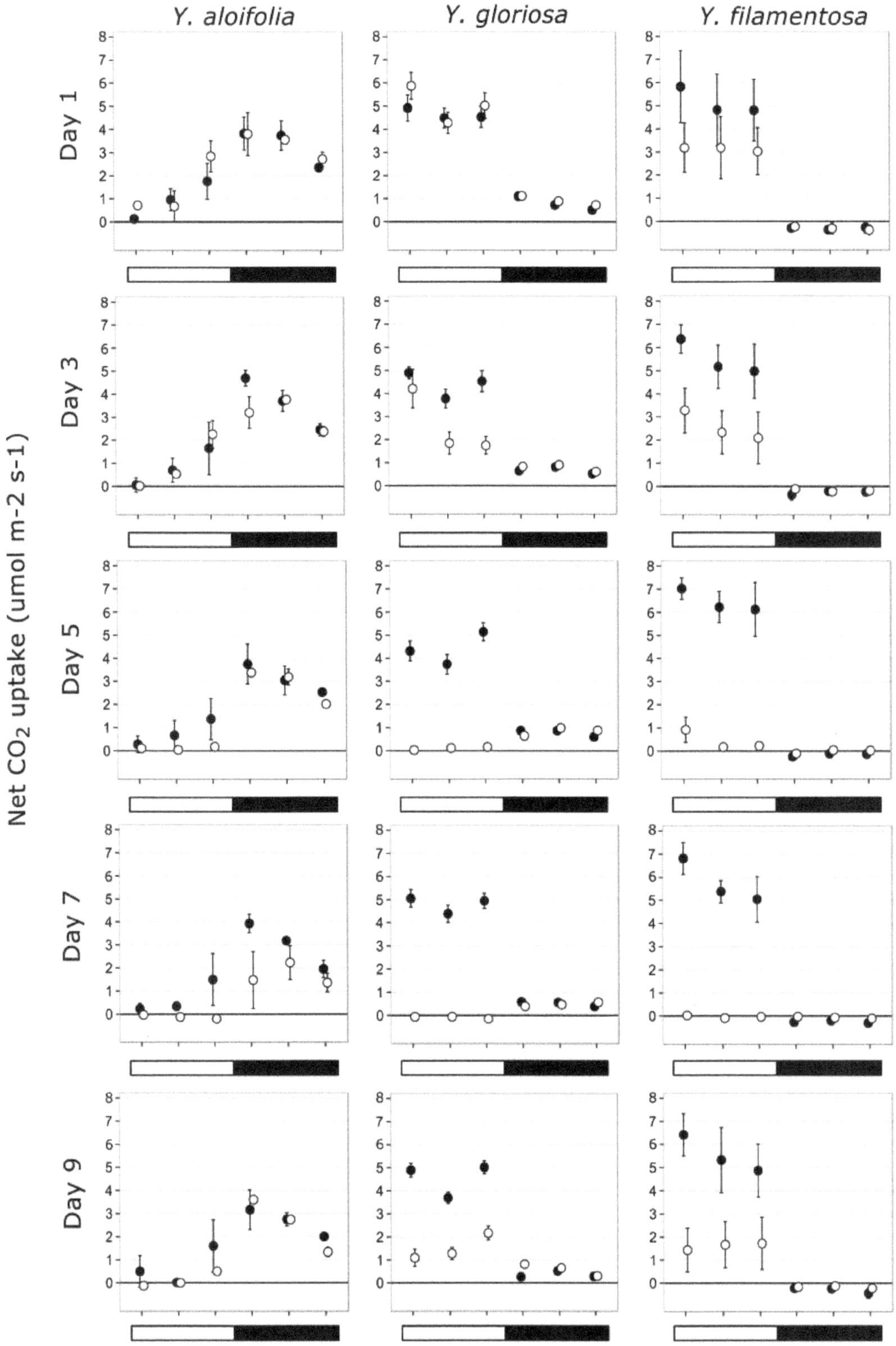

Figure 2. Gas exchange patterns for *Y. aloifolia*, *Y. filamentosa*, and *Y. gloriosa* across all days of the growth chamber dry-down. Filled circles indicate the clone kept under well-watered conditions, open circles indicate clones which were subjected to dry-down starting after day 1. The open bar indicates hours under light; the filled bar indicates time when lights were off. Points represent averages across all genotypes measured for a given species, with standard errors (too small to be visible for some points).

All genotypes of the parental species were consistent in the general pattern of gas exchange expected for each photosynthetic pathway. Although overall rates of photosynthetic activity may have varied between genotypes, the CAM and C_3 patterns were maintained in all genotypes of *Y. aloifolia* and *Y. filamentosa*, respectively. *Yucca gloriosa* had somewhat more variable responses to drought, though nearly all genotypes used night-time carbon uptake at low levels even under well-watered conditions. Some genotypes of the hybrid had a more pronounced level of night-time carbon uptake under drought, though most maintained well-watered levels even while stressed.

Titratable Acidity

Yucca aloifolia had the highest levels of acid accumulation (Table 1) as indicated by ΔH^+ and was the only species to show a significant effect of drought on the degree of leaf acidity (day 7, Student's *t*-test, df=8,$P<0.001$). No sample of *Y. filamentosa* ever required titration, demonstrating a complete lack of acid accumulation in this species. *Yucca gloriosa* had variable acid accumulation across time, with a species average value significantly greater than zero on days 1,7, and 9 but not 5 (Table 1). Some genotypes of *Y. gloriosa* had no acid accumulations, while *Y. aloifolia* had variable but positive accumulation across all genotypes and all treatments.

Table 1. Night-time acid accumulation (ΔH^+, μequivalents g^{-1}) derived from difference in dusk and dawn titratable acidity measurements, averaged across samples for all three species with standard error

Species and conditions	Day 1		Day 5		Day 7		Day 9	
	ΔH⁺	Sig	ΔH⁺	Sig	ΔH⁺	Sig	ΔH⁺	Sig
Y. aloifolia								
Well watered	144.2±16.0	***	106.7±16.2	***	133.5±14.6	***	125.5±7.3	***
Drought stressed	135.1±6.9	***	119.8±6.7	***	48.0±14.2	**	99.5±9.1	***
Y. gloriosa								
Well watered	2.4±0.97	*	0.44±0.26	n.s.	2.0±0.71	*	1.6±0.67	*
Drought stressed	3.7±1.4	*	1.79±0.87	n.s.	1.7±1.0	n.s.	2.4±0.77	**
Y. filamentosa								
Well watered	0±0	n.s.	0±0	n.s.	0±0	n.s.	0±0	n.s.
Drought stressed	0±0	n.s.	0±0	n.s.	0±0	n.s.	0±0	n.s.

Averages for plants kept well watered for the duration of the experiment or under drought stress (which began on day 3, not sampled, and ended after day 7). Significance values (Sig) indicate species by treatment measurements that are significantly different from zero by Student's *t*-test: ***$P<0.001$, **$P<0.01$, *$P<0.05$, n.s., not significant.

Leaf Anatomy

Cross sections of each of the three species were easily distinguishable (Fig. 3) and almost all anatomical traits measured were significantly different between all three species (Table 2). *Yucca aloifolia* had the most succulent leaves, followed by *Y. gloriosa,* with *Y. filamentosa's* leaves being the least succulent. Leaf IAS was significantly smaller in *Y. aloifolia*and was the largest in *Y. filamentosa*. Similarly, *Y. aloifolia* had the largest cells. Across species, average cell size was negatively correlated to IAS (Fig. 4, R^2=0.4122, $P<0.01$), but this relationship did not hold within a species; in *Y. gloriosa*, for example, average cell size was not correlated significantly to IAS (R^2= 0.0737, P=0.42). However, there was no significant difference in the number of stomata on either adaxial or abaxial sides of the leaf between species, although *Y. gloriosa* has a weakly significant difference (P=0.04955) on the adaxial leaf surface compared to *Y. filamentosa.* The average distance between major veins was not significantly different between *Y. gloriosa* and *Y. filamentosa*, and*Y. gloriosa* did not differ in distance between minor veins from either parent, although the parental phenotypes were significantly different from each other. All three species showed similar propensities for 3D venation and there were no significant difference between them (Table 1).

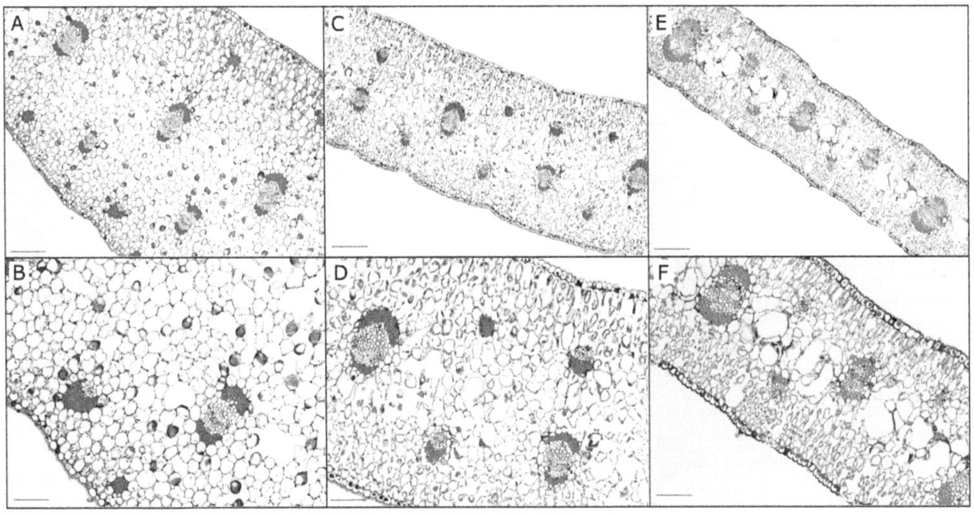

Figure 3. Cross sections of *Y. aloifolia*(A and B), *Y. gloriosa* (C and D), and *Y. filamentosa* (E and F) at 5× (A, C, E) and 10× (B, D, F) magnification. Scale bars for 5×=200 μm and at 10×=100 μm.

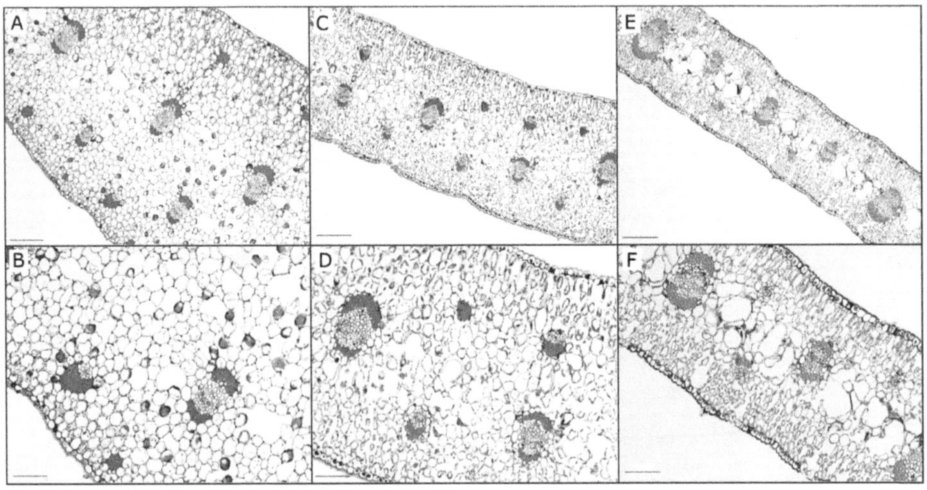

Figure 4. Intercellular air space vs. cell size, with standard errors around each genotype average. *Yucca aloifolia* is represented by black filled circles, *Y. gloriosa* by open unfilled circles, *Y. filamentosa* by grey circles.

Table 2. Mean and standard error for traits in the three species and significance of pairwise comparisons

Trait	*Y. aloifolia*	*Y. gloriosa*	*Y. filamentosa*	YA vs. YF	YA vs. YG	YG vs. YF
Succulence (g cm^{-2})	0.1172±0.0027	0.0699±0.0022	0.0396±0.0022	***	***	***
Stomata, ad (mm^{-2})	75.02±5.78	125.11±17.14	79.06±5.60	n.s.	n.s.	*
Stomata, ab (mm^{-2})	93.06±13.07	125.12±17.15	92.88±8.63	n.s.	n.s.	n.s.
Thickness (μm)	1331.29±90.97	748.23±24.20	520.00±27.67	**	**	***
IAS (%)	14.5±0.54	21.5±1.28	28.7±2.38	**	**	*
Cell size, ad (μm)	2124.28±169.19	1153.99±95.21	696.31±47.85	**	**	***
Cell size, ab (μm)	1960.99±242.04	985.08±88.52	634.18±27.50	**	**	**
Dist major vein (μm)	701.56±34.37	535.59±37.08	450.86±61.73	*	*	n.s.
Dist minor vein (μm)	673.01±8.04	551.02±46.42	460.37±38.87	**	n.s.	n.s.
Planes of veins (μm)	2.46±0.30	2.55±0.19	2.55±0.14	n.s.	n.s.	n.s.

YA, *Y. aloifolia*; YF, *Y. filamentosa*; YG, *Y. gloriosa*. Sample sizes are 4, 10, and 7 genotypes of *Y. aloifolia*, *Y. gloriosa*, and *Y. filamentosa*, respectively, except for succulence, which had n=6, 12, and 8. IAS is shown as a percentage of mesophyll area. Stomatal and cell size values calculated separately for adaxial (ad) and abaxial (ab) surface areas. ***P< 0.001, **P<0.01, *P<0.05, n.s. not significant.

The PCA (Fig. 5) of all phenotypic data shows two ends of trait space defined by *Y. aloifolia* and *Y. filamentosa*. Further, *Y. gloriosa* appears slightly closer to *Y. filamentosa* but falls between the two clusters of parental species. The five traits with highest loading scores for the first principal component include adaxial and abaxial cell size (loading scores of –0.30 and –0.31, respectively), average proportion of dark CO_2 uptake under watered conditions (–0.32), maximum dark CO_2 uptake (–0.32), and leaf thickness (–0.31). The traits with highest loading scores for the second component include adaxial and abaxial stomatal density (–0.53 and –0.55, respectively), the average distance between major veins (0.44), succulence (0.22), and leaf IAS (0.21). The large loading scores for stomatal densities on PC2 are driven largely by a single genotype of *Y. filamentosa*. A number of traits were correlated, including positive relationships between leaf density (succulence, thickness, and cell sizes) and maximum dark CO_2 uptake rates and proportion of CO_2 taken up at night (Fig. 6). ΔH^+ under watered conditions was positively correlated to proportion of dark CO_2 uptake both under well-watered and drought conditions, but negatively correlated to maximum CO_2 uptake in the light. Negative correlations existed between IAS and the maximum dark CO_2 uptake rate, as well as IAS and leaf thickness. Only two traits—proportion of night-time CO_2 uptake in watered and drought conditions—were not unimodal across species; correlations between proportions of CO_2 uptake and any other trait should therefore be treated with caution.

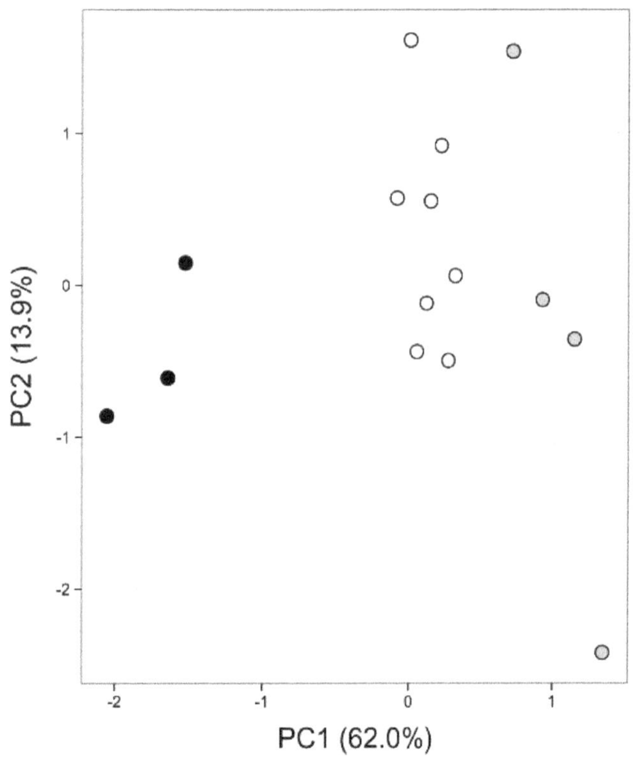

Figure 5. PCA of phenotypic data.*Yucca aloifolia* is represented by black filled circles, *Y. gloriosa* by open circles, and *Y. filamentosa*by grey circles. Phenotypes include those listed in Table 2: succulence, stomatal densities for adaxial and abaxial sides, leaf thickness, %IAS, average cell size on adaxial and abaxial sides, average distance between major veins, and average distance between minor veins, and additionally maximum dark CO_2 uptake rate, maximum light CO_2uptake rate, ΔH^+ well watered, ΔH^+ drought conditions, and proportion of CO_2 taken up at night well watered and under drought conditions.

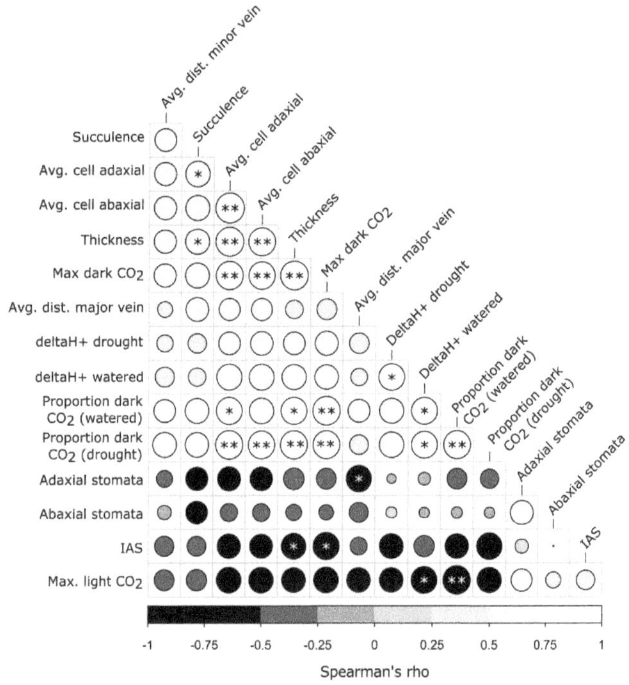

Figure 6. Correlation matrix of phenotypic data, based on Spearman's rho. Rho of –1 is black, rho of 1 is white. The size of the circles indicates the absolute value of the correlation, with larger circles referencing larger absolute correlations. Holm–Bonferoni corrected*P*-values: **P*≤0.05, ***P*≤0.01. Trait names that are italicized on the vertical axis are traits that were not unimodal according to Hartigan's dip test.

DISCUSSION

Photosynthesis in *Yucca*

Using a combination of gas exchange measurements, titratable acidity, and leaf anatomy, the use of the CAM and C_3 photosynthetic pathways by*Y. aloifolia* and *Y. filamentosa*, respectively, was verified. These species of*Yucca* crossed to form a natural hybrid species, *Y. gloriosa* (Rentsch and Leebens-Mack, 2012), in which previous natural surveys showed nightly acid accumulation but no detectable nocturnal gas exchange (Martin *et al.*, 1982). Our results show that *Y. gloriosa* is an intermediate C_3–CAM species, with the ability to uptake CO_2 nocturnally but with relatively low levels of nightly acid accumulation (Fig. 2 and Table 1). Anatomically, the leaves of the three species are distinct, and *Y. gloriosa* has intermediate phenotypes for a variety of traits measured (Table 2). Trait values reported here for different *Y. gloriosa* genotypes are not only intermediate relative to the parental species, but also 'fill' the phenotypic space between the extremes of C_3 and CAM. Individuals of *Y. gloriosa* sampled from natural populations are segregating for parental markers and none of the nine *Y. gloriosa* genotypes used in our analyses are F1s (Fig. 1). These results are in agreement with an earlier study suggesting that *Y. gloriosa* is genetically distinct from either parental species and is largely evolving independent of *Y. aloifolia* and *Y. filamentosa* (Rentsch and Leebens-Mack, 2012). The range of hybrid index scores for genotypes sampled in this and the earlier study raises the possibility that populations are segregating for CAM-related traits, but additional replication of individual *Y. gloriosa*genotypes is needed to assess intraspecific variation for physiological traits.

Gas exchange patterns for *Y. aloifolia* and *Y. filamentosa* are representative of their photosynthetic pathways, as are ΔH^+ values. For *Y. gloriosa*, the photosynthetic machinery is intermediate; the hybrid has daytime CO_2 uptake levels that are comparable to the C_3 parent, and while it has the ability to use the CAM pathway at low levels, its night-time rate of CO_2 uptake and acid accumulation never reached the levels in *Y. aloifolia.* Non-zero but low CO_2 uptake rates at night are common in facultative CAM plants, as the facultative upregulation of the CAM cycle is often coincident with abiotic stress (Winter and Holtum, 2014). Since water stress decreased the magnitude of night-time CAM in *Y. aloifolia* (Fig. 2), it is unsurprising that drought stress likewise limited nocturnal CO_2uptake in *Y. gloriosa*. More important than the magnitude of night-time CO_2 uptake is the proportion of daily carbon acquired at night relative to total carbon gain; for drought-stressed *Y. gloriosa* plants, 100% of carbon acquisition occurred during the night.

Acid accumulation in the hybrid was highly variable, though more acid accumulated in *Y. gloriosa* leaves than the C_3 parent, paralleling its ability to use the CAM cycle at night. The lack of an increase in acidification under drought stress in *Y. gloriosa* is correlated to little increase in net CO_2 uptake rates in drought-stressed plants relative to well watered. The variability among genotypes and between time points in acid accumulations may be partly due to leaf heterogeneity and the inability to sample exact replicates in terms of leaf age and position. ΔH^+ levels may also be so low in the hybrid that they were not detected consistently with titration methods used. Alternatively, the ability to accumulate acid in the leaves—a proxy for CO_2 fixation by PEPC—may be segregating in genotypes of the hybrid, despite a relatively consistent ability across genotypes to acquire carbon via stomatal opening at night.

In species that are facultatively CAM, the use of the CAM pathway has been shown to relate to environment or seasonality (Guralnick *et al.*, 1984; Ball*et al.*, 1991; Borland *et al.*, 1992; Lüttge, 2006; Winter and Holtum, 2007). For example, in the annual *Mesmbryanthemum crystallinum,* induction of CAM occurs at the start of the dry season, with C_3 photosynthesis being the primary mode of carbon acquisition during the wet period (Bloom and Troughton, 1979; Winter and Holtum, 2007). Similarly, *Clusia uvitana*, a weak CAM species, was shown to increase the proportion of carbon acquired via CAM from 27% in the wet season to 42% during the dry season (Zotz and Winter, 1994). In theory, seasonal or environmentally induced photosynthetic switching should allow plants to grow rapidly by using C_3 photosynthesis during favourable conditions, and allow these plants to continue to grow by utilizing CAM during dry conditions or seasons. Even under extreme drought conditions, plants with CAM phenotypes can shut stomata completely and keep carbon metabolism primed by recycling respired CO_2. This process, known as CAM 'cycling', produces no net growth as net CO_2 intake is zero, but allows the photosynthetic machinery to stay active until conditions become more favourable. While drought stress forces *Y. gloriosa* from mostly C_3 carbon gain to 100% CAM carbon gain, how this plant responds to drought in its natural setting is not known. *In situ* studies are required to better describe seasonal and environmental impacts on the frequency of CAM use in *Y. gloriosa*.

Leaf Anatomy

Yucca aloifolia possessed traits expected for CAM plants, including increased succulence and decreased IAS. Previous work in *Clusia, Annanas,*and *Kalanchoe* (Nelson and Sage, 2008) showed similar relationships between IAS and strong CAM. Zambrano *et al.* (2014) found that CAM*Clusia* species have thicker leaves, are more succulent, and have lower IAS than their C_3 counterparts. The correlation between leaf thickness or succulence, internal air space, and CAM across different plant lineages indicates that these traits represent fundamental requirements for the CAM cycle. An increase in succulence, and the corresponding decrease in IAS, has been hypothesized to limit conductance of gases through the mesophyll. For CAM plants that generate very high concentrations of CO_2in the cells during daytime decarboxylation of malic acid, this limitation in conductance serves to keep captured carbon in the leaf. The intermediate levels of succulence and IAS found in *Y. gloriosa*, however, present a unique challenge to this species, as the majority of its carbon is assimilated during the day under well-watered conditions (Fig. 2). Movement of CO_2 through the leaf is imperative for efficient C_3photosynthesis, which is typically inhibited by leaf CO_2 levels. The proportion of IAS in *Y. gloriosa* measured here approaches levels found in the fully CAM *Y. aloifolia*, which would limit C_3 photosynthesis during the day in *Y. gloriosa*. In addition, stomatal density was nearly identical in *Y. aloifolia* and *Y. gloriosa*, precluding the hybrid from using a greater number of stomata to compensate for lower leaf CO_2 movement. How *Y. gloriosa* is able to conduct CO_2 at the same level as *Y. filamentosa* could be addressed by stomatal aperture size, but the sunken nature and large subsidiary cells in stomata of *Yucca* make measuring aperture size difficult. In addition, IAS in *Y. gloriosa* is higher than in *Y. aloifolia*, and may conversely be limiting the hybrid's ability to use the CAM pathway to as great a degree as is found in *Y. aloifolia*.

While research into the vascular architecture of CAM plants is lacking, a wealth of information about venation and photosynthetic efficiency comes from the C_4 literature. An increase in leaf vein density is the basis of Kranz anatomy for C_4 plants, enabling a high mesophyll to bundle sheath cell ratio required for efficient spatial concentration of CO_2. For CAM, it is unlikely that vein density plays a direct role in photosynthetic efficiency as it does in C_4 species; rather, the increase in succulence in CAM plants likely leads to modified venation patterns to maintain hydraulic connectivity. While increasing succulence is predicted to be correlated to a decrease in vein density and the distance between veins (Noblin *et al.*, 2008), succulent species have been shown to circumvent limitations to hydraulic connectivity by evolving 3D venation (Ogburn and Edwards, 2013). All three species of *Yucca*, including the C_3 *Y. filamentosa*, had more than one plane of veins in their cross sections, indicating that the tendency toward 3D venation may be ancestral in *Yucca*, perhaps as a response to arid environments, and may have allowed for the further evolution of succulent CAM species in this genus (Griffiths, 2013). If propensities for thick leaves and 3D venation allow for repeated independent origins of CAM in a lineage, understanding why *Y. filamentosa* is not CAM will be important for describing the evolutionary trajectory from C_3 to CAM.

Evolutionary Implications

Recent work on transitions from different photosynthetic states has focused on using intermediate plants that have varying propensities or similarities towards one state or another. *Flaveria*, which has C_3, C_4, and C_3–C_4 intermediate species, has allowed for detailed studies on the trajectory of anatomical, physiological, and genetic changes required to evolve C_4 from C_3 (Huxman and Monson, 2003; McKown and Dengler, 2007). Similarly, evolutionary studies within the grasses have described an anatomical progression from C_3 to C_4, and have shown that certain grass lineages possess pre-requisites for C_4 photosynthesis (Christin *et al.*, 2013). While there are many evolutionary model systems in CAM plants, including facultative species in *Clusia, Mesembryanthemum, Kalanchoe*, and the Orhicdaceae, *Yucca* holds particular promise due to closely related species that are C_3, CAM, and C_3–CAM intermediates. In addition, molecular analysis has shown that individuals of *Y. gloriosa* are later generation hybrids that segregate for molecular markers (Rentsch and Leebens-Mack, 2012, this study); photosynthetic phenotypes may segregate as well, but this hypothesis needs to be tested. While this study does not have the necessary replication of genotype gas exchange patterns to conclusively shown variation in the ability to use CAM, leaf anatomical traits do differ between genotypes, and indicate that parental species physiology is likewise segregating among individuals of the hybrid. Future replicated analyses of gas exchange patterns across time points for more genotypes of *Y. gloriosa* will elucidate the degree of variation and genetic architecture of CAM-related traits within this species.

Finally, the intermediate nature of *Y. gloriosa* indicates it should be classified as a 'C_3–CAM species' according to Winter *et al.* (2015) and raises the question of whether such intermediacy represents a stable state or a point on an evolutionary trajectory toward C_3 or CAM. The traditional classifications of intermediate CAM include facultative species and CAM cyclers, although recent work in the Orchidaceae has described the prevalence of 'weak' CAM, a largely C_3 plant with low levels of night-time acid accumulation (Silvera *et al.*, 2005). *Yucca gloriosa* does not fit into any of these traditional categories, and therefore prompts questions about the defining features of CAM plants (Winter *et al.*, 2015). Certainly, the ability of *Y. gloriosa* genotypes to obtain carbon entirely via the CAM cycle under drought stress should place it onto the CAM spectrum, but where along the continuum is less clear. Using *Y. gloriosa* and other intermediate species as models for evolution from C_3 to CAM will help clarify the distribution of anatomical, physiological, and molecular traits along a CAM continuum. Better characterization of this continuum will ultimately inform understanding of CAM evolution and the potential for engineering CAM in a C_3 species (Borland *et al.*, 2011, 2014; Borland and Yang, 2013).

CONCLUSIONS

Assessment of photosynthetic pathway and leaf anatomy reveals *Y. gloriosa* as an intermediate C_3–CAM species, possessing the ability to rely entirely on nocturnal carbon uptake during drought stress. This species' ability to use CAM may be limited by a number of anatomical features, including smaller cells that prohibit accumulation of malic acid and greater intercellular airspace that might promote rapid loss of nocturnally fixed CO_2 during the day. *Yucca gloriosa*'s intermediate nature poises it as an ideal system to study the evolution of CAM photosynthesis from a C_3ancestor, and future work will focus on understanding variation among hybrid genotypes.

ACKNOWLEDGEMENTS

Special thanks to the State Parks of North Carolina and South Carolina for permission and help in plant collection, Jeremy Rentsch for plant locality information, Jeremy Ray for assistance with titrations, and Beth Richardson for help with imaging. Funding was provided by the National Science Foundation (DEB 1442199 and IOS 0922742 to JLM), the University of Georgia Graduate School Innovative and Interdisciplinary Research Grant to KH, and the Botanical Society of America Graduate Student Research Award to KH.

REFERENCES

1. Ball E, Hann J, Kluge M, Lee HSJ, Luttge U, Orthen B, Popp M, Schmitt A, Ting IP. 1991.Ecophysiological comportment of the tropical CAM-tree *Clusia* in the field II: Modes of hotosynthesis in trees and seedlings. New Phytologist 117, 483–491.
2. Bloom AJ, Troughton JH. 1979. High productivity and photosynthetic flexibility in a CAM plant. Oecologia 38, 35–43.
3. Borland AM, Barrera Zambrano VA, Ceusters J, Shorrock K. 2011. The photosynthetic plasticity of crassulacean acid metabolism: an evolutionary innovation for sustainable productivity in a changing world. New Phytologist 191,619–633.
4. Borland AM, Griffiths H, Maxwell C, Broadmeadow MSJ, Griffiths NM, Barnes JD. 1992. On the ecophysiology of the Clusiaceae in Trinidad: expression of CAM in*Clusia minor* L. during the transition from wet to dry season and characterization of three endemic species. New Phytologist 122, 349–357.
5. Borland AM, Hartwell J, Weston DJ, Schlauch KA, Tschaplinski TJ, Tuskan GA, Yang X, Cushman JC. 2014. Engineering crassulacean acid metabolism to improve water-use efficiency. Trends in Plant Science 19, 327–338.
6. Borland AM, Yang X. 2013. Informing the improvement and biodesign of crassulacean acid metabolism via system dynamics modelling. New phytologist 200,946–949.
7. Christin P-A, Osborne CP, Chatelet DS, Columbus JT, Besnard G, Hodkinson TR, Garrison LM, Vorontsova MS, Edwards EJ. 2013. Anatomical enablers and the evolution of C_4 photosynthesis in grasses. Proceedings of the National Academy of Sciences of the United States of America 110, 1381–1386.
8. Cockburn W, Ting IP, Sternberg LO. 1979. Relationships between stomatal behavior and internal carbon dioxide concentration in crassulacean acid metabolism plants. Plant Physiology 63, 1029–1032.
9. Cushman JC. 2001. Crassulacean acid metabolism. A

plastic photosynthetic adaptation to arid environments. Plant Physiology 127, 1439–1448.

10. Dodd AN, Borland AM, Haslam RP, Griffiths H, Maxwell K. 2002. Crassulacean acid metabolism: plastic, fantastic. Journal of Experimental Botany 53, 569–580.
11. Doyle J. 1987. A rapid DNA isolation procedure for small quantities of fresh leaf tissue. Phytochemical Bulletin 19, 11–15.
12. Flatz R, Yoder JB, Lee-Barnes E, Smith CI. 2011. Characterization of microsatellite loci in *Yucca brevifolia* (Agavaceae) and cross-amplification in related species. American Journal of Botany 98, e67–e69.
13. Garcia TM, Heyduk K, Kuzmick E, Mayer JA. 2014. Crassulacean acid metabolism biology. New Phytologist 204, 738–740.
14. Gibson AC. 1982. The anatomy of succulence. In: Ting IP Gibbs M , eds.Crassulacean Acid Metabolism: Proceedings of the fifth annual symposium in botany. Rockville, MD: American Society of Plant Physiologists, 1–17.
15. Gompert Z, Buerkle AC. 2010. INTROGRESS: a software package for mapping components of isolation in hybrids. Molecular Ecology Resources 10, 378–384.
16. Griffiths H. 2013. Plant venation: from succulence to succulents. Current Biology23, R340–341.
17. Guralnick LJ, Rorabaugh PA, Hanscom Z. 1984. Seasonal shifts of photosynthesis in *Portulacaria afra* (L.) Jacq. Plant Physiology 76, 643–646.
18. Hattersley PW. 1984. Characterization of C_4 type leaf anatomy in grasses (Poaceae). Mesophyll: bundle sheath area ratios. Annals of Botany 53, 163–180.
19. Heyduk K, Trapnell DW, Barrett CF, Leebens-Mack J. 2015. Phylogenomic analyses of species relationships in the genus *Sabal* (Arecaceae) using targeted sequence capture. Biological Journal of the Linnean Society, doi: 10.1111/bij.12551.
20. Huxman TE, Monson RK. 2003. Stomatal responses of C_3, C_3-C_4 and C_4 *Flaveria*species to light and intercellular CO_2 concentration: implications for the evolution of stomatal behaviour. Plant, Cell and Environment 26, 313–322.
21. Kearse M, Moir R, Wilson A et al. 2012. Geneious Basic: an integrated and extendable desktop software platform for the organization and analysis of sequence data. Bioinformatics 28, 1647–1649.
22. Lee HSJ, Griffiths H. 1987. Induction and repression of CAM in *Sedum telephium*L. in response to photoperiod and water stress. Journal of Experimental Botany 38,834–841.
23. Lüttge U. 2004. Ecophysiology of crassulacean acid metabolism (CAM). Annals of Botany 93, 629–652.
24. Lüttge U. 2006. Photosynthetic flexibility and ecophysiological plasticity: questions and lessons from *Clusia*, the only CAM tree, in the neotropics. New Phytologist 171, 7–25.
25. Martin CE, Lubbers AE, Teeri JA. 1982. Variability in Crassulacean acid metabolism: A survey of North Carolina succulent species. Botanical Gazette 143,491–497.
26. McKown AD, Dengler NG. 2007. Key innovations in the evolution of Kranz anatomy and C_4 vein pattern in *Flaveria* (Asteraceae). American Journal of Botany 94,382–399.
27. Nelson EA, Sage RF. 2008. Functional constraints of CAM leaf anatomy: tight cell packing is associated with increased CAM function across a gradient of CAM expression. Journal of Experimental Botany 59, 1841–1850.
28. Nelson EA, Sage TL, Sage RF. 2005. Functional leaf anatomy of plants with crassulacean acid metabolism. Functional Plant Biology 32, 409–419.
29. Noblin X, Mahadevan L, Coomaraswamy IA, Weitz DA, Holbrook NM, Zwieniecki MA. 2008. Optimal vein density in artificial and real leaves. Proceedings of the National Academy of Sciences of the United States of America 105, 9140–9144.
30. Ogburn RM, Edwards EJ. 2013. Repeated origin of three-dimensional leaf venation releases constraints on the evolution of succulence in plants. Current Biology 23, 722–726.
31. Osmond CB. 1978. Crassulacean acid metabolism: A curiosity in context. Annual Review of Plant Physiology 29, 379–414.
32. R Core Team. 2013. R: A language and environment for statistical computing.Vienna, Austria: R Foundation for Statistical Computing.
33. Rentsch JD, Leebens-Mack J. 2012. Homoploid hybrid origin of *Yucca gloriosa*: intersectional hybrid speciation in *Yucca* (Agavoideae, Asparagaceae). Ecology and Evolution 2, 2213–2222.
34. Schuelke M. 2000. An economic method for the fluorescent labeling of PCR fragments. Nature Biotechnology 18, 233–234.
35. Silvera K, Neubig KM, Whitten WM, Williams NH, Winter K, Cushman JC. 2010. Evolution along the crassulacean acid metabolism continuum. Functional Plant Biology 37, 995–1010.
36. Silvera K, Santiago LS, Winter K. 2005. Distribution of crassulacean acid metabolism in orchids of Panama: evidence of selection for weak and strong modes. Functional Plant Biology 32, 397–407.
37. Štorchová H, Hrdličková R, Chrtek J, Tetera M, Fitze D, Fehrer J. 2000. An Improved method of DNA isolation from plants collected in the field and conserved in saturated NaCl/CTAB solution. Taxon 49, 79–84.
38. Ting IP. 1985. Crassulacean acid metabolism. Annual Review of Plant Physiology36, 595–622.
39. Ueno O, Kawano Y, Wakayama M, Takeda T. 2006. Leaf vascular systems in C_3 and C_4 grasses: a two-dimensional analysis. Annals of Botany 97, 611–621.
40. Winter K, Garcia M, Holtum JAM. 2011. Drought-stress-induced up-regulation of CAM in seedlings of a tropical cactus, *Opuntia elatior*, operating predominantly in the C_3 mode. Journal of Experimental Botany 62, 4037–4042.

41. Winter K, Holtum JAM. 2007. Environment or development? Lifetime net CO_2exchange and control of the expression of Crassulacean acid metabolism in*Mesembryanthemum crystallinum* . Plant Physiology 143, 98–107.

42. Winter K, Holtum JAM. 2014. Facultative crassulacean acid metabolism (CAM) plants: powerful tools for unravelling the functional elements of CAM photosynthesis. Journal of Experimental Botany 65, 3425–3441.

43. Winter K, Holtum JAM, Smith JAC. 2015. Crassulacean acid metabolism: a continuous or discrete trait? New phytologist 208, 73–78.

44. Winter K, Smith JAC (eds). 1996. Crassulacean acid metabolism. Biochemistry, ecophysiology and evolution. Berlin, Germany: Springer.

45. Winter K, Ziegler H. 1992. Induction of crassulacean acid metabolism in*Mesembryanthemum crystallinum* increases reproductive success under conditions of drought and salinity stress. Oecologia 92, 475–479.

46. Zambrano VAB, Lawson T, Olmos E, Fernández-García N, Borland AM. 2014. Leaf anatomical traits which accommodate the facultative engagement of crassulacean acid metabolism in tropical trees of the genus *Clusia* . Journal of Experimental Botany 65, 3513–3523.

47. Zotz G Winter K . 1994. A one-year study on carbon, water and nutrient relationships in a tropical C_3-CAM hemi-epiphyte, *Clusia uvitana* Pittier. New Phytologist 127, 45–60.

CHAPTER 13 MORPHOLOGY, STRUCTURE, AND ONTOGENY OF TRICHOMES OF THE GRAPE GENUS (VITIS, VITACEAE)

Zhi-Yao Ma, Jun Wen, Stefanie M. Ickert-Bond, Long-Qing Chen, and Xiu-Qun Liu

Keywords: *Vitis*, trichome, morphology, structure, ontogeny, SEM, TEM

INTRODUCTION

Trichomes usually originate from epidermal cells and are ubiquitous in many plant families, showing great diversity in morphology, cellular structure and function (Uphof, 1962). These epidermal appendages are unicellular or multicellular, branched or unbranched, glandular or non-glandular (Levin, 1973; Werker, 2000; Yang and Ye, 2013). They have been observed on all vegetative and reproductive organs in angiosperms (Werker, 2000; Wagner et al., 2004). These special structures can protect from herbivory by large mammals, insect and pathogens attacks, increase light reflectance, regulate temperature and decrease water loss (Levin, 1973; Wagner et al., 2004).

Grapes have been widely recognized for their agronomic and economic importance as fresh fruits, and sources of wine and raisins (Wen, 2007). Other species of the grape genus (*Vitis* L.) are important germplasm resources for the wine grape (*V. vinifera* L.), and have been bred to improve resistance of cultivated varieties against many fungal diseases and to enhance cold tolerance (Staudt and Kassemeyer, 1995; Wang et al., 1995; Staudt, 1997; Brown et al., 1999; Wan et al., 2007; Zhang et al., 2012; Gerrath et al., 2015).

Vitis includes ca. 70 species mostly in the temperate regions of the Northern Hemisphere (Chen et al., 2007; Wen, 2007; Zecca et al., 2012; Moore and Wen, 2016). *Vitis* has been shown to be part of the *Ampelocissus–Vitis* clade, one of the five major clades in Vitaceae (Soejima and Wen, 2006; Liu et al., 2013, 2016; Wen et al., 2013). The genus *Vitis* has been widely studied over the past 10 years and confirmed to include two subgenera (subgenus *Muscadinia* and subgenus *Vitis*) (Mullins et al., 1992; Soejima and Wen, 2006; Aradhya et al., 2008; Tröndle et al., 2010; Péros et al., 2011; Zecca et al., 2012; Miller et al., 2013; Wan et al., 2013; Liu et al., 2016). Subgenus *Muscadinia* includes only two species: *V. popenoei* J. H. Fennel from Mexico and *V. rotundifolia* Michx. from the south-eastern U.S.A. (Brizicky, 1965;Moore, 1991; Wen, 2007), while, subgenus *Vitis* includes the Eurasian and the New World clades (Péros et al., 2011;Zecca et al., 2012; Wan et al., 2013; Liu et al., 2016). The phylogeny and taxonomy within the subgenus *Vitis* have not been well resolved due to the highly similar morphology, overlapping distribution, introgression and interspecific hybridization among species within the clade (Aradhya et al., 2008; Moore and Wen, 2016). Many *Vitis* species are differentiated and delimited mainly by the types, distribution and density of trichomes (Moore, 1991; Chen et al., 2007).

However, trichomes of *Vitis* have rarely been studied systematically. Detailed documents of trichome types and distribution in vegetative parts of *Vitis* have not been shown in previous studies (Galet, 1988; Moore, 1991; Chen et al., 2007). He et al. (1994) and Cheng et al. (2013) have explored trichome diversity and some anatomy in *Vitis* trichomes. Nonetheless, they did not show clear and detailed anatomical structures and distribution of trichomes and did not perform any developmental studies of trichomes. The lack of structural and ontogenetic studies on *Vitis* trichomes has limited the exploration of their function and the significance as the basis for inferring the evolutionary relationships among the species in *Vitis*.

Vitis davidii is the only species with prickles on stems and petioles in *Vitis*. The unique character was considered to distinguish *V. davidii* from other species in *Vitis* (Galet, 1988; Chen et al., 2007). However, very little is known about its structure, ontogeny, function and the relationship with other trichome types.

The primary purpose of this study is to: (1) document types and distribution of *Vitis* trichomes to provide available and reliable information for taxonomy, and (2) to explore the

Citation: Ma Z-Y, Wen J, Ickert-Bond SM, Chen L-Q and Liu X-Q (2016) Morphology, Structure, and Ontogeny of Trichomes of the Grape Genus (Vitis, Vitaceae).Front. Plant Sci. 7:704. doi: 10.3389/fpls.2016.00704.

anatomical structure, ultrastructure and ontogeny of *Vitis*trichomes, especially prickles of *V. davidii*, to infer the relationship among trichomes and prickles.

MATERIALS AND METHODS

Plant Materials

This study sampled 34 *Vitis* species covering three different distribution areas: 27 from Asia, six from North America and one from Europe (Table 1).

Table 1. The distribution of trichomes on different vegetative organs in *Vitis*.

	Species	Plant parts									Number of plants
		Current year's branchlets	Young leaves	Mature leaves							
				Petioles	Leaf surfaces					Leaf margins	
					Adaxial		Abaxial				
					Leaf epidermis	Leaf veins	Leaf epidermis	Leaf veins	Leaf vein axils		
ASIAN SPECIES											
1	*V. adenoclada*[1,2,3]	G+ R	G R+ S	G R		R S	R+	R S			3
2	*V. amurensis*	R	R+ S	S*				R* S*	S		4
3	*V. balansana*	R	R+				R*				3
4	*V. bellula* var. *pubigera*	R	R+ S	S		S	R+	R S		S	3
5	*V. betulifolia*	R	R+ S	R		R S	R	R S			1
6	*V. bryoniifolia*	R	R+ S	R S		R S	R	R S		S	4
7	*V. chunganensis*[1,2,3]	G* R*	G* R*	G*				G*			3
8	*V. chungii*		R* S*			S*		S*		S*	5
9	*V. davidii*[1,2,3]	P+	P R S	P		S		R S	S	S	3
10	*V. erythrophylla*		R S	S		R S		R S		S	3
11	*V. flexuosa*[1]	S	R S	S		S		S	S		5
12	*V. hancockii*[1]	S+	R S+	R S+		S		R S+	S	S	4
13	*V. heyneana*	R	R+ S	R		R S	R+	R S			3
14	*V. heyneana* subsp. *ficifolia*	R	R+ S	R		R S	R+	R S			3
15	*V. hui*	R S	R+ S	R S	S+	R S	R+	R S		S	3
16	*V. lanceolatifoliosa*	R	R+ S	R		R S	R+	R S			3
17	*V. piasezkii*	R* S*	R S	R* S*	S+	S	R*	S+		S	3
18	*V. piasezkii* var. *pagnucii*		R S	S		S		R S	S		3
19	*V. pseudoreticulata*	R	R+ S	R S		R S		R S	S	S*	5
20	*V. retordii*[1,2]	R	R+ S+	R S*	S+	S	R+	R S		S	3
21	*V. romanetii*[1,2,3]	G+ R S+	G+ R+ S+	G R S+		S	R	G R S+		S	3
22	*V. ruyuanensis*		S	S		S		S		S	4
23	*V. shenxiensis*[1,2]	G R	G R S	G R S		S		S	S	S	1

24	*V. sinocinerea*	R S	R+ S	R S	S	R+	R S		S	3
25	*V. tsoii*	S	R S	R S	R S	R*	R S		S	3
26	*V. wilsoniae*[12]	R	R+ S	R S	S		R S	S	S	3
27	*V. wuhanensis*	R	R+ S	R S	R S		R S		S	3
EUROPEAN SPECIE										
28	*V. vinifera*	R	R S*	R S*		R*	R S*			4
NORTH AMERICAN SPECIES										
29	*V. labrusca*	G* R	R+ S	G* R	S	R+	R S		S	1
30	*V. mustangensis*	R	R+	R		R+	R			1
31	*V. riparia*		S	S	S		S	S	S	1
32	*V. rupestris*		R S	S	S		S	S	S	1
33	*V. shuttleworthii*	R	R+	R		R+	R			1
34	*V. rotundifolia*	G* S	R S	S	S		S	S		1

Abbreviations in table: R, ribbon trichomes; G, glandular trichomes; P, prickles; S, simple trichomes. The blank spot indicate that the organ is glabrous. The numbers "1," "2," "3" indicate the species were examined with light, scanning electron and transmission electron microscopy respectively. "+" indicate that the trichome type is stable present and forms remarkable dense indumentum on the vegetative organ. "" indicate that the trichome type is variable and not stable present on the vegetative organ. The unmarked indicate that the trichome type is stable present, but not forms remarkable dense indumentum on the vegetative organ.*

Stereomicroscope

Herbarium specimens and fresh vegetative parts (current year's branchlets, young leaves, petioles, adaxial and abaxial leaf surfaces of mature leaves, and leaf margins) of *Vitis* were examined to characterize the types and distribution of trichomes in 34 *Vitis* species using a Nikon SMZ800 stereomicroscope. At least three plants were sampled for most species, except *V. betulifolia*, *V. shenxiensis* and all North American species for which there was only one individual sampled. Young leaves were taken from shoot tips, while mature ones were taken from nodes after the 9th node (counting from the top) to assure their mature condition. Five young and five mature leaves from each plant were sampled respectively. The presence/absence and types of trichomes on different vegetative organs were documented.

The detailed anatomy, ontogeny and ultrastructure studies of *Vitis* trichomes were conducted using light microscopy (LM), scanning electron microscopy (SEM) and transmission electron microscopy (TEM).

Light Microscopy

To examine the anatomical structure of trichomes, 33 samples of stems, petioles and leaf mid-veins, from nine *Vitis*species (*V. adenoclada*, *V. chunganensis*, *V. davidii*, *V. flexuosa*, *V. hancockii*, *V. retordii*, *V. romanetii*, *V. shenxiensis*, and *V. wilsoniae*), were fixed in 70% FAA for 24 h. Samples were dehydrated in an ethanol series (Johansen, 1940) and embedded in paraffin wax. Transversal and longitudinal sections of 6–8 μm were cut with a Leica RM2235 rotary microtome, mounted on slides and stained with 1% safranin O and 0.5% fast green.

Scanning Electron Microscopy

Thirty-eight samples including stems, petioles and leaves from representatives of eight species (*V. adenoclada*, *V. chunganensis*, *V. davidii*, *V. heyneana*, *V. retordii*, *V. romanetii*, *V. shenxiensis*, and *V. wilsoniae*) were fixed in 2.5% glutaraldehyde in 0.1M phosphate buffer, pH7.2. The fixed samples were dehydrated in an ethanol series and critical point dried. Samples were placed with double-sided tape on stubs and sputter-coated with a thin gold layer (Robards, 1978). The types and morphology of trichomes were examined with a JEOL JSM-6390LV Scanning Electron Microscope. Based on the SEM images, we measured the heights/lengths of different trichome types and the sizes of parts of glandular trichomes or prickles.

Transmission Electron Microscopy

Four samples of glandular trichome or young prickle from *V. adenoclada*, *V. chunganensis*, *V. davidii*, and *V. romanetii*were fixed in 1% osmium tetroxide, in 0.1M phosphate buffer, pH7.2, and processed following standard techniques (Roland and Vian, 1991). Ultra-thin sections of both glandular trichomes and the prickles were made and viewed under a Hitachi H-7650 Transmission Electron Microscope.

Histochemical Lignin Estimation

The histochemical lignin tests for prickles of different growth stages of *V. davidii* were performed in fresh hand-cut sections. The sections were soaked in a fresh phloroglucinol solution, and then a few drops of 35% HCl were added (Pillonel et al., 1991). The positive result will shows an intense pink color.

RESULTS

Trichome Types

We used the terminology of Payne (1978) to describe the trichomes of *Vitis* in this study. Two general trichome types in*Vitis* are documented: non-glandular and glandular. Non-glandular trichomes include ribbon and simple trichomes.

Non-glandular Trichomes

In *Vitis* the distribution and densities of non-glandular trichomes vary among different vegetative organs including current year's branchlets, young leaves and mature leaves, as well as on petioles, adaxial and abaxial leaf surfaces and leaf margins (Figures 1A–I). Almost all species sampled bear the ribbon and simple trichomes on different plant organs (Table 1).

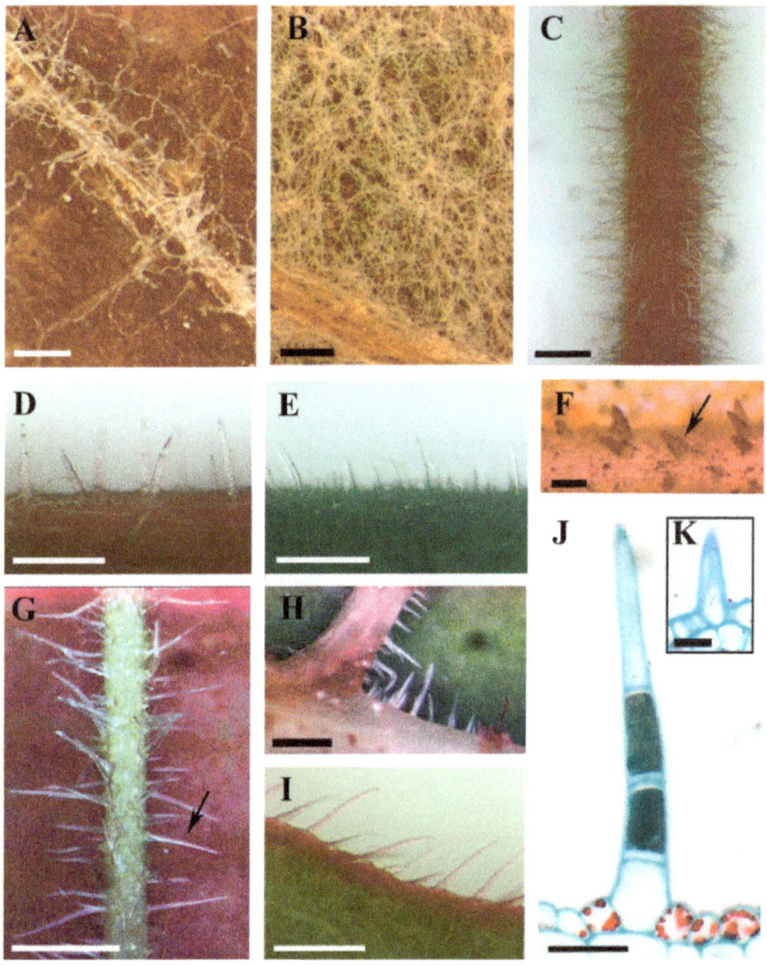

Figure 1. Distributions and structure of ribbon and simple trichomes. (A,B) Ribbon trichomes on different plant organs. Ribbon trichomes on adaxial leaf surfaces of *V. bryoniifolia* (A) and on abaxial leaf surfaces of *V. adenoclada* (B). (C–I) Simple trichomes on different plant organs. Simple trichomes on branchlets of *V. hancockii*(C), petioles of *V. pseudoreticulata* (D), adaxial leaf surfaces of *V. retordii* (E), adaxial leaf veins (F) and abaxial leaf veins of *V. hancockii* (G), abaxial leaf vein axils of *V. shenxiensis* (H) and leaf margins of *V. piasezkii* (I). Arrow in(F) shows a simple unicellular trichome and in (G) shows a simple uniseriate multicellular trichome. (J)Photomicrograph of microtome section of a simple trichome with uniseriate cells of *V. hancockii*. (K) A simple unicellular trichome of *V. davidii*. Scale bars: (A,B) = 0.25 mm; (C,E) = 1 mm; (D,G–I) = 0.5 mm; (F) = 100 μm;(J) = 50 μm; (K) = 20 μm.

Ribbon trichomes are flattened, slender, greatly elongated, twisted and unicellular (Figures 1A,B, 2A,B). Ribbon trichomes can be as long as several millimeters. In total, 32 of the 34 species sampled bear the ribbon trichomes (Table1). Twenty-three of the 34 species sampled have ribbon trichomes on current year's branchlets. On mature leaves, we observed ribbon trichomes on the petioles of 21 of the 34 species sampled. In addition, on the adaxial leaf surfaces of some species (e.g., *V. adenoclada* and *V. bryoniifolia*) only the leaf veins bear sparse ribbon trichomes (Figure 1A; Table1). On the abaxial leaf surfaces, ribbon trichomes are found in 26 species, of which 11 species have non-deciduous dense arachnoid indumentum composed of this trichome type (Figure 1B; Table 1; *V. adenoclada*, *V. bellula* var. *pubigera*, *V. heyneana*, *V. heyneana* subsp. *ficifolia*, *V. hui*, *V. lanceolatifoliosa*, *V. retordii*, *V. sinocinerea*, *V. labrusca*, *V. mustangensis*, and *V. shutterworthii*).

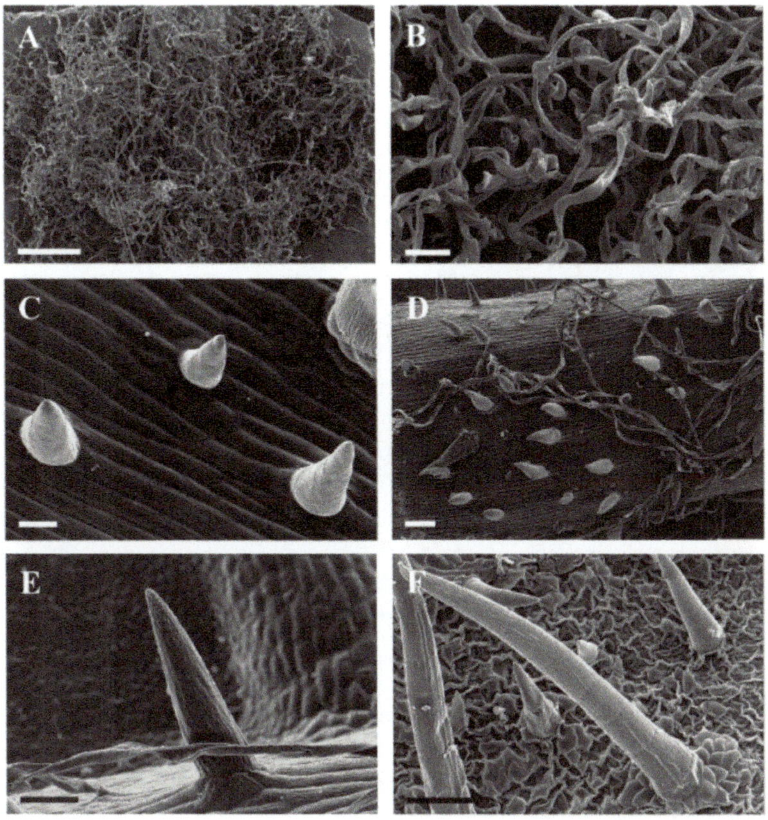

Figure 2. SEM micrographs of ribbon and simple trichomes. (A,B) Ribbon trichomes on abaxial leaf surfaces of a young leaf in *V. wilsoniae*. (C–F) Simple trihcomes: (C) young simple trichomes on veins of *V. romanetii*; (D,E)simple trichomes on veins of young leaves in *V. davidii*; (F) simple trichomes on adaxial leaf surfaces of *V. retordii*. Scale bars: (A) = 500 µm; (B,E) = 50 µm; (C) = 20 µm; (D,F) =100 µm.

Simple trichomes are unbranched, uniseriate, or unicellular (Figures 1C–J, 2C–F), and originate from the epidermis (Figure 2C). Simple trichomes are found in 30 of the 34 *Vitis* species examined, but their distribution on different plant organs varies (Table 1). On mature leaves, only eight species have simple trichomes on the current year's branchlets (Figure 1C), and 22 species have this trichome type on petioles (Figure 1D), while 28 species bear simple trichomes on adaxial leaf surfaces (Figures 1E,F). Simple trichomes are present on the veins of abaxial leaf surfaces in 30 species (Figure 1G), and 11 species (*V. amurensis*, *V. davidii*, *V. flexuosa*, *V. hancockii*, *V. piasezkii* var. pagnucii, *V. pseudoreticulata*, *V. shenxiensis*, *V. wilsoniae*, *V. riparia*, *V. rupestris*, and *V. rotundifolia*) bear simple trichomes on the abaxial leaf vein axils (Figure 1H). On leaf margins, 20 species bear simple trichomes (Figure 1I). We observed dramatic length differences in simple trichomes ranging from 50–200 µm in length on veins of adaxial and/or abaxial leaf surfaces in *V. adenoclada*, *V. davidii*, *V. hancockii*, *V. heyneana*, and *V. wilsoniae* (Figures 1F, 2D,E), while some of simple trichomes are up to ca. 750–900 µm long in *V. romanetii* and *V. retordii* (Figure 2F).

Almost all species examined (30/34) show an indumentum composed of ribbon trichomes on young leaves (Figures 3A–E), which may become sparse or glabrescent on mature leaves (Table 1). Ribbon trichomes are only occasionally observed on the young leaves of some species (*V. chunganensis* and *V. chungii*) (Table 1). Thirty species sampled were covered with simple trichomes on young leaves (Figure 3F; Table 1) and some species sampled are nearly glabrous (Figure 3G). Twenty-eight species sampled bear both ribbon and simple trichome types (Figure 3A, Table 1).

Figure 3. Young leaves with ribbon and/or simple trichomes. (A–E) Dense ribbon trichomes on abaxial leaf surfaces of *V. romanetii* (A), *V. lanceolatifoliosa* (B) and *V. retordii* (C) and adaxial leaf surfaces of *V. wuhanensis*(D) and *V. wilsoniae* (E). (F) Dense simple trichomes on abaxial leaf surfaces of *V. hancockii*. (G) *V. chungii* nearly glabrous on the abaxial leaf surfaces. Scale bars: (A–E,G) = 3 mm; (F) = 4 mm.

The non-glandular trichome density, especially the ribbon trichome density, changed on leaves of some species of *Vitis*(e.g., *V. amurensis* and *V. wilsoniae*) during development of leaves. The ribbon trichomes become lost with development of the leaves in some species of *Vitis*.

Glandular Trichomes and Prickles

Glandular trichomes are observed in six species (*V. romanetii*, *V. adenoclada*, *V. chunganensis*, *V. labrusca*, *V. rotundifolia*, and *V. shenxiensis*), but prickles are only observed in *V. davidii*. Glandular trichomes vary in shape, size, density, distribution and color among *Vitis* species in this study (Figures 4A–J, 5A–I). All of them are multicellular, non-vascularized and composed of both epidermis and subjacent layers (Figures 4B,E,G,J). The glandular trichomes of these species are composed of apparent glandular heads with multiple layers of epidermal cell and stalks with one layer of epidermal cell (Figure 4J). Stomata are frequently observed on the upper region of glandular heads in most *Vitis*species sampled except in *V. chunganensis* (Figures 5C,F,H,I).

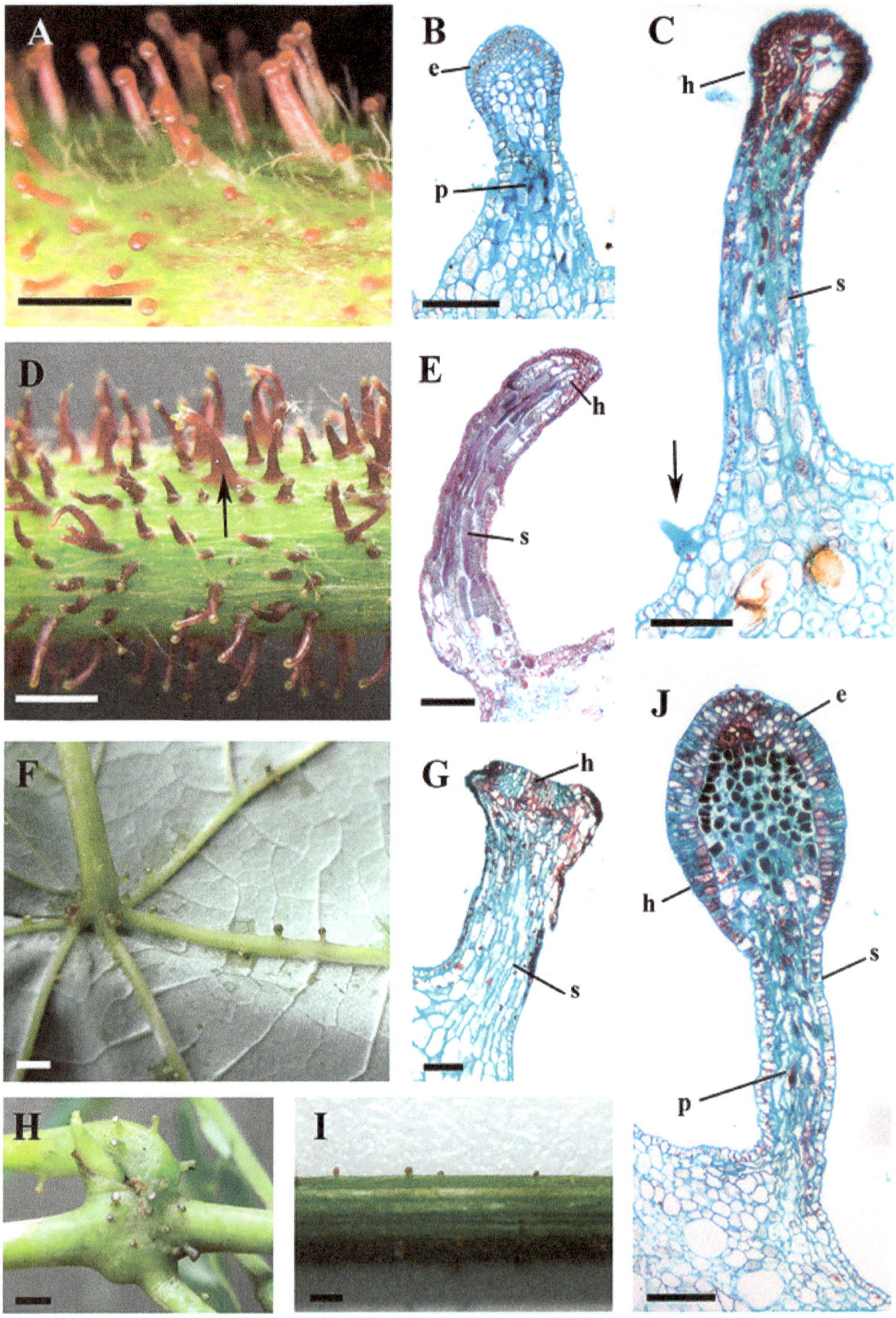

Figure 4. Structures and distributions of glandular trichomes in *Vitis*. (A–C) Glandular trichomes of *V. romanetii*: (A) glandular trichomes on stems; (B,C) structures of glandular trichomes and simple trichomes on the epidermis of stalks of glandular trichomes (arrow). (D,E) Glandular trichomes on the stems of *V. adenoclada*: glandular trichomes on stems and some of these trichomes form complexes (arrow) (D). (F–H) Glandular trichomes of *V. chunganensis*: glandular trichomes on basal parts of veins of abaxial leaf surfaces (F) and nodes (H). (I,J)Glandular trichomes on stems of *V. shenxiensis*. Abbreviations: e, epidermal cells; p, parenchymatic cells; h, head; s, stalk. Scale bars: (A,D,F,H,I) = 1 mm; (B,C,E,G,J) = 100 μm.

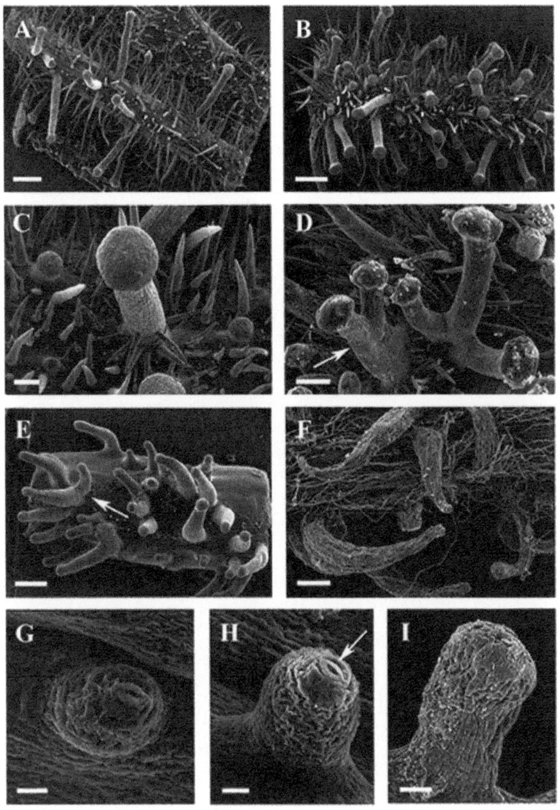

Figure 5. SEM micrographs of glandular trichomes of *Vitis*. (A,B) Glandular trichomes and simple trichomes on abaxial leaf surfaces (A) and petioles (B) of *V. romanetii*. (C) Simple trichomes on the epidermis of stalks of glandular trichomes of *V. romanetii*. (D) Trichome complexes of glandular trichomes (arrow) of *V. romanetii*. (E)Complexes (arrow) of prickles on petioles of *V. davidii*. (F) Glandular trichomes with hooked stalks on stems of *V. adenoclada*. (G,H) Emerging cell masses (G) and a glandular head without stalk and an obvious stoma (arrow) on the top of glandular head (H) in *V. shenxiensis*. (I) A glandular trichome on nodes of *V. chunganensis*. Scale bars:(A,B,E) = 500 μm; (C) = 100 μm; (D,F) = 200 μm; (G,H) =20 μm; (I) = 50 μm.

Developmentally, we can recognize three stages of glandular trichome growth (Figures 6A–E, 7A–F). Initially, cell masses emerge from the epidermis where stomata may develop with rapid proliferation and elongation of upper region epidermal cells (Figures 5G, 7A,B), forming a sessile glandular head (Figures 5H, 6A–C, 7C). Secondly, rapid proliferation and expansion of epidermal cells and underlying parenchymatous cells cause a stalk to develop (Figures 6D, 7D,E). Finally, glandular trichomes fully mature (Figures 6E, 7F) and proceed to the senescence period in which they appear shrunken (Figure 5I).

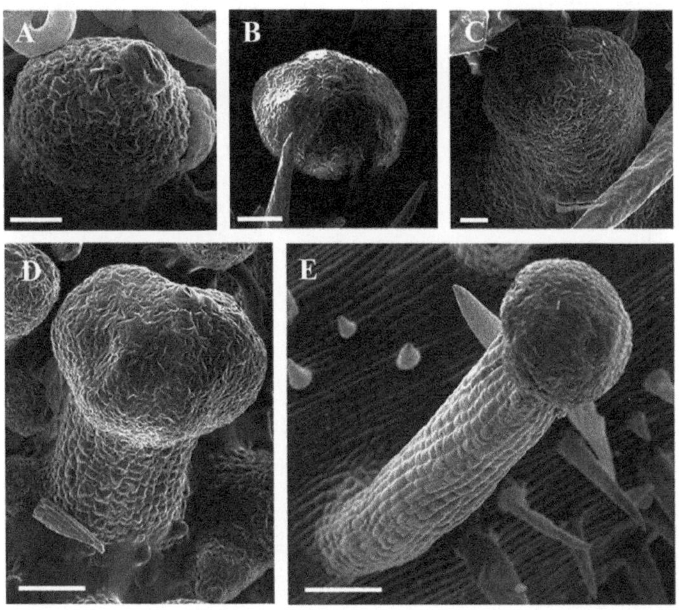

Figure 6. SEM micrographs of development of glandular trichomes of *V. romanetii*. (A–C) Glandular heads without apparent stalks. (D) Young glandular trichome with short stalk. (E) Mature glandular trichome with long stalk. Scale bars: (A) = 20 μm; (B) = 50 μm; (C) = 20 μm; (D) = 50 μm; (E) = 100 μm.

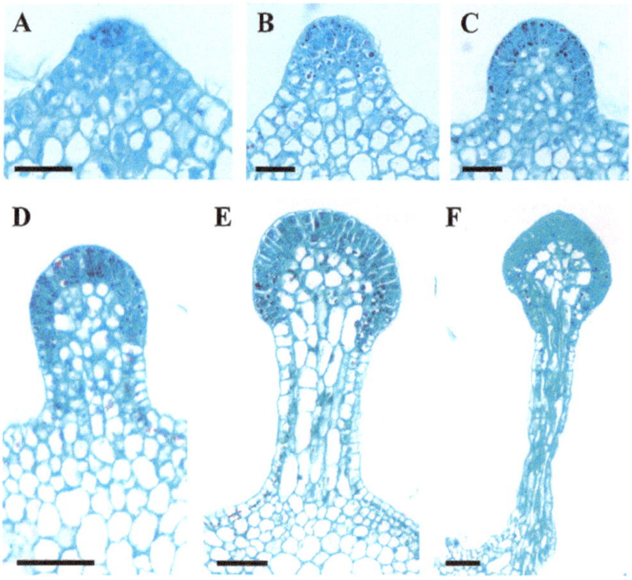

Figure 7. Development of glandular trichomes of *V. romanetii*. (A) Emerging cell masses. (B) Original gland head with rapid proliferation and elongation of upper region epidermal cells. (C) Emerging glandular head without stalk. (D) Young glandular trichome with short stalk. (E) Elongation of stalk of young glandular trichome. (F) Mature glandular trichome. Scale bars: (A–C) = 50 μm; (D–F) = 100 μm.

The predominant distributions of dense glandular trichomes of *V. romanetii* are on branchlets, petioles and veins of abaxial leaf surfaces alongside dense and long simple trichomes (Figures 4A–C, 5A–D). The glandular trichomes are pink/yellow, erect and ca. 1200 μm long. Diameters of their glandular heads are ca. 200 μm. They emerge and continue to grow with the extension of stalks until maturity. Interestingly, adjacent glandular trichomes may produce trichome complexes which may be formed after fusion or branching, meanwhile, this similar structure is also observed in *V. adenoclada* and *V. davidii* (Figures 4D, 5D,**E**). Simultaneously, many simple trichomes are observed on the epidermis of the stalks of glandular trichomes (Figures 4C, 5C,D).

The glandular trichomes of *V. romanetii* are mucilaginous and emit volatile substance based on our observation. Many insects are often glued to their glands (Figures 8A–C).

Figure 8. Insects were glued to the glandular trichomes of *V. romanetii* (A–C). Arrows indicate stuck insects. Scale bars: (A–C) = 3 mm.

Glandular trichomes of *V. adenoclada* have wine-red and hooked stalks and faint-yellow glandular heads (Figures 4D,E,5F). They are commonly observed on branchlets and petioles. The height of these mature trichomes is up to 1100 μm and the diameter of the glandular heads is ca. 80–100 μm.

The glandular trichomes of *V. chunganensis* are present on basal veins of abaxial leaf surfaces, petioles and at along stem nodes (Figures 4F,H) and have short stalks and inverted trapezoid heads (Figures 4G, 5I). The height of these trichomes is only 80–550 μm and the diameter of the glandular heads is 100–350 μm.

Glandular trichomes of *V. shenxiensis* bear similar structures to those of *V. romanetii*, but are less dense on branchlets (Figures 4I,J). Furthermore, these trichomes are ca. 650 μm long and have large glandular heads with diameters up to 250 μm.

The prickles of *V. davidii* are similar to those of other glandular trichomes in *Vitis*. They are also multicellular, non-vascularized and composed of both epidermis and subjacent layers. The prickles have stalks with one layer of epidermal cells and apparent glandular heads with multiple layers of epidermal cells on which stomata are frequently observed. Similar to early developmental stages of glandular trichomes, initially, the cell masses emerge from the epidermis with notable stomata at ca. 200 μm in diameter (Figure 9A); then the glandular heads grow without stalks (Figure 9B); and finally, the stalks of prickles form with rapid proliferation and expansion of epidermal and parenchymatous cells (Figures 9C,D, 10A). However, the prickles increase in both height and basal width during development (Figure 9E). They become conical and glaucous (Figure 10B) and lignified at maturity (Figure 10C). Some glandular heads of prickles shrink with lignification (Figures 9F,G). Finally, they become hard and prickly (positive result for histochemical lignin estimation) (Figures 9H and 10D).

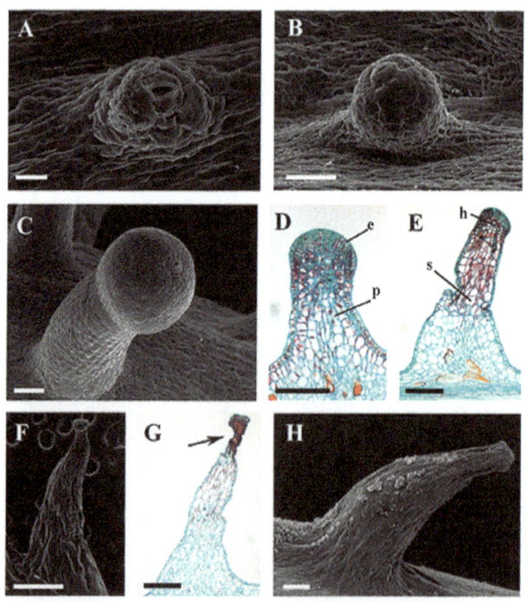

Figure 9. Development and structure of prickles of *V. davidii*. (A) Emerging cell masses with a notable stoma and (B) a glandular head without stalk. (C) Young prickle with stalk. (D) Structure of the young prickle. (E) Mature prickle. (F) Shrinking prickle. (G) Senescent prickle, arrow indicating the shrinking top of gland. (H) The lignifying prickle. Abbreviations: e, epidermal cells; p, parenchymatic cells;

h, head; s, stalk. Scale bars: (A) = 20 μm; (B,C) = 50 μm; (D) = 100 μm; (E,H) = 200 μm; (F) = 500 μm; (G) =250 μm.

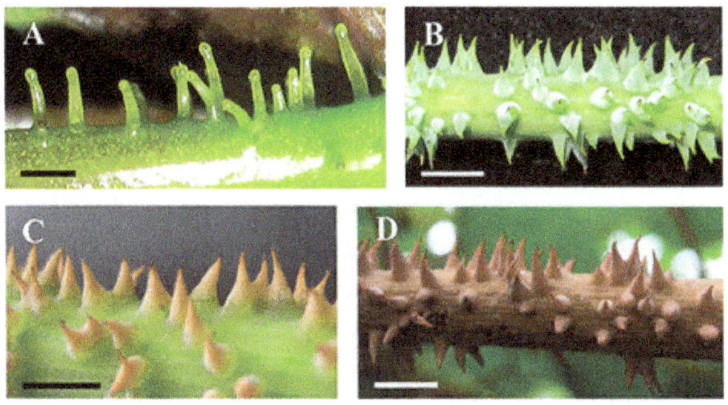

Figure 10. Development and morphology of prickles of *V. davidii*. (A) Young prickles on petiole. (B) Mature prickles on stem. (C) The withering and lignifying terminal part of prickles. (D) Fully lignifying prickles. Scale bars:(A) = 1 mm; (B–D) = 5 mm.

Ultrastructure

The glandular heads of *V. davidii*, *V. adenoclada*, *V. chunganensis*, and *V. romanetii* showed similar ultrastructure. There are numerous osmiophilic granular and irregular masses found in the dense cytoplasm and vacuoles in the epidermal cells of glandular heads of all species sampled (Figures 11A–E). The intercellular spaces of the epidermal cells contain granular and/or thready material in *V. davidii*, *V. adenoclada*, and *V. chunganensis* (Figures 11A–D).

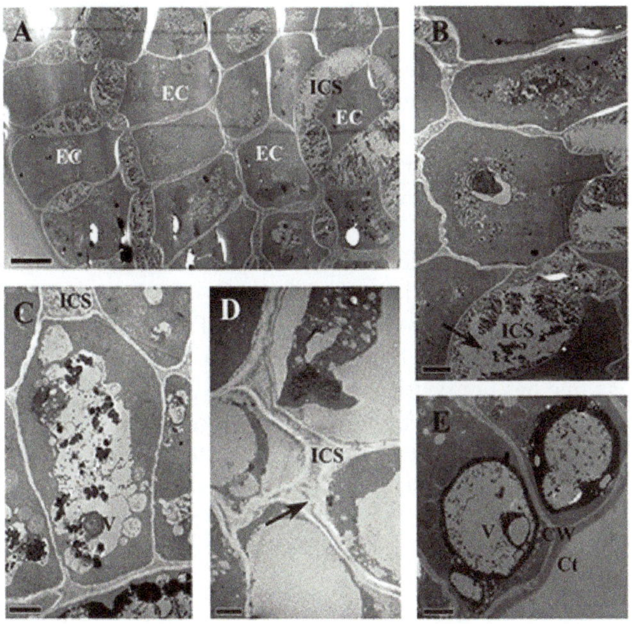

Figure 11. TEM images of the glandular trichomes and young prickles. (A) Epidermal cells of gland head in*V. davidii*, showing containing of dense thready and granular material in vacuoles and intercellular spaces. (B) Detail of the epidermal cells of *V. davidii*. Arrow indicate dense granular material. (C) Numerous irregular masses of osmiophilic material accumulating in vacuoles of epidermal cells of gland head in *V. adenoclada*. (D) Dense thready material (arrow) in the intercellular spaces of *V. chunganensis*. (E) Vacuoles are filled with osmiophilic material. Abbreviations: CW, cell wall; Ct, cuticle; EC, epidermal cells; ICS, intercellular spaces; V, vacuole. Scale bars: (A) = 5 μm; (B,C,E) = 2 μm; (D) = 1 μm.

DISCUSSION

The trichome types include ribbon, simple and glandular trichomes in representatives of *Vitis* on different vegetative plant organs. These trichomes may play different roles in plant physiology and ecology with variable morphological, mechanical and phytochemical characteristics (Kortekamp and Zyprian, 1999; Wagner et al., 2004).

Ribbon Trichomes

Ribbon trichomes are flattened and twisted with variable densities on the different vegetative organs among species of*Vitis* (Figures 1A,B; Table 1). The young leaves of most *Vitis* species are covered by ribbon trichomes, which may serve as a mechanical barrier against harsh spring temperature, herbivores and pathogens (Kortekamp and Zyprian, 1999;Werker, 2000) and restrict insect activities on the leaf surfaces (Wagner et al., 2004). They may also play a photo-protective role in decreasing light stress (Liakopoulos et al., 2006). Dense ribbon trichomes (Figures 3A–E) may be a significant defense mechanism against abiotic and biotic stresses for young plant organs of *Vitis*. These trichomes may also help decrease water loss and facilitate acclimation to xeric environments (Wagner et al., 2004). During development of these young leaves, in some *Vitis* species sampled (e.g., *V. flexuosa*, *V. rupestris*, *V. rotundifolia*, and *V. shenxiensis*), the ribbon trichomes are deciduous, becoming glabrescent at maturity (Table 1). This phenomenon may be due to the promotion of photosynthetic competence and establishment of phytochemical defense mechanism of leaves when they mature, at which point they no longer need the protection of ribbon trichomes.

Ribbon trichomes are frequently observed in both subgenera of *Vitis* (subgenus *Muscadinia* and subgenus *Vitis*) distributed in North America, East Asia, and Europe (Moore, 1991; Chen et al., 2007; our observations). Interestingly, ribbon trichomes also occur in other genera of the *Ampelocissus–Vitis* clade (e.g., *Ampelocissus* and *Pterisanthes*) (Liu et al., 2013, 2016; Wen et al., 2013) in Vitaceae according to the previous studies (Latiff, 1982; Jackes, 1984; Ickert-Bond et al., 2015) and our observations, while, as far as we know they are not observed in any other genera of Vitaceae (e.g., *Ampelopsis, Rhoicissus, Tetrastigma, Cayratia, Cissus, Parthenocissus*, and *Yua*). Our results support that *Vitis, Ampelocissus* and *Pterisanthes* have similar morphological characteristics, confirming the hypothesis that *Vitis* is most closely related to *Ampelocissus* and *Pterisanthes* based on the molecular evidence (Liu et al., 2013, 2016; Wen et al., 2013) and inflorescence morphology (Ickert-Bond et al., 2015). We infer that the ribbon trichome may be a morphological synapomorphy for the *Ampelocissus–Vitis* clade, which should be tested in conjunction with a fully resolved phylogeny of the *Ampelocissus–Vitis* clade in the future.

Simple Trichomes

In most *Vitis* species, simple trichomes differ in size, density, cell number and growth positon, and they are all unbranched, uniseriate or unicellular (Figures 1J,K). The development of uniseriate trichomes in the majority of flowering plants is achieved by initial enlargement and a series of periclinal divisions of an epidermal cell (Werker, 2000; Yang and Ye, 2013).

A dense indumentum composed of simple trichomes usually acts as a filter protecting plant tissues against damage of ultraviolet-B radiation and as a deterrent to insect activity, including oviposition and feeding in many species (Levin, 1973; Karabourniotis et al., 1995; Liakoura et al., 1997; Manetas, 2003; Yan et al., 2012). Simple trichomes in wine grape leaves (*V. vinifera* L.) are also considered as a defense against photo-damage during leaf development by modifying and filtering both ultraviolet and visible lights (Karabourniotis et al., 1999; Liakopoulos et al., 2006).Kortekamp and Zyprian (1999) demonstrated that the indumentum of grape leaves with both simple and ribbon trichomes acts as a protective barrier and could restrict the development of fungi and prevent the fungal hyphae from reaching the leaf cuticle.

However, the density of leaf trichomes in grapevine cultivars has no influence on the activity and oviposition of leafhoppers, on the contrary, the density of simple trichomes on the leaf veins is negatively correlated with the parasitism rate by *Anagrus* spp., which is a biological control agent to parasitize the eggs of leafhoppers (Pavan and Picotti, 2009). Short simple trichomes did not affect the oviposition of the leafhoppers, but instead reduced the parasitism rate of *Anagrus* spp., while the long simple trichomes could influence the oviposition of the leafhopper on different herbaceous plants (Pavan and Picotti, 2009). We thus infer that different sizes of insects are influenced by the different height of simple trichomes in *Vitis*, with the short simple trichomes (e.g. in *V. davidii* and *V. vinifera*) impeding activities of smaller insects, and the long simple trichomes (e.g., in *V. retordii* and *V. romanetii*) may protect against bigger insects.

Glandular Trichomes and Prickles

Glandular trichomes show great variation in morphology and cell structure (Werker, 2000). They may have uni- or multiserate stalks and uni- or multicellular glandular heads. They can be derived from the epidermis (e.g., capitate glandular trichomes) or from both the epidermis and subjacent layers. Some types of glandular trichomes, which are solely derived from the epidermis are found in many species and are particularly prominent in Asteraceae, Cucurbitaceae, Cannabaceae, Lamiaceae, and Solanaceae (Akers et al., 1978; Werker et al., 1985; Nielsen et al., 1991;Ascensão and Pais, 1998; Kolb and Müller, 2004; Tissier, 2012; Celep et al., 2014).

However, all glandular trichomes and the prickles of *Vitis* which were examined histologically in our study are derived from both the epidermis and subjacent layers, which are similar to the glandular trichomes on the leaves of *Drosera*(Fahn, 1979) and those on *Passiflora foetida* stipules (Durkee et al., 1984), as well as the food bodies, which are globose structures attached to the plant by a short peduncle for promoting mutualism between plants (e.g., in Vitaceae) and ants (Buono et al., 2008; Paiva et al., 2009).

Glandular trichomes can secrete terpenes, phenolics, alkaloids, lipophilic or other substances deterring or poisoning herbivores and pathogens (Levin, 1973; Werker, 2000; Tissier, 2012) or secrete resin to protect the developing tissues from cold temperatures (Lapinjoki et al., 1991). Secreted mucilage from digestive glands of *Drosera* acts as a sticky trap for insects (Outenreath and Dauwaldert, 1986).

The glandular trichomes of *V. romanetii* can secret mucilage and emit volatile substance so that many insects are often glued to glands based on our observation (Figures 8A–C). This trichome type may act as a defense mechanism against insects. The composition of mucilage and volatile substance will be explored in future. However, the glandular trichomes of other species lacked apparent secretions.

Akers et al. (1978) reported that accumulation of the secretions of glandular trichomes of *Nicotiana tabacum* can be observed both within extraplasmic and intercellular spaces. These extracellular spaces for the preservation of the secretions were observed in other glandular trichomes as well (Vermeer and Peterson, 1979; Ascensão et al., 1997;Gravano et al., 1998; Turner et al., 2000). In the secretory cells of glandular trichomes of *Passiflora foetida*, dense masses of granular or fibrillar material accumulate in vacuoles, intercellular spaces and beneath the cuticle (Durkee et al., 1984). Similarly, the ultrastructure of the glandular trichomes and the young prickle of *Vitis* show that there are numerous osmiophilic granular, thread-like and irregular masses in the cytoplasm, vacuoles and/or the intercellular spaces in the epidermal cells of glandular heads. These results suggest that the glandular trichomes and the young prickle might excrete substances and the intercellular spaces may be the places of temporary storage of these secretions in the *Vitis* species examined. There is always one stoma at the apex of glandular trichome (except *V. chunganensis*) based on our observations.

Classification of glandular trichomes is based on morphological and ultrastructural differences (Werker, 2000). In *Vitis*the prickles are only present on stems and petioles of *V. davidii*. Considering the morphology, ultrastructural and development of glandular trichomes of other species in *Vitis*, the prickles of *V. davidii* and the glandular trichomes have similar morphological and ultrastructural characteristics and share similar early developmental stages. Our studies reveal that the prickles of *V. davidii* are homologous to glandular trichomes. The prickles undergo early developmental stages like other glandular trichomes (Figures 9A–D, 10A) and then increase in both height and basal width (Figures9E–G, 10B), lignification (Figure 10C) and harden (Figures 9H, 10D). Our findings are consistent with the developmental process of prickles on raspberries and roses (Kellogg et al., 2011). Furthermore, trichomes as the precursors to prickles have been mentioned in many other species (Delbrouck, 1875; Leelavathi and Ramayya, 1983). The prickles of *V. davidii* may facilitate climbing and deter large herbivores from feeding on this species. Since birds are important seed dispersers of grapes (Hardie and Obrien, 1988), the prickles may protect fruits from frugivores as well.

Glandular trichomes and the prickles are here observed on Asian *Vitis* species, and they do not usually occur on species from North America and Europe. Moore and Wen (2016) suggest that only *V. labrusca* and *V. rotundifolia* occasionally show few glandular trichomes (Table 1). In addition, glandular trichomes are present on the stems of *Ampelocissus martini* and *A. arachnoidea* and on the pedicel of *A. xizangensis* (Latiff, 1982, 2001; Chen et al., 2007). Conical prickles are also present on the stems of *A. aculeata* and *A. acetosa* in Papua New Guinea and Northern Australia (Jackes, 1984;Latiff, 2001). Further study is needed to elucidate the origin of the glandular trichomes and prickles of *Vitis* and *Ampelocissus*.

Taxonomic Utility of Trichomes in *Vitis*

The trichome types and density of *Vitis* are influenced by different regional environment and growth stages based on our observation (Table 1). For example, not all *V. chunganensis* which were collected from different localities have glandular trichomes. Some plants of *V. piasezkii* only have ribbon trichomes while the others bear simple trichomes along the current years' branchlets. The ribbon trichomes are dense to nearly glabrous on leaves of *V. wilsoniae* during the growing season. The ribbon trichomes with variable density are present on different leaves of *V. balansana*.

Ribbon and simple trichomes are known from most species of *Vitis*, except for some nearly glabrous species (Table 1). Nineteen and three species sampled have dense ribbon and simple trichomes on young leaves respectively, which are useful taxonomic characters. Meanwhile, young leaves of many species often differ in their types and density of trichomes compared to mature leaves on some vegetative organs (Table 1), due to the loss of trichomes during development of leaves. So it is important for identification to examine mature leaves. For example, in *V. wilsoniae*, the young leaves bear very dense ribbon trichomes (Figure 3E), while, there are hardly any ribbon trichomes on the mature leaves (only a few of trichomes on the veins of abaxial leaf surfaces were observed).

Eleven species sampled in this study show a non-deciduous dense arachnoid indumentum composed of ribbon trichomes on the abaxial leaf surfaces, which distinguish these species from any of the other *Vitis* species examined. *Vitis retordii* shows a dense indumentum composed of long simple trichomes on adaxial leaf surfaces, which can easily distinguish this species from any other *Vitis* species examined. Simple trichomes on abaxial leaf vein axils or leaf margins are important taxonomic characters for some species (e.g., *V. flexuosa*, *V. piasezkii*, *V. piasezkii* var. *pagnucii*, and *V. shenxiensis*).

Glandular trichomes and prickles are only documented in seven species and these species can be further distinguished by the morphological differences of the glandular trichomes. The prickles on stems and petioles of *V. davidii* are considered an unique character that distinguishes this species from any other *Vitis* species (Chen et al., 2007). The glandular trichomes of *V. romanetii* are dense, long, pink/yellow and columnar on branchlets, petioles and veins of abaxial leaf surfaces, while those of *V. adenoclada* are dense, long, wine-red and hooked on stems and petioles. In comparison the glandular trichomes of *V. chunganensis* are less dense, short, green and have inverted trapezoid heads on basal veins of abaxial leaf surfaces, petioles and along stem nodes. The distribution of the glandular trichomes of *V. shenxiensis* is very sparse, and the glandular trichomes are short and the glandular heads are oval in *V. shenxiensis*. Details of the distribution of the three types of trichomes in different species in this study (Table 1) can contribute to increase our understanding of *Vitis* systematics and taxonomy.

AUTHOR CONTRIBUTIONS

ZM and XL designed the study, ZM performed the research and analyzed the data. ZM, JW, SI, LC, and XL discussed the data and wrote the manuscript.

CONFLICT OF INTEREST STATEMENT

The authors declare that the research was conducted in the absence of any commercial or financial relationships that could be construed as a potential conflict of interest.

ACKNOWLEDGMENTS

We thank Prof. C.H. Liu and Dr. J.F. Jiang (Zhengzhou Fruit Research Institute, Chinese Academy of Agricultural Science) who gave us some materials of *Vitis* and we would like to thank three master candidates (L.J. Duan, L. Dong and, L.L. Gui) in Huazhong Agricultural University for collecting additional samples.

REFERENCES

1. Akers, C. P., Weybrew, J. A., and Long, R. C. (1978). Ultrastructure of glandular trichomes of leaves of *Nicotiana tabacum* L., cv. Xanthi. *Am. J. Bot.* 65, 282–292. doi: 10.2307/2442269
2. Aradhya, M., Koehmstedt, A., Prins, B. H., Dangl, G. S., and Stover, E. (2008). Genetic structure, differentiation, and phylogeny of the genus *Vitis*: implications for genetic conservation. *Acta Hort.* 799, 43–49. doi: 10.17660/ActaHortic.2008.799.4
3. Ascensão, L., Marques, N., and Pais, M. S. (1997). Peltate glandular trichomes of *Leonotis leonurus* leaves: ultrastructure and histochemical characterization of secretions. *Int. J. Plant Sci.* 158, 249–258. doi: 10.1086/297436
4. Ascensão, L., and Pais, M. S. (1998). The leaf capitate trichomes of *Leonotis leonurus*: histochemistry, ultrastructure and secretion. *Ann. Bot.* 81, 263–271. doi: 10.1006/anbo.1997.0550
5. Brizicky, G. K. (1965). The genera of Vitaceae in the southeastern United States. *J. Arnold Arbor.* 46, 48–67.
6. Brown, M. V., Moore, J. N., McNew, R. W., and Fenn, P. (1999). Inheritance of downy mildew resistance in table grapes. *J. Amer. Soc. Hort. Sci.* 124, 262–267.
7. Buono, R. A., De Oliveira, A. Í. D. B., and Paiva, E. A. S. (2008). Anatomy, ultrastructure and chemical composition of food bodies of *Hovenia dulcis*(Rhamnaceae). *Ann. Bot.* 101, 1341–1348. doi: 10.1093/aob/mcn052
8. Celep, F., Kahraman, A., Atalay, Z., and Doğan, M. (2014). Morphology, anatomy, palynology, mericarp and trichome micromorphology of the rediscovered Turkish endemic *Salvia quezelii* (Lamiaceae) and their taxonomic implications. *Plant Syst. Evol.* 300, 1945–1958. doi: 10.1007/s00606-014-1020-1
9. Chen, Z. D., Ren, H., and Wen, J. (2007). "Vitaceae," in *Flora of China*, Vol. 12, eds C. Y. Wu, D. Y. Hong, and P. H. Raven (Beijing: Science Press, and St. Louis: Missouri Botanical Garden Press), 173–222.
10. Cheng, D. W., Jiang, J. F., Fan, X. C., Zhang, Y., Zhang, G. H., and Liu, C. H. (2013). Diversity analysis of Chinese wild grape species. *J. Plant Genet. Resour.* 14, 996–1012.
11. Delbrouck, C. (1875)."Die Pflanzen-Stacheln," in

Botanische Abhandlungen, Vol. 2, ed. J. Hanstein (Bonn: Adolph Marcus), 1–119.

12. Durkee, L. T., Baird, C. W., and Cohen, P. F. (1984). Light and electron microscopy of the resin glands of *Passiflora foetida* (Passifloraceae). *Am. J. Bot.*71, 596–602. doi: 10.2307/2443335
13. Fahn, A. (1979). *Secretory Tissues in Plants*. London: Academic Press.
14. Galet, P. (1988). *Cépages et Vignobles de France. Tome 1. Lesvignes Américaines.* Montpellier: Pierre Galet.
15. Gerrath, J., Posluszny, U., and Melville, L. (2015). *Taming the Wild Grape. Botany and Horticulture in the Vitaceae*. Heidelberg: Springer.
16. Gravano, E., Tani, C., Bennici, A., and Gucci, R. (1998). The ultrastructure of glandular trichomes of *Phillyrea latifolia* L.(Oleaceae) leaves. *Ann. Bot.* 81, 327–335. doi: 10.1006/anbo.1997.0562
17. Hardie, W. J., and Obrien, T. P. (1988). Considerations of the biological significance of some volatile constituents of grape (*Vitis* spp.). *Aust. J. Bot.* 36, 107–117. doi: 10.1071/BT9880107
18. He, Y. H., Li, C. L., and Cao, Y. L. (1994). Comparative anatomy of vegetative organs in the genus *Vitis* L. and its systematic significance. *Acta Phytotaxon. Sin.* 32, 154–164.
19. Ickert-Bond, S. M., Gerrath, J. M., Posluszny, U., and Wen, J. (2015). Inflorescence development in the *Vitis–Ampelocissus* clade of Vitaceae: the unusual lamellate inflorescence of *Pterisanthes*. *Bot. J. Linn. Soc.* 179, 725–741. doi: 10.1111/boj.12348
20. Jackes, B. R. (1984). Revision of the Australian Vitaceae, 1. *Ampelocissus* Planchon. *Austrobaileya* 2, 81–86.
21. Johansen, D. A. (1940). *Plant Microtechnique*. New York, NY: McGraw Book Co.
22. Karabourniotis, G., Bornman, J. F., and Liakoura, V. (1999). Different leaf surface characteristics of three grape cultivars affect leaf optical properties as measured with fibre optics: possible implication in stress tolerance. *Aust. J. Plant Physiol.* 26, 47–53. doi: 10.1071/PP98052
23. Karabourniotis, G., Kotsabassidis, D., and Manetas, Y. (1995). Trichome density and its protective potential against ultraviolet-B radiation damage during leaf development. *Can. J. Bot.* 73, 376–383. doi: 10.1139/b95-039
24. Kellogg, A. A., Branaman, T. J., Jones, N. M., Little, C. Z., and Swanson, J. D. (2011). Morphological studies of developing *Rubus* prickles suggest that they are modified glandular trichomes. *Botany* 89, 217–226. doi: 10.1139/b11-008
25. Kolb, D., and Müller, M. (2004). Light, conventional and environmental scanning electron microscopy of the trichomes of *Cucurbita pepo* subsp. pepo var. styriaca and histochemistry of glandular secretory products. *Ann. Bot.* 94, 515–526. doi: 10.1093/aob/mch180
26. Kortekamp, A., and Zyprian, E. (1999). Leaf hairs as a basic protective barrier against downy mildew of grape. *J. Phytopathol.* 147, 453–459. doi: 10.1111/j.1439-0434.1999.tb03850.x
27. Lapinjoki, S. E., Elo, H. A., and Taipale, H. T. (1991). Development and structure of resin glands on tissues of *Betula pendula* Roth. during growth. *New Phytol.* 117, 219–223. doi: 10.1111/j.1469-8137.1991.tb04902.x
28. Latiff, A. (1982). Studies in Malesian Vitaceae, I–IV. *Fed. Mus. J.* 27, 46–93.
29. Latiff, A. (2001). Diversity of the Vitaceae in the Malay Archipelago. *Malayan Nat. J.* 55, 29–42.
30. Leelavathi, P., and Ramayya, N. (1983). Structure, distribution and classification of plant trichomes in relation to taxonomy III. Papilionoideae. *Proc. Plant Sci.* 92, 421–441.
31. Levin, D. A. (1973). The role of trichomes in plant defense. *Q. Rev. Biol.* 48, 3–15. doi: 10.1086/407484
32. Liakopoulos, G., Stavrianakou, S., and Karabourniotis, G. (2006). Trichome layers versus dehaired lamina of *Olea europaea* leaves: differences in flavonoid distribution, UV-absorbing capacity, and wax yield. *Environ. Exp. Bot.* 55, 294–304. doi: 10.1016/j.envexpbot.2004.11.008
33. Liakoura, V., Stefanou, M., Manetas, Y., Cholevas, C., and Karabourniotis, G. (1997). Trichome density and its UV-B protective potential are affected by shading and leaf position on the canopy. *Environ. Exp. Bot.* 38, 223–229. doi: 10.1016/S0098-8472(97)00005-1
34. Liu, X. Q., Ickert-Bond, S. M., Chen, L. Q., and Wen, J. (2013). Molecular phylogeny of Cissus L. of Vitaceae (the grape family) and evolution of its pantropical intercontinental disjunctions. *Mol. Phylogenet. Evol.* 66, 43–53. doi: 10.1016/j.ympev.2012.09.003
35. Liu, X. Q., Ickert-Bond, S. M., Nie, Z. L., Zhou, Z., Chen, L. Q., and Wen, J. (2016). Phylogeny of the *Ampelocissus–Vitis* clade in Vitaceae supports the New World origin of the grape genus. *Mol. Phylogenet. Evol.* 95, 217–228. doi: 10.1016/j.ympev.2015.10.013
36. Manetas, Y. (2003). The importance of being hairy: the adverse effects of hair removal on stem photosynthesis of *Verbascum speciosum* are due to solar U-VB radiation. *New Phytol.* 158, 503–508. doi: 10.1046/j.1469-8137.2003.00768.x
37. Miller, A. J., Matasci, N., Schwaninger, H., Aradhya, M. K., Prins, B., Zhong, G. Y., et al. (2013). *Vitis* phylogenomics: hybridization intensities from a SNP array outperform genotype calls. *PLoS ONE* 8:e78680. doi: 10.1371/journal.pone.0078680
38. Moore, M. O. (1991). Classification and systematics of eastern North American *Vitis* L. (Vitaceae) north of Mexico. *SIDA* 14, 339–367.
39. Moore, M. O., and Wen, J. (2016). "Vitaceae," in *Flora of North America, North of Mexico*, Vol. 12, eds Flora of North America Editorial Committee (New York, NY; Oxford: Oxford University Press).
40. Mullins, M. G., Bouquet, A., and Williams, L. E. (1992). *Biology of the Grapevine*. Cambridge: Cambridge University Press.
41. Nielsen, M. T., Akers, C. P., Järlfors, U. E., Wagner, G. J., and Berger, S. (1991). Comparative ultrastructural features of secreting and nonsecreting glandular trichomes of two genotypes of *Nicotiana tabacum* L. *Bot. Gaz.* 152, 13–22. doi: 10.1086/337858

42. Outenreath, R. L., and Dauwaldert, M. (1986). Ultrastructural and radioautographic studies of the digestive gland cells of *Drosera capensis* II. Changes induced by stimulation. *J. Ultrastruct. Mol. Struct. Res.* 95, 164–174. doi: 10.1016/0889-1605(86)90038-8

43. Paiva, E. A. S., Buono, R. A., and Lombardi, J. A. (2009). Food bodies in *Cissus verticillata* (Vitaceae): ontogenesis, structure and functional aspects.*Ann. Bot.* 103, 517–524. doi: 10.1093/aob/mcn237

44. Pavan, F., and Picotti, P. (2009). Influence of grapevine cultivars on the leafhopper *Empoasca vitis* and its egg parasitoids. *Biocontrol* 54, 55–63. doi: 10.1007/s10526-008-9151-3

45. Payne, W. W. (1978). A glossary of plant hair terminology. *Brittonia* 30, 239–255. doi: 10.2307/2806659

46. Péros, J. P., Berger, G., Portemont, A., Boursiquot, J. M., and Lacombe, T. (2011). Genetic variation and biogeography of the disjunct *Vitis* subg. *Vitis*(Vitaceae). *J. Biogeogr.* 38, 471–486. doi: 10.1111/j.1365-2699.2010.02410.x

47. Pillonel, C., Mulder, M. M., Boon, J. J., Forster, B., and Binder, A. (1991). Involvement of cinnamyl-alcohol dehydrogenase in the control of lignin formation in *Sorghum bicolor* L. Moench. *Planta* 185, 538–544. doi: 10.1007/bf00202964

48. Robards, A.W. (1978). "An introduction to techniques for scanning electron microscopy of plant cells," in *Electron Microscopy and Cytochemistry of Plant Cells*, ed J. L. Hall (New York, NY: Elsevier), 343–403.

49. Roland, J. C., and Vian, B. (1991). "General preparation and staining of thin sections," in *Electron Microscopy of Plant Cells*, eds J. L. Hall, and C. Hawes (London: Academic Press), 1–66.

50. Soejima, A., and Wen, J. (2006). Phylogenetic analysis of the grape family (Vitaceae) based on three chloroplast markers. *Am. J. Bot.* 93, 278–287. doi: 10.3732/ajb.93.2.278

51. Staudt, G. (1997). Evaluation of resistance to grapevine powdery mildew (*Uncinula necator* [Schw.] Burr., anamorph *Oidium tuckeri* Berk.) in accessions of *Vitis* species. *Vitis* 36, 151–154.

52. Staudt, G., and Kassemeyer, H. H. (1995). Evaluation of downy mildew resistance in various accessions of wild *Vitis* species. *Vitis* 34, 225–228.

53. Tissier, A. (2012). Glandular trichomes: what comes after expressed sequence tags? *Plant J.* 70, 51–68. doi: 10.1111/j.1365-313X.2012.04913.x

54. Tröndle, D., Schröder, S., Kassemeyer, H. H., Kiefer, C., Koch, M. A., and Nick, P. (2010). Molecular phylogeny of the genus *Vitis* (Vitaceae) based on plastid markers. *Am. J. Bot.* 97, 1168–1178. doi: 10.3732/ajb.0900218

55. Turner, G. W., Gershenzon, J., and Croteau, R. B. (2000). Development of peltate glandular trichomes of peppermint. *Plant Physiol.* 124, 665–680. doi: 10.1104/pp.124.2.665

56. Uphof, J. C. T. (1962). "Plant hairs," in *Encyclopedia of Plant Anatomy*, Vol. 5, eds W. Zimmermann and P. G. Ozenda (Berlin: Gebrüder Borntrager), 1–206.

57. Vermeer, J., and Peterson, R. L. (1979). Glandular trichomes on the inflorescence of *Chrysanthemum morifolium* cv. Dramatic (Compositae). II. Ultrastructure and histochemistry. *Can. J. Bot.* 57, 714–729. doi: 10.1139/b79-091

58. Wagner, G. J., Wang, E., and Shepherd, R. W. (2004). New approaches for studying and exploiting an old protuberance, the plant trichome. *Ann. Bot.* 93, 3–11. doi: 10.1093/aob/mch011

59. Wan, Y., Schwaninger, H. R., Baldo, A. M., Labate, J. A., Zhong, G. Y., and Simon, C. J. (2013). A phylogenetic analysis of the grape genus (*Vitis* L.) reveals broad reticulation and concurrent diversification during Neogene and Quaternary climate change. *BMC Evol. Biol.* 13:141. doi: 10.1186/1471-2148-13-141

60. Wan, Y., Schwaninger, H., He, P., and Wang, Y. (2007). Comparison of resistance to powdery mildew and downy mildew in Chinese wild grapes. *Vitis* 46, 132–136.

61. Wang, Y., Liu, Y., He, P., Chen, J., Lamikanra, O., and Lu, J. (1995). Evaluation of foliar resistance to *Uncinula necator* in Chinese wild *Vitis* species. *Vitis*34, 159–164.

62. Wen, J. (2007). "Vitaceae," in *The families and Genera of Vascular Plants*, Vol. 9, ed. K. Kubitzki (Berlin: Springer), 466–478.

63. Wen, J., Xiong, Z., Nie, Z. L., Mao, L., Zhu, Y., Kan, X. Z., et al. (2013). Transcriptome sequences resolve deep relationships of the grape family. *PLoS ONE*8:e74394. doi: 10.1371/journal.pone.0074394

64. Werker, E. (2000). Trichome diversity and development. *Adv. Bot. Res.* 31, 1–35. doi: 10.1016/S0065-2296(00)31005-9

65. Werker, E., Ravid, U., and Putievsky, E. (1985). Structure of glandular hairs and identification of the main components of their secreted material in some species of the Labiatae. *Israel J. Bot.* 34, 31–45.

66. Yan, A., Pan, J. B., An, L. Z., Gan, Y. B., and Feng, H. Y. (2012). The responses of trichome mutants to enhanced ultraviolet-B radiation in *Arabidopsis thaliana*. *J. Photochem. Photobiol. B* 113, 29–35. doi: 10.1016/j.jphotobiol.2012.04.011

67. Yang, C. X., and Ye, Z. B. (2013). Trichomes as models for studying plant cell differentiation. *Cell. Mol. Life Sci.* 70, 1937–1948. doi: 10.1007/s00018-012-1147-6

68. Zecca, G., Abbott, J. R., Sun, W. B., Spada, A., Sala, F., and Grassi, F. (2012). The timing and the mode of evolution of wild grapes (*Vitis*). *Mol. Phylogenet. Evol.* 62, 736–747. doi: 10.1016/j.ympev.2011.11.015

69. Zhang, J., Wu, X., Niu, R., Liu, Y., Liu, N., Xu, W., et al. (2012). Cold-resistance evaluation in 25 wild grape species. *Vitis* 51, 153–160.

CHAPTER 14

THE ORGANIC PRODUCTION OF CASH CEREALS AND PSEUDOCEREALS

Franc Bavec

INTRODUCTION

Organic Production of Major Cereal Crops and Maize

General Characteristic

Monocotyledonous maize (*Zea mays* L.) and spending winter or spring sown cereals such as common wheat (*Triticum aestivum* L.), barley (*Hordeum vulgare* L.), rye (*Secale cereale* L.), oat (*Avena sativa*L.), and triticale (*xTriticosecale* Wittm. & Camus) provide the main staple foods for the world's human and animal populations. They are conventionally grown on approximately 30 % of the total world arable land, and about 50% of cereals (data from FAO for 2012), provide almost one half of total nutrition energy. Over 500 important industrial products and byproducts may be obtained from maize, and also the similar amount may be obtained from cereals. Their excellent adaptability and wide distribution belongs to the spread family *Poaceae*. For example, wheat is grown from 40° s. l. (Southern latitude) to 60° n. l. (Northern latitude); barley even to 70° n. l. Rye is less sensitive to climate changes than wheat and barley and could be sown at altitudes up to 1200 m under European conditions. Similarly, maize is grown from 53° (optimum 45°) s.l. to 35° n.l., especially wide-spread in areas with a (sub) tropical climate. Especially new crops in the region, like sweet maize in temperate climate needs to be studied carefully [1].

The aim of this chapter was to analyse data mainly from Web Sci and Sci Direct and integrate actual organic cultivation knowledge as a suggestions for further development of maize and cereal production.

The use of organic farming technologies has certain advantages in some situations and for certain crops such as maize [2], but like in case of the yield performances of 24 organic wheat yields were low and variable 3.5 +/- 1.4 t ha^{-1} [3]. The proportion of cereals incl. maize produced organically is relatively high compared to vegetable and fruits (FAO, 2002), due to their less efficient production and more pretentious marketing. In the USA, 47% of the certified organic area was devoted to field and fodder crops in 2003, compared to 34% to pastures and 8% to fruit and vegetable crops (Economic Research Service, USDA).

In many cases farmers introduce into organic farming the old varieties and land race populations of cereals, because of their better resistance on plant diseases and due to their low fertilizer inputs, but also the yields are lower. Instead of common wheat often grown are spelt, emmer and einkorn in Austria, Switzerland and Slovenia. In spite of high inputs and low resistance on plant diseases, the new (modern) cultivars have changed many of their characteristics (grain size and leaf area have increased almost twenty times, ageing of top leaves has become slower and the period of assimilate accumulation in grains has been extended, better expressed productive tillering by cereals, cereals were bred to produce short-stem varieties to prevent lodging, ...).

The competitive strength to weeds is highest for rye, less so for winter barley, oat and triticale, while wheat and summer barley are the least competitive species, due to their height of steams. Monocotyledonous weeds like couch grass (*Elytrigia repens* L.), wild oat (*Avena fatua* L.), hair grass (*Apera spica – venti* L.) and some dicotyledonous weeds like catch weed (*Galium aparine* L.) and perennial weeds like field bind weed (*Convolvulus arvensis* L.), bull thistle (*Cirsium arvense* L.), etc. are associated in cereals. Maize is associated with other group of wide-leaved – to the high temperatures sensitive annual weeds like amaranths (*Amaranthus* sp.), white goose

Citation: Franc Bavec (2014). The Organic Production of Cash Cereals and Pseudocereals, Organic Agriculture Towards Sustainability, Prof. Vytautas Pilipavicius (Ed.), InTech, DOI: 10.5772/57475.

foot (*Cheopodium album* L.), and perennial weeds. Weeds in cereals result in difficult harvesting and cause an intensive spreading of weed seeds and decrease the yield; and additional influence of weeds in maize is suppression of growth in the first leaf stages (Maier code 21-26, [4]).

In organic farming, especially early in the spring at the tillering stage (EC 22-23, see growth stages of cereals described by Zadoks [5] of winter wheat and barley, lack of available nitrogen in the soil is problematic, due to nitrogen leaching and lack of mineralization caused by low temperatures. Rye and oat are less demanding, and thus more suitable for organic production. Also in case of maize the lack of nitrogen in the spring time is a usual problem.

Position in a Crop Rotation

Success of cereal crop production primarily depends on the choice of the previous crop, especially in less fertile soil. In conventional farming systems, there is a possibility to compensate for a less suitable foregoing crop by using higher rates of fertilizers and pesticides, but in organic farms, residual soil fertility from crop residues primarily determines the potential yield of a cereal crop. Rotation is also essential to prevent the build-up of pests and diseases; in particular foot and root rot diseases.

In mixed farming systems (incl. animal and crop production) where a grass/clover ley is part of the rotation, cereals can be grown after incorporation of the grass/clover mixture or after a more N-demanding intermediate crop grown after grass-clover incorporation, such as potatoes. In stockless farming systems (without livestock), legumes, cereal-legume mixtures, root crops, or oil seed crops are suitable foregoing crops, provided sufficient N is left from the root- or oil seed crops. In exceptional cases, cereals can be grown for two consecutive years, for example a winter and summer cereal. In that case, it is best to use a less demanding crop (rye or oat in cool climates, maize or millet in warmer climates) as the second crop. Compared to conventional rotations, organic crop rotations generally have a lower proportion of cereals, not exceeding 50%. Above this rate, only crops with low nutrient demands (rye, oat, triticale) can be used. If a legume such as red clover or alfalfa is used as foregoing crop, phytotoxicity od the decomposing residues may be a problem for the germinating seeds of cereal crops [6]. Therefore these foregoing crops should be ploughed under at least three weeks before the cereal is sown.

Because the maize is more demanding for N than temperate cereals, maize is usually grown after a leguminous green manure or fodder crop [7]. Maize is often only rotated with soybeans in conventional production systems, but a corn-soybean rotation does not provide adequate N in an organic system, as most of the N is removed with the soybean. Thus, organic rotations need to include additional leguminous crops besides soybeans or use additional animal manure. A common organic rotation used in the Midwestern USA consists of maize, soybeans, a cereal, legume green manure, and a fifth crop that is varied from year to year [8]. The effects of 2- versus 4-year rotations on corn and soybean yields were studied in organic and conventional systems in Minnesota [9]. Maize in the organic four-year maize-soybean-oat/alfalfa-alfalfa rotation yielded only 7-9% less than in the conventional two-year rotation, and almost twice the yield in the organic 2-year rotation. In another study in Iowa, yields of organic and conventional corn were equivalent in a similar 4-year rotation [10].

Cereal crops also have a useful function for the following crops in an organic rotation, due to their fibrous root systems that may loosen deeper soil layers, depending on the particular crop (Table 1). Moreover, monocotyledonous crops generally are not susceptible to diseases of dicotyledonous crops and, thus, provide a break in the build-up of diseases of dicotyledonous crops.

Table 1. Properties of cereals when used as a foregoing crop (arranged according to Demo, Bielek *et al.* [11].

Crop	Root		Tolerance to yourself	Foregoing crop quality (generally)	Land cover-age	Pertinence like cover crop
	Amount	Depth				
Winter wheat	+	++	+	+	+	+++
Spring wheat	+	+	+	-	+	++
Winter rye	++	++	++	++	++	++
Winter barley	+	+	+	+++	+	+++
Spring barley	+	+	++	++	+	++
Oat	++	++	++	-	+	++
Millet	++	+++	+	-	+	++

Explanatory notes: +++ good; ++ proper; + poor; - improper

Crop rotation is also essential for regulation of weeds in cereal crops. Summer cereals should be preferentially grown when winter weeds, such as mouse foxtail (*Alopecurus myosuroides* Huds.), scentless mayweed (*Matricaria maritima* L.), catch weed (*Galium aparine* L.), hair grass (*Apera spica – venti* L.), field poppy (*Papaver rhoeas* L.) are problematic, while winter crops can better be grown when summer weeds (such as wild oats or wild rape seed) occur more frequently. Cereal crops show different competitive strengths against weeds. In general, winter cereals are more competitive than summer cereals, rye more than wheat, and oat more than summer barley. In the low-stem cereals under sown cover crops (from genus *Trifolium*) can be included, because they compete generally well with weeds after harvest and represents the further leguminous in the crop rotation.

Crop rotation also contributes to the regulation of pests and diseases. Black stalk or take-all disease (*Gaeumannomyces graminis)* and eyespot disease (*Pseudocercosporella herpotrichoides*) represent major fungal diseases of cereal crops that are strongly affected by rotation. Take-all is commonly suppressed by antibiotic-producing *Pseudomonas fluorescens* bacteria when wheat is grown without rotation [12], but this is not an option for organic wheat production. In organic farming systems, take-all can be reduced by using soil-improving, non-susceptible annual crops in the rotation, because the pathogens do not persist in soil for a long time period. Suitable *G. graminis*-interrupting crops are maize, oat, potatoes, sugar beet, legumes, rape, or flax. High microbial activity and diversity, promoted by a complex rotation, also contribute to take-all control [12]. Cereal crops need to be interrupted by other crops for at least 2-3 years to limit eyespot occurrence. The most suitable crops for this purpose are clover or alfalfa for 2 - 3 years. Very suitable are sequences of these crops, for example maize - oat, potatoes - oat, potatoes - legumes etc. However, to prevent the build-up of smut diseases, such as *Tilletia controversy* on wheat and *Ustilago maydis* on maize, longer rotations of 6-8 years are required. This is especially true when organically produced seeds are used, as the common seed treatments are not allowed. Too many cereal crops in the rotation may also enhance the risk of foliar diseases such as powdery mildew (*Erysiphe graminis*). Due to low resistance of cereals to the diseases, one of the most effective measures for stable yielding is intercropping of different species of cereals [13] or just different varieties (with the same harvesting time) of one cereal species [14].

The most frequent cereal pests are aphides, damaging the assimilatory organs and spikes. As cereal aphids are quite specific, it is important to rotate with non-cereal crops. In Europe and Western Asia, the corn ground beetle (*Zabrus gibbus*) can be very damaging, especially on wheat [15]. The most damaging pests in maize are corn borer (*Ostrinia nubilalis* L.) and Western corn rootworm (*Diabrotica virgifera* L.), becoming new pests in Eastern Europe. Rotation of cereals with legumes and beet is a sufficient measure against these pests, which' larvae damage green grains and adults feed on crop flowers and grains during milk ripe stages. Wire worms (larvae of the click beetle) can be a major pest in organic crop production, especially after incorporation of large amounts of crop residues, for example perennial fodder crops or a cereal crop with under sown grass-clover. Populations of click beetles are generally reduced by cultivation operations. As root crops and rape seed require multiple cultivation operations, these crops are more suitable for rotation.

Cereal Crops Management Practice

Variety Selection

The main criteria for selection of cereal crops and varieties for organic production are: suitability for the certain location, yielding ability, stability and plasticity, large rooting and nutrients uptake ability (especially of nitrogen), resistance to fungal diseases, high competitive ability towards weeds, and stress tolerance (drought, water logging, extreme temperatures). The lower internodes should be relatively short to prevent lodging, but the upper internodes long to support higher ability of assimilation during the period of grain forming, even if the leaves are damaged by fungal diseases. Mid-height and bearded varieties seem to be optimal as these produce good grain yields with fewer productive stalks but with more grains per spike. It is important to select varieties with resistance or tolerance to fungal diseases prevailing at the particular location, as synthetic fungicides are prohibited in organic cereal production, and copper fungicides are generally not used on these crops. In case of maize the special attention must be done to the sowing appropriate FAO groups of hybrids, depending on geographical latitudes (FAO classification valid for Corn Belt in USA, number of growing days for each group growth out of this latitude are longer) and other production demands, and also to the target type of grains (dent, flint, semi dent or flint, pop, sweet) due to it's potential use.

Older varieties of cereals usually have favourable features to low pretentiousness of production inputs, with possible respect to grain quality, but lower productivity. Such varieties can be used for particular contracts with processors, where the price of a specific product balances lower yields.

Preparatory Soil Cultivation

Soil cultivation is primarily needed for destruction of crop debris and stubble of a previous crop, weed control and mineralization of nutrients, but also to enhance porosity for aeration and water and root penetration. Several cultivation operations can be distinguished, each with their own objectives, for example: ploughing to turn the soil so that crop residues, weed seeds and plant pathogens are buried below the immediate rooting zone, tillage to loosen the soil without turning it, harrowing to prepare a loose seedbed (4-5 cm deep), and hoeing to remove weeds.

The timing of cultivation is extremely important, because the wrong timing can destroy the soil structure, and nullify the benefits of proper organic matter management. This was shown for a biodynamic farm, where soil had been ploughed in a too wet condition, leading to soil compaction. The soil structure was not significantly better than that of a neighbouring conventional farm in comparison to the much better soil structure of permanent grassland. For winter cereals, the soil is obviously prepared for seeding in the fall, but for summer cereals and maize there is a choice of fall or spring soil preparation, depending on the climate. When springs are generally wet, for example in land climates with heavy snow fall, fall ploughing may be the only option, even for spring crops.

The depth of cultivation is another important choice. Common ploughing should not be deeper than usuall, because of live and well-structured layer of the soil (suggested 20-25 cm). Even higher ploughing must be step-by step i.e. cm by cm. Recent research has shown that soil organic carbon can be preserved by minimum tillage (shallow tillage, 10-15 cm deep). This also results in minimal disturbance of soil life, particularly fauna and fungi. Many organic farmers opt for shallow tillage, using the so-called eco-plough. After potato for winter cereals or annual cover crops for maize, direct sowing is possible also in organic production. This has resulted in considerable reduction in erosion and nutrient losses, and in improvement of soil structure. However, no-till is hardly an option in organic farming, as herbicides cannot be used, and hand-weeding and mechanical in-row hoeing are not feasible for field crops. Minimum tillage may also result in an increased occurrence of weeds, especially hair grass (*Apera spica – venti* L.), camomile (*Chamomilla*), couch-grass (*Elytrigia repens* L.), creeping thistle (*Cirsium arvense* L.), barnyard grass (*Echinochloa crus – galli* L.), and pig-weed (*Amaranthus*L.), which thrive on compacted soils.

The number of operations and their intervals ultimately determine weed infestation rates. To avoid weed problems in the beginning of crop growth, soils are sometimes harrowed an extra time before final seedbed preparation, so that weeds germinate, start to grow, but are killed by the second harrowing. The first operation is called preparation of a 'false seedbed'. A certain level of weed cover could be tolerated when crops are past their most sensitive stage [16], but noxious weeds should be removed before setting seed. Too frequent tillage operations lead to soil erosion and nutrient and carbon losses, and ultimately to poor soil structure.

Nutrition of Cereals and Maize

The uptake of nutrients with aboveground maize yields can reach up to 278 kg N ha^{-1}, 46 kg P ha^{-1} and 171 kg K ha^{-1} [17], in cereals approximately up to half of the amount. Target fertilization rate depends on available nutrients in the soil and target yield. Target fertilization rate depends also on climatic conditions, genotype, and production system like plant populations [18, 19]. According to the EU regulations limited input with organic fertilizers is 170 kg N ha^{-1} as an average farm^{-1}. It means that is impossible input the same levels of nutrients in organic production comparing with conventional agriculture. Deria et al. [20] considered that several factors causes grain yield differences (organically grown wheat was: increased at one site by 17%; decreased at three sites by an average of 27%, and not changed at another three sites) between the organic and conventional wheat: like lower nitrogen supply, and lower extractable-P of the organic wheat.

Cereals absorb nutrients mainly from the upper soil layer. Their development depends on available nutrients, especially phosphorus and nitrogen, released from organic fertilizers or crop residues by mineralization.

Nitrogen nutrition is often not limited by the total nitrogen in soil, but by a deficiency in certain growth stages, especially at the tillering stage and generative stages of winter cereals. Release of mineral nitrogen is often maximal in mid-summer, often resulting in too high nitrogen levels in brewing barley, but still fairly low nitrogen (and gluten) levels in bread wheat.

Besides crop residues, organic cereal crops may be fertilized with animal manure, especially on poor soils, at rates of 4-10 t ha^{-1} (up to content of 180 kg N ha^{-1} year^{-1}). Also legumes as preceding crops improve nitrogen nutrition on subsequent winter wheat, e. g. the Nitrogen Nutrition Index at flowering (NNIf) = 0.51 +/- 0.12 for a crop rotation with a rate of legumes over 37% vs. 0.41 +/- 0.11 for a crop rotation with a rate of legumes under 25% [3]. In the case of comparing conventional animal production, organic animal production, and organic cereal production residual fertility effects from the preceding crop produced higher yields in organic animal production winter and spring wheat than in conventional. Cultivation of winter wheat in organic animal production was a more efficient use of nitrogen resources than conventional [21]. The findings that microbial properties and N availability with 182-285% increase in potentially mineralizable N, for thus the yields of crops differed under different organic input (regimes cotton gin trash, animal manure and rye/vetch green manure), [22].

Organic fertilizers may also be combined as phosphorus and potassium fertilizers which are allowed (depending on the certification agency).

An additional nutrient dressing can be applied at the generative stages (EC 20–35) by dispersion of dung (5-7 t ha^{-1}) or slurry (20-40 m^3 ha^{-1}). Higher rates of dung or slurry can cause expansion of ruderal weeds (docks, goosefoot, hardship, scentless may-weed, barnyard grass, etc.). Harmonious fertilization maintains the competitive strength of crops, promoting fast foliage growth and better soil coverage.

Additional supply of nitrogen is possible by intercropped legumes. The data showed that nitrogen content of the wheat grain and whole plant biomass significantly increased when the density of beans in the intercrops increased; this was reflected in a significant increase in grain protein at harvest of organically grown wheat [23]. The clear and significant differences in wheat yield and protein content between the organic and conventional systems suggest a limited supply of available N in the organic fertility management system which is also supported by the significant interaction effect of the preceding crop on protein content [24].

Seed Selection and Sowing

According to EU regulations, if available, certified organically grown seed is required for certified organic crop production (EU Regulation No. 834/2007). As sufficient certified organic seeds of cereal crops and maize are on the market, European organic growers are bound to use these seeds. The situation is different in North America and Canada, where grower-saved seeds are often used, or commercial, conventionally grown seeds without fungicide coating.

The selected sowing rate should ensure an optimum stand density that results in soil shading and weed suppression. Sowing at high density at a narrow row distance is more effective to control weeds in a cereal crop than wide row spacing combined with mechanical weed control. Early sowing is not as effective for weed control as sowing after repeated harrowing (preparing a false seedbed to kill germinated weeds). Finally, the sowing depth is crucial: not too deep to avoid root and foot diseases, and not too shallow to avoid drying of the seed.

Both, cereals and maize can be intercropped by legumes; in the case of maize intercropping with climbing bean is suggested and in case of cereals with pea and beans. Sowing of above mentioned intercropped crops is separate, depending on suitable periods for seed emergence and growth.

Weed Control

High weed pressure is one of the main problems of organic cereal and maize production. Weed seedbank size was larger under the organic system of maize production than conventional system; and for example differences among maize management systems depends mainly upon weed control efficacy rather than upon tillage effects [25].

The most problematic weed species in cereal production are couch-grass and creeping thistle. Application of herbicides is impossible and thus different ways of weed regulation must be used. The aim of such regulation is not to completely destroy weeds, but to keep their coverage under the damage threshold. The emphasis is put on prevention of introduction and spread of regenerative organs of weeds (seeds, stolons, etc.), and on indirect control like promoting optimal growing conditions for the crops that enhance their competitive strength against weeds. Among the preventive measures belong selection of a location that is suited to a crop's demands, a varied and well-balanced crop rotation including fodder crops, proper soil cultivation, use of barnyard manure, harmonic fertilization, proper choice of species and varieties, proper sowing, prevention of weed introductions, optimum time and way of harvesting as well as post-harvest operations. Finally, care must be taken that noxious weeds don't spread from field margins and surrounding natural vegetation.

If the preventive measures are not effective enough, direct weed control treatments must be applied: mechanical, physical and biological methods. Harrowing using different harrow types (weeders, net harrow, and spike harrow) can be used for deep-rooted crops (maize, sorghum) before the plants sprout but after germination of weeds. A second harrowing can be applied shortly after the stage of 2-3 true leaves (Maier code 21-23). Line weeding (with brush cultivators or deer-tongue cultivators) can be applied in wide-row cereal crops on heavy soils (row spacing over 15 cm) or to combat overgrown weeds when cultivation was delayed, for example due to wet conditions. Thermal weed control (using propane-butane burners) can be used to eliminate all weeds before sprouting of the crop, or to put down dicotyledonous weeds in wide-row crops like maize. Biological weed control by plant pathogens or insect pests is still hardly used in organic agriculture.

Harvesting, Post-Harvest Treatments and Storage

Organic grains need to be harvested and stored separately from conventional grains. Perfect threshing and pre-cleaning of the grain is a basic expectation for proper harvesting of cereals. Grain needs to be dried to 13.5 - 14% moisture. Drying of grain is performed gradually, removing only about 2% water at a time if it originally had moisture content over 20%. Organically grown grain should be dried like seed material, preventing too high temperatures, because sprouted grains can be used for human consumption.

Grains of cereals and maize can be infected by fungi that produce mycotoxins (primarily *Fusarium*species), which can seriously affect human and animal health. Therefore mouldy grain should not be used, not even for feeding purposes. To prevent mould infection, grain temperature, moisture, smell and pest invasion must be monitored, and the storage silos should be aerated. The air temperature should be 5 °C lower than the grain temperature. The optimum temperature for grain storage is 5 – 10 °C, and the temperature should not exceed 20 °C. Storage pests, especially grain weevils (*Sitophilus granarius*), grain moths (*Nemapogon granellus*), and flour mites (*Acari*) can be a serious problem for organic grain, as they can only be controlled within the limits given by regulations. The storage silos need to be cleaned before use. Natural pyrethrums can be used to control invading pests, but these substances have a short-term effect only (5-6 hours), so the application has to be done repeatedly. Natural pyrethrums do not affect warm-blooded animals. If these disinfection procedures are followed, limited occurrence of pests can be expected.

Labour Use and Economics

Costs and Prices

The yields of organic cereals and maize are usually somewhat lower than those achieved under conventional management, mainly due to limited nitrogen availability. However, in a long-term experiment with a five year rotation, yields of organic maize were very similar to those of conventional: 6.4 versus 6.5 tons/ha [26]. In drought years, organic maize yielded even more (28-34%) than conventional maize, thanks to the greater water-holding capacity of the organically managed soil. But in drought conditions in case of organic maize yielded just 38% and 137% relative to conventional were formed [27].

The direct costs for organic grain production are frequently lower than those for conventional production, as there are no expenditures for pesticides, growth regulators (for small grain cereals only) and synthetic fertilizers. These lower costs may sometimes compensate for the lower yields, even at conventional market prices, as observed for maize production in the Corn Belt in the USA [28]. However, for calculation of net income from a crop, opportunity costs need to be included, as less profitable leguminous rotation crops are needed to provide sufficient nitrogen for a corn crop [26, 29]. In addition, more expensive seed, more frequent cultivation and the associated fuel costs, and higher costs for manual labor for preparation of natural fertilizers and weed control detract from the profitability of organic small grain and corn production [26]. Taking the lower yields per ha and all costs per ha into consideration (Fig. 1), the production of cereals and maize would mostly not be economic at conventional prices for the products. Premium prices are essential for a positive balance [26].

Price premiums vary considerably, from 15 to 50% for organic cereals (Sullivan, 2003), and from 20 to 50% for organic corn [29]. Even a premium price of 200% of organic over conventional maize has been reported [30]. The height of the price premium depends, of course, on the kind of cereal, the quality, the extent of processing, the kind of market, and supply versus demand. Especially for maize, specialty crops (sweet maize, pop corn, conserved grains) fetch higher prices than regular feed maize. The premiums will likely remain fairly high as long as the demand for organically grown grains is increasing at the same pace or faster compared to the increase in organic grain production.

However, the economic net return was higher for conventional farming system, but when the economic subsidies from EU were considered, the integrated net return was higher than under organic farming system for wheat and soya [31]. As the demand for organic meat is rising, this should also increase the demand for certified organic feeds, and possibly the premium prices (as long as the number of certified organic grain growers is not expanding as fast or faster). Altogether, it is difficult to predict future organic grain prices [29].

Markets

Depending on the quality of the grain, it can be used as animal feed or for human consumption; grading standards for grains for human consumption are higher than for feed grains [32]. In mixed organic farms, grains are commonly used as livestock feed on the farm itself. Fodder grains can also be sold on the regional, national or international market. However, organic feed grain is generally not transported over great distances. Although it is more difficult to grow organic corn and cereal grains for human consumption due to the higher quality standards, they result in higher net profits than feed grains. Organic cereal grains for human consumption are commonly marketed in different ways than conventional grains, since the buyers are usually smaller-scale than buyers of conventional grain [33]. Organic grain is often produced on a contract, and producers may have more grain than a typical small-scale buyer would need. Thus, organic grain often needs to be stored, and it may be beneficial for an organic grower to have his/her own storage facilities [33]. Some organic farmers process their grain into flour and bread or other cereal products, and sell it directly to consumers at home or at the local market. The greater the share of the production chain is controlled by the farmers themselves, the more profitable the production of cereals will be.

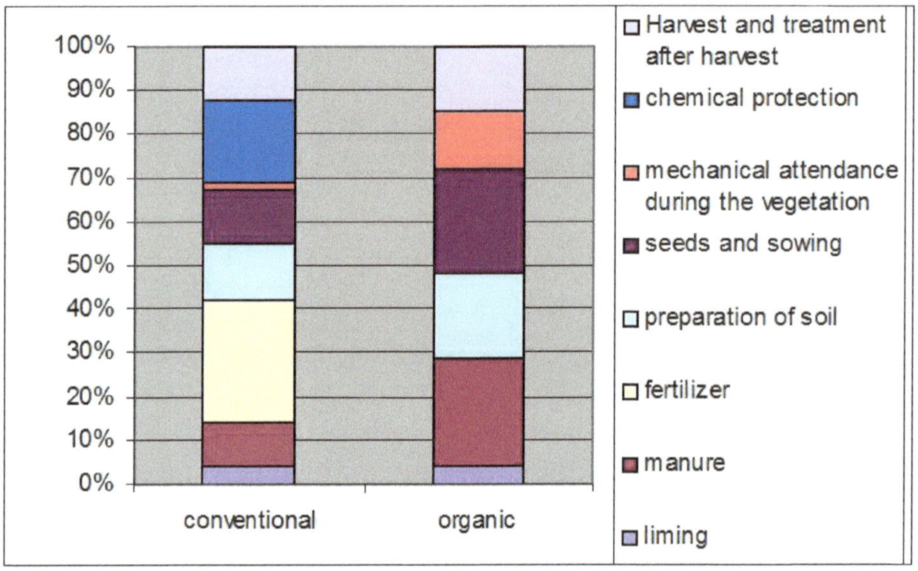

Figure 1. Comparison costs of convencional and organic farming (growing wheat) – variable cost and fixed cost of machines (Moudrý, 2005) – not published data.

Ecological Impact

The results show a markedly reduced ecological footprint of the organic systems (icluding biodynamic) in production of wheat (*Triticum aestivum* L. 'Antonius') and spelt (*Triticum spelta* L. 'Ebners rotkorn'), mainly due to the absence of external production factors. When yields were also considered, the organic systems again had a reduced overall footprint per product unit and increased ecological efficiency of production. Thus, this farming systems present viable alternatives for reducing the impact of agriculture on environmental degradation and climate change. Nevertheless, room for improvement exists in the area of machinery use in all systems studied and yield improvement in the organic farming system [33].

ORGANIC PRODUCTION OF PSEUDOCEREALS

General Characteristics

Pseudocereals described in this book include dicotyledonous field crops of various botanical families i.e. common buckwheat (*Fagopyrum esulentum* Moench) family *Polygonancae,* quinoa (*Chenopodium quinoa* Willd.) family *Chenopodiace* and grain amaranths (*Amaranthus caudatus, A. cruentus, A. hypohondriacus*) family *Amaranthaceae*, which were traditional food in Asia from 10th to11th BC, of Inca's and Aztec civilization, respectively. Their production (Table 2) has become more and more interesting, due to grains nutritional and healthy value, especially if they are gown organically. For those the own experiences at the farm and in research are again evaluated within the scope of new findings cited in research databases.

These crops have an extensive diversification among genotypes, which were utilized and adapted under specific environments; consequently their increasing introductions in the new environments differ in expression of genetic potential of yield and duration of vegetation period to reach full maturity. The main limitations for production under European temperate climate and its above sea altitudes are frost free days during vegetation period and necessary effective temperatures for active photosynthesis (assimilation), except in buckwheat where the temperatures exceeding 24°C cease pollination and stop growth due to short day-length in some genotypes. The main preferences are well adaptation of quinoa and amaranths to dry conditions, soil pH and salinity, and also C_4 pathway in amaranths.

Table 2. Pseudocereals production characteristics and environmental requirements

	Buckwheat	Quinoa	Amaranths
Production			
Origin of utilizatio^{n}a	Asia, highlands	Andes mountains	South America
Main producers	Russia, China, USA, Japan, E^{U}b	Peru, Bolivia, Argentin^{a}c	USA, Peru, Bolivi^{a}b
Yield (kg grain ha^{-1}) - approximate	800-100^{0}b	400-1200^{b}, 640-92^{0}c	400^{b}-150^{0}a
minimum, maximum	500-220^{0}b	435-659^{1}b	10^{d}-4000, 500^{0}b
Requirements			
Growing season (days)	100	120-240^{e}, 14^{0}f	105-160^{b}, 150^{f} 90^{g,h}-10^{0}h
Min. germination temperature (°C)	^{7}f	5^{f}-^{7}b	12, 1^{5}i
Frost resistance to (°C)	-1^{f}, (-1.3 to -2.9$^{)}$j	-3 do -1^{5}k	0
Temperature stop assimilation (°C)	10	8	15
Transpiration coefficient (kg water kg^{-1} of dry plant matter)	500-60^{0}b	400	200-33^{3}k

Sources: [a] [35], [b] [36], [c][37], [d] [38], [f] [39], [g] [40], [h] [41], [i][42], [k][43], [j][44], [k][45]

The grains vary in morphological characteristics, shape and weight; for example 1000-g seed weight in buckwheat varies from 18 to 32 (38) g, in quinoa from 2 to 6 g, and in grain amaranths from 0.3 to 1 g. These pseudocereals have a great possibility of niches organic products with high nutritional and healthy value, suitable to diabetics and people with celiac disease, and prohibited milk proteins, etc. In comparison with cereals, grains of these pseudocereals are gluten free food, suitable for people with celiac disease, with higher protein content and their rich amino acid composition, rich in fibers, polisaturated (gamma) fatty acids incl. squalene in amaranths and minerals, etc. (Table 3), however composition depends also on milling fractions and food processing. All three pseudocereals are rich in lysine (the first limiting essential amino acid in cereals); quinoa also in histidine, isoleucine and metionine + cystine, its' consumption less than 50 g of grain per adults' daily need, it is also ideal reference of essential amino acid pattern for children according to FAO/WHO/UNO standards (except for valine); amaranths are a substitute for meat meals, especially due to rich content of methionine, cisteine and arginine [45, 36]. For those mixtures with cereals supports balanced meals in amino acid composition, especially for vegetarians.

Table 3. Nutritional composition of pseudocereals (% of dry matter)

	Buckwheat				Amaranth	Quinoa
	Seed	Groa^{t}b	Dark flou^{r}a	White flou^{r}a	Variation	Average^{f}(variation)
Protein	11-15,12.^{3}a	16.8	14.1	6.4	13.3^{c}-17.^{9}d	14.7 (9.6-22.1)
Starch	73.^{3}a	67.8	68.6	79.5	6^{2}g	58.2 (46.9-77.4)
Fat	2.^{3}a	11	3.5	1.2	5.1-7.7	7.2 (1.8-8.2)
Fibre	10.^{9}a	0.6	8.3	0.5	^{8}g	6.4 (1.1-13.4)
Ash	2.^{1}a	2.2	1.8	0.9	^{3}g	3.4 (2.4-9.7)

Sources: [a] [47], [b] [48], [c] [46], [d][49], [f] average data from Fleming and Galwey [50], [g][39].

In addition pseudocereals are characterized by many health-important substances like antioxidants – flavonoids: rutin, isovitexin, etc., flavones, phytosterols, fagopyrins and thiamine binding proteins and prebiotic activity in buckwheat. Consumption of buckwheat can improve diabetes, obesity, hypercholesterolemia, constipation as well as reduce blood pressure and cellular proliferation. The most important component in amaranths is squalene, which affected synthesis of cholesterol, possesses antisclerotic properties and may be used in the prophylaxis against cancer (see references in Bavec and Bavec[36]. From quinoa grains leached bitter saponins represent natural substances used in organic detergents, soaps, shampoos, cosmetics, etc.; the high-growth cultivars of amaranth are suitable for energetic purposes.

Among pseudocereals buckwheat is the most weed resistant crop which results from its fast development in early growth stages and might due to allelopathy [51].

In buckwheat do not exist disease and pest infestation, which might cause economically important damages, except emerging plants by nematodes. Also quinoa and amaranths should not require special pest and disease control. But several species of cosmopolitan polyphagos pests (e.g., *Agrotis ipsilon*Hufnagel, Lepidoptera: nuctuidae) may cause economic losses. Some of widespread pests endemic to the Andean region may cause up to 50% loss of yield in quinoa, especially two species of moths (*Eursacca quinoae* Povolyny, *E. melanocampta* Meyrick) [52]. Emerging plants may also be damaged by flea beatles and caterpillars. Diseases caused by *Peronospora* sp., *Sclerotinium* sp., *Phoma* sp.,*Botrytis* sp, and *Pseudomonas* sp. may have an influence on yielding of quinoa. However, viruses found on spinach or beets have been observed also in quinoa fields.

Position in Rotation

Because of previously mentioned plant diseases and possible polyphagous pests which could appear, and increase of weeds, each pseudocereal may not be planted again on the same field for at least 3 years; 5-6 years is suggested in organic farming. A negative consequence of incorrect buckwheat crop rotation is also increased occurrence of necrosis and the appearance of some disease signs, especially root diseases caused by *Gaeumannomyces* sp. In spite of amaranths, quinoa seeds also do not exhibit dormancy and they germinate when conditions are suitable; the plant itself though, in wild form, may remain in soil for 2 to 3 years without germinating.

According to different requirements about nitrogen needs in amaranths (high nitrogen accumulative C_4plant), recommended rest of available nitrogen (after precrop and/ or additional fertilization with organic manure) is high in comparison with buckwheat, due to lodging in the case of high value of available nitrogen in the soil [36]. However, suitable precrops crops are cereals, grain legumes and perennial legumes. Buckwheat appears as a very suitable precrop for fiber flax due to its strong competitive strength against cough-grass and dicotyledonous weeds. Considering its shorter growth season can be used as stubble crop after early harvested previous crops (early potatoes, barley, crop-legumes mixtures for green feeding). Because amaranths and quinoa growth slowly than buckwheat during early growth stages, the field in quinoa, and especially in amaranths, must be free of weed competition, for those precrop must suppressed the weeds. Buckwheat, sown as crop, cover crops or for green manure, may produce allelophatic substances that could inhibit weeds [51] and also follows crop in rotation.

Crop Management

Species and Varieties Selection

In common buckwheat exist more than 4500 different varieties and land race populations, but in every climate special attention should be given to choosing appropriate cultivars, because of different photoperiodical reactions.

From more than 60 wild species of genus *Amaranthus*, four of them were selected like grain amaranths. They are different among each others in germination [42], growth, habitus, inflorescence, ripening and seed colour [53].

The genus *Chenopodium* consists of 150 species, classified into 16 sections. Few plants from section *Chenopodium* are used for human consumption. Quinoa is the most important species from genus *Chenopodium*. In the first stage, cultivated genotypes originated from 6 ecotypes and 4 plant populations from different regions [36]. Also in amaranths and quinoa introduction of new cultivars must based on previous analyses of trials results in every new climate.

Fertilization

Pseudocereals requirements for nutrients need vary from low to very high. Organic fertilizers that are left over from the preceding crop or given stable manure in autumn provide adequate yielding in the organic production system. Modest available level of nutrients results in low yields, except in case of some heavy metals in amaranths (important remediate plant), and Ca-bound phosphorus in buckwheat, which efficiency might be influenced by acidify of the rhizosphere [54] compared to spring wheat. In buckwheat in case of more than 20 g NO_3-N kg^{-1}dry soil to the depth of 0.3 m fertilization with nitrogen manure is not allowed, due to possible lodging problems; it should instead be used for the previous crop in the crop rotation [36]. Although amaranth yield is responsive to available nitrogen in the soil like maize, but a high level of available nitrogen can negatively affect grain harvest in terms of increased lodging, and delayed crop maturity [55]. Nitrogen application up to an Nmin target value of 140kg N ha^{-1} raised protein concentration in grain of amaranths, and maintained the content of essential amino acids in protein. Essential amino acids in grain fertilized to the target value 140 kg N ha^{-1} was higher (397 g kg^{-1}) than the standard requirement for preschool children (339 g kg^{-1}). Among essential amino acids, only valine concentration responded to nitrogen supply. Leucine was the limiting amino acid in grain protein [56, 57]. In quinoa, compensation is suggested by organic fertilizers at an application rate of about 120 kg of N ha^{-1} [58] for the expected enhancement of grain yield.

Sowing

Buckwheat is often sown as stubble crop, amaranths and quinoa are main crops. Different sowing methods, including sowing rates variy among buckwheat (mainly suggested 200 – 300; row spacing is from 12.5 to 25 cm, sowing depth is about 2-3 cm, sowing rate is 1,0 1,5 mil. seeds, i.e. 40-60 kg per ha. Late sowing, for example on weedy fields, is recommended to be done using closer row spacing and higher sowing rates. Greenhouse experimental data in quinoa showed that optimal plant populations varies from 30 plants m^{-2} [58] to an optimal plant population of 140 plants m^{-2} [59], to 200 plants m^{-2}in the 2.3 field trials (Jacobsen et al., 2005) with interrow spacing 0.4 to 0.8 m. In amaranths 0.12 to 08 m interrow spacing, 150 to 200 seeds m^{-2} and final plant population about 40 plants m^{-2}. From variable data we can stress that in *A. cruentus* cultivars would be recommended above 30-cm row spacing and populations of 74000 and 272000 plants ha^{-1} [40].

Yields of *A. cruentus* cv. 'G6' were affected by growing season and date of sowing. There was a higher protein content (165 g kg^{-1}) and lower (39.3 g kg^{-1}) in grain of plants sown in May and June, respectively [57].

Harvesting, Post-Harvest Treatments and Stocking

According to botanical distinction harvesting, postharvest techology and food processing [60] vary in comparison with cereals, except seeding machines, harvesters and mills in buckwheat. Buckwheat is harvested when about 75% of the seeds have reached ripeness, in quinoa when at the maturity stage, plants become lighter and more yellow, and the leaves fall off and in amaranths the appropriate harvesting period is indicated by the yellowing of bottom leaves and dry seed [36]. After harvesting, which can be done with special adaptation of cereals or clover threshers, must follow immediate winnowing, after-drying to 10-12% water content and placing in a slot-floor storages equipped with an air ventilation system. Storage facilities must be dry and airy.

Labour Use and Economics

Pseudocereals as low input crops (labor, fertilizers) have a petty differences between yield in conventional and organic farming system, where production should be more profitable according to higher prices of organic products.

CONCLUSIONS

Cereals and maize are the most important food and fodder crops also in organic farming, they are essential rotation crops, as long as they are not taking up more than 50% of the crops in the rotation. Cereals are more suitable for cool and humid climates than maize, due to the slow initial growth caused by low temperatures and less expressed weed competition. Well-structured soils and enough available nutrients in the soil based on crop rotation and inputs of organic manures, and systematic use of other preventive and corrective measures for pest, disease, and weed control will contribute to stabilization of yield at about 80-90% in comparison with conventional crops, depending on climate and soil type. Using organic price premiums for organic crops, net incomes from organic cereals and maize are higher than those of their conventional counterparts.

Buckwheat, grain amaranths and quinoa are very suitable crops for organic farming, especially for small scale farms. They extend agro-biodiversity and range of niches organic products, also for special uses. For an effective production selection of varieties and appropriate growing conditions should be studied and knowledge of post-harvest technology and food processing must be detailed and up to dated.

REFERENCES

1. Fekonja M, Bavec F, Grobelnik-Mlakar S, Bavec F. Growth Performance of Sweet Maize under Non-Typical Maize Growing Conditions. Biol. Agric. Hortic. 2011;(27) 147-164.
2. Pimentel D. Economics and energetics of organic and conventional farming. J. Agric. Environ. Ethics, 1993;63 53-60.
3. David C, Jeuffroy MH, Henning J, Meynard JM. Yield variation in organic winter wheat: a diagnostic study in the Southeast of France. Agron. Sust. Develop., 2005; (25) 213-223
4. Shütte F., Maier U. Entwicklungstadien des Mais zum Gebrauch fuer das Versuchwesen, die Beratung und die Praxis in der Landwirtshaft. Biol. Bundesanst. F. Land-u Fortwirtsch., Merkblat. 1981;27 p.
5. Zadoks J. C., Chang T. T., Konzak C. F. A decimal code for growth stages of cereals. Weed Res., 1974;14 415-421.
6. Urban J., Šarapatka B. Ekologicke zemedelstvi (Organic farming)-ucebnice pro skoly i praxi. I. Dil, MZP Praha, 2003;280 p.
7. Drinkwater, L.E., Wagoner, P., and Sarrantonio, M. Legume-based cropping systems have reduced carbon and nitrogen losses. Nature. 1998;396 262—264.
8. Frerichs R. Organic Food-grade Corn, 2003 – http://web.aces.uiuc.edu/value/factsheets/corn/fact-organic-corn.htm acssed 10.11.2013
9. Porter P.M., Huggins D.R., Perillo C.A., Quiring S.R. and Crookston R.K. Organic and other management strategies with two- and four-year crop rotations in Minnesota. Agron. J. 2003;95 233-244.
10. Delate K, Cambardella CA. Agroecosystem Performance during Transition to Certified Organic Grain Production. Agron J. 2004;96 1288-1298.
11. Demo Bielek et al. Regulačné technológie v produkčnom procese polnohospodárských plodín, SPU Nitra, 2000, 667p.
12. Hiddink GA, van Bruggen AHC., Termorshuizen AJ, Raaijmakers JM, Semenov AV. Effect of organic management of soils on suppressiveness to Gaeumannomyces graminis var. tritici and its antagonist, Pseudomonas fluorescens. Europ. J. Plant Pathol. 2005;113 417-435.
13. Sobkowitcz P., Tendziagolska E. Competition and productivity in mixture of oats and wheat. J. Agr. And Crop Sci, 2005;191 377-385.
14. Walsh E. J., Noonan M. G. Agronomic and quality performance of variety mixtures in spring wheat (Triticum aestivum L.) under Irish conditions. Cereal Res. Commun., 1999;26 427-432.
15. Pests, (http://www.agroatlas.spb.ru/pests/Zabrus_tenebrioides_en.htm), accessed 01.12.2013.
16. Mortensen D. A., Byastiaans L., Sattin M. The role of ecology in the development of weed management systems: an autlook. Weed Res., 2000;40 49-62.
17. Bavec F., Bavec M. Effect of maize plant double row spacing on nutrient uptake, leaf area index and yield. Rost. Vyroba, 2001;47 135-140.
18. Bavec F, Bavec M. Effects of plant population on leaf area index, cob characteristics and grain yield of early maturing maize cultivars (FAO 100-400). Eur. J. Agron., 2002;16 151-159.
19. Bavec M, Bavec F, Varga B, Kovačević V. Relationship among yield component in winter wheat (Triticum aestivum L.) cultivars affected by seeding rates. Bodenkultur, 2002;53 151-159.
20. Deria AM, Bell RW, O›Hara GW. Organic wheat production and soil nutrient status in a Mediterranean climatic zone. J. Sust. Agric. 2003;21 21-47.
21. L-Baeckstrom G., Hanell U., Svensson G. Nitrogen use efficiency in an 11-year study of conventional and organic wheat cultivation. Commun. Soil Sci. Plant Anal., 2006;37 417-449.

22. Tu C, Ristaino JB, Hu SJ. 2006. Soil microbial biomass and activity in organic tomato farming systems: Effects of organic inputs and straw mulching, Soil Biol. Biochem. 2006;38 247-255.

23. Bulson HAJ, Snaydon RW, Stopes CE. Effects of plant density on intercropped wheat and field beans in an organic farming system. J. Agric. Sci. 1997;128 59-71, Part1

24. Bilsborrow P, Cooper J, Tetard-Jones C, Srednicka-Tober D, Baraniski M, Eyre M, Schmidt C, Shotton P, Volakakis N, Cakmak I, Ozturk L, Leifert C, Wilcockson S. The effect of organic and conventional management on the yield and quality of wheat grown in a long-term field trial. Eu. J. Agron., 2013;51 71-80

25. Barberi P, Cozzani A, Macchia M, Bonari E. Size and composition of the weed seedbank under different management systems for continuous maize cropping. Weed Res. 1989;38 319-334.

26. Pimentel D., Hepperly P., Hanson J., Douds D., Seidel R. Environmental, Energetic, and Economic Comparisons of Organic and Conventional Farming Systems, BioScience 2005;55 573-582.

27. Lotter DW, Seidel R, Liebhardt W.2003. The performance of organic and conventional cropping systems in an extreme climate year. American J. Alt. Agric., 2003;18 146-154

28. Lockeretz, W., Shearer, G. and Kohl, D. 1981. Organic farming in the Corn Belt. Science 1981;211 540-547.

29. Kuepper G. Organic firld corn production. 2002, 16p.

30. Delate K, van Dee K. Evaluation of Corn, Soybean and Barley Varieties for Certified Organic Production-Crawfordsville Trial, 2000

31. Sartori L, Basso B, Bertocco M, Oliviero G. Energy use and economic evaluation of a three year crop rotation for conservation and organic farming in NE Italy. Biosyst. Engin. 2005;91 245-256.

32. Stearns L., David L. W. Northern Plains Organic Crops Marketing Analysis:Wheat, Oats, Sunflower. Agricultural Economics Report No. 293. Department of Agricultural Economics-Agricultural Experiment Station. North Dakota State University. 1993

33. Sullivan P. Organic Small Grain Production. ATTRA. http://www.attra.ncat.org/attra-pub/smallgrain.html, 2003, accessed 15.10.2013

34. Bavec M, Narodoslawsky M, Bavec F, Turinek M. Ecological impact of wheat and spelt production under industrial and alternative farming systems. Renew. Agric Food Syst. 2012; 27 242-250.

35. Williams, J.T., Ed. Cereals and Pseudocereals, Chapman and Hall, London, 1995

36. Bavec F, Bavec M. Organic Production and Use of alternative Crops, Taylor and Francis; Boca Raton, New York, London, 2006;250 p.

37. Mujica, A. et al. Quinoa (Chenopodium quinoa Willd.): Ancestral cultivo Andino, alimento del presente y futuro, project FAO presentation, Santiago, Chile, 2001;303 pp.

38. Kaul HP et al. The suitability of amaranth genotypes for grain and fodder use in Central Europe, Die Bodenkultur, 1996;47 173-181.

39. Aufhamer W, Lee JH, Kúbler E, Kun M, Wagner S. Anbau und Nutzung der Pseudocerealien Buchweizen (Fagopyrum esculentum), Reismelde (Chenopodium quinoa) und (Amaranthus spp.) als Kornefruchtarten. Die Bodenkultur, 1995;2 1-16.

40. Henderson TL, Johnson BL, Schneiter AA. Row spacing, plant population, and cultivar effects on grain amaranth in the northern Great Plains, Agron. J., 2000;92 329-334.

41. Teixeira, D.L., Spehar, C.R. and Souza L.A.C. Agronomic characterization of amaranth for cultivation in the Brazilian Savannah, Pesquisa Agropecuaria Brasileira, 2003;38 45-51.

42. Bavec F, Grobelnik-Mlakar S. Effects of soil and climatic conditions on ermegence of grain amaranths, Eur. J. Agron., 2002;17 93-103.

43. Jacobsen SE., Jorgensen, I., and Stolen, O. Cultivation of quinoa (Chenopodium quinoa) under temperate climatic conditions in Denmark, J. Agric. Sci., 2004;122 47-52.

44. Kalinova J, Moudry J. Evaluation of frost resistance in varieties of common buckwheat (Fagopyrum esculentum Moench), Plant Soil and Environ., 2003;49 410-413

45. Nabwile Omami, E. Response of amaranth to salinity stress. PhD. Thesis, University of Pretoria, 2005; p102, 235 p.

46. Matuz, Kuepper, G. Organic Field Corn Production. 2002. ATTRA. http://www.attra.org/attra-pub/PDF/fieldcorn.pdf. accessed 20.1.2013

47. Robinson, R.G. The Buckwheat Crop in Minnesota, Agr. Exp. Sta. Bul., Univ.. Minnesota, St. Paul, 1980;539 p.

48. Steadman, K.J. Minerals, phytic acid, tannin and rutin in buckwheat seed milling fractions, J. Sci. Food Agric., 2001;81 1094-1100.

49. Muchova, Z., Cukova, L., and Mucha, R. Seed protein fractions of amaranth (Amaranthus sp.), Rostl. Vyroba, 2000;46 331-336.

50. Fleming JE, Galwey NW. Quinoa (Chenopodium quinoa), in Underutilized Crops: Cereals and Pseudocereals, Williams, J.T., Ed., Chapman & Hall,London, 1995. p3.-85.

51. Iqbal Z et al. Allelopathic activity of buckwheat: Isolation and characterization of phenolics, Weed Sci., 2003;51 657-662.

52. Rasmussen, C., Lagnaoui, A., Esbjerg, P. Advances in the knowledge of quinoa pests, Food Reviews Int., 2003;19 61-75.

53. Jamriska P. The effect of variety and row spacing on seed yield of amaranth (Amaranthus ssp.), Rostl. Vyroba, 1998;44 71-76.

54. Zhu, Y.G. et al. Buckwheat (Fagopyrum esculentum Moench) has high capacity to take up phosphorus (P) from a calcium (Ca)-bound source, Plant Soil., 2002;239 1-8.

55. Myers, R.L. Nitrogen fertilizer effect on grain amaranth, Agron. J., 1998;90 597-602

56. Grobelnik-Mlakar S. Agronomic characteristics, potential utilisation and quality of grain amaranth Amaranthus cruentus L. Doct. Thesis. Univ. Maribor, 2012.

57. Grobelnik-Mlakar S, Jakop M, Turinek M; Robačer M, Bavec M, Bavec F. Protein concentration and amino acid composition in grain amaranth (Amaranthus cruentus L.) as affected by sowing date and nitrogen fertilization. African J. Agric. Res. 2012;7 5238-5246.

58. Erley GSA et al. Yield and nitrogen utilization efficiency of the pseudocereales amaranth, quinoa, and buckwheat under differing nitrogen fertilization, Eur. J. Agron., 2005;22 95-100.

59. Johnson DL, Ward S. Quinoa, in New Crops, Janick, J. and Simon, J.E.,Eds, Wiley, New York, 1996;222-227.

60. Belton P, Taylor J, Eds. Pseudocereals and Less Common Cereals: Grain Properties and Utilization Potential, Springer-Verlag, Berlin, 2002;269 pp.

CHAPTER 15

FACTORS INFLUENCING DEOXYNIVALENOL ACCUMULATION IN SMALL GRAIN CEREALS

Stephen N. Wegulo

Keywords: Fusarium head blight, deoxynivalenol, small grain cereals, environmental factors, growth stage, inoculum dosage, pathogen aggressiveness, chemotype, lodging, tillage system, cultivar resistance, fungicide application, regulatory/advisory standards

INTRODUCTION

Deoxynivalenol (DON), also known as vomitoxin, is a trichothecene mycotoxin produced by the fungal plant pathogens *Fusarium graminearum* (sexual stage: *Gibberella zeae*) and *F. culmorum* [1,2]. Both pathogens and several other species of *Fusarium* and its allies cause a disease known as Fusarium head blight (FHB) in wheat, barley, and other small grain cereals [1,2,3,4]. *F. graminearum* also causes ear and stalk rots in maize [5]. In addition to DON, the trichothecenes nivalenol (NIV) and T-2 toxin and the sterol zearalenone (ZEA) are produced by FHB-causing pathogens [6,7]. These mycotoxins are harmful to humans and animals. Although DON is the least toxic of them, it is the most commonly detected *Fusarium* mycotoxin [8] and is the predominant and most economically important mycotoxin in small grain production [1]. Therefore, this review examines the factors that influence the accumulation of DON in small grain cereals under field conditions. An understanding of these factors can be useful in devising strategies aimed at reducing the amounts of DON that accumulate in grain before harvest and, ultimately, in mitigating the human and animal health risks associated with DON contamination of food and feed.

During the growing season in the field, FHB causes premature bleaching of spikes of small grain cereals. In wheat, the bleaching starts with one or more spikelets on a spike (Figure 1) and can continue until the entire spike is whitened. In a disease-favorable growing season, numerous spikes can be seen randomly scattered in the field (Figure 2). Bleached spikelets are sterile or contain shriveled and/or chalky white or pink kernels known as *Fusarium*-damaged kernels (FDK), scabby kernels, or “tombstones” [9] (Figure 3). Kernels that appear to be healthy also may be infected, especially if infection occurred late in kernel development. Infected kernels contain mycotoxins, primarily DON. In controlled experiments, DON was detected as early as 36 hours after inoculation of wheat spikelets with *F. culmorum* [10].

DON concentration in *Fusarium*-damaged grain generally increases with the percentage of damaged grain in a given sample. To demonstrate this, the author mixed wheat *Fusarium*-damaged kernels (FDK) with healthy kernels in 5% (by weight) increments from 0% FDK, 100% healthy kernels to 100% FDK, 0% healthy kernels. The FDK were obtained by separating them from grain collected from winter wheat fields and grain elevators in 2007 and 2008 when there were severe FHB epidemics in Nebraska, USA. Subsamples of grain from each mixture were ground to flour and submitted to the North Dakota Veterinary Diagnostic Laboratory for DON content determination using gas chromatography with electron capture detection (GC/ECD) [11]. In both years, DON concentration increased with the percentage of FDK in grain samples (Figure 4).

The acetylated derivatives of DON, 3-ADON and 15-ADON, are commonly detected in contaminated grain [12]. Miller *et al.* [13] showed that DON-producing isolates of *F. graminearum* collected from different regions of the world differed in their production of 3-ADON and 15-ADON. Some isolates produced one or the other type of acetylated derivative, whereas others produced both types. For those isolates that produced either 3-ADON or 15-ADON but not both, there was a correlation between geographical origin and isolate chemotype. Isolates from China produced predominantly 3-ADON whereas isolates from Mexico and North America produced mainly 15-ADON.

Citation: Stephen N. Wegulo. Factors Influencing Deoxynivalenol Accumulation in Small Grain Cereals, Toxins, 2012; 4(11), 1157-1180. DOI:10.3390/toxins4111157.

Figure 1. A wheat spike with a few spikelets bleached following natural infection by *Fusarium graminearum* in the field.

Figure 2. A wheat field affected by a severe Fusarium head blight (FHB) epidemic in south central Nebraska, USA in 2008.

In humans, food poisoning characterized by diarrhea, nausea, vomiting, abdominal pain, headache, dizziness, and fever has been associated with consumption of *Fusarium*-infested cereals or food products [1,14,15]. Desjardins [1] discussed some historical case studies of *Fusarium* mycotoxin-associated illnesses. They include human alimentary toxic aleukia in Russia and Central Asia, described in 1932 by Russian scientists. The disease was characterized by the food poisoning symptoms described above in mild cases and often fatal skin rashes and necrotic lesions of the alimentary tract and other organs in serious cases. It was associated with consumption of wheat, barley, and other cereal grains left in the field during the winter and harvested in the spring. At about the same time, Japanese scientists linked outbreaks of a human toxicosis to the consumption of cereal grains, including wheat and barley, discolored by a red mold disease known as akakabi-byo. In the United States, swine feed refusal accompanied by vomiting, weakness and emaciation occurred when the animals were fed wheat, barley and oat grain harvested from fields that were affected by a severe FHB epidemic in the central states in 1928. *F. graminearum* was the predominant *Fusarium* species isolated from the grain.

Trichothecenes have multiple effects on eukaryotic cell functions; the primary effect is believed to be inhibition of protein synthesis [16]. Although DON is the least toxic of the trichothecene mycotoxins produced by *Fusarium* spp. in small grains, it can cause significant harm to humans and animals when ingested in large quantities. DON has been shown to inhibit the absorption of certain nutrients by human intestinal epithelial cells [17]. In animals, clinical signs of trichothecene toxicosis include feed refusal and weight loss, emesis, hemorrhage, and cellular necrosis of mitotically active tissues such as the intestinal mucosa, skin, and bone marrow [16].

In plants, DON and 3-ADON have been shown to be phytotoxic. Wang and Miller [18] showed that of several *F. graminearum* metabolites, DON and 3-ADON most strongly inhibited growth of wheat coleoptile tissue. Growth of the coleoptile tissue of most of the 14 spring wheat cultivars tested was inhibited at a DON and 3-ADON concentration of 10^{-6}M, and inhibition was much stronger at higher concentrations. Bruins *et al.* [19] exposed seedlings, coleoptile segments, anther-derived callus, and anther-derived embryos to DON and 3-ADON. They found that DON inhibited growth of all four types of plant material. Shimada and Otani [20] found DON to strongly inhibit root growth in seedlings of seven wheat cultivars and suggested, based on their observation, that DON might be useful in selecting cells resistant to *F. graminearum* in cell cultures. Because DON is water soluble [21], it can be translocated to other parts of the plant where it can exert physiological effects. Kang and Buchenauer [10] found DON and 3-ADON in mycelium-free wheat plant tissues distant from *F. culmorum*-inoculated spikelets. They concluded that the toxins can be translocated upwards via xylem vessels and phloem sieve tubes, and downward via phloem sieve tubes.

Figure 3. Wheat grain damaged by *Fusarium* (left) and healthy grain.

OCCURRENCE AND ECONOMIC IMPORTANCE OF FHB AND DON IN SMALL GRAIN CEREALS

FHB occurs worldwide and especially in temperate regions where small grain cereals (wheat, barley, triticale, oats, rye) are grown [22,23]. It is caused by several species of *Fusarium* and its allies, notably *F. graminearum, F. culmorum, F. avenaceum, F. poae,* and *Microdochium nivale* [4,23]. *F. graminearum* and *F. culmorum* are the most pathogenic and most common [22]. These fungi produce mycotoxins, mainly trichothecenes, which contaminate grain. Their geographical distribution appears to be influenced by climate. For example, *F. graminearum* most commonly occurs in North America, Southern China, and Eastern Europe whereas *F. culmorum* is common in cooler areas such as Western Europe [2,13,22,24].

FHB can cause significant economic losses. Johnson *et al.* [25] estimated that FHB caused direct losses in wheat and barley totaling more than $1.3 billion in the U.S. during the period from 1991 to 1997. They estimated the total economic impact in rural communities and businesses related to grain production and marketing to be three to four times this amount. Direct losses are manifested as reduced yield due to FDK or spikelet sterility; low test weight of harvested grain; grain cleaning costs to remove FDK; and price discounts at grain elevators due to DON contamination. In addition, germination of *Fusarium*-infected grain can be reduced to unacceptable levels [26] and if the grain is used as seed for the next season's crop, seedling blights and root and crown rots can significantly reduce stand establishment and subsequently yield. In years in which severe and widespread FHB epidemics occur, there can be scarcity of seed needed to plant the next season's crop. Indirect losses result from poor quality food products (e.g., bread and beer) made from *Fusarium*-damaged and DON-contaminated grain; reduced productivity of livestock due to toxicosis or feed refusal; costs incurred in the treatment of toxicosis in humans and animals; costs associated with fungicide seed treatment to control seedling and root and crown rot diseases; and loss of profits or failure of businesses related to grain production and marketing. Worldwide, losses due to FHB can be staggering in years when environmental conditions favor disease development.

FHB DISEASE CYCLE

Because some of the factors that influence the accumulation of DON in small grain cereals include or are related to pathogen growth stages, a brief description of the disease cycle of FHB is provided. *F. graminearum* overwinters as chlamydospores or mycelia in the soil or in host crop residues which serve as a source of primary (initial) inoculum in the spring [3]. The fungus can also survive on wheat seed. Primary inoculum mainly consists of ascospores produced in perithecia (sexual fruiting structures), which form on crop residues in the spring as temperatures warm up. In the spring, ascospores and/or conidia are released from crop residues and spread by wind and/or splashing water from rain or irrigation. They land on spikes and during wet, warm weather they germinate and infect glumes, flower parts, or other parts of the spike. Infections occur mostly during anthesis (stage at which anthers rupture and shed pollen during flowering) partly because the anthers contain stimulants for spore germination and pathogen growth [27]. Spikes are susceptible from anthesis until the soft dough stage. Infections that occur during anthesis are the most damaging. During warm temperatures (25°C to 30°C) and wet conditions, blight symptoms develop within 2 to 4 days after infection. Therefore, an apparently healthy crop can show symptoms suddenly. Later in the growing season or after harvest, perithecia may form on spikes. FHB is considered a monocyclic or one cycle disease, that is, after the initial or primary infection, little or no secondary infection occurs by conidia formed on infected spikes.

FACTORS INFLUENCING DON ACCUMULATION IN GRAIN

The amount of DON produced by *F. graminearum* is positively correlated with fungal biomass. Using real-time polymerase chain reaction (PCR) analysis, Demeke *et al.* [28] quantified *F. graminearum* DNA in grain of two classes of Canadian wheat (western red spring and western red winter). They also measured DON in the same grain using gas chromatography with mass selective detection. Linear regression of DON on fungal DNA revealed a strong, positive relationship ($R^2 \geq 0.90$) between the two variables in both classes of wheat. Sneller *et al.* [29] similarly showed that in *F. graminearum*-susceptible wheat genotypes, fungal biomass had a strong, positive correlation with DON. Therefore, when discussing the factors that affect DON accumulation in grain, it is assumed that factors that promote pathogen growth in a susceptible host also promote DON production. Pathogen growth and expression of pathogenicity in the host are quantified indirectly as disease intensity (FHB incidence, severity, or index).

Field studies have demonstrated a positive, linear relationship between FHB intensity and DON [30,31,32]. In 2008, when there were severe FHB epidemics in wheat fields in Nebraska and other states in the FHB-prone regions of the U.S., the author generated, in a field experiment, different levels of FHB intensity (measured as index (%) = [incidence (%) × severity (%)]/100) by applying or not applying the fungicide Prosaro (prothioconazole + tebuconazole) to two FHB-susceptible winter wheat wheat cultivars, '2137' and 'Jagalene' at early anthesis. All plots were spray-inoculated with *F. graminearum*spores (1×10^5 spores/mL) 24 hours after fungicide application. In addition, maize kernels colonized by *F. graminearum* had been spread on the soil surface (0.42 g/m^2) in the plots one week before anthesis. Following rehydration of the maize kernels by natural rain, mycelial growth of the fungus culminated in sporulation, which provided additional spore inoculum during anthesis. The experimental design was a split plot in randomized complete blocks with six replications. The main plot and subplot treatments were cultivar and fungicide application (Prosaro or no Prosaro), respectively. Index was measured 21 days after spray-inoculation. Following harvest, a grain subsample from each plot was ground to flour and submitted to the North Dakota Veterinary Diagnostic Laboratory for DON content determination using gas chromatography with electron capture detection (GC/ECD) [11]. Regression of DON concentration on index using plot data revealed a significant, positive linear relationship between the two variables (Figure 5). Similar results were obtained from replicated field studies using the same methodology conducted over three years in Kansas and Nebraska, USA [30].

Other field studies have similarly shown a positive, linear relationship between DON concentration and FHB intensity. Hernandez Nopsa *et al.* [32] tagged spikes of two winter wheat cultivars in increasing FHB severity categories and, following harvest, submitted ground subsamples of grain for DON content determination as described above. They found significant, positive, linear relationships (correlation coefficients ranging from 0.57 to 0.77) between FHB severity and DON concentration in both cultivars in all the three years during which the study was conducted. Paul *et al.* [31] used meta-analysis to analyze 163 published and unpublished field studies. They found more than 65% of correlation coefficients between measures of FHB intensity (incidence, severity, index) to be >0.50. The analysis determined that the relationship between diseased spike severity and DON concentration was linear and positive with a mean correlation coefficient of 0.53. The results from these studies indicate that FHB intensity is a major factor influencing DON accumulation in small grain cereals. Therefore, factors that influence FHB development will also influence DON accumulation in grain.

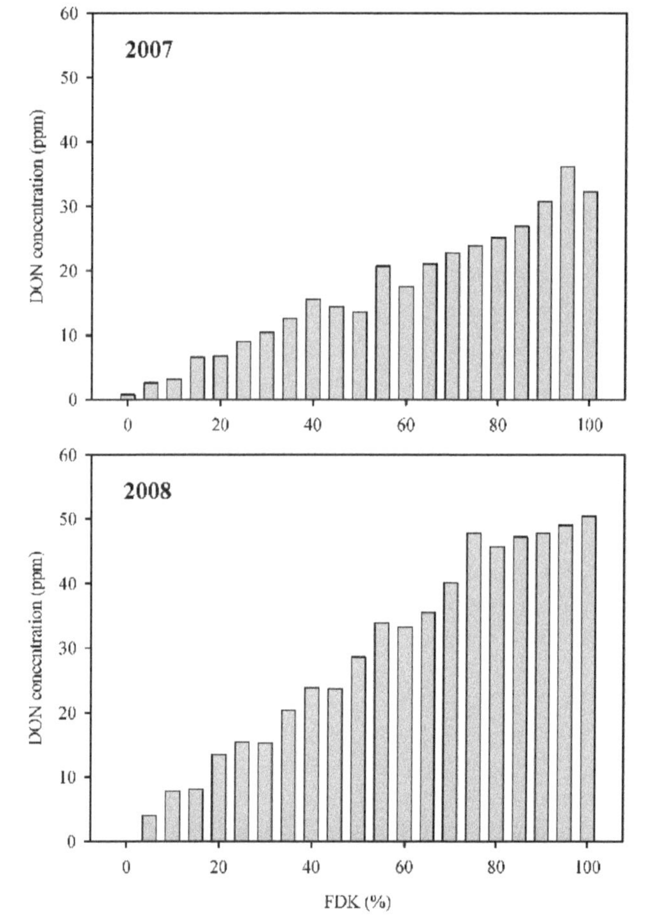

Figure 4. Deoxynivalenol (DON) concentration in wheat grain samples with an increasing proportion (by weight) of *Fusarium*-damaged kernels.

Environmental Factors

Temperature, moisture, and relative humidity (RH) are the most important environmental factors influencing the development of FHB and therefore DON accumulation in small grain cereals. Climate change can greatly influence these factors by causing excessive rainfall or drought [33]. Andersen [34] conducted detailed experiments to determine the effects of temperature and moisture on FHB development in spring wheat. Plants were inoculated with conidia of *F. graminearum* at the past flowering growth stage (kernels nearly filled and in the milk to soft dough stage) and incubated at temperatures ranging from 15 to 30°C with exposure to continuous wetness for periods ranging from 24 to 60 hours. Results showed that the optimum temperature for FHB development was 25°C regardless of the duration of continuous wetness. Longer periods of exposure to continuous wetness resulted in higher disease severity (percent of spikelets killed).

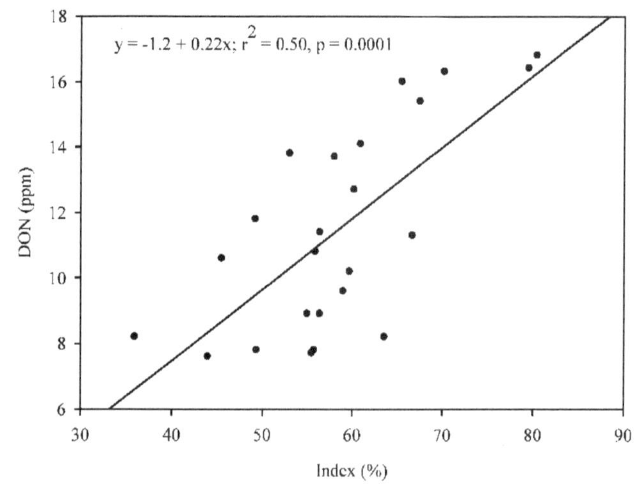

Figure 5. Relationship between FHB index and DON concentration in grain from a field experiment in which different levels of FHB intensity were generated by applying or not applying the fungicide Prosaro to two winter wheat cultivars.

Using non-parametric correlation analysis and stepwise logistic regression analysis of environmental data collected at 50 location-years in four states representing three wheat production regions in the U.S., De Wolf *et al.* [35] identified combinations of weather variables and their occurrence and duration relative to anthesis that accurately predicted FHB outbreaks in the field. They found the most useful predictor variables to be the duration (hours) of rainfall 7 days before anthesis, the duration (hours) when temperature was between 15 and 30°C 7 days before anthesis, and the duration (hours) that temperature was between 15 and 30°C and RH was equal to or greater than 90%. In Belgium, Chandelier *et al.* [36] analyzed climatic data in relation to DON accumulation in winter wheat over a 7-year period. During a 24-day period centered on the mean flowering date, they found a strong positive correlation between the number of days with a mean RH above 80% and mean annual DON content in grain.

Kriss *et al.* [37] analyzed the relationships between environmental factors and biological variables (e.g., FHB intensity, fungal biomass, and DON) associated with FHB in wheat using measurements obtained from 150 location-years originating in three European countries. They determined that in the period from 18 days before anthesis to harvest, moisture-related variables such as average RH and hours of RH above 80% had the highest positive correlations with biological variables. They found significant semi-partial correlations between RH variables and DON in harvested grain. A similar study in the U.S. [38] but with longer window lengths (time of planting to crop maturity) showed that moisture- or wetness-related variables such as daily average RH and total daily precipitation were positively correlated with FHB intensity.

Cowger *et al.* [39] investigated the effects of post-anthesis moisture on DON accumulation in soft red winter wheat under field conditions. Cultivars moderately resistant or susceptible to FHB were inoculated with conidia of *F. graminearum* at mid anthesis (50% of spikes with extruded anthers) and then subjected to different durations of mist ranging from 0 to 30 days. Misting for 10 and 20 days significantly increased DON compared to no misting in both moderately resistant and susceptible cultivars.

Hernandez Nopsa *et al.* [32] observed that in a year in which there was continuous rainfall starting from before anthesis and continuing through grain maturation, DON accumulation in two winter wheat cultivars was much higher than in a year when there was rainfall beginning from before anthesis and continuing through anthesis followed by a dry period after anthesis. Similarly, there was less DON accumulation in the two cultivars in a year in which it was dry prior to and during anthesis followed by rainfall after anthesis. The author compared total rainfall in the months of May and June (the period of highest risk for FHB infections and DON accumulation in wheat in southeastern Nebraska, USA) and total DON in an FHB integrated management study conducted in 2008 and repeated in 2009 [30]. Results showed that total DON was much higher in 2008 when it was excessively wet compared to 2009 when it was relatively dry in the 2-month period (Figure 6).

The results from these studies indicate that moisture plays an important role in the accumulation of DON in small grain cereals. However, free moisture can also leach out DON from *F. graminearum*-infected wheat spikes. In field experiments, Culler *et al.* [40] showed that DON levels were lower in wheat grain from plots that were subjected to extended irrigation (from anthesis to harvest) compared to grain from plots that were irrigated from anthesis to the early dough stage. In greenhouse experiments, Gautam and Dill-Macky [41] showed that regardless of cultivar, DON levels were lower in winter wheat plants whose spikes were subjected to a single wetting event lasting 6 hours compared to plants that were not wetted. They also detected DON in runoff water from the wetted plants, confirming that DON can leach out from wheat spikes exposed to water such as that from irrigation or rainfall.

Growth Stage

FHB in small grain cereals develops from infections that occur during the heading growth stages. In wheat, the most damaging infections occur during flowering or shortly after flowering. However, infections can continue as grain maturation progresses. Miller and Young [42] inoculated spikes of winter wheat with spores of *F. graminearum* in field plots followed by misting at 4 and 16 hours, then measured DON concentration in spikes (grain and chaff) over time for up to 9 weeks after inoculation (w.a.i). They found that DON concentration increased up to 9.5 ppm 6 w.a.i then declined thereafter to 2.5 ppm by 9 w.a.i. They attributed the decline in DON concentration after Week 6 to its breakdown by plant enzymes and suggested, based on these observations, that the timing of harvest may influence the amount of DON in grain.

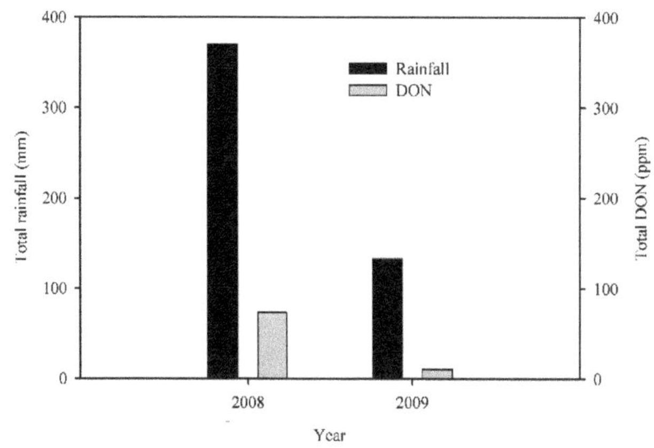

Figure 6. Total rainfall for the months of May and June and total DON in harvested grain in an integrated management field experiment conducted in Nebrasaka, USA in 2008 and repeated in 2009 [30].

Infections that occur late in grain development can lead to accumulation of DON in apparently healthy grain. In field experiments in North Carolina, USA, Cowger and Arrellano [43] inoculated spikes of eight soft red winter wheat cultivars with spores of *F. graminearum* at 0, 10, or 20 days after anthesis (daa) and then misted the plots for 0, 10, 20, or 30 days starting at anthesis. FDK and DON were found to be higher in the 0-and 10-daa inoculations and in the 0- to 20-daa misted treatments than in later inoculation and misting treatments. However, DON still accumulated to unacceptable levels in the late inoculation treatments. The investigators found that high percentages of low FDK (symptomless kernels) with high DON concentrations occurred under conditions of low disease intensity involving late infection and concluded that late infection is an important factor that results in low FDK and high DON in grain.

In Japan, Yoshida and Nakajima [44] similarly found from greenhouse experiments that *F. graminearum* infections that occurred as late as 20 daa resulted in DON and NIV contamination of grain without visible FHB symptoms on wheat spikes. In a greenhouse study, Del Ponte *et al.* [45] inoculated spikes of spring wheat cultivar Norm with a DON-producing isolate of *F. graminearum* at each of six growth stages from flowering to hard dough. They found that high incidences of kernel infection and significant concentrations of DON resulted from inoculations as late as the hard dough stage, with no corresponding reductions in grain weight. They concluded that FHB-favorable environments well past the flowering growth stage could lead to late infections of wheat spikes and kernels by *F. graminearum*.

In greenhouse studies, Andersen [34] inoculated spring wheat with conidia of *F. graminearum* at boot (spike still enclosed in the leaf sheath), before flowering (spikes emerged but no anthers extruded), flowering (flowers in anthesis or anthers already extruded), after flowering (small immature kernels present), and past flowering (kernels nearly filled and in the milk to soft dough stage). Plants were then subjected to continuous wetness for 48 hours at 25°C. Results showed that three days after inoculation, no infection of spikelets occurred in plants inoculated at the boot stage and only 6% of spikelets were infected in plants inoculated at the before flowering stage. However, 79, 98, and 100% of spikelets were infected in plants inoculated at the flowering, after flowering, and past flowering stages, respectively. Five days after inoculation, there was no death of spikelets in plants inoculated at the boot stage and only 4% of spikelets in plants inoculated after spike emergence but before flowering were killed. In contrast, 84, 87, and 100% of spikelets were killed in plants inoculated at the flowering, after flowering, and past flowering stages, respectively.

Susceptibility of the wheat spike to infection by *F. graminearum* at flowering has been attributed to the anthers, which are considered the main site of initial or primary infections [46]. Strange and Smith [27] showed that extruded anthers promoted invasion of the spike by *F. graminerum*. They further demonstrated that extracts of anthers stimulated growth of *F. graminearum in vitro* more than extracts of any other wheat organ, and infection occurred on non-emasculated but not emasculated plants when spikes were inoculated with ascospores of the pathogen. Greater infection of wheat spikes resulted when drops of inoculum were placed on anthers rather than glumes within seven days of anthesis and inoculation of spikes after pollen shed (six weeks after anthesis) did not result in fungal growth or infection of spikelets and spread within the rachis [27]. Stimulants of *F. graminearum* found in wheat anthers were identified as the quaternary ammonium compounds betaine and choline [47,48]. Several studies have shown that susceptibility of the wheat spike to infection by *F. graminearum*declines sharply after the soft dough stage [34,49,50]. Therefore, under field conditions more DON is likely to result from infections that occur from the beginning of flowering to the soft dough stage compared to those that occur at other growth stages.

Inoculum Dosage

In greenhouse experiments, Andersen [34] demonstrated that FHB severity (number of spikelets killed) on *F. graminearum*-inoculated wheat spikes increased with increasing spore concentration, reaching 100% severity at a concentration of two million spores per pot of 10 plants. This implies that in a susceptible cultivar, the higher the concentration of *F. graminearum*spores, the higher the severity of FHB and, therefore, the greater the amount of DON that accumulates in grain. In greenhouse experiments, Stein *et al.* [51] inoculated spikes of a hard red spring wheat cultivar susceptible to FHB with conidia of a highly aggressive isolate of *F. graminearum*. Inoculation was done at the anthesis growth stage and conidial concentration ranged from 0 to 100,000 conidia per ml. Fifteen days after inoculation, FHB incidence (percentage of diseased spikes and severity (percentage of diseased spikelets) were determined and spikes were oven-dried. DON analysis was then done on grain only and whole spikes (grain and chaff). Results showed that FHB incidence and severity increased exponentially and DON concentration in grain only and in whole spikes increased linearly with inoculum concentration. Disease severity did not increase further above a spore concentration of 5,000 spores per spike. However, the level of disease severity reached at this spore concentration differed among replicate experiments. On the other hand, DON concentration continued to increase linearly up to the maximum spore concentration of 25,000 spores per spike reported in the study.

Results from these studies imply that in growing seasons during which environmental conditions (moderate to warm temperatures and prolonged rainfall) favor ascospore production on crop residues, more severe FHB will develop leading to higher accumulation of DON in grain compared to seasons with unfavorable (dry) conditions for disease development. Locally, inoculum concentrations will be higher in fields in which crop residue (maize and wheat stubble) is retained on the soil surface, which is common in reduced or no-tillage systems [23,52,53]. Under these conditions, unacceptably high levels of DON can be expected in grain at the end of the cropping season.

Aggressiveness and Chemotype of Pathogen Isolate

Aggressiveness or virulence is the relative ability of a pathogen to colonize and cause damage to plants [54]. Isolates of a given pathogen can vary in aggressiveness on the same host species. Desjardins *et al.* [55] showed that a trichothecene-nonproducing mutant of *F. graminearum* obtained by disrupting *Tri5*, the gene ecoding trichodiene synthase, was less virulent than a wild-type, trichothecene-producing strain. They inoculated individual florets on wheat spikes in the field at mid anthesis and measured FHB incidence and severity, seeds per spike, total yield per spike, and individual seed weight. Inoculation with the trichothecene-nonproducing mutant resulted in significantly lower disease incidence and severity, more seeds per spike, greater total yield per spike, and greater individual seed weight compared to the wild-type strain.

In Canada, Langevin *et al.* [56] point-inoculated the spikes of six small grain species (bread wheat, durum wheat, triticale, rye, barley, and oats) with two isogenic strains of *F. graminearum* that consisted of a trichothecene- and a non-trichothecene-producing strain. They found that the trichothecene-producing strain was generally more aggressive than the non-producing strain, but the aggressiveness varied with crop species.

In North Dakota, USA, Puri and Zhong [57] investigated the chemotype and aggressiveness on spring wheat of old (collected from 1980 to 2000) and new (collected in 2008) isolates of *F. graminearum*. They found a 15-fold increase in 3-ADON isolates in the new compared to the old collection. Under greenhouse conditions, single-floret inoculation of spikes showed that in a susceptible and a moderately resistant cultivar, the 3-ADON isolates were more aggressive (higher FHB severity) and produced more DON in grain than the 15-ADON isolates. In addition, the 3-ADON isolates produced more DON in rice culture and sporulated more on agar media compared to the 15-ADON isolates.

In Germany, Gang *et al.* [58] inoculated, under field conditions, a susceptible population of winter rye with 42 isolates of*F. culmorum* collected from nine European countries and Australia. They also incubated the same 42 isolates *in vitro* on rye grain. In both assays, the isolates were found to widely differ in the amounts of DON they produced. In the field assay, more DON was produced in grain by the more aggressive isolates, a result similar to that obtained from the *F. graminearum*study by Puri and Zhong [57].

These studies show that the more aggressive isolates of *F. graminearum* and *F. culmorum* produce more DON than the less aggressive isolates, and that the acetylated DON chemotype of a given isolate can influence the level of aggressiveness. Under natural field conditions, it is expected that there will be a mixture of isolates varying widely in aggressiveness. However, based on previous studies [13,59] that have shown a correlation between geographical origin and isolate chemotype, a given geographic location is likely to have isolates that are predominantly of one chemotype (3-ADON or 15-ADON). It should be noted that under natural field conditions, environmental conditions (especially moisture amount and duration) may play a more important role in determining the amount of DON produced than the chemotype of pathogen isolates present in a given location.

Lodging

In the field, lodging is caused mainly by wind, but other factors such as rain, soil type, and nutrient levels (especially nitrogen) in the soil can play a role. Few studies have been done to investigate the effect of lodging on DON accumulation in small grain cereals. Langseth and Stabbetorp [60] conducted field experiments to determine the effect of lodging and time of harvest on DON contamination in barley and oats. They collected grain samples from 19 fields planted separately with a barley or an oat cultivar. Two cultivars of each crop species were included in the study. Half of each field was artificially lodged with a drum roller about three weeks after spike emergence. Grain was harvested at the mealy ripe growth stage and at 15 and 30 days after the mealy ripe growth stage and analyzed for DON content. Results showed that overall mean DON content in grain from the lodged sections of fields (1369 ppb) was nearly twice that from the non-lodged sections (745 ppb). Overall, DON content in grain declined with delayed harvesting.

Nakajima *et al.* [61] investigated the effects of lodging on DON and NIV accumulation in wheat, barley, and rice naturally or artificially infected with *F. graminearum*. They analyzed a total of 66 grain sample sets from fields in which the crops were naturally infected. Each sample set consisted of grain from lodged and non-lodged plants. They also analyzed grain samples from wheat fields in which part of an inoculated area of the field was completely lodged by trampling. In both cases (natural infection and lodging and artificial inoculation and lodging) they found the concentrations of DON and NIV to be significantly higher in grain from lodged compared to non-lodged plants.

These studies indicate that lodging can significantly increase DON concentration in small grain cereals. This increase in DON due to lodging can be attributed to a moist and more humid environment near the soil surface compared to higher locations in the plant canopy where air circulation limits the duration of free moisture and reduces RH. Moisture, RH, and temperature are the three most important environmental factors influencing the growth of *F. graminearum* and the development of FHB (see the *Enviromental Factors* subsection above). Within individual wheat fields, the author has observed more severe FHB in low-lying areas with poor drainage compared to higher ground with good drainage. In addition to moisture and humidity, spikes on lodged plants are in closer proximity to pathogen inoculum on the soil surface, especially in no-till fields that have abundant crop residue. This can result in an increase in infections per spike and therefore an elevated concentration of DON in grain.

Tillage System and Crop Sequence

Tillage system and crop sequence can significantly influence FHB development and therefore DON accumulation in small grain cereals. Over the last several decades, emphasis has been placed on conservation tillage systems that leave crop residues on the soil surface. These systems conserve soil and water resources and thus contribute to sustainable crop production to meet the food and fiber needs of a growing world population [62]. However, surface residues can provide a suitable habitat for the survival, growth, and multiplication of plant pathogens [63]. Pereyra and Dill-Macky [64] examined the presence of *Fusarium* spp. in residues of various crops common in wheat and barley cropping systems. The residues were collected from no-tillage and reduced-tillage field plots in Uruguay and included residues of wheat, barley, maize, sunflower, pasture, and gramineous weed species. *F. graminearum* was recovered from most of the residues, but most frequently from wheat and barley residues. *F. graminearum* colonized wheat and barley residues longer in no-till than in reduced tillage production systems, suggesting that these residues may be major contributors to FHB inoculum.

Maize and wheat residues are particularly suitable for survival of *F. graminearum*. In Canada, Khonga and Sutton [65] infested maize ears and stems and wheat stems, spikes, and grain with *F. graminearum* and then placed them above, on, or below the soil surface. They observed perithecia formation and spore production on the residues placed on or above the soil surface for up to three years. In Minnesota, USA, Dill-Macky and Jones [66] investigated the effects of previous crop residues and tillage practices on FHB in spring wheat. They determined that FHB intensity was higher when wheat followed maize and lower when wheat followed soybean. They also found that moldboard plowing resulted in lower FHB intensity compared to no-till or chisel plowing. DON concentration in harvested grain was highest when wheat followed maize and lowest when wheat followed soybean. In Italy, Maiorano *et al.* [67] quantified maize residues present on the soil surface and in the top 10 cm of soil in tilled and no-till fields. They then evaluated the influence of the residues on *Fusarium* infection of winter wheat and DON contamination of grain. They found a strong correlation between the total amount of residues and DON contamination of grain. DON levels were higher in no-till fields and in residues on the soil surface compared to tilled fields and buried residues. The investigators concluded that residues on the soil surface played a major role in *Fusarium*infection of wheat and DON contamination of grain.

In Ohio, USA, Selby and Manns [68] found that FHB was more severe in areas where wheat was planted continuously. Hoffer *et al.* [69] and Adams [70] noted that observations in Indiana, Pennsylvania, and Wisconsin, USA showed a greater prevalence of FHB in fields where wheat followed maize in the crop sequence. These observations and the studies mentioned above demonstrate that tillage systems and crop sequence can play a major role in DON accumulation in small grain cereals. Locally, DON concentrations in harvested grain will generally be higher in no-till or reduced tillage systems compared to clean tillage systems and in crop sequences in which FHB-susceptible small grains follow maize or are planted continuously. Recent research by Bergstrom *et al.* [71], however, suggests that even though crop residue (specifically maize residue) can play an important role in DON accumulation in grain locally, regional atmospheric inoculum is the strongest contributor to FHB infections and, therefore, DON accumulation.

Cultivar Resistance

Resistance to FHB in small grains is a complex, quantitative trait [72]. Five categories of resistance have been described [50,72,73,74,75]. They are resistance to initial infection (Type I), resistance to pathogen spread in infected tissue (Type II), resistance to kernel infection (Type III), tolerance (Type IV), and resistance to toxins (Type V). Both native and exotic sources of resistance have been identified in the FHB-prone regions of the world [72,76,77,78,79,80]. Resistance in the Chinese wheat line Sumai 3 and its derivatives is used extensively in breeding programs worldwide [72,77,81]. Many studies have shown that cultivars or genotypes of small grain cereals differ in their resistance to FHB and DON accumulation [29,30,32,82,83,84,85,86]. The majority of these studies have shown that in general, FHB-susceptible cultivars (based on phenotype) accumulate more DON than susceptible ones. However, some studies [32,87] have shown that phenotypically moderately resistant cultivars are susceptible to DON accumulation, suggesting that in some cultivars, resistance to FHB may be independent of resistance to DON accumulation.

The mechanisms by which resistance to FHB and DON accumulation is expressed in small grain cereals are complex and have not been fully elucidated [72]. Miller *et al.* [88] infected nine, five, and six cultivars of spring wheat, rye, and triticale, respectively, with *F. graminearum* and compared fungal biomass (determined as ergosterol) and DON concentration in kernels and chaff. They found that FHB-susceptible cultivars contained higher concentrations of DON in kernels than resistant cultivars and the concentration of DON in chaff in both susceptible and resistant cultivars was approximately eight times that in the kernels. Susceptible cultivars contained higher ergosterol concentrations, suggesting that these cultivars were more susceptible to hyphal invasion. Ergosterol:DON ratios were much higher in resistant compared to susceptible cultivars, implying that resistant cultivars had mechanisms that prevented synthesis or promoted degradation of DON. In a similar study in which 22 wheat genotypes differing in resistance to FHB were infected with *F. graminearum*, Snijders and Krechting [89] also found ergosterol and DON concentrations in the chaff and kernels to be higher in susceptible than in resistant cultivars and concluded that susceptible cultivars were more susceptible to hyphal invasion. They observed that DON was translocated from the chaff to young kernels followed by fungal colonization and that in resistant cultivars, this translocation was inhibited, leading to little fungal colonization in kernels. They attributed the limited fungal colonization of kernels in resistant cultivars to membrane-based trichothecene tolerance.

Sneller *et al.* [29] assessed the genetic variation for resistance to kernel infection and resistance to DON accumulation in 32 wheat genotypes by measuring DON concentration and fungal biomass in grain from spikes with varying degrees of visually estimated FHB intensity. They found some genotypes to consistently have low fungal biomass in grain despite increasing FHB intensity, implying that there was resistance to kernel infection in these cultivars. Some genotypes consistently had low DON concentrations in grain despite increasing FHB intensity, suggesting resistance to DON accumulation. There was lack of significant correlations between resistance to kernel infection and FHB intensity and between resistance to DON accumulation and FHB intensity leading the authors to conclude that these resistances may be independent of Type I and Type II resistances which determine the degree of FHB intensity.

The results shown in Figure 4 suggest that small grain cultivars susceptible to *Fusarium* kernel damage are likely to accumulate more DON than cultivars resistant to kernel damage. Existence of resistance to kernel damage has been demonstrated in wheat. Bonin and Kolb [90] identified three kernel damage quantitative trait loci (QTLs) in chromosomes 2B, 4B, and 6B in a winter wheat recombinant inbred line population. The 4B QTL explained seven and 12.3% of the phenotypic variation for kernel damage in greenhouse and field experiments, respectively. Because the 4B QTL had previously been identified for FDK [91], the authors concluded that this QTL may contribute to the reduction of FDK in wheat.

Fungicide Application

Several studies have demonstrated the efficacy of fungicides in reducing FHB and DON in small grain cereals under field conditions [30,83,85,92,93,94,95]. Of the two most commonly used fungicide classes (triazoles and strobilurins) for control of diseases in small grain cereals, the triazoles are more effective in controlling FHB and DON than the strobilurins, in part because strobilurins have been shown to be associated with elevated levels of DON in grain. In Italy, Blandino *et al.* [92] found that in field trials, mixtures of triazoles and strobilurins effectively controlled FHB in winter wheat but increased DON in grain (compared to the non-treated check) when conditions were wet. In Hungary, Mesterházy *et al.* [83] observed an increase in DON levels in grain over the non-treated check when azoxystrobin and carbendazim were applied to winter wheat to control FHB in field plots. In China, Zhang *et al.* [95] similarly found increased DON concentrations in grain when azoxystrobin and carbendazim were applied to winter wheat to control FHB under field conditions.

A multivariate meta-analysis of over 100 FHB uniform fungicide studies across 11 years and 14 states in the U.S. showed that among the triazoles, prothioconazole + tebuconazole (Prosaro) and metconazole (Caramba) were the most effective in controlling FHB and DON [94]. The two fungicides reduced FHB intensity and DON by up to 52 and 45%, respectively. For maximum efficacy in reducing FHB and DON, optimum timing (at anthesis) of fungicide application, using fungicide rates recommended on the label, and thorough coverage of spikes are essential.

Research has shown that combining fungicide application and cultivar resistance is more effective in reducing FHB and DON than using either strategy alone. Wegulo *et al.* [30] showed that fungicide efficacy in reducing FHB and DON was higher in moderately resistant compared to susceptible cultivars. Willyerd *et al.* [85] used data from over 40 trials and 12 states in the U.S. to evaluate the integration of host resistance and application of the fungicide prothioconazole + tebuconazole to manage FHB and DON in small grain cereals. They found that relative to susceptible cultivars not treated with fungicide, the highest mean percent control of FHB (76%) and DON (71%) occurred in moderately resistant cultivars treated with fungicide whereas the lowest mean percent control (43% for FHB and 30% for DON) occurred in moderately susceptible cultivars not treated with fungicide.

REGULATORY/ADVISORY STANDARDS FOR DON

Due to the harmful effects of mycotoxins in humans and animals, many countries have established guidelines or regulatory limits for the maximum levels of mycotoxins in food and feed. In 2003, the Food and Agriculture Organization (FAO) compiled these guidelines and limits in various countries around the world [96]. In the U.S., the Food and Drug Administration (FDA) has established advisory levels for DON as follows [97]: one part per million in finished wheat products for human consumption; 10 ppm in grain and grain byproducts and 30 ppm in distillers grains and brewers grains destined for ruminating beef and feedlot cattle older than 4 months (total ration not to exceed 10 ppm DON), ruminating dairy cattle older than 4 months (total ration not to exceed 5 ppm DON), and chickens (DON-contaminated feed not to exceed 50% of the diet); 5 ppm for grain and grain byproducts destined for swine (DON-contaminated feed not to exceed 20% of the diet); and 5 ppm for grain and grain byproducts for all other animals (DON-contaminated grain not to exceed 40% of the diet). In Canada, maximum DON levels permitted in wheat are 2 ppm in uncleaned soft wheat for use in non-staple foods and 1 ppm in uncleaned soft wheat for use in baby foods [98]. The Canadian Grain Commission has set tolerances for the percentage of *Fusarium*-damaged grain and Agriculture and Agri-Food Canada has established the following DON feeding guidelines: one part per million for swine, dairy cattle, and horses and 5 ppm for poultry and growing beef cattle and sheep [99]. In the European Union, the European Commission has set maximum limits for DON in small grain cereals as 1.25 ppm in unprocessed cereals other than durum wheat and oats; 1.75 ppm in unprocessed durum wheat and oats; 0.75 ppm in cereals for direct human consumption, cereal flour, and bran and germ products marketed for direct human consumption; 0.75 ppm for dry pasta; 0.5 ppm for bread, pastries, biscuits, cereal snacks, and breakfast cereals; and 0.2 ppm for processed cereal-based foods and baby foods for infants and young children [100].

CONCLUSIONS

In FHB-prone regions of the world, the disease and the associated mycotoxins, especially DON, continue to pose health risks for humans and animals, as well as economic uncertainty for farmers of small grain cereals and businesses related to grain production and marketing. An understanding of the factors that favor FHB development and DON accumulation in grain is helpful in devising management strategies for the disease and the mycotoxin. These strategies include an integrated approach that combines the use of resistant cultivars, fungicide application at anthesis, FHB and mycotoxin forecasting models, practices that minimize lodging, crop residue management, and a crop rotation sequence that minimizes disease development and mycotoxin accumulation in grain. Additional reduction in DON contamination of harvested grain can be achieved by harvesting severely affected sections of a field separately from the rest of the field, adjusting the combine travel speed and the fan speed and/or the shutter opening of the combine during harvesting to blow out severely damaged kernels which are lighter than healthy kernels [101], using cleaning equipment to remove FDK from harvested grain, and blending*Fusarium*-damaged grain with healthy grain in proportions that will lower DON concentration to levels below regulatory/advisory limits. Regulations that limit the levels of DON and other mycotoxins in food and feed are desirable because they mitigate the health risks in humans and animals associated with the mycotoxins.

REFERENCES

1. Desjardins, A.E. *Fusarium Mycotoxins. Chemistry, Genetics, and Biology*; American Phytopathological Society: St. Paul, MN, USA, 2006.
2. Miller, J.D. Epidemiology of Fusarium ear diseases of cereals. In *Mycotoxins in Grain. Compounds Other than Aflatoxin*; Miller, J.D., Trenholm, H.L., Eds.; Eagan Press: St. Paul, MN, USA, 1994; pp. 19–36.
3. Dill-Macky, R. Fusarium head blight (scab). In *Compendium of Wheat Diseases and Pests*, 3rd; Bockus, W.W., Bowden, R.L., Hunger, R.M., Morrill, W.L., Murray, T.D., Smiley, R.W., Eds.; American Phytopathological Society: St. Paul, MN, USA, 2010; pp. 34–36.
4. Liddell, C.M. Systematics of *Fusarium* species and allies associated with Fusarium head bligh. In *Fusarium Head Blight of Wheat and Barley*; Leonard, K.J., Bushnell, W.R., Eds.; American Phytopathological Society: St. Paul, MN, USA, 2003; pp. 35–43.
5. Payne, G.A. Ear and kernel rots. In *Compendium of Corn Diseases*, 3rd; White, D.G., Ed.; American Phytopathological Society: St. Paul, MN, USA, 1999; pp. 44–47.
6. McCormick, S. The role of DON in pathogenicity. In *Fusarium Head Blight of Wheat and Barley*; Leonard, K.J., Bushnell, W.R., Eds.; American Phytopathological Society: St. Paul, MN, USA, 2003; pp. 165–183.
7. Gale, L.R. Population biology of *Fusarium* species causing head blight of grain crops. In *Fusarium Head Blight of Wheat and Barley*; Leonard, K.J., Bushnell, W.R., Eds.; American Phytopathological Society: St. Paul, MN, USA, 2003; pp. 120–143.
8. Villar, D.; Carson, D.L. Trichothecene mycotoxins. In *Clinical Veterinary Toxicology*; Pumlee, K.H., Ed.; Mosby: St. Louis, MO, USA, 2004; pp. 270–275.
9. McMullen, M.; Jones, R.; Gallenberg, D. Scab of wheat and barley: a re-emerging disease of devastating impact. *Plant Dis.* 1997, *81*, 1340–1348.
10. Kang, Z.; Buchenauer, H. Immunocytochemical localization of fusarium toxins in infected wheat spikes by *Fusarium culmorum*. *Physiol. Mol. Plant P.* 1999, *55*, 275–288.
11. Tacke, B.K.; Casper, H.H. Determination of deoxynivalenol in wheat, barley, and malt by column cleanup and gas chromatography with electron capture

detection. *J. Assoc. Off. Anal. Chem.* 1996, *79*, 472–475.

12. Mirocha, C.J.; Xie, W.; Filho, E.R. Chemistry and detection of Fusarium mycotoxins. In *Fusarium Head Blight of Wheat and Barley*; Leonard, K.J., Bushnell, W.R., Eds.; American Phytopathological Society: St. Paul, MN, USA, 2003; pp. 144–164.
13. Miller, J.D.; Greenhalgh, R.; Wang, Y.; Lu, M. Trichothecene chemotypes of three *Fusarium* species. *Mycologia* 1991, *83*, 121–130.
14. Pestka, J.J. Deoxynivalenol: toxicity, mechanisms and animal health risks. *Anim. Feed Sci. Tech.* 2007, *137*, 283–298.
15. Sobrova, P.; Adam, V.; Vasatkova, A.; Beklova, M.; Zeman, L.; Kizek, R. Deoxynivalenol and its toxicity. *Interdisc. Toxicol.* 2010, *3*, 94–99.
16. Mostrom, M.S.; Raisbeck, M.F. Trichothecenes. In *Veterinary Toxicology*, 1st; Gupta, R.C., Ed.; Elsevier: New York, NY, USA, 2007; pp. 951–976.
17. Maresca, M.; Mahfoud, R.; Garmy, N.; Fantini, J. The mycotoxin deoxynivalenol affects nutrient absorption in human intestinal epithelial cells. *J. Nutr.* 2002, *132*, 2723–2731.
18. Wang, Y.Z.; Miller, J.D. Effects of *Fusarium graminearum* metabolites on wheat tissue in relation to Fusarium head blight resistance. *J. Phytopathology* 1988, *122*, 118–125.
19. Bruins, M.B.M.; Karsai, I.; Schepers, J.; Snijders, C.H.A. Phytotoxicity of deoxynivalenol to wheat tissue with regard to*in vitro* selection for Fusarium head blight resistance. *Plant Sci.* 1993, *94*, 195–206.
20. Shimada, T.; Otani, M. Effects of *Fusarium* mycotoxins on the growth of shoots and roots at germination in some Japanese wheat cultivars. *Cereal Res. Commun.* 1990, *18*, 229–232.
21. Snijders, C.H.A. Resistance in wheat to Fusarium infection and trichothecene formation. *Toxicol. Lett.* 2004, *153*, 37–46.
22. Miller, J.D. Aspects of the ecology of Fusarium toxins in cereals. In *Mycotoxins and Food Safety*; DeVries, J.W., Trucksess, M.W., Jackson, L.S, Eds.; Kluwer Academic/Plenum Publishers: New York, NY, USA, 2002; pp. 19–28.
23. Parry, D.W.; Jenkinson, P.; McLeod, L. *Fusarium* ear blight (scab) in small grain cereals - a review. *Plant Pathol.* 1995,*44*, 207–238.
24. Snijders, C.H.A.; Perkowski, J. Effects of head blight caused by *Fusarium culmorum* on toxin content and weight of wheat kernels. *Phytopathology* 1990, *80*, 566–570.
25. Johnson, D.D.; Flaskerud, G.K.; Taylor, R.D.; Satyanarayana, V. Quantifying economic impacts of Fusarium head blight in wheat. In *Fusarium Head Blight of Wheat and Barley*; Leonard, K.J., Bushnell, W.R., Eds.; American Phytopathological Society: St. Paul, MN, USA, 2003; pp. 461–483.
26. Argyris, J.; van Sanford, D.; TeKrony, D. *Fusarium graminearum* infection during wheat seed development and its effect on seed quality. *Crop Sci.* 2003, *43*, 1782–1788.
27. Strange, R.N.; Smith, H. A fungal growth stimulant in anthers which predisposes wheat to attack by *Fusarium graminearum*. *Physiol. Plant Pathol.* 1971, *1*, 141–150.
28. Demeke, T.; Gräfenhan, T.; Clear, R.M.; Phan, A.; Ratnayaka, I.; Chapados, J.; Patrick, S.K.; Gaba, D.; Lévesque, C.A.; Seifert, K.A. Development of a specific TaqMan® real-time PCR assay for quantification of *Fusarium graminearum* clade 7 and comparison of fungal biomass determined by PCR with deoxynivalenol content in wheat and barley. *Int. J. Food Microbiol.* 2010, *141*, 45–50.
29. Sneller, C.; Guttieri, M.; Paul, P.; Costa, J.; Jackwood, R. Variation for resistance to kernel infection and toxin accumulation in winter wheat infected with *Fusarium graminearum*. *Phytopathology* 2012, *102*, 306–314.
30. Wegulo, S.N.; Bockus, W.W.; Hernandez Nopsa, J.; De Wolf, E.D.; Eskridge, K.M.; Peiris, K.H.S.; Dowell, F.E. Effects of integrating cultivar resistance and fungicide application on Fusarium head blight and deoxynivalenol in winter wheat. *Plant Dis.* 2011, *95*, 554–560.
31. Paul, P.A.; Lipps, P.E.; Madden, L.V. Relationship between visual estimates of Fusarium head blight intensity and deoxynivalenol accumulation in harvested wheat grain: A meta-analysis. *Phytopathology* 2005, *95*, 1225–1236.
32. Hernandez Nopsa, J.; Baenziger, P.S.; Eskridge, K.M.; Peiris, K.H.S.; Dowell, F.E.; Harris, S.D.; Wegulo, S.N. Differential accumulation of deoxynivalenol in two winter wheat cultivars varying in FHB phenotype response under field conditions. *Can. J. Plant Pathol.* 2012, *34*, 380–389.
33. Miller, J.D. Mycotoxins in small grains and maize; old problems, new challenges. *Food Addit. Contam.* 2008, *25*, 219–230.
34. Andersen, A.L. The development of *Gibberella zeae* headblight of wheat. *Phytopathology* 1948, *38*, 599–611.
35. De Wolf, E.D.; Madden, L.V.; Lipps, P.E. Risk assessment models for wheat Fusarium head blight epidemics based on within-season weather data. *Phytopathology* 2003, *93*, 428–435.
36. Chandelier, A.; Nimal, C.; André, F.; Planchon, V.; Oger, R. *Fusarium* species and DON contamination associated with head blight in winter wheat over a 7-year period 92003-2009) in Belgium. *Eur. J. Plant Pathol.* 2011, *130*, 403–414.
37. Kriss, A.B.; Paul, P.A.; Xu, X.; Nicholson, P.; Doohan, F.M.; Hornok, L.; Reitini, A.; Edwards, S.G.; Madden, L.V. Quantification of the relationship between the environment and Fusarium head blight, Fusarium pathogen density, and mycotoxins in winter wheat in Europe. *Eur. J. Plant Pathol.* 2012, *133*, 975–993.
38. Kriss, A.B.; Paul, P.A.; Madden, L.V. Relationship between yearly fluctuations in Fusarium head blight intensity and environmental variables: A window-pane analysis. *Phytopathology* 2010, *100*, 784–797.
39. Cowger, C.; Patton-Özkurt, J.; Brown-Guedira, G.;

Perugini, L. Post-anthesis moisture increased Fusarium head blight and deoxynivalenol levels in North Carolina winter wheat. *Phytopathology* 2009, *99*, 320–327.

40. Culler, M.D.; Miller-Gavin, J.E.; Dill-Macky, R. Effect of extended irrigation and host resistance on deoxynivalenol accumulation in *Fusarium*-infected wheat. *Plant Dis.* 2007, *91*, 1464–1472.
41. Gautam, P.; Dill-Macky, R. Free water can leach mycotoxins from *Fusarium*-infected wheat heads. *J. Phytopathol.* 2012,*60*, 484–490.
42. Miller, J.D.; Young, J.C. Deoxynivalenol in an experimental *Fusarium graminearum* infection of wheat. *Can. J. Plant Pathol.* 1985, *7*, 132–134.
43. Cowger, C.; Arrellano, C. Plump kernels with high deoxynivalenol linked to late *Gibberella zeae* infection and marginal disease conditions in winter wheat. *Phytopathology* 2010, *100*, 719–728.
44. Yoshida, M.; Nakajima, T. Deoxynivalenol accumulation in wheat infected with *Fusarium graminearum* during grain development. *Phytopathology* 2010, *100*, 763–773.
45. Del Ponte, E.M.; Fernandes, J.M.C.; Bergstrom, G.C. Influence of growth stage on fusarium head blight and deoxynivalenol production in wheat. *J. Phytopathology* 2007, *155*, 577–581.
46. Shaner, G. Epidemiology of Fusarium head blight of small grain cereals in North America. In *Fusarium Head Blight of Wheat and Barley*; Leonard, K.J., Bushnell, W.R., Eds.; American Phytopathological Society: St. Paul, MN, USA, 2003; pp. 84–119.
47. Strange, R.N.; Smith, H.; Majer, J.R. Choline, one of two fungal growth stimulants in anthers responsible for the susceptibility of wheat to *Fusarium graminearum. Nature* 1972, *238*, 103–104.
48. Strange, R.N.; Majer, J.R.; Smith, H. The isolation and identification of choline and betaine as two major components in anthers and wheat germ that stimulate *Fusarium graminearum in vitro. Physiol. Plant Pathol.* 1974, *4*, 277–290.
49. Atanasoff, D. Fusarium blight (scab) of wheat and other cereals. *J. Agric. Res.* 1920, *20*, 1–32.
50. Shroeder, H.W.; Christensen, J.J. Factors affecting resistance of wheat to scab caused by *Gibberella zeae. Phytopathology*1963, *53*, 831–838.
51. Stein, J.M.; Osborne, L.E.; Bondalapati, K.D.; Glover, K.D.; Nelson, C.A. Fusarium head blight severity and deoxynivalenol concentration in wheat in response ot *Gibberella zeae* inoculum concentration. *Phytopathology* 2009, *99*, 759–764.
52. Sutton, J.C. Epidemiology of wheat head blight and maize ear rot caused by *Fusarium graminearum. Can. J. Plant Pathol.*1982, *4*, 195–209.
53. Bai, G.; Shaner, G. Scab of wheat: prospects for control. *Plant Dis.* 1994, *78*, 760–766.
54. Shurtleff, M.C.; Averre, C.W., III. *Glossary of Plant-Pathological Terms*; APS Press: St. Paul, MN, USA, 1997.
55. Desjardins, A.E.; Proctor, R.H.; Bai, G.; McCormick, S.P.; Shaner, G.; Buechley, G.; Hohn, T.M. Reduced virulence of trichothecene-nonproducing mutants of *Gibberella zeae* in wheat field tests. *Mol. Plant Microbe In.* 1996, *9*, 775–781.
56. Langevin, F.; Eudes, F.; Comeau, A. Effect of trichothecenes produced by *Fusarium graminearum* during Fusarium head blight development in six cereal species. *Eur. J. Plant Pathol.* 2004, *110*, 735–746.
57. Puri, K.D.; Zhong, S. The 3ADON population of *Fusarium graminearum* found in North Dakota is more aggressive and produces a higher level of DON than the prevalent 15ADON population in spring wheat. *Phytopathology* 2010, *100*, 1007–104.
58. Gang, G.; Miedaner, T.; Schumacher, U.; Schollenberger, M.; Geiger, H.H. Deoxynivalenol and nivalenol production by *Fusarium culmorum* isolates differing in aggressiveness toward winter rye. *Phytopathology* 1998, *88*, 879–884.
59. Kim, H.-S.; Lee, T.; Dawlatana, M.; Yun, S.-H.; Lee, Y.-W. Polymorphism of trichothecene biosynthesis genes in deoxynivalenol- and nivalenol-producing *Fusarium graminearum* isolates. *Mycol Res.* 2003, *107*, 190–197.
60. Langseth, W.; Stabbetorp, H. The effect of lodging and time of harvest on deoxynivalenol contamination in barley and oats. *J. Phytopathology* 1996, *144*, 241–246.
61. Nakajima, T.; Yoshida, M.; Tomimura, K. Effect of lodging on the level of mycotoxins in wheat, barley, and rice infected with the *Fusarium graminearum* species complex. *J. Gen. Plant Pathol.* 2008, *74*, 289–295.
62. Unger, P.W. Residue production and uses-an introduction to managing agricultural residues. In *Managing Agricultural Residues*; Unger, P.W., Ed.; Lewis Publishers: Boca Raton, FL, USA, 1994; pp. 1–6.
63. Watkins, J.E.; Boosalis, M.G. Plant disease incidence as influenced by conservation tillage systems. In *Managing Agricultural Residues*; Unger, P.W., Ed.; Lewis Publishers: Boca Raton, FL, USA, 1994; pp. 261–283.
64. Pereyra, S.A.; Dill-Macky, R. Colonization of the residues of diverse plant species by *Gibberella zeae* and their contribution to Fusarium head blight inoculum. *Plant Dis.* 2008, *92*, 800–807.
65. Khonga, E.B.; Sutton, J.C. Inoculum production and survival of *Gibberella zeae* in maize and wheat residues. *Can. J. Plant Pathol.* 1988, *10*, 232–239.
66. Dill-Macky, R.; Jones, R.K. The effect of previous crop residues and tillage on Fusarium head blight of wheat. *Plant Dis.* 2000, *84*, 71–76.
67. Maiorano, A.; Blandino, M.; Reyneri, A.; Vanara, F. Effects of maize residues on the *Fusarium* spp. infection and deoxynivalenol (DON) contamination of wheat grain. *Crop Prot.* 2008, *27*, 182–188.
68. Selby, A.D.; Manns Thos, F. Studies in diseases of cereals and grasses. *Ohio Agric. Exp. Sta. Bull.* 1909, *203*, 187–236.
69. Hoffer, G.N.; Johnson, A.G.; Atanasoff, D. Corn root-rot and wheat-scab. *J. Agric. Res.* 1918, *14*, 611–612.
70. Adams, J.F. Observations on wheat scab in Pennyslvania and its pathological histology. *Phytopathology* 1921, *11*, 115–125.

71. Bergstrom, G.C.; Waxman, K.D.; Bradley, C.A.; Hazelrig, A.L.; Hershman, D.E.; Nagelkirk, M.; Sweets, L.E.; Wegulo, S.N. Effects of local corn debris management on FHB and DON levels in seven U.S. wheat environments in 2011. In Proceedings of the 2011 National Fusarium Head Blight Forum; Canty, S., Clark, A., Anderson-Scully, A., Van Sanford, D., Eds.; U.S. Wheat & Barley Scab Initiative and ASAP Printing, Inc.: Lansing, MI, USA, 2011; pp. 119–121.
72. Bai, G.; Shaner, G. Management and resistance in wheat and barley to Fusarium head blight. *Annu. Rev. Phytopathol.*2004, *42*, 135–161.
73. Wang, Y.Z.; Miller, J.D. Screening techniques and sources of resistance to Fusarium head bligh. In *Wheat Production Constraints in Tropical Environments*; CIMMYT: Lisboa, Mexico, 1988.
74. Mesterházy, Á. Types and components of resistance to Fusarium head blight of wheat. *Plant Breeding* 2006, *114*, 377–386.
75. Mesterházy, Á. Breeding wheat for Fusarium head blight resistance in Europe. In *Fusarium Head Blight of Wheat and Barley*; Leonard, K.J., Bushnell, W.R., Eds.; American Phytopathological Society: St. Paul, MN, USA, 2003; pp. 211–240.
76. Gosman, N.; Bayles, R.; Jennings, P.; Kirby, J.; Nicholson, P. Evaluation and characterization of resistance to fusarium head blight caused by Fusarium culmorum in UK winter wheat cultivars. *Plant Pathol.* 2007, *56*, 264–276.
77. Yu, J.-B.; Bai, G.-H.; Cai, S.-B.; Dong, Y.-H.; Ban, T. New Fusarium head blight-resistant sources from Asian wheat germplasm. *Crop Sci.* 2008, *48*, 1090–1097.
78. McKendry, A. Native resistance: an essential building block for accelerating the development of scab resistant soft red winter wheat. *Cereal Res. Commun.* 2008, *36*, 135–137.
79. Zwart, R.S.; Muylle, H.; Van Bockstaele, E.; Roldan-Ruiz, I. Evaluation of genetic diversity of Fusarium head blight resistance in European winter wheat. *Theor. Appl. Genet.* 2008, *117*, 813–828.
80. Ma, H.; Ge, H.; Zhang, X.; Lu, W.; Yu, D.; Chen, H.; Chen, J. Resistance to Fusarium head blight and deoxynivalenol accumulation in Chinese barley. *J. Phytopathology.* 2009, *157*, 166–171.
81. Anderson, J.A.; Stack, R.W.; Liu, S.; Waldron, B.L.; Fjeld, A.D.; Coyne, C.; Moreno-Sevilla, B.; Mitchel Fetch, J.; Song, Q.J.; Cregan, P.B.; *et al.* DNA markers for Fusarium head blight resistance QTLs in two wheat populations. *Theor. Appl. Genet.* 2001, *102*, 1164–1168.
82. Mesterházy, Á.; Bartok, T.; Mirocha, C.G.; Komoroczy, R. Nature of wheat resistance to Fusarium head blight and the role of deoxynivalenol for breeding. *Plant Breeding* 1999, *118*, 97–110.
83. Mesterházy, Á.; Bartok, T.; Lamper, C. Influence of wheat cultivar, species of *Fusarium*, and isolate aggressiveness on the efficacy of fungicides for control of Fusarium head blight. *Plant Dis.* 2003, *87*, 1107–1115.
84. Bai, G.-H.; Shaner, G. Variation in *Fusarium graminearum* and cultivar resistance to wheat scab. *Plant Dis.* 1996, *80*, 975–979.
85. Willyerd, K.T.; Li, C.; Madden, L.V.; Bradley, C.A.; Bergstrom, G.C.; Sweets, L.E.; McMullen, M.; Ransom, J.K.; Grybauskas, A.; Osborne, L.; Wegulo, S.N.; *et al.* Efficacy and stability of integrating fungicide and cultivar resistance to manage Fusarium head blight and deoxynivalenol in wheat. *Plant Dis.* 2012, *96*, 957–967.
86. Gautam, P.; Dill-Macky, R. Fusarium head blight development and trichothecene accumulation in point inoculated Fusarium infected wheat heads. *World Mycotoxin J.* 2012, *5*, 45–55.
87. Arseniuk, E.; Foremska, E.; Goral, T.; Chelkowski, J. Fusarium head blight reactions and accumulation of deoxynivalenol (DON) and some of its derivatives in kernels of wheat, triticale, and rye. *J. Phytopathol.* 1999, *147*, 577–590.
88. Miller, J.D.; Young, J.C.; Sampson, D.R. Deoxynivalenol and Fusarium head blight resistance in spring cereals. *J. Phytopathol.* 1985, *113*, 359–367.
89. Snijders, C.H.A.; Krechting, C.F. Inhibition of deoxynivalenol translocation and fungal colonization in Fusarium head blight resistant wheat. *Can. J. Botany* 1992, *70*, 1570–1576.
90. Bonin, C.M.; Kolb, F.L. Resistance to Fusarium head blight and kernel damage in a winter wheat recombinant inbred line population. *Crop Sci.* 2009, *49*, 1304–1312.
91. Abate, Z.A.; Liu, S.; McKendry, A.L. Quantitative trait loci associated with deoxynivalenol content and kernel quality in the soft red winter wheat 'Ernie'. *Crop Sci.* 2008, *49*, 1408–1418.
92. Blandino, M.; Minelli, L.; Reyneri, A. Strategies for the chemical control of Fusarium head blight: effect on yield, alveographic parameters and deoxynivalenol contamination in winter wheat grain. *Europ. J. Agronomy* 2006, *25*, 193–201.
93. Paul, P.A.; Lipps, P.E.; Hershman, D.E.; McMullen, M.P.; Draper, M.A.; Madden, L.V. A quantitative review of tebuconazole effect on Fusarium head blight and deoxynivalenol content in wheat. *Phytopathology* 2007, *97*, 211–220.
94. Paul, P.A.; Lipps, P.E.; Hershman, D.E.; McMullen, M.P.; Draper, M.A.; Madden, L.V. Efficacy of triazole-based fungicides for Fusarium head blight and deoxynivalenol control in wheat: a multivariate meta-analysis.*Phytopathology* 2008, *98*, 999–1011.
95. Zhang, Y.J.; Fan, P.S.; Zhang, X.; Chen, C.J.; Zhou, M.G. Quantification of *Fusarium graminearum* in harvested grain by real-time polymerase chain reaction to assess efficacies of fungicides on Fusarium head blight, deoxynivalenol contamination, and yield of winter wheat. *Phytopathology* 2009, *99*, 95–100.
96. van Egmond, H.P.; Jonker, M.A. *Worldwide Regulations for Mycotoxins in Food and Feed*; FAO: Rome, Italy, 2003. Available online: ftp://ftp.fao.org/docrep/fao/007/y5499e/y5499e00.pdf (accessed on 24 September 2012).
97. Guidance for industry and FDA: Advisory levels for deoxynivalenol (DON) in finished wheat products for human consumption and grains and grain by-

products used for animal feed. Food and Drug Administration: Washington, DC, USA, 2010. Available online: http://www.fda.gov/downloads/Food/GuidanceComplianceRegulatoryInformation/GuidanceDocuments/NaturalToxins/UCM217558.pdf (accessed on 20 October 2012).

98. Canadian standards (maximum levels) for various chemical contaminants in foods. Health Canada: Ottawa, Ontario, Canada, 2012. Available online: http://www.hc-sc.gc.ca/fn-an/securit/chem-chim/contaminants-guidelines-directives-eng.php (accessed on 20 October 2012).

99. Grading tolerances for fusarium-damaged grain and DON feeding guidelines. Canadian Grain Commision: Winnipeg, Manitoba, Canada, 2012. Available online: http://www.grainscanada.gc.ca/guides-guides/don/don-1-eng.htm(accessed on 20 October 2012).

100. Commission Regulation (EC) No 1881/2006 of 19 December 2006 setting maximum levels for certain contaminants in foodstuffs. The European Commission: Brussels, Belgium, 2006. Available online: http://eur-lex.europa.eu/LexUriServ/LexUriServ.do?uri=CONSLEG:2006R1881:20090701:EN:PDF (accessed on 20 October 2012).

101. Salgado, J.D.; Wallhead, M.; Madden, L.V.; Paul, P.A. Grain harvesting strategies to minimize grain quality losses due to Fusarium head blight in wheat. *Plant Dis.* 2011, *95*, 1448–1457.ss

CHAPTER 16

DISTINGUISHING BETWEEN YIELD ADVANCES AND YIELD PLATEAUS IN HISTORICAL CROP PRODUCTION TRENDS

Patricio Grassini, Kent M. Eskridge, and Kenneth G. Cassman

INTRODUCTION

The first decade of the new millennium saw an abrupt reversal of long-term trends in declining grain prices since the onset of the green revolution in the mid-1960s, and an increase in expansion of land area used for crop production. Whether this recent expansion in crop area is transitory or permanent will depend in large part on trajectories in grain and oilseed prices, which in turn depend on trends in crop yields. Estimating these trends with a high degree of confidence is therefore essential to inform development of appropriate agricultural policies and priorities for agricultural research to ensure future food security and minimize conversion of carbon-rich and biodiverse natural ecosystems to cropland, which contributes substantially to anthropogenic greenhouse gas emissions and climate forcing[1,2,3]. In fact, agricultural production, including indirect emissions associated with land-use change, accounts for 15–25% of the total anthropogenic greenhouse gas emissions[4,5]. Hence, a critical question is whether current yield trajectories are adequate to achieve the needed production increases on existing farmland because econometric projections of grain prices and land use change are highly sensitive to underpinning assumptions about future crop yield increases.

Here, by applying a robust, statistical framework to evaluate historical yield trends of major crop-producing countries since 1965, we find that all trends can be described by linear functions with or without an abrupt decrease in rate of gain or an upper yield plateau. We conclude that estimates of future crop production and land use must consider both historical yield trends and biophysical yield ceilings to improve forecasting capability.

Citation: Patricio Grassini, Kent M. Eskridge, and Kenneth G. Cassman (2013). Distinguishing between yield advances and yield plateaus in historical crop production trends, Nature Communications 4, Article number: 2918. DOI:10.1038/ncomms3918.

RESULTS

Recent Trends in Land Use for Crop Production

At a global level, changes in land use for crop production have been driven in large part by increases in land area devoted to the three major cereals (rice, wheat and maize). During the first 16 years of the green revolution, for example, expansion of area for the major cereals represented >70% of land use increase for all crops, followed by two decades in which both total crop area and area in major cereals remained relatively constant (Fig. 1). This period of stability came to an abrupt end in 2002. Since then, crop production area has increased at nearly ten million hectares per year and 60% of this expansion is due to increased production of the major cereals. An additional 25% is attributable to expansion of soybean area. Nearly all of the increased crop area since 2002 has occurred in South America, Asia and Africa.

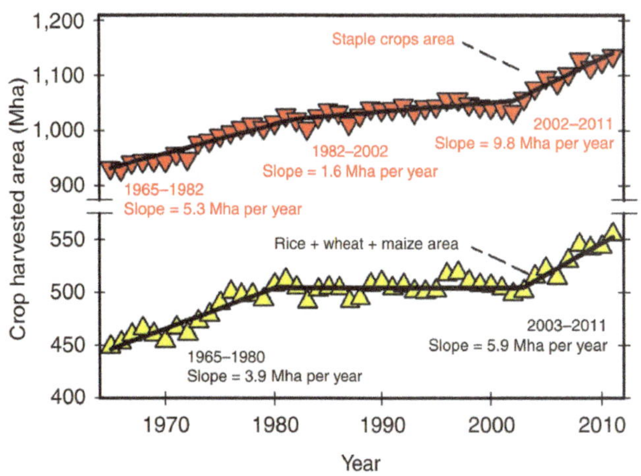

Figure 1: Trends in total harvested area of staple crops and three major cereal crops. Staple crops include cereal, oil, sugar, pulses, fibre, tuber plus root crops. The three major cereal crops are rice, wheat and maize. Slopes of the fitted trilinear models are shown when significant (Student's *t*-test; $P<0.01$, $n=47$ years of yield data).

Previous Analyses of Historical Crop Yield Trends

Global rates of yield increase have been decidedly linear for most major cereal crop species since the start of the green revolution in the 1960s[6,7]. Some projections of global food security assume that these linear rates of increase will continue unchanged during the next 40 years[8,9,10,11]. Other projections, including several based on computational partial equilibrium models that evaluate both food demand and supply, assume compound annual rates of yield increase without recognizing biophysical limits to crop yields[12,13,14,15,16,17,18] (Fig. 2). In a compound rate model, annual yield gain represents a constant proportion of the current yield and thus the magnitude of absolute gain increases each year. Although exponential increase in crop yields can occur over short time periods of one or two decades, such growth rates are not feasible over the long term because average farm yields eventually approach a yield potential ceiling determined by biophysical limits on crop growth rates and yield[19,20,21,22,23]. For irrigated crops, yield potential is defined as the yield of a crop cultivar when growth is only limited by solar radiation, temperature and carbon dioxide supply from the atmosphere, and also by the amount and distribution of precipitation in rainfed cropping systems[23,24].

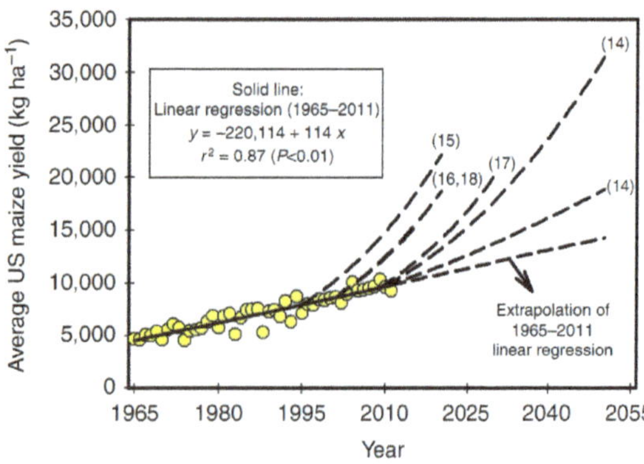

Figure 2: Historical trend in average US maize yield and reported projected trajectories based on compound rates of yield gain. Historical trend (1965–2011, n=47 years of yield data) is indicated with the solid line and yellow data points, and associated linear-regression equation, coefficient of determination (r^2) and Student's t-test P-value are shown. Trajectories reported in publications that evaluated future food production potential based on these projected yield trajectories are indicated with the dashed lines. Numbers associated with each trajectory indicate the reference in which this exponential rate was used. The trajectory based on extrapolation of the 1965–2011 linear regression is also shown.

To account for the expected deceleration in crop yields as they approach their yield potential ceiling, several recent projections of food security have considered a declining compound rate over time, which aims to mimic the shape of an asymptotic trend[25,26,27]. Another recent study suggests that yields have stagnated in a large portion of the world's crop production area, and that current yield trends are not sufficient to meet the demand on existing farmland[28]. However, this study is based on yield trend estimates and their pattern over time using gridded, historical crop yield data derived from average yields reported at coarser levels of spatial resolution. In addition, many of the highlighted yield trends came from regions with negligible crop production (e.g., maize in Moldova, soybean in Congo, rice in Greece). The statistical analysis was limited to only three models, including the cubic model to detect yield plateaus—which has no biophysical justification—and model selection was not robust because regression analysis assumptions were not tested. Only one recent study followed a robust, statistically sound method to describe yield trends, but that study was limited to one crop (wheat) and only two statistical models were evaluated[29].

Exaggerated yield gain rates, sometimes justified by unverifiable 'record' crop yields, have been used to support optimistic projections of grain production[12,13,14,15,16,17,18] (Fig. 2). For instance, a widely publicized goal for average US maize yield of 20 Mg ha^{-1} by 2030 would require an annual yield gain of 506 kg ha^{-1} per year (or 3.6% per year compounded)[17]. This rate is four times greater than the rate of increase in US maize yield from 1965 to 2011 (114 kg ha^{-1} per year). Another recent study predicted a 2.8% per year rate of gain in US maize yield (equivalent to 460 kg ha^{-1} per year as a linear rate) under the assumption that farmers would respond to the incentive of higher grain prices caused by increased global demand, partly as a result of the US bioethanol mandates[18].

Improved Analytical Framework for Estimating Historical Yield Trends

Trajectories in national average yields are driven by changes in crop management practices, crop genetic improvement through conventional breeding and genetic engineering, climate and interactions among these factors, under influence of surrounding social, economic and political environments[30]. Depending on country and time period, the mathematical form of historical yield trend for a specific crop in a given country can be linear, exponential, parabolic, linear plateau or flat. Statistical trend analysis is the only objective method to determine the mathematical model that best fits observed data. And although food security projections and yield gain rates of major cereal crops are extremely sensitive to mathematical form, rigorous statistical trend analysis to identify the most appropriate statistical model (hereafter called 'best-fit model') have not been used to inform previous projections of global food security as published in the papers cited herein.

Here we develop a framework for statistical trend analysis of historical crop yields and use it to analyse yield trends of the major cereals since the start of the green revolution in 36 countries and regions, which together account for 84, 56 and 71% of global rice, wheat and maize production, respectively. These three crops are the focus of our analysis because rice, wheat and maize together account for ~85% of global cereal production and contribute a majority of human calories eaten directly as staple foods or indirectly through consumption of livestock fed with grain. Six statistical models, widely used in the literature for describing time series trends in crop yields, provided the basis for a comprehensive analysis of historical crop yield trends (Fig. 3). Testing all six models based on statistically sound criteria provided an objective approach for identifying the most appropriate shape of historical yield trend (Fig. 4). From this analysis, we estimate the proportion of global grain production from countries and regions in which yields are flat, rising, declining or plateauing, consider the most plausible explanations for cases of accelerating or plateauing yields and posit implications of these findings for studies of future food production capacity and land area required to produce it.

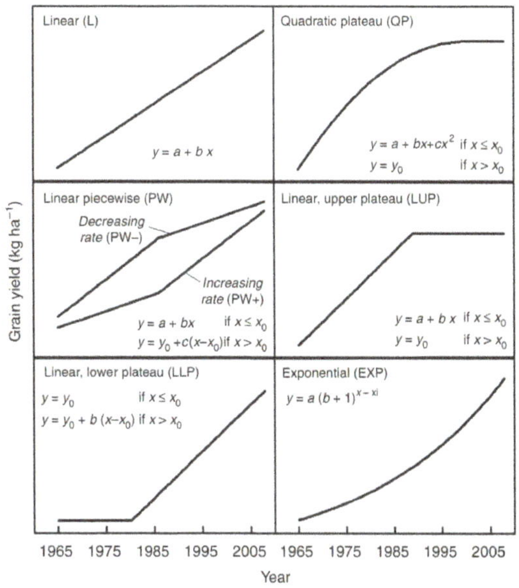

Figure 3: Six statistical models evaluated for their performance to fit observed crop yield trends. For piecewise and linear-plateau models, x_0 is the breakpoint year and y_0 is the yield value of the upper (LUP) or lower plateau (LLP), or in the breakpoint year of the PW model.

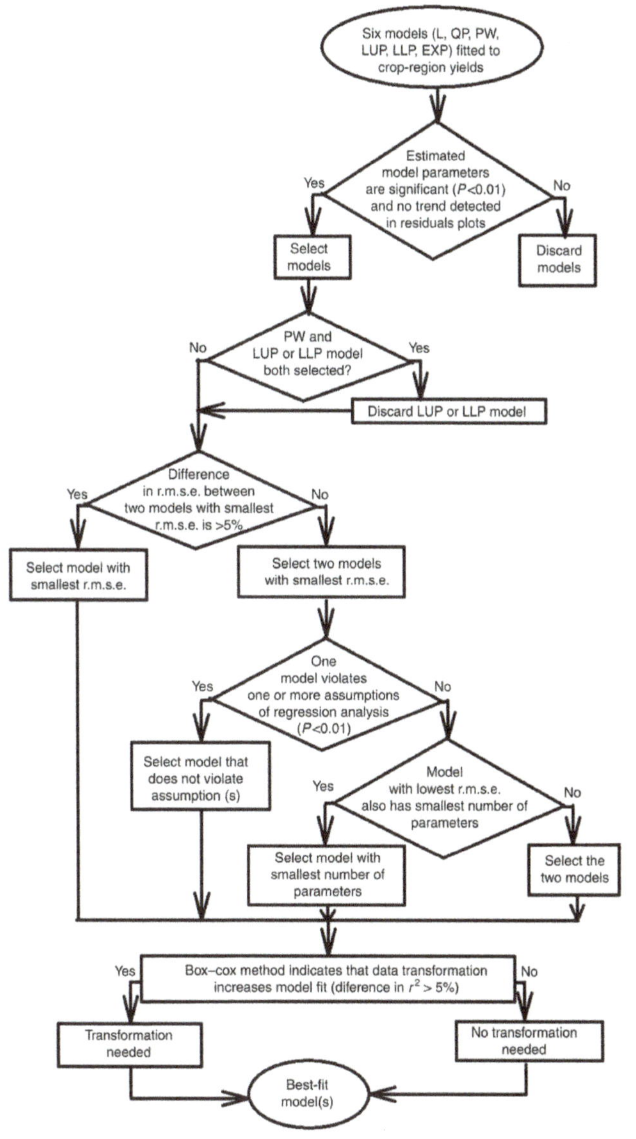

Figure 4: Decision tree used to identify best-fit models for each crop–country or –region case. Note the description of statistical tests and detailed explanation of the cases exhibiting significant serial correlation, as detected with the Durbin–Watson test and the residual plots, in the Methods section. r.m.s.e., root mean square error; r², coefficient of determination.

Rate of Increase in Cereal Crop Yield Is Generally Linear

Best-fit trends for selected crop–country and –region cases are shown in Fig. 5. Linear models, with or without a discontinuous breakpoint, adequately described all of the yield trends (Table 1 and Fig. 5). In 10 of 36 cases, a linear rate of gain throughout the entire time series provided the best fit. Dividing the rate of gain by the trendline yield provides the relative rate of gain as a percentage of the estimated yield for a given year. For example, average rice yield in India was ~1,500 kg ha^{-1} in 1970, whereas the rate of gain was 45 kg ha^{-1} per year, which gives a relative rate of gain of 2.9%. By 2010, average rice yield had risen to ~3,300 kg ha^{-1}, which means the relative rate of gain had fallen to 1.3%. Thus, in 28% of the cases evaluated, rate of increase remains constant such that relative rates of gain decline throughout the entire time series.

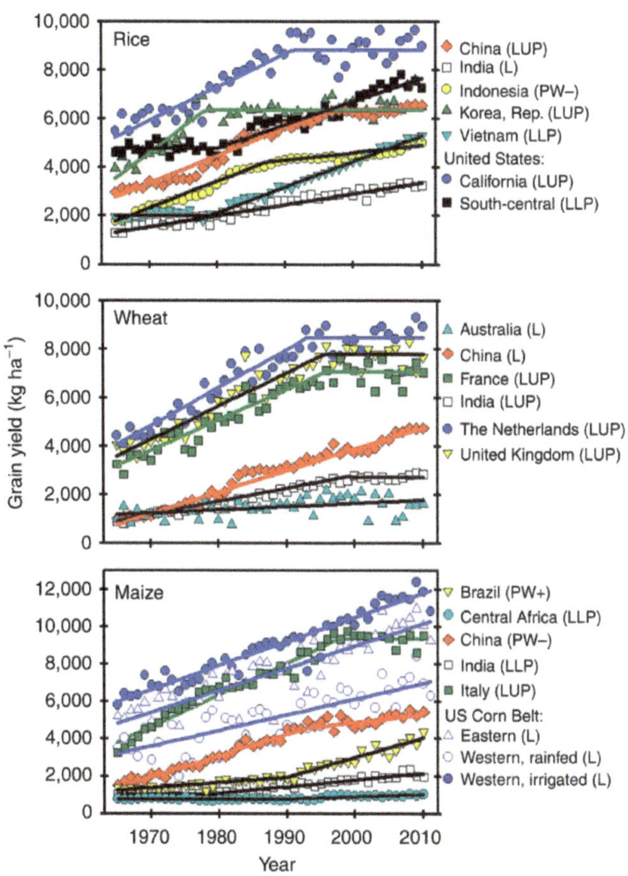

Figure 5: Trends in grain yield of the three major cereal crops for selected regions since the start of the green revolution in the 1960s. Fitted model for each crop–region case is indicated in parenthesis. L, linear; QP, quadratic plateau; PW, piecewise with (+) increasing or (–) decreasing rate after breakpoint year; LUP or LLP, linear with upper or lower plateau; EXP, compound exponential.

Table 1: Best-fit models identified for each crop and country or region.

Country/region	**Crops**		
	Rice	**Wheat**	**Maize**
Africa			
East			LUP*
Central			LLP
West			LLP
America			
Argentina		L, EXP	PW+, EXP
Brazil			PW+
Canada		LLP	
United States			
California	LUP		
South-central	LLP		
South Great Plains		LUP	
Central Great Plains		L, EXP	
North Great Plains		LLP, EXP	
Eastern corn belt			L, EXP
Western Corn Belt			
Irrigated			L, EXP
Rainfed			L, EXP
Asia			
Bangladesh	PW+		
China	LUP	L	PW−
India	L	LUP	LLP
Indonesia	PW−		
Japan	L, LUP		
Korea	LUP *		
Philippines	L		
Thailand	PW+		
Vietnam	LLP		
Europe			
Denmark		LUP	
France		LUP	LUP
Germany		LUP	
Italy			LUP
The Netherlands		LUP	
United Kingdom		LUP	
Oceania			
Australia		L, EXP	

L, linear (no plateau); QP, quadratic plateau; PW, piecewise with (+) increasing or (−) decreasing rate after breakpoint year; LUP or LLP, linear with upper or lower plateau; EXP, compound exponential.

*Box–Cox procedure indicates that reciprocal transformation improves model fit.

Decreasing absolute rates of gain were observed for rice in Indonesia and maize in China, where yield gain during recent decades (31 and 42 kg ha^{-1} per year, respectively) are much slower than rates at the beginning of the time series (110 and 115 kg ha^{-1} per year, respectively). In contrast, abrupt, transitory increases in both absolute and relative rates of yield gain were observed in 11 cases. These cases corresponded to crops and countries where little yield increase occurred at the beginning of the time series and average yield remained low for a long time period, such as rice in Vietnam and maize in Brazil (Fig. 5). For rice in Vietnam, socioeconomic stabilization following the Vietnam War, introduction of modern indica rice varieties and increased use of fertilizer supported an abrupt upward yield trend beginning in 1979. In Brazil, the development of improved management practices for acid, infertile soils in the Cerrado and the introduction of hybrid seed led to a rapid rise in maize yields after 1990. However, even in these cases, subsequent yield gain after the abrupt upturn was remarkably linear and therefore relative rates of gain decline.

Strong Evidence of Yield Plateaus in Some of the World's Most Intensive Cropping Systems

A major concern is the observation that yields in some major cereal-producing regions have not increased for long periods of time following an earlier period of steady linear increase, hereafter called 'upper yield plateaus' (Table 1 and Fig. 5). Upper yield plateaus were observed in 14 of the 36 cases. Moreover, there was no case of a return to the previous rising yield trend after a statistically significant upper yield plateau occurred. Taken together, cases with statistically significant upper yield plateaus represent 33% of global rice and 27% of global wheat production (Table 2). Given the importance of these intensive systems to global food supply, identification of underpinning causes for yield plateaus in these systems is fundamental for estimating future food production capacity. A hypothesis that can explain the occurrence of yield plateaus is that average farm yields approach a biophysical yield ceiling for the crop in question, which is determined by its yield potential in the regions where the crop is produced[19,20,21]. This seems to be the case in high-yield systems for rice in East Asia (China, Republic of Korea and Japan), wheat in Northwest Europe (United Kingdom, France, Germany, The Netherlands, Denmark) and India, and maize in South Europe (Italy and France). Other factors may also contribute to observed yield plateaus, including cyclical weather patterns, land degradation, shift in the location of production area to regions with poorer soils and climate, policies on the use of fertilizers and pesticides, and insufficient or poorly oriented investment in agricultural research and development (R&D)[27,29,31].

Table 2: Percentage of global crop production under different crop yield trajectories.

Crop species	% of global production*	Increasing rate (%)†	Constant rate (%)	Decreasing rate (%)	Upper yield plateaus (%)
Rice	84	19	23	9	33
Wheat	56	5	24	0	27
Maize	71	13	33	20	5
Total rice, wheat and maize	70	12	27	10	21

*Percentage of global production of rice, wheat and maize, and the three crops together accounted for by the 36 crop–country cases analysed in this study.

†The 36 crop–country cases were grouped according to the pattern of change in rates of crop yield gains during the 1965–2010 interval and total production was calculated for each group and for each crop.

At issue is how many years of plateau are needed before the trend becomes statistically significant. Analysis of cases in this study indicates that 4–18 years of 'flat' yields are needed to detect a statistically significant yield plateau, depending on the degree of year-to-year variability in yield along the entire time series. In favourable irrigated or rainfed environments where yield variability due to weather and water supply is small, ~8 years are needed to identify a significant yield plateau (for example, irrigated rice in China and California, rainfed wheat in Northwestern Europe and maize in South Europe). Longer time series (13–18 years) are needed to identify yield plateaus in harsher rainfed cropping systems due to the high year-to-year variability in yield, such as for wheat in the southern US Great Plains where it appears there has been little improvement in yields of recent varietal releases[22].

Length of the ascending linear phase is also important. For example, a recent study of wheat yield trends in three regions in The Netherlands reports that average farm yield continues to rise at a linear rate[32], in contrast to our analysis (Table 1 and Fig. 5). The reason for this different conclusion is that the Dutch study uses a shorter time series, which begins in 1979, and thus has only 15 years of ascending linear phase, based on the breakpoint year (1993) found in the present study. In our analysis, the time series begins in 1965 and thus has 29 years to establish the linear phase.

In contrast to plateaus in high-yield systems, yield stagnation at low-yield levels or very low rates of yield gain are observed in countries or regions where farmers lack access to agricultural inputs, infrastructure, capital, markets and extension services, such as for maize in Africa. These regions exhibit the highest potential for intensification because substantial increases in crop production can be achieved without expansion of current cropland area, through modest increases in current yield gain rates. For example, Africa accounts for 15% of global maize harvested area but produces <5% of current global maize. With current rates of yield gains of 0, 13 and 29 kg ha^{-1} per year (East, Central and West Africa, respectively), total maize production in Sub-Saharan Africa will increase only by 9% in the next 10 years which, in turn, will put pressure on expanding cropland area as observed during the last decade (Fig. 1). In contrast, if current yield gains could be increased to a modest rate of 80 kg ha^{-1} per year, similar to the rates observed for maize in other harsh environments such as the western US Corn Belt, total maize production in Sub-Saharan Africa would increase by 53% in the next 10 years, helping to decrease pressures to expand cropland area and food imports.

No Support Exists for Use of Compound, Exponential Rates of Yield Increase to Project Yield Trends

Exponential models underperformed compared with linear models and exhibited large biases in the distribution of yield residuals (Figs 6, 7, and 8). Even in the few cases where exponential models performed well at describing yield trends, their residual error was comparable to that of the best-fit linear model (for example, maize and wheat yields in USA). Hence, there are three reasons why the use of exponential models to describe or project yield trajectories is not appropriate. First, where yields are increasing, absolute rates of yield increase are linear. Second, there is no case in which relative rates of yield gain have consistently increased or remained constant over time. Third, there are a number of cases where yield plateaus or a discontinuous break to slower rates of yield gain suggest the possibility of a biophysical 'ceiling' yield. In contrast, much of the published literature on yield projections is based on exponential rates of yield increase (Fig. 2), which, as shown here, do not occur in the real world (Fig. 5). For example, a previous published study projected an average 1.4% per year compound rate of yield increase for irrigated rice in California from 2010 to 2050 (ref. 14) even when the average actual yield of California rice has not increased since 1990 (Fig. 5). Predicted irrigated rice yield in California by 2050, based on the above compound rate, is ~15,000 kg ha^{-1}, which is >70% above the current average yield of ~9,000 kg ha^{-1}.

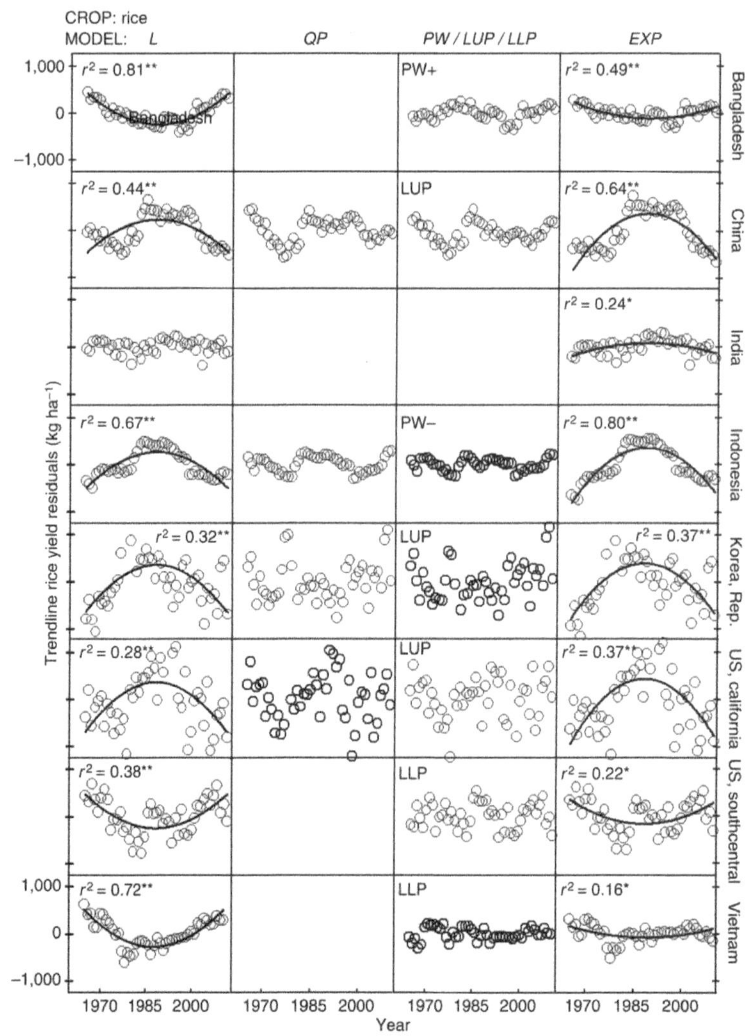

Figure 6: Trendline residuals for rice yield plotted against year for selected regions. Selected regions are indicated in the right axis labels. Only one residual plot is shown for either the piecewise (PW+, PW−) or linear plateau with upper/lower plateau (LUP/LLP) because these models are mutually exclusive. The *y* axis scale is the same across all panels and is shown only in the upper and bottom panels. Asterisks indicate significance at *F*-test *P<0.01 or **P<0.001 of the fitted regression line to the residuals over time (n=46 years of yield data, except for USA (n=47)). L, linear; QP, quadratic plateau; PW, piecewise with (+) increasing or (−) decreasing rate after breakpoint year; LUP/LLP, linear with upper/lower plateau; EXP, compound exponential.

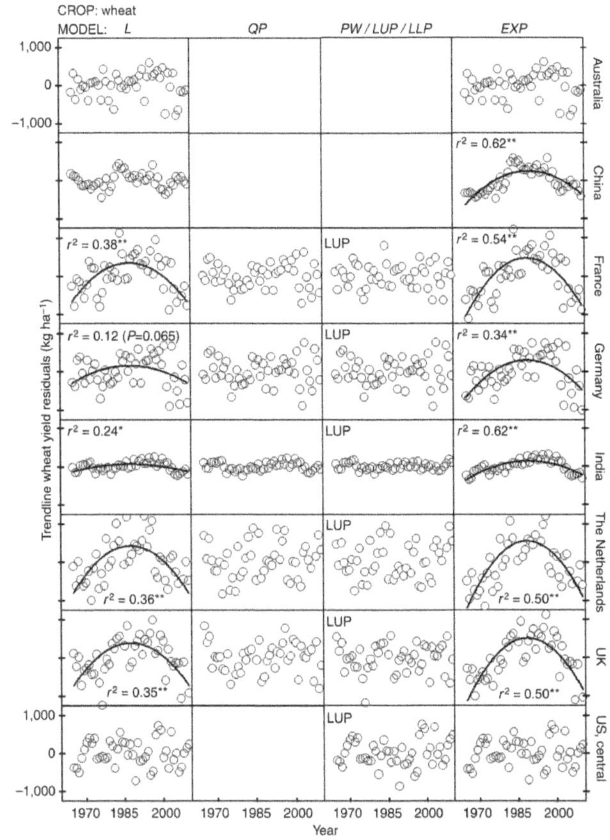

Figure 7: Trendline residuals for wheat yield plotted against year for selected regions. Selected regions are indicated in the right axis labels. Only one residual plot is shown for either the piecewise (PW+, PW−) or linear plateau with upper/lower plateau (LUP/LLP) because these models are mutually exclusive. The *y* axis scale is the same across all panels and is shown only in the upper and bottom panels. Asterisks indicate significance at *F*-test $*P<0.01$ or $**P<0.001$ of the fitted regression line to the residuals over time (n=46 years of yield data, except for USA (n=47)). L, linear; QP, quadratic plateau; PW, piecewise with (+) increasing or (−) decreasing rate after breakpoint year; LUP/LLP, linear with upper/lower plateau; EXP, compound exponential.

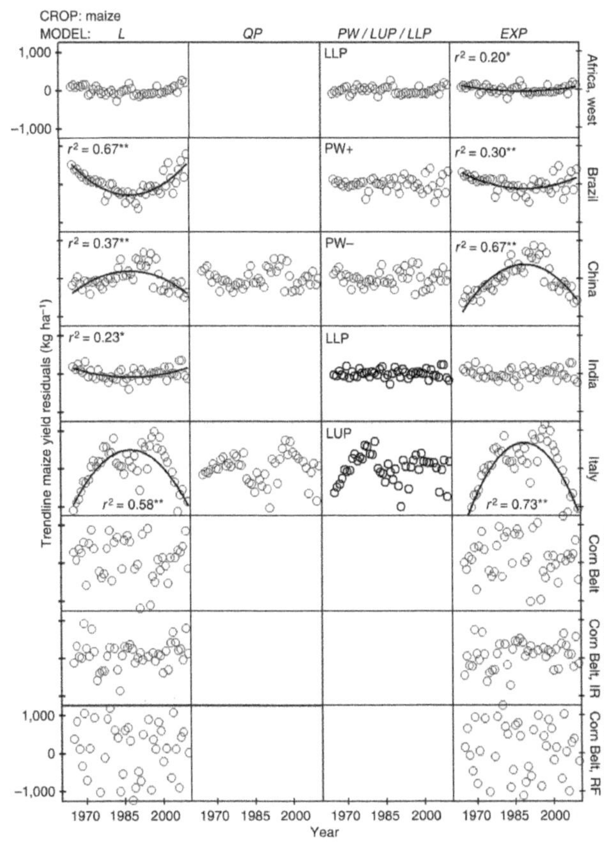

Figure 8: Trendline residuals for maize yield plotted against year for selected regions. Selected regions are indicated in the right axis labels. Only one residual plot is shown for either the piecewise (PW+, PW−) or linear plateau with upper/lower plateau (LUP/LLP) because these

models are mutually exclusive. The *y* axis scale is the same across all panels and is shown only in the upper and bottom panels. Asterisks indicate significance at *F*-test $*P<0.01$ or $**P<0.001$ of the fitted regression line to the residuals over time (n=46 years of yield data, except for USA (n=47)). L, linear; QP, quadratic plateau; PW, piecewise with (+) increasing or (−) decreasing rate after breakpoint year; LUP/LLP, linear with upper/lower plateau; EXP, compound exponential; R, rainfed; I, irrigated.

DISCUSSION

Results from our analysis suggest that projections of crop yield trajectories based on extension of historical trends of the past five decades should be viewed with caution because these past trends were driven by rapid adoption of green revolution technologies that were largely one-time innovations. These include the development of semi-dwarf wheat and rice varieties, first widespread use of commercial fertilizers and pesticides, and large investments to expand irrigation infrastructure. A concern is that despite the increase in investment in agricultural R&D and education during this period[27,33], the relative rate of yield gain for the major food crops has decreased over time together with evidence of upper yield plateaus in some of the most productive domains. For example, investment in R&D in agriculture in China has increased threefold from 1981 to 2000 (ref. 33). However, rates of increase in crop yields in China have remained constant in wheat, decreased by 64% in maize as a relative rate and are negligible in rice (Fig. 5). Likewise, despite a 58% increase in investment in agricultural R&D in the United States from 1981 to 2000 (sum of public and private sectors), the rate of maize yield gain has remained strongly linear (Fig. 5), implying that the marginal yield increase per unit of research investment has decreased substantially over time and highlighting the need for increasing the level of investment on agricultural R&D to sustain current and future increases in crop yields[33].

These findings are consistent with the notion that as farmers' yields move up towards the yield potential threshold, it becomes more difficult to sustain further yield gain because it requires fine tuning of many different facets of management in the production system[34]. Such fine tuning is often difficult to achieve in farmer's fields, and the associated marginal costs, labour requirements, risks and environmental impacts may outweigh the benefits[34]. Therefore, substantial increases in future grain production will require significant increases in average crop yields in countries where current yield gaps are large as others have noted[35]. Paradoxically, many of the countries exhibiting the largest gaps have the poorest access to technology, infrastructure and capital required for agricultural development[36].

To summarize, we found widespread deceleration in the relative rate of increase of average yields of the major cereal crops during the 1990–2010 period in countries with greatest production of these crops, and strong evidence of yield plateaus or an abrupt drop in rate of yield gain in 44% of the cases (Table 1), which, together, account for 31% of total global rice, wheat and maize production (Table 2). The results strongly support the proposition that estimates of future cereal production should be derived from yield projections based on linear models, with breakpoints and plateaus to reflect the linear nature of the crop yield gains in an ascending phase during which modern crop management practices are adopted, and the existence of a biophysical upper limit for grain yield best estimated by robust crop simulation models[3]. Approaches that rely on compound rates of yield increase or constant linear rates with no upper limit to yield growth are not supported by the analysis of historical yield trends and current understanding of crop physiology, and they are likely to overestimate future increases in crop yields by a large margin (Fig. 2). In turn, overestimating trajectories in crop yields leads to estimates of land requirements for crop production that are too low and diminish capacity for effective strategic planning and research prioritization to ensure future food security and conservation of natural resources.

METHODS

Overview

The time period of 1965–2010 over which yield trends were evaluated is the most relevant for use in estimating historical rates of yield gain to inform projections of future food security because it represents a contemporary period when public and private sectors in many developed and developing countries have invested heavily in application of modern genetics, agronomy and supporting basic sciences and information technologies to develop and gain adoption of technological advances in crop production.

Evaluated mathematical models were as follows: linear (L), quadratic with upper plateau (QP), linear piecewise (PW), linear with upper (LUP) or lower plateau (LLP), and compound-rate exponential (EXP) (Fig. 3). Except for the United States and Africa, where yield trends were analysed at the subnational and regional levels, respectively, the present study focuses on average yield at the national level because this is the spatial scale at which most projections of food security are made. Criteria for the selection of the best-fit trend are based on the assumption that the most appropriate statistical model is one that can describe the observed time trend in yield with the lowest error and minimal bias in distribution of residuals, and which uses the smallest possible number of estimated parameters while meeting the assumptions of normality, independence and homogenous variance in regression analysis[37] (Fig. 4).

Analysis of Trends in Cropland Harvested Area

Long-term (1965–2011) data in crop-harvested area were retrieved from FAOSTAT and used to analyse changes in cropland area of staple crops (cereal, oil, sugar, fibre, pulses, tuber and root crops) and three major cereal crops (rice, wheat and maize; FAOSTAT Database–Agricultural Production (http://faostat.fao.org/). A three-phase linear model was fitted to the observed trends using the PROC NONLIN procedure in SAS Version 9.1 (SAS Institute, 9.1 Foundation for 64-bit Microsoft® Windows®, SAS Institute, Cary, NC):

$$\begin{aligned} y &= a + bx && \text{if } x < x_1 \\ y &= (a + bx_1) + c(x - x_1) && \text{if } x_1 \le x \le x_2 \\ y &= (a + bx_1) + c(x_2 - x_1) + d(x - x_2) && \text{if } x > x_2 \end{aligned} \tag{1}$$

where y is crop-harvested area (Mha), x is year and $x_{1,2}$ are the breakpoint years. The fitted trilinear models have coefficient of determination (r^2) of 0.97 and 0.93 (root mean square error (r.m.s.e.)=9.5 and 6.9 Mha) for area of staple crops and three major cereal crops, respectively. A total of 47 years of yield data were used in the regression analysis (1965–2011 time period). All estimated parameters were significant (Student's *t*-test, $P<0.01$), except for parameter *c* in the fitted trend for cropland area of the three major cereal crops (Student's *t*-test, $P=0.20$). An initial period of increase in harvested area occurred until 1980, followed by a period of little or no increase in harvested area of staple and major cereal crops, respectively, which lasted until early 2000s (Fig. 1). This period was followed by an unprecedented rate of expansion in harvested cropland area during the last decade, as indicated by the statistically significant higher rates of increase in harvested area of staple and major cereal crops during the last decade (parameter *d* in (equation 1)) compared with earlier rates of increase during the first two decades of the green revolution (parameters *b* and *c*in (equation 1); Student's *t*-test; $P<0.01$).

Change in crop-harvested area during the last 10 years (2002–2011) was calculated for staple crops and, separately, for four crops (rice, wheat, maize and soybean) that accounted, altogether, for 83% of observed increase in staple crop area:

$$\text{Net change} = \text{HA}_{2010-2011} - \text{HA}_{2002-2003} \quad (2)$$

$$\text{Relative change} = \left(\text{Net change}/\text{HA}_{2002-2003}\right) * 100 \quad (3)$$

where net and relative changes are the absolute (Mha) and relative (%) differences in harvested area between two periods (Mha), and $\text{HA}_{2002\text{–}2003}$ and $\text{HA}_{2010\text{–}2011}$ are the calculated 2-year average harvested area for the 2002–2003 and 2010–2011 intervals, respectively. A net change of 85 Mha was observed for staple crops between 2002–2003 and 2010–2011 intervals, indicating that an 8% increase in global cropland area has occurred in only 10 years. This change was mostly accounted for by increased harvested area of maize and rice in Africa (+9 Mha), soybean in South America (+15 Mha) and maize, rice and soybean in South and Southeastern Asia (+16 Mha). In addition, remarkable was the little relative change of harvested crop areas (negative in some cases) observed for North and South Africa, West Asia, Europe and North America.

Published Projections in Grain Yields of Cereal Crop Yields

Previously published projections on yield trajectories, based on exponential compound rates, were retrieved from the literature[12,13,14,15,16,17,18]. In many of these projections, countries were grouped into categories according to their geographic location or degree of economic development, and a crop-specific exponential yield gain rate was assumed for each category. Other studies simplified this approach by assuming a single, worldwide exponential yield gain rate for each crop. A common feature of previous studies is the lack of recognition of a biophysical limit on crop yields determined by solar radiation, temperature and water supply from both rainfall and irrigation. To illustrate the discrepancy between reported projections and observed historical trends, reported projected trajectories for the US average maize yield were plotted and compared against the projected yields based on the historical (1965–2011) yield trend (Fig. 2).

Data on Grain Yield Data for Cereal Crops

Average annual yield data from 1965 to 2010 were retrieved for rice, wheat and maize in selected countries and regions, resulting in a total of 36 (crop–country or –region) cases that include a wide range of production environments and yield levels (FAOSTAT Database–Agricultural Production (http://faostat.fao.org/; National Agricultural Statistics Service—Crops US state and county databases (http://www.nass.usda.gov/index.asp). Available wheat and maize yield data in the USA also includes 2011 National Agricultural Statistics Service—Crops US state and county databases (http://www.nass.usda.gov/index.asp). Hence, a total of 46 years of yield data were used for the yield trend analysis, except for the United States where a total of 47 years of yield data were used. In the case of the United States, trends were analysed for major producing regions: rice in California and south-central, wheat in the southern, central and northern Great Plains, and maize in the eastern and western Corn Belt. These regions accounted for 77, 47 and 85% of total US production of rice, wheat and maize, respectively. Separate analyses were performed for rainfed and irrigated maize yields in the western US Corn Belt, where both irrigated and rainfed production are important (54 and 46% of total maize production in western US Corn Belt, respectively). The analysis of maize yield trends in Africa was conducted for three regions: East, Central and West Africa. The average yield for a given region in the United States or Africa was calculated as the sum of total crop production in all the states (the United States) or countries (Africa) within the region divided by total harvested area.

Statistical Analysis of Yield Trends

Data on average annual grain yield were plotted against year for each crop–region case. Six statistical models, extensively used in the literature to describe yield trajectories, were evaluated for their performance of fitting trends in grain yields since the onset of the green revolution (Fig. 3):

$$\text{Linear(L)} \quad y = a + bx \quad (4)$$

$$\text{Quadratic with plateau (QP)} \quad \begin{array}{ll} y = a + bx + cx^2 & \text{if } x < x_0 \\ y = y_0 & \text{if } x \geq x_0 \\ x_0 = 0.5b/c & \end{array} \quad (5)$$

$$\text{Linear piecewise (PW)} \quad \begin{array}{ll} y = a + bx & \text{if } x < x_0 \\ y = (a + bx_0) + c(x - x_0) & \text{if } x \geq x_0 \end{array} \quad (6)$$

$$\text{Linear with upper plateau (LUP)} \quad \begin{array}{ll} y = a + bx & \text{if } x \leq x_0 \\ y = y_0 & \text{if } x > x_0 \end{array} \quad (7)$$

$$\text{Linear with lower plateau (LLP)} \quad \begin{array}{ll} y = y_0 & \text{if } x \leq x_0 \\ y = y_0 + b(x - x_0) & \text{if } x > x_0 \end{array} \quad (8)$$

$$\text{Compound exponential (EXP)} \quad y = a(b+1)^{x - x_i} \quad (9)$$

where *y* is grain yield (kg ha^{-1}), *x* is year, x_i is the initial year of the time series (1965 in the present study) and y_0 is the yield plateau level (kg ha^{-1}).

SAS Version 9.1 programmes and procedures were used for all statistical analyses. The six models were fitted to observed yields using the PROC MIXED, PROC REG and PROC NONLIN procedures. Estimates of model coefficients (and associated confidence intervals), coefficient of determination (r^2) and the r.m.s.e. were calculated for each crop–region model combination. In the case of nonlinear models, r^2 was calculated as the square correlation between predicted and observed values[38]. Violations of assumptions of regression analysis were identified by performing Shapiro–Wilks test for normality, Levene's test for variance homogeneity and the Durbin–Watson's (D) test for serial correlation[39]. Residuals were plotted against year and the significance of linear- and quadratic models fitted to these plots was evaluated to detect potential biases along the time series (Figs 6, 7, and 8). Finally, the Box–Cox method was used to identify suitable power transformations that can improve models performance[40].

The following criteria were used to identify the best-fit model for each crop–region yield trend: highly significant (Student's *t*-test; $P<0.01$) model parameters, highest r^2 and lowest r.m.s.e. compared with other models, and independent, normally distributed yield residuals with homogenous variance and unbiased distribution when residuals were plotted against year (Fig. 4). An additional criteria was that none of the applied data transformations improve model fit by >5% based on comparison of calculated r^2 among models based on transformed and untransformed data[37] (Fig. 4). In the process of selecting the best-fit model, the PW and LUP or LLP models were mutually exclusive, that is, the LUP or LLP models were discarded if estimated parameters of the PW model were all significant and no bias was detected in the distribution of yield residuals. In 22 out of the 36 cases, one statistical model clearly outperformed the others by a difference in r.m.s.e. ≥5% and this was the chosen best-fit model. When differences in r.m.s.e. among two or more models were <5% (14 out of 36 cases), the two models with lowest r.m.s.e. were selected as best-fit models, unless there was a justification to discard one of them, for example, failure to meet assumptions of regression analysis (EXP in Canada) or when the model with higher r.m.s.e. also had the larger number of parameters (QP for wheat in the Netherlands and maize in Italy).

Selected best-fit models exhibited highly significant coefficients (Student's *t*-test; $P<0.01$) and higher goodness of fit compared with other fitted models (higher r^2 and smaller r.m.s.e.). Out of 45 selected best-fit models, only 1 and 4 did not meet assumptions of variance homogeneity (LLP model for rice in Vietnam) and normality (L and EXP models for maize in the United States), respectively (Levene and Shapiro–Wilks tests; $P<0.01$). In contrast, ten selected best-fit models exhibited significant positive serial autocorrelation: rice in Bangladesh, China, Indonesia, Republic of Korea, Philippines, Vietnam, wheat in China and maize in central and west Africa and Italy (D-test; $P<0.01$). Residual plots of the best-fit models did not exhibit any obvious trends over time, except for four cases with severe positive serial autocorrelation: rice and wheat in China and rice in Indonesia and Philippines ($0.45<D<0.80$), which are possibly related to vigorous public sector varietal improvement and agronomic research programmes that promote rapid adoption of improved rice varieties and associated fertilizer and pest management practices and/or the frequency of the yield survey (Figs 6, 7, and 8). Serial correlation affects the estimated variances but does not affect the value of the estimated model coefficients, which results in the estimators looking more accurate than they actually are[38]. To test how serial correlation may have affected the significance of the estimators, we assumed the estimated variance to be 60% of the true variance, which is expected for a first-order serial correlation coefficient (AR(1)) of 0.5 (ref. 39), similar to the estimated AR(1) of 0.6 found for the above four crop–country cases. Results indicate that the parameter estimates of the best-fit models were still highly significant despite serial correlation (Student's *t*-test; $P<0.01$). EXP models exhibited a remarkable bias on their residuals plots in 75% of the 36 cases (with significant positive autocorrelation in 58% of the cases) and did not meet assumptions of regression analysis in 31% of the cases (Figs 6, 7, and 8).

Other evidence of the robustness of selected best-fit models was that their fit, based on the original untransformed data, was not improved after data transformation using the Box–Cox method[40], except for rice in Republic of Korea and maize in east Africa. In both cases, model fitness increased slightly after applying a reciprocal transformation to the yield data ($r^2=0.90$ versus 0.84 (Republic of Korea) and 0.57 versus 0.49 (East Africa), with and without data transformation, respectively). None of the non-selected models fit to the transformed data outperformed the fit of the selected models shown in Table 1. And although the presence of abnormally low- or high-yield years during the last years of the time series can potentially hinter the identification of yield plateaus, this is not a large concern in the present study because all cases that exhibited an upper yield plateau have ≥10 years of yield data after the breakpoint year, except for maize in France with 7 years. Perhaps, more important, the 99% confidence interval of the breakpoint year (x_0) was within the 1965–2010 interval in all cases. In addition, visual inspection of the crop–country trends with statistically significant yield plateaus did not have unusually high or low yields around the breakpoint year. Therefore, it is unlikely that the upper yield plateaus identified in the present study are the consequence of a few yield outliers at the end of the time series. The presence of autocorrelation and changes in variance over time can also hinder the identification of yield plateaus[29]. However, out of the 21 selected LUP or LLP models in the present study, variance was not homogenous in only one case (rice in Vietnam) and autocorrelation was significant in six cases (rice in China, Republic of Korea and Vietnam, and maize in central and west Africa, and Italy). And even in the six cases with significant autocorrelation or non-homogenous variance, visual inspection of the residuals clearly indicated that LUP or LLP models outperform linear models (Figs 6, 7, and 8). Although the yield residual versus time relationship of the linear model exhibited a strong trend, there was no detectable pattern in the yield residuals of the LUP or LLP models, with the only exception of maize in West Africa where there was no detectable trend in both linear and LLP models.

Yield trends from cases in which there was evidence of yield plateaus were re-analysed to determine the number of years after the breakpoint year (x_0) that were needed to identify a yield plateau (hereafter called x_n). The LUP model was successively fitted to yield trends in which data-years after x_0 were removed and then added one by one until estimated x_0 became statistically significant (Student's *t*-test; $P<0.01$). Values of x_n varied across cases, depending on the magnitude of year-to-year variation in yields along the entire trend line. In fact, there was a negative relationship between x_n and r^2 of the fitted LUP model to each crop–country case (y=20−14.9 x; r^2=0.63; Student's t-test; P<0.01). A conservative value of x_n=7 years can be taken as the minimum number of years needed, after the breakpoint year, to detect a statistically significant yield plateau in production systems that exhibited high r^2 (>0.85) such as rice in China and California, wheat in northwestern Europe and maize in Italy and France. In production systems with larger variability in yield along the yield trends, a greater number of years is required, ranging from 9 years in Japan (r^2=0.63) to 17 years in East Africa (r^2=0.49). The total number of years with 'flat' yields required to detect a statistically significant yield plateau, including the breakpoint year, was calculated as x_n+1 and this is the value reported in the Results section.

Calculation of Absolute and Relative Yield Gain Rates

Absolute yield gain rates (kg ha^{-1} per year) were calculated from the first derivative of the best-fit model, for each crop–region case, at three points in the time interval: 1970, 1990 and 2010. For those cases in which there were two best-fit models, the model with lowest r.m.s.e. was chosen to describe the yield trend. Despite the slightly better performance of the EXP model, the L model was chosen to describe trends in maize yield in the United States because fit of the EXP and L models was very similar (difference in r.m.s.e. ≤2%), there was no evidence of changes in the absolute yield gain rate between 1965 to 2011 as tested using piecewise with one, two or three breakpoint years and the EXP model underperformed compared with other models in all the other cases. Moreover, using an exponential model would imply predicted average maize yields by 2030 that are 35 and 37% higher than the 2010 yield for maize in the eastern and (irrigated) western US Corn Belt, which will require average (2010–2030) yield gain rates of 180 and 220 kg ha^{-1} per year. These rates are 50 and 70% higher than the observed yield gain rates during the 1965–2010 interval. This is unlikely to occur given the slowdown in yield gain rates observed in other intensive cropping systems and lack of increase observed for irrigated maize in contest winners and farmers' fields in the western US Corn Belt[3,19]. In fact, although no yield plateau was detected for irrigated maize in the western US Corn Belt for the 1965–2011 interval, the linear regression for the last 10-year period (2002–2011) has a slope indistinguishable from zero (Student's t-test; P=0.26, n=10).

Relative rates of yield gain (% per year) were calculated for 1970, 1990 and 2010 as the ratio between the absolute yield gain rate and the yield-trend-predicted yield in the particular year, expressed as percentage. Exponential compound rates (% per year) were also derived for each crop–region case from the parameter b in (equation 9), expressed as percentage. The objective was to highlight the limitations of using compound rates of yield gain to predict future yield trajectories: out of a total of 29 cases that have compound rate >1% per year, there was evidence of yield plateaus or decreasing yield gain rates in 41% of the cases, decreasing relative rates of yield gain during the entire time series in 66% of the cases and decrease in the relative rate of yield gain in the time interval between 1990 and 2010 in all cases.

REFERENCES

1. Royal Society of London. *Reaping the Benefits: Science and the Sustainable Intensification of Global Agriculture* Royal Society: London, (2009).
2. FAO. *The State of The World's Land and Water Resources for Food and Agriculture (SOLAW)–Managing Systems at Risk* FAO: Rome and Earthscan, London, (2011).
3. Van Ittersum, M. K. *et al.* Yield gap analysis with local to global relevance—a review. *Field Crops Res.* 143, 4–17 (2013).
4. Burney, J., Davis, S. J. & Lobell, D. B. Greenhouse gas mitigation by agricultural intensification. *Proc. Natl Acad. Sci. USA* 107, 12052–12057 (2010).
5. Vermeulen, S. J., Campbell, B. M. & Ingram, J. S. I. Climate change and food systems. *Annu. Rev. Environ. Resource* 37, 195–222 (2012).
6. Fischer, R. A. & Edmeades, G. O. Breeding and cereal yield progress. *Crop Sci.* 50, 85–98 (2010).
7. Hafner, S. Trends in maize, rice, and wheat yields for 188 nations over the past 40 years: a prevalence of linear growth. *Agri. Ecosyst. Environ.* 97, 275–283 (2003).
8. Searchinger, T. R. *et al.* Use of U.S. croplands for biofuels increases greenhouses gases through emissions from land-use change. *Science* 319, 1238–1240 (2008).
9. Jaggard, K. W., Qi, A. & Ober, E. S. Possible changes to arable crop yields by 2050. *Phil. Trans. R Soc. B* 365, 2835–2851 (2010).
10. Dyson, T. World food trends and prospects to 2025. *Proc Natl Acad Sci U.S.A.* 96, 5929–5936 (1999).
11. Ray, D. K., Mueller, N. D., West, P. C. & Foley, J. A. Yield trends are insufficient to double global crop production by 2050. *PLoS ONE*8, e66428 (2013).
12. Evenson, R. & Rosegrant, M. W. *Productivity Projections for Commodity Market Modeling* Yale University Economic Growth Center: New Haven, (1995).
13. Evenson, R. E. Global and local implications of biotechnology and climate change for future food supplies. *Proc. Natl Acad. Sci. USA*96, 5921–5928 (1999).
14. Nelson, G. C. *et al. Food Security, Farming, and Climate Change to 2050* IFPRI: Washington, DC, (2010).
15. Reilly, J. M. & Fuglie, K. O. Future yield growth in field crops: what evidence exists? *Soil Till. Res.* 47, 275–290 (1998).
16. Heisey, P. W. *Amber Waves* USDA-ERS Publication (2009)http://webarchives.cdlib.org/sw1vh5dg3r/http://ers.usda.gov/AmberWaves/December09/Features/USCornYields.htm.
17. Edgerton, M. D. Increasing crop productivity to meet global needs for feed, food, and fiber. *Plant Phys.* 149, 7–13 (2009).
18. Hertell, T. W. *et al.* Global land use and greenhouse gas emissions impacts of maize ethanol: the role of market-mediated responses.*Biosci.* 60, 223–231 (2010).
19. Duvick, D. N. & Cassman, K. G. Post-green revolution trends in yield potential of temperate maize in the North-Central United Sates. *Crop Sci.* 39, 1622–1630 (1999).

20. Specht, J. E., Hume, D. J. & Kumudini, S. V. Soybean yield potential - A genetic and physiological perspective. *Crop Sci.* 39, 1560–1570 (1999).
21. Peng, S., Cassman, K. G., Virmani, S. S., Sheehy, J. & Khush, G. S.Yield potential trends of tropical rice since the release of IR8 and the challenge of increasing rice yield potential. *Crop Sci.* 39, 1552–1559 (1999).
22. Graybosch, R. A. & Peterson, C. J. Genetic improvement in winter wheat yields in the Great Plains of North America, 1959–2008.*Crop Sci.* 50, 1882–1890 (2010).
23. Lobell, D. B., Cassman, K. G. & Field, C. B. Crop yield gaps: their importance, magnitude, and causes. *Annu. Rev. Environ. Resource*34, 1–26 (2009).
24. Evans, L. T. *Crop Evolution, Adaptation, and Yield* Cambridge University Press: Cambridge, (1993).
25. Bruinsma, J. *The Resource Outlook to 2050: By How Much do Land, Water Use and Crop Yields Need to Increase by 2050?* ed Conforti P. chapter 6FAO: Rome, (2011).
26. Alexandratos, N. & Bruinsma, J. *World Agriculture Towards 2030/2050: the 2012 Revision. ESA Working Paper No. 12-03* FAO: Rome, (2012).
27. Alston, J. M., Beddow, J. M. & Pardey, P. G. Agricultural research, productivity, and food prices in the long run. *Science* 325, 1209–1210 (2009).
28. Ray, D. K. *et al.* Recent patterns of crop yield growth and stagnation. *Nat. Commun.* 3, 1293 (2012).
29. Lin, M. & Huybers, P. Reckoning wheat yield trends. *Environ. Res. Lett.* 7, 024016 (2012).
30. Anderson, J. R. & Hazell, P. B. R. *Implications for Agricultural Research and Policy in Developing Countries* The Johns Hopkins University Press: Baltimore and London, (1989).
31. Brisson, N. *et al.* Why are wheat yields stagnating in Europe? A comprehensive data analysis for France. *Field Crops Res.* 119, 201–212 (2010).
32. Rijk, B., van Ittersum, M. & Withagen, J. Genetic progress in Dutch crop yields. *Field Crop Res.* 149, 201.
33. Pardey, P. G., Beintema, N., Dehmer, S. & Wood, S. *Agricultural Research. A Growing Global Divide?* International Food Policy Research Institute: Washington, DC, (2006).
34. Cassman, K. G., Dobermann, A., Walters, D. T. & Yang, H. S.Meeting cereal demand while protecting natural resources and improving environmental quality. *Annu. Rev. Environ. Resource* 28, 315–358 (2003).
35. Foley, J. A. *et al.* Solutions for a cultivated planet. *Nature* 478, 337–342 (2011).
36. Pardey, P. G., Alston, J. M. & Piggott, R. R. *Agricultural R&D in the Developing World: Too Little, Too Late?* International Food Policy Research Institute: Washington, DC, (2006).
37. Archontoulis, S. V. & Miguez, F. A. Nonlinear regression models and application in agricultural research. *Agron. J.* 105, 1–13 (2013).
38. Gallant, A. R. *Nonlinear Statistical Models* John Willey: New York, (1989).
39. Johnston, J. & Di Nardo, J. *Econometric Methods* McGraw Hill: New York, (1997).
40. Box, G. E. P. & Cox, D. R. An analysis of transformations. *J. R. Stat. Soc. B Met.* 26, 211–252 (1964).

CHAPTER 17

ANTIOXIDANT SECONDARY METABOLITES IN CEREALS: POTENTIAL INVOLVEMENT IN RESISTANCE TO FUSARIUM AND MYCOTOXIN ACCUMULATION

Vessela Atanasova-Penichon, Christian Barreau, and Florence Richard-Forget

Keywords: *Fusarium*, mycotoxins, cereals, antioxidants, resistance

INTRODUCTION

Fusarium Head Blight (FHB) of small-grain cereals such as wheat and barley and *Gibberella* Ear Rot (GER) and *Fusarium* Ear Rot (FER) of maize are three devastating fungal diseases affecting crops worldwide. Both FHB and GER are caused by the same *Fusarium* species on wheat and maize respectively, *Fusarium graminearum* and *Fusarium culmorum* being the most predominant in Europe (Bottalico and Perrone, 2002). FER is caused by *Fusarium* species belonging to the *Gibberella fujikuroi* complex, including *Fusarium proliferatum* and *Fusarium verticillioides*. These three fungal diseases lead to huge economic losses, resulting from reduced yields, deteriorated grain quality and contamination of grains with mycotoxins.

F. graminearum and *F. culmorum* can produce zearalenone and type B trichothecenes (TCTB). TCTB include deoxynivalenol (DON) and its two acetylated forms, 3-acetyl-deoxynivalenol (3-ADON) and 15-acetyl-deoxynivalenol (15-ADON), as well as nivalenol (NIV) and its acetylated form 4-acetylnivalenol or fusarenon X (FX). *F. proliferatum*and *F. verticillioides* are major sources of maize contamination with fumonisins (FB) among which fumonisin B1 (FB1), FB2 and FB3 are predominant. All these *Fusarium* toxins exhibit various acute and chronic effects on humans and animals (Bennett and Klich, 2003). Consequently, thresholds for maximal DON, FB1+FB2 and zearalenone content in foodstuffs have been set up in Europe: Commission regulation published in 2005 (EC number 856/2005) and amended in July 2007 (EC number 1126/2007) for DON and zearalenone and Commission regulation number 1126/2007 for FB1+FB2. Published surveys on the mycotoxin status of European cereals and derived products clearly show that mycotoxins produced by *Fusarium* species are ubiquitously present and that contamination levels exceeding the EU maximum levels or guidance values are likely to occur, leading to significant economic losses (Streit et al., 2012;Schatzmayr and Streit, 2013; Nordkvist and Haggblom, 2014). There are no data on the economic costs of mycotoxins in Europe with the exception of one study in Hungary where these costs were estimated to be 100 million euros in 1998, consecutively to a severe FHB outbreak (Milicevic et al., 2010). Economic costs directly result from: (1) yield loss due to fungal diseases, (2) reduced crop value resulting from mycotoxin contamination, (3) losses in animal productivity and (4) trade impacts. Additional costs include the cost of management linked to prevention, sampling, analysis, mitigation, litigation, and research. For instance, the annual cost for monitoring aflatoxin alone in the US is estimated to be 30–50 million dollars.

Production of DON, FB and zearalenone by *Fusarium* spp. occurs during infection of crops. Structurally, DON is defined as an epoxide containing sesquiterpenoid skeleton. The epoxide group at position 12–13 allows DON to bind to ribosomes leading to the activation of various protein kinases, the modulation of gene expression, the inhibition of protein synthesis and cell toxicity (Maresca, 2013). The chemical structure of FB consists of an aminopentol backbone with one tricarballylic acid on each side chain and one or more hydroxyl groups (Bezuidenhout et al., 1988). Due to their structural similarity with sphinganine, FB may act as specific inhibitors of sphingolipid biosynthesis, which are major constituents of cell membranes and important components of many signaling pathways (Merrill et al., 2001). Zearalenone is a phenolic resorcyclic acid lactone and its toxicity is mainly related to its ability to competitively bind to estrogen receptors. Excellent reviews describing the detailed biosynthesis pathway for DON and FB have been recently published (Brown et al., 2007; Alexander et al., 2009) whilst there are still significant

Citation: Atanasova-Penichon V, Barreau C and Richard-Forget F (2016). Antioxidant Secondary Metabolites in Cereals: Potential Involvement in Resistance to Fusarium and Mycotoxin Accumulation. Front. Microbiol. 7:566. DOI:10.3389/fmicb.2016.00566.

knowledge gaps in the understanding of zearalenone biosynthesis. During the last decades, tremendous progress has also been made in identifying the environmental factors that significantly impact the regulation of DON and FB biosynthesis during the colonization of plant tissues (Picot et al., 2010; Merhej et al., 2011; Montibus et al., 2015). Temperature, water availability, pH variations, nutrient sources and plant defense metabolites were pointed out as key factors regulating DON and FB production.

TCTB and FB toxins are heat-stable molecules that are not fully eliminated during food and feed processing (Hazel and Patel, 2004; Humpf and Voss, 2004). As a result, the best way to reduce or avoid contamination of food and feed is to control the biosynthesis of these mycotoxins at the field level during plant cultivation. Three major factors have been reported to significantly influence fungal development and mycotoxin accumulation in grains: (i) environmental conditions, (ii) agricultural practices, and (iii) susceptibility of cereal genotypes (Edwards, 2004). Several cultural practices such as crop rotation, tillage, use of chemicals as well as breaking the fungal disease cycle by adapting the sowing period or using resistant hosts have been shown to reduce efficiently the level of primary pathogen inoculum (Pirgozliev et al., 2003). More recently, integrated management studies have demonstrated the improvements that can be gained by combining multiple control strategies (Blandino et al., 2012). Plant breeding strategies are among the most promising and performing approaches to durably fight against *Fusarium* diseases and the contamination of cereals with mycotoxins. Undoubtedly, such strategies will be among the most important pillars of any integrated disease management programs (Terzi et al., 2014).

Plant resistance to *Fusarium* and mycotoxin accumulation is a highly complex mechanism. Five major types of resistance have been classified for wheat, and are transferable for barley and maize. However, mechanisms associated with one of these five types can be host specific. In wheat and barley, type I resistance operates against initial infection of the floret (Schroeder and Christensen, 1963), and in maize, it may be associated with silk resistance. Type II resistance limits spreading of the infection within the host. Unlike in wheat, fungal infection in barley usually does not spread from initially infected spikelets to adjacent spikelets. Type II resistance has therefore little meaning for barley. Type III concerns resistance to grain infection; type IV, tolerance and ability to maintain yields and finally type V resistance gathers all mechanisms of resistance to mycotoxin accumulation (Miller et al., 1985; Mesterhazy, 1995, 2002). Boutigny et al. (2008) proposed to divide the type V resistance into two components. The first one, called type V-1, represents resistance to toxin accumulation operated by metabolic transformation involving biochemical modification catalyzed by enzymes such as UDP-glycosyltransferases, gluthatione-S-transferases or cytochrome P450 mono-oxygenases (Karlovsky, 2011; De Boevre et al., 2014). The second one (type V-2) corresponds to resistance *via*inhibition of mycotoxin biosynthesis through the action of plant endogenous compounds. These compounds include both constitutively synthetized compounds and those induced in response to pathogen infection.

In addition to genetic approaches aiming at identifying and characterizing Quantitative Trait Loci (QTL) for FHB, FER and GER resistance, recent biochemical studies have been attempted to decipher the biochemical defenses that contribute to FHB, FER, and GER resistance and low mycotoxin accumulation. Mainly based on comparative approaches of metabolite composition of resistant and susceptible varieties, challenged or not with *Fusarium*, these attempts have implemented targeted analytical approaches and non-targeted global metabolomic developments (Siranidou et al., 2002; Bollina et al., 2011; Atanasova-Penichon et al., 2012; Picot et al., 2013; Sampietro et al., 2013;Gunnaiah and Kushalappa, 2014). A large set of metabolites potentially acting in cereals to counteract toxigenic *Fusaria*and reduce mycotoxin accumulation has been highlighted by these studies. These metabolites derive from primary and secondary plant metabolism and can be roughly classified in six major groups: fatty acids, amino acids and derivatives, carbohydrates, amines and polyamines, terpenoids and phenylpropanoids (Gauthier et al., 2015). Plant secondary metabolites with antioxidant properties, mainly terpenoids and phenylpropanoids are among the most frequently reported for their potential involvement in plant defense against fungal pathogens (Balmer et al., 2013). In addition to their key role as plant defense mediators and their participation to cell wall reinforcement, these compounds display antifungal properties and some of them can interfere with mycotoxin biosynthesis (Gauthier et al., 2015).

Here, we review the latest scientific advances that support the potential contribution of grain antioxidant secondary metabolites to cereal resistance to *Fusarium* and mycotoxin accumulation focusing on (i) *in vitro* studies on the effect of antioxidants on fungal development and mycotoxin production by *Fusarium*, (ii) identification of the major antioxidant metabolites that *Fusarium* can encounter during ear infection process, from anthesis to grain maturity, and (iii) relation between resistance to *Fusarium* and antioxidant content in cereals.

PRINCIPAL ANTIOXIDANT SECONDARY METABOLITES IN CEREALS

In cereals, the main secondary metabolites with antioxidant activity belong to three groups including phenolic compounds, carotenoids and tocopherols (Boutigny et al., 2008). An additional group, consisting of benzoxazinoid derivatives, less abundant in grains but with multiple recognized biological activities, needs also to be addressed.

Phenylpropanoids

Phenolics are considered the major contributors to total antioxidant capacity of cereal grains (Awika et al., 2003;Gorinstein et al., 2008). Phenolic compounds derive from the phenylpropanoid pathway and are divided into two groups: flavonoid phenylpropanoids including flavones, flavonols, flavanones, flavanols, anthocyanins and chalcones, and non-flavonoid phenylpropanoids such as stilbenes, lignans, and phenolic acids.

Non-Flavonoid Phenylpropanoids

Among non-flavonoid phenylpropanoids, phenolic acids are predominant in cereals (Dykes and Rooney, 2007; Gauthier et al., 2015). Phenolic acids are derivatives of either benzoic or cinnamic acids. In cereals, benzoic acid derivatives include gallic, *p*-hydroxybenzoic, vanillic, syringic, and protocatechuic acids while cinnamic acid derivatives include caffeic, chlorogenic, *p*-coumaric, sinapic, and ferulic acids. Phenolic acids found in cereals exist in both soluble (free) and insoluble (cell-wall-bound) forms. Soluble phenolic acids are either free acids or esterified to sugar conjugates. Insoluble phenolic acids are linked to various polysaccharides and to lignin through ester and ether bonds. Soluble forms are compartmentalized within the plant cell vacuoles and insoluble forms are distributed in cell walls. Phenolic compounds are concentrated in the outer layers of the grain, the pericarp and the aleurone, and in the germ, and are less abundant in the endosperm (Bily et al., 2003; Das and Singh, 2015).

Studies comparing composition of phenolic compounds in cereals reveal significant differences between cereal types, within varieties as well as within grain fractions (Adom and Liu, 2002; Ndolo and Beta, 2014; Pihlava et al., 2015). This variability associated to the large set of extraction protocols and analytical procedures that can be used when addressing the phenolic composition of grains explains the frequent discrepancies observed in published data. For instance, in the study of Adom and Liu (2002), maize grains were reported as the richest in total phenolic acids, followed by wheat, oat and rice while, in the report of Irakli et al. (2012), the highest levels in both free and bound phenolic acids were found in oat, followed by maize, wheat and rice. Nevertheless, in all published studies, the major portion of phenolics in grains exists as bound forms: 85% in maize, 75% in oat and wheat and 62% in rice (Adom and Liu, 2002; Boz, 2015; Das and Singh, 2015).

Among free phenolic acids, ferulic acid is by far predominant, followed by *p*-coumaric and vanillic acids (Adom and Liu, 2002; Bakan et al., 2003; Santiago et al., 2007). Caffeic, *p*-hydroxybenzoic and sinapic acids are also present but at very low concentrations (0.5–1.5 μg/g) (Irakli et al., 2012). In addition to phenolic acid monomers, hydroxycinnamic polyamines such as *p*-coumaroyl-feruloylputrescine (CFP) and diferuoylputrescine (DFP) have been quantified in significant amounts in free phenolic maize extracts. Their concentrations can reach 330 μg equiv. 8-5′-benzofuran-diferulic acid/g (Moreau et al., 2001; Atanasova-Penichon et al., 2012; Figure 1A).

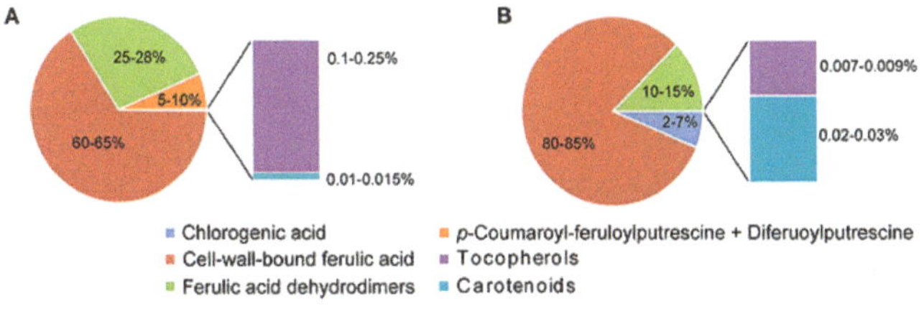

Figure 1. Main antioxidant secondary metabolites quantified in maize at (A) maturity and at (B) early stage when mycotoxin biosynthesis is initiated (milk-dough stage). This model is based on maize results obtained by Atanasova-Penichon et al. (2012), Atanasova-Penichon and Richard-Forget (2014) and Picot et al. (2013).

Among cell-wall-bound phenolic acids, ferulic acid is the most abundant one in common cereals and represents up to 90% of the total phenolic compounds (Adom and Liu, 2002; Boz, 2015). Concentrations of this compound can reach 3000 μg/g for some maize varieties (Li et al., 2007). Ferulic acid and its oxidatively coupled products named ferulic acid dehydrodimers or diferulic acids (DiFA) are found in greater concentrations in cereal brans. Ferulic acid dehydrodimers are potent antioxidants and are ester-linked to the cell wall polymers. The most important ferulic acid dehydrodimers in common cereals are: 8–5 DiFA (open form), 5-5′DiFA, 8-O-4′DiFA and 8-5′benDiFA (benzofuran form) and their sum represents 20–30% of the total phenolic acids (Atanasova-Penichon et al., 2012; Boz, 2015) (Figure1A). The highest levels of ferulic acid dehydrodimers in cereals ranged between 250 and 475 μg/g (Jilek and Bunzel, 2013).

Flavonoid Phenylpropanoids

The second group of phenolic compounds with significant concentrations in cereal grains is the class of flavonoids, located in the pericarp and the germ (Dykes and Rooney, 2007; Das and Singh, 2015). Flavonoids are also major active ingredient in corn silks (Hu et al., 2010). As phenolic acids, most grain flavonoids are found in the cell-wall-bound fraction: 93% of the total flavonoids in wheat, 91% in maize, 65% in rice and 61% in oat (Adom and Liu, 2002). According to Adom and Liu (2002), maize grains contain the highest level in total flavonoids followed by wheat, oat and rice. The most frequently cited flavonoids in cereal grains are the flavonols kaempferol and quercetin for maize (Das and Singh, 2015), the flavanone naringenin and its glycosylated forms and the flavanols catechin and epicatechin for barley (Bollina et al., 2010, 2011; Zilic et al., 2011), the flavones vitexin and luteolin for rye (Pihlava et al., 2015) and the anthocyanins in colored grains (Dykes and Rooney, 2007). According to the report of Reid et al. (1992), corn silks are characterized by high concentrations of flavones including luteolin and apigenin and flavone glycosides such as maysin, iso-orientin, and iso-vitexin.

Lipophilic Compounds

In cereals, the major lipophilic secondary metabolites with antioxidant properties include tocols (or commonly referred to as tocopherols) and carotenoids. The latest group consists of carotenes, of which α-carotene and β-carotene are the major representatives, and xanthophylls, mostly lutein and zeaxanthin. According to the results of two-year field studies, lipophilic antioxidant secondary metabolites represent less than 0.25% of total antioxidant secondary metabolites in mature maize grains (Atanasova-Penichon et al., 2012; Picot et al., 2013; Figure 1A). Concentrations in carotenoids in grains significantly vary according to the cereal type, from 1.8 μg/g in oat to 18.2 μg/g in maize (Ndolo and Beta, 2013). The major carotenoids in maize are concentrated in the endosperm fraction, ranging from 14.2 to 31.2 μg/g of endosperm, while the major carotenoids in small-grain cereals are found in germ and range from 3.2 to 14.8 μg/g of germ (Ndolo and Beta, 2013). In wheat, barley and oat grains, lutein was reported as the major xanthophyll and zeaxanthin as the minor one (Ndolo and Beta, 2013).

Tocol composition of cereals includes tocopherols (α-, β-, δ- and γ-tocopherol) and tocotrienols (α-, β-, δ- and γ-tocotrienols). The α-forms are predominant (Gutierrez-Gonzalez et al., 2013). Tocopherols are mainly present in the germ fraction while tocotrienols are present in the pericarp and endosperm fractions (Falk et al., 2004). In small-grain cereals such as oat, barley and wheat, tocotrienols are the main tocols and their concentrations range between 40 and 60 µg/g depending on the cereal type and the variety (Falk et al., 2004). Conversely, maize grains contain more tocopherols than tocotrienols, with concentrations ranging between 34–70 µg/g and 20–25 µg/g respectively (Das and Singh, 2015).

Benzoxazinoid Derivatives

Benzoxazinoids are a group of secondary metabolites found in maize, rye, wheat and triticale, but not in sorghum and rice (Niemeyer, 2009; Andersson et al., 2014). The mono and dihexose conjugates of cyclic hydroxamic acid DIBOA (2,4-dihydroxy-1,4-benzoxazin-3-one) and the corresponding lactam HBOA (2-hydroxy-1,4-benzoxazin-3-one) are the major benzoxazinoids in rye, while in wheat and maize, DIMBOA (2,4-dihydroxy-7-methoxy-1,4-benzoxazin-3-one) and its glycosylated derivative dominate (Etzerodt et al., 2015; Pihlava et al., 2015). Rye is by far the crop characterized by the highest content in benzoxazinoids, with concentration values more than 20-fold higher to that found in whole grain wheat (Andersson et al., 2014; Pihlava et al., 2015). Benzoxazinoids are located in all fractions of seeds, with greater concentrations in bran and germ (Pihlava et al., 2015).

ANTIOXIDANTS MODULATE FUNGAL DEVELOPMENT AND MYCOTOXIN PRODUCTION BY *FUSARIUM* SPP.

The biosynthetic pathways that lead to the production of TCTB and FB by *Fusarium* species have been well established and characterized by several oxygenation steps (Proctor et al., 2003; Desjardins, 2007). Therefore, changes in the oxidative parameters of the medium are likely to interfere with the fungal secondary metabolism and to modulate the level of mycotoxin production (Ponts, 2005; Montibus et al., 2015). Due to the antioxidant properties of cereal secondary metabolites, several studies have been devoted to their anti-fungal and anti-mycotoxin effect.

Antifungal Properties of Cereal Antioxidants

Phenolic acids are toxic toward many fungi including *Fusarium* species (Guiraud et al., 1995; Ponts et al., 2011; Gauthier et al., 2016). Their fungicidal efficiency has been characterized against different *Fusarium* species and IC_{50} values (concentration that inhibits 50% of fungal growth) ranging between 0.7 to >10 mM have been reported (Table 1). The comparison of IC_{50} values has to be considered with caution as these values are method and condition dependent.

Table 1. IC_{50}^{a} values of non-flavonoids and flavonoids against different *Fusarium* species.

Compound	*Fusarium* species	IC_{50}, mM	References
NON FLAVONOIDS: CINNAMIC ACID DERIVATIVES			
Ferulic acid	*F. graminearum*	0.7–2.2	Ponts et al., 2011
	F. graminearum	1.8	Bollina et al., 2010
	F. graminearum	2.4	Kumaraswamy et al., 2011a
	F. graminearum	2.3, 3.4	McKeehen et al., 1999
	F. culmorum	1.7	McKeehen et al., 1999
	F. coeruleum, *F. moniliforme*, *F. solani*	5.2–>5.2	Guiraud et al., 1995
p-Coumaric acid	*F. graminearum*	1–4.2	Ponts et al., 2011
	F. graminearum	1.2	Bollina et al., 2010
	F. graminearum	1.9, 4.8	McKeehen et al., 1999
	F. culmorum	3.4	McKeehen et al., 1999
Caffeic acid	*F. graminearum*	4.0–7.1	Ponts et al., 2011
	F. graminearum	2.5	Kumaraswamy et al., 2011a
	F. graminearum	6.7–10.1	Gauthier et al., 2016
	F. culmorum	8.8–> 10.0	Gauthier et al., 2016
Chlorogenic acid	*F. graminearum*	>10.0	Gauthier et al., 2016
	F. culmorum	>10.0	Gauthier et al., 2016
NON FLAVONOIDS: BENZOIC ACID DERIVATIVES			
p-Hydroxybenzoic acid	*F. graminearum*	6.6–> 15.0	Ponts et al., 2011
Syringic acid	*F. graminearum*	3.5–6.2	Ponts et al., 2011
	F. coeruleum, *F. moniliforme*, *F. solani*	>5.0	Guiraud et al., 1995
Protocatechuic acid	*F. coeruleum*, *F. moniliforme*, *F. solani*	>6.5	Guiraud et al., 1995
Vanillic acid	*F. coeruleum*, *F. moniliforme*, *F. solani*	>6.0	Guiraud et al., 1995
FLAVONOIDS: FLAVONES			
Quercetin	*F. graminearum*	2.9	Bollina and Kushalappa, 2011
Kaempferol	*F. graminearum*	4.8	Bollina et al., 2010
FLAVONOIDS: FLAVANONES			
Flavanone	*F. graminearum*, *F. culmorum*, *F. poae*	<0.2	Silva et al., 1998
	F. avenaceum, *F. nivale*	<0.8	Silva et al., 1998
Naringenin	*F. graminearum*	1.6	Bollina et al., 2010
FLAVONOIDS: FLAVONES			
Flavone	*F. graminearum*, *F. culmorum*, *F. poae*	<0.2	Silva et al., 1998
	F. avenaceum	<0.8	Silva et al., 1998
	F. nivale	<0.05	Silva et al., 1998
4′-methylflavone	*F. graminearum*, *F. nivale*, *F. poae*	<0.2	Silva et al., 1998
	F. culmorum, *F. avenaceum*	>0.8	Silva et al., 1998
4′-methyoxyflavone	*F. graminearum*, *F. culmorum*, *F. avenaceum*, *F. nivale*, *F. poae*	>0.8	Silva et al., 1998

[a]*Concentration that inhibits 50% of growth.*

Compared to IC_{50} values ascribed to phenolic acids, those determined for flavones and flavanones against different*Fusarium* species including *F. culmorum* and *F. graminearum* are substantially weaker (Table 1). According to Treutter (2005), the efficiency to inhibit fungal growth directly results from the ability of flavonoids to irreversibly combine with nucleophilic amino acids in fungal proteins. Among the different groups of the flavonoid subclass, i.e., flavanones, flavones and flavanols, data reported in Table 1 suggest that unsubstituted flavones and flavanones (IC_{50} between < 0.05 to 1.6 mM) display a more efficient antifungal activity than hydroxylated flavones, i.e., flavonol (IC_{50} between 2.9 to 4.8 mM). The promising ability of flavonoids to inhibit spore development and restrain mycelium hyphae elongation of plant pathogens have been the subject of numerous investigations (Treutter, 2006; Mierziak et al., 2014). Flavonoids are also the subject of intensive medical research, with the aim of identifying alternatives to synthetic drugs for counteracting human fungal pathogens that increasingly display resistance to commonly used antifungal agents such as triazole ones (Cushnie and Lamb, 2005).

IC_{50} values gathered in Table 1 also illustrate the great variability in phenolic acid sensitivity between *F. graminearum*strains. Similar variations seem to occur among *F. culmorum* strains; however there is insufficient data to assert this hypothesis. When comparing results obtained in the same conditions by Gauthier et al. (2016) and Ponts et al. (2011) for *F. graminearum* and *F. culmorum*, it also appears that *F. culmorum* strains (IC_{50} between 8.8 to >10.0 mM) could be less susceptible to caffeic acid than *F. graminearum* strains (IC_{50} between 4 to 10.1 mM). Phenolic acids could be ranked in ascending order of toxicity toward *F. graminearum* as follows: chlorogenic acid < *p*-hydroxybenzoic acid < caffeic acid < syringic acid < *p*-coumaric acid < ferulic acid (Table 1). Chlorogenic acid, a cinnamic-derived phenolic acid, displays a lower fungicidal activity than *p*-hydroxybenzoic and syringic acid, which is contradictory with the assumption that cinnamic-derived phenolic acids are roughly more toxic than benzoic acid-derived ones (Beekrum et al., 2003; Ponts et al., 2011). Chlorogenic acid displays however weak lipophilic properties, which, according to Guiraud et al. (1995) and Ponts et al. (2011) are primary factors in the antifungal efficiency of phenolic acids.

As regards to benzoxazinoids, their antifungal activities have been the subject of numerous publications (Glenn et al., 2001; Martyniuk et al., 2006). According to the results of Glenn et al. (2001), *Fusarium* species responsible for GER and FER in maize show a wide range of sensitivity to 6-methoxybenzoxazolin-2(3H)-one (MBOA) and benzoxazolin-2(3H)-one (BOA), with the most tolerant being *F. verticillioides, Fusarium subglutinans* and *F. graminearum*. As demonstrated by Glenn et al. (2001), differences in tolerance can be ascribed do different abilities to metabolize and therefore detoxify these antimicrobial compounds.

Cereal Antioxidants Inhibit Mycotoxin Biosynthesis by *Fusarium*

In addition to displaying antifungal properties, several antioxidant secondary metabolites of cereals can modulate the production of mycotoxins by various fungal pathogens. According to the report of Boutigny (2007), cinnamic acid derivatives such as sinapic, caffeic, *p*-coumaric, chlorogenic, and ferulic acids are efficient inhibitors of TCTB production by *F. graminearum* and *F. culmorum* while benzoic acid derivatives, with the exception of syringic acid, have an activating effect. It is noteworthy that the effect of phenolic compounds is strain and molecule dependent (Boutigny et al., 2009; Gauthier et al., 2016). Increasingly, phenolic acids are becoming the subject of anti-mycotoxin research and many groups have demonstrated their efficiency to modulate *in vitro* the biosynthesis of various mycotoxins, including type A trichothecenes (Ferruz et al., 2016), fumonisins (Beekrum et al., 2003; Samapundo et al., 2007; Atanasova-Penichon et al., 2014), ochratoxin (Palumbo et al., 2007), and aflatoxins (Norton, 1999).

Similarly, several studies illustrated the potential impact flavonoids could exert on mycotoxin production. Recently, rutin was demonstrated as a potent inhibitor of aflatoxin B1 production by *Aspergillus flavus* (Norton, 1999; Chitarrini et al., 2014) and naringin, hesperidin and some glucosides were characterized for their capacity to restrain the production of patulin by *Penicillium expansum, Aspergillus terreus*, and *Byssochlamys fulva* (Salas et al., 2012). As regards to TCTB, effects of flavonoids on their biosynthesis have been poorly documented with exception of the publication of Desjardins et al. (1988) describing an inhibitory effect of flavones on the biosynthetic step that catalyzes the conversion of trichodiene (the first chemical intermediate in trichothecene biosynthesis) to oxygenated trichothecenes. In addition to phenolic compounds, carotenoids and tocopherols are potent cereal antioxidant compounds, but their antifungal and antimycotoxin activities against *Fusarium* are poorly documented. Recent works have shown that sub-lethal doses of α-tocopherol significantly affected fumonisin production (Picot et al., 2013) and that 50 μg/ml of β-carotene added to the culture medium led to a significant decrease (close to 50%) in TCTB accumulation (Boutigny, 2007). A few additional studies investigated the impact of carotenoids on other mycotoxin production but they led to opposite results, depending on the mycotoxin targeted. While capsanthin (a major carotenoid in paprika) has been shown to inhibit aflatoxin yield (Masood et al., 1994), more recent results demonstrated its lack of inhibitory effect on ochratoxin production (Santos et al., 2010).

The toxin suppressive effects of benzoxazinoids have also been addressed in several publications. This effect was first suggested by Miller et al. (1996) who reported that 4-acetyl-benzoxazolin-2-one (4-ABOA) and related compounds present in an active maize fraction were able to reduce trichothecene and aflatoxin productions by *F. culmorum* and *A. flavus*, respectively. Anti-mycotoxin activities of benzoxazinoids were recently confirmed by Etzerodt et al. (2015) who demonstrated that a 250 μM concentration of DIMBOA caused 50% inhibition of 15-ADON production by *F. graminearum*.

Mechanisms of Fungal Toxicity and Inhibition of Mycotoxin Production

Regardless the phytochemical considered (phenolic compound, carotenoid, tocopherol, and benzoxazinoid), the exact mechanisms by which fungal growth and mycotoxin production are inhibited remain unclear. While few published studies have focused on plant fungal pathogens, the mechanism of action of phenolic acids on human pathogens and mainly on *candida* species have been the subject of extensive research. Last and most significant insights on the anti-adhesion, anti-biofilm effects of phenolic acids together with their inhibitory activity on morphogenesis and fungal exoenzymes production have been recently gathered in the review of Teodoro et al. (2015). Interestingly, phenolic acids have been evidenced for their ability to breakdown the fungal membrane permeability barrier, probably through a perturbation of the lipid bilayers causing the leakage of ions and other chemicals as well as the formation of pores and modification of the electric potential of membranes (Sung and Lee, 2010).

As regards to the effects of phenolic acids against plant fungal pathogens, Guiraud et al. (1995) and more recently,Boutigny (2007) and Ponts et al. (2011) indicated that the toxicity of phenolic compounds is related to their lipophilicity as well as their strong antioxidant properties. Accordingly, Pani et al. (2014) and Roleira et al. (2010) suggested that the balance among lipophilicity and antioxidant activity can be a key factor to predict the capacity of a phenolic to inhibit mycotoxin production. However, it is essential to keep in mind that fungal cultures are multi-component systems, where the media can be considered as both lipidic and emulsion systems and that in such biological media, several physicochemical parameters including pH, light or temperature can affect the lipophilicity and antioxidant capacity of phytochemicals. Thus, correlating theoretical antioxidant potential and lipophilicity values with experimental data is not a straightforward approach. Nonetheless, the hypothesis that antioxidant properties of cereal metabolites can be primary factors for their antimycotoxin activity is highly consistent with the assumed activating effect of oxidative stress on the biosynthesis of mycotoxins. Indeed, an increasing body of work, recently gathered in the review of Montibus et al. (2015), emphasizes the modulation of fungal secondary metabolism by oxidative stress and the enhancement of mycotoxin production, including DON and FB, after exposure to reactive oxygen species. Thus, due to their capacity to quench oxygen free radicals, antioxidant metabolites may reduce or suppress upstream signals, such as oxidative stress, that modulate toxin biosynthesis. According to Guiraud et al. (1995), toxicity of phenolic acids can also be linked to their interaction with various intra or extracellular fungal enzymes, including phenol oxidases and several hydrolytic activities (El Modafar et al., 2000; Paul et al., 2003). Moreover, Passone et al. (2009) mentioned that antioxidant compounds interfere with mycotoxin production probably indirectly *via* their capacity to perturb the membrane function and modify its permeability. Lastly, the results of Boutigny et al. (2009) and Etzerodt et al. (2015) that indicate a downregulation of the expression of the genes involved in DON biosynthesis by *F. graminearum* when ferulic acid and DIMBOA is added to *in vitro* culture media are in accordance with a transcriptional control exerted by phenolic acids and benzoxazinoids. A similar conclusion was evidenced by the study of Sanzani et al. (2009) that proved that quercetin and umbelliferon reduced patulin accumulation by acting on the transcription level of biosynthetic genes.

ANTIOXIDANT SECONDARY METABOLITES ENCOUNTERED BY *FUSARIUM* SPP. DURING THE EAR INFECTION PROCESS

To date, most of the attempts aiming at clarifying the contribution of cereal secondary metabolites to the *in planta*control of *Fusarium* mycotoxin accumulation have targeted mature grains. However, during plant development, grain antioxidant composition is likely to be dramatically modified. *Fusarium* commonly infects cereal ears shortly after anthesis, and the compounds the fungus has to face at the onset of infection are certainly extremely different from those found in the mature grain. There are very few dynamical studies that have addressed the composition of the grain in the early stages of grain development, when the biosynthesis of mycotoxin is initiated. In recent field experiments on maize inoculated with *F. graminearum* or *F. verticillioides*, the kinetics of fungal development and the accurate stage at which mycotoxin production is initiated were established (Picot et al., 2011, 2013; Atanasova-Penichon et al., 2012). *In planta*TCTB and FB accumulation were found to start between 10 and 20 days after flowering, i.e., at the milk-dough stage (Figure 2). Major free and bound antioxidant secondary metabolites present at the milk-dough stage were quantified and are detailed in Figure 1B. Cell-wall-bound ferulic acid, which represents 80–85% of the analyzed antioxidants, is the predominant compound, followed by ferulic acid dehydrodimers (10–15%) and free chlorogenic acid (2–7%). A particular attention was paid to free antioxidants, particularly to chlorogenic acid that represents almost 80% of the total free phenolic acids in the early stages of maize grain development. Indeed, free antioxidant compounds are more likely to interfere first with *Fusarium.* Phenolic compounds present in kernels at early stages are likely to alleviate fungal infection in a manner similar to that observed in the *in vitro* inhibition studies. In maize grains at the milk-dough stage, lipophilic antioxidants such as carotenoids and tocopherols are present at much lower levels than phenolic acids and represent only 0.02–0.03% of the total antioxidant content (Figure 1B). However, because their antioxidant properties are much higher than that of phenolic compounds, it cannot be excluded that, despite their low concentrations, they also significantly contribute *in planta* to the inhibition of *Fusarium* toxin biosynthesis. In addition to highlight the milk-dough stage as a critical step, the studies of Atanasova-Penichon et al. (2012) and Picot et al. (2013) provided information on the evolution of antioxidant secondary metabolites during maize ear ripening (Figure3). Similar evolution patterns of phytochemicals were reported by the previous authors for the two years of experimentation, suggesting that they may correspond to an intrinsic characteristic of maize genotypes not dependent on environmental factors. Free and cell-wall-bound phenolic acid as well as carotenoid and tocopherol contents show large fluctuations during the ripening of maize grains. Composition in free phenolic acid evolves qualitatively over time whereas the composition in cell-wall-bound phenolic acids, carotenoids and tocopherols remains unchanged and only shows quantitative variation at the different grain stages. Kinetic of free chlorogenic acid, cell-wall-bound ferulic acid and ferulic acid dehydrodimers as well as xanthophylls, carotenes and tocopherols during maize ear ripening is presented in Figure 3. Except for tocopherols, all antioxidant secondary metabolites are found at higher concentrations in the grain at early stages, suggesting that these compounds are the main antioxidants that *F. graminearum* and *F. verticillioides* potentially encounter when their mycotoxin production is initiated. Figure 3 indicates that, after a rapid increase from anthesis to the silking-blister stage (with exception of chlorogenic acid), levels of cell-wall-bound monomers represented by ferulic acid, of free phenolic acids represented by chlorogenic acid, of xanthophylls and of carotenes decrease to reach traces at maturity. As regards the ferulic acid dehydrodimers, their concentration exhibits a pattern similar to monomeric phenolic acids in the first stages of grain development and then increases until the mature stage. This increase reflects the contribution of ferulic acid dehydrodimers to cell wall structure through their role in forming bridges between hemicellulose chains. A similar pattern for evolution of bound ferulic acid has also been reported for wheat (Shewry et al., 2012) and rice (Lin and Lai, 2011). Similarly, a decrease in free ferulic acid in rice (Lin and Lai, 2011) and total free phenols in oat (Alfieri and Redaelli, 2015) has been described, supporting old data on soft and durum wheat grain (McCallum and Walker, 1990; Régnier and Macheix, 1996; McKeehen et al., 1999). However, considering that most of the studies mentioned above were conducted with few genotypes, caution should be taken in generalizing the results.

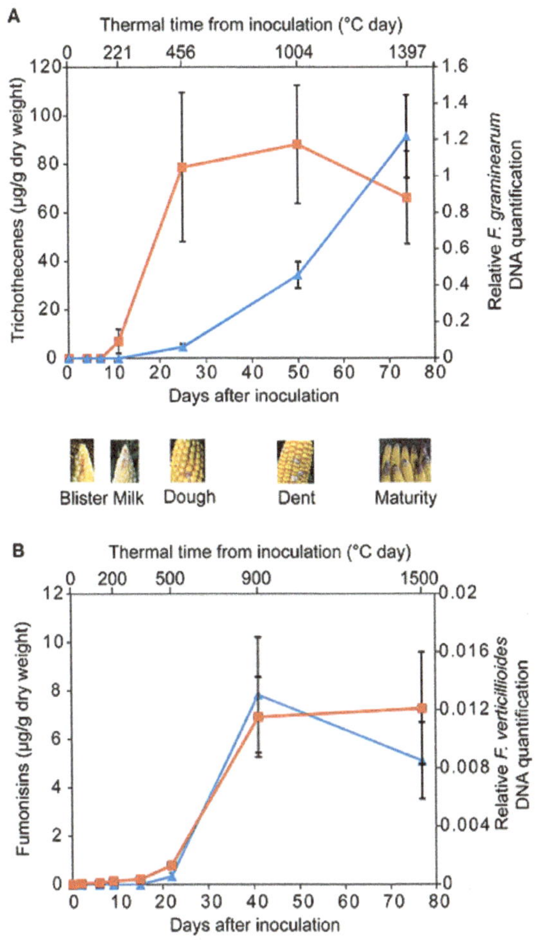

Figure 2. (A) Relative quantification of fungal DNA in maize kernels, expressed as a log10 (*F. graminearum*DNA/maize DNA) ratio of a susceptible variety (blue triangles, right Y-axis), and level of trichothecenes accumulated in maize kernels (red squares, left Y-axis) after silk inoculation with *F. graminearum*. Vertical bars show standard error of the mean. Top X-axis: thermal time from inoculation (mean value of 2 years), bottom X-axis: days after inoculation for each sampling (mean value of 2 years). Data from the 2 years and two repetitions were pooled (mean values ± SEM,$n = 4$). (B) Relative quantification of fungal DNA in maize kernels, expressed as a log10 *F. verticillioides* DNA/log10 maize DNA ratio of a susceptible variety (blue triangles, right Y-axis), and level of fumonisins accumulated in maize kernels (red squares, left Y-axis) after silk inoculation with *F. verticillioides*. Vertical bars show standard deviations. Top X-axis: thermal time from inoculation, bottom X-axis: days after inoculation for each sampling. Data from the two sowing date treatments were pooled. Kinetics are established with data published by Atanasova-Penichon et al. (2012)and Picot et al. (2011).

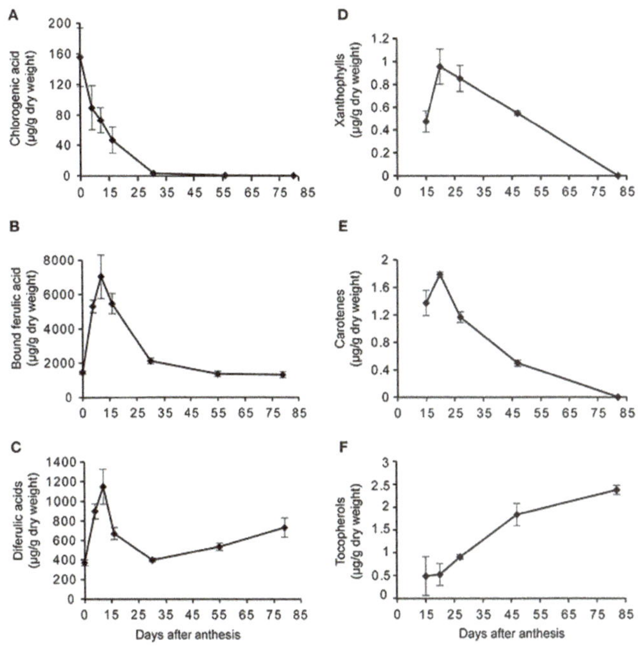

Figure 3. Kinetics of antioxidant compounds during maize grain development: (A) chlorogenic acid; (B) cell-wall-bound ferulic acid; (C) ferulic acid dehydrodimers; (D) xanthophylls; (E) carotenes and (F) tocopherols. Kinetics are established with data published by Atanasova-Penichon et al. (2012), Atanasova-Penichon and Richard-Forget (2014) and Picot et al. (2013).

The declining concentrations of phenolic acids during grain ripening can be ascribed to several rationales. First, the activity of phenyl-alanine ammonia-lyase and L-tyrosine ammonia-lyase, two crucial enzymes for the initial committed step in the biosynthesis of phenylpropanoids, have been shown to be maximal only during the early stages of grain development (McCallum and Walker, 1990; Régnier and Macheix, 1996). Second, the rate of endosperm development surpasses the rate of synthesis of the outer coverings during grain ripening which leads to a dilution of the overall phenolic constituents within the grain. Third, the decrease in phenolic acids can also result from their oxidative degradation involving phenoloxidases and peroxidases, induced by the breakdown of cellular structure in the pericarp at the end of the milk stage and during further maturation (Régnier and Macheix, 1996). Finally, the decrease in cell wall-bound phenolic acid contents can be correlated with the formation of alkali-resistant bounds occurring in cross-linked polymers in cell walls not extractable with the method commonly used to analyze phenolic acids (Iiyama et al., 1994).

Figure 3 also indicates a decrease in xantophylls and carotenes from the dough stage until maturity, in accordance with the pattern reported by Rodríguez-Suárez et al. (2014) and Sreenivasulu et al. (2010) for carotenoids in durum wheat and barley. Reduced levels in carotenes and xanthophylls during ripening may be due to their oxidation sensitivity, as a result of the high degree of unsaturation present in their structure. According to Mellado-Ortega and Hornero-Mendez (2016) and Sandmann et al. (2006), carotenes are likely to be more prone to oxidation than xanthophylls. This oxidation is caused by reactive oxygen species and especially singlet oxygen or free radicals generated by enzymatic systems such as lipoxygenases. Lipoxygenases catalyze the hydroperoxidation of polyunsaturated fatty acids, preferentially non-esterified polyunsaturated fatty acids, to form conjugated diene hydroperoxides (Loiseau et al., 2001). These hydroperoxides react with carotenoids, breaking down the carbon backbone into smaller compounds, including volatile molecules and apocarotenoids (e.g., epoxyaldehydes, ketones; Mellado-Ortega and Hornero-Mendez, 2016). Lipoxygenase is widely distributed in cereals and located in the germ and bran of the grain (Loiseau et al., 2001). A second explanation to the declining levels of carotenoids during grain ripening, could be linked to their esterification with fatty acids that produce mono and/or diesters and is catalyzed by xanthophyll acyltransferase enzymes. However, this second hypothesis is unlikely to occur based on the fact that xanthophyll esters seem to be absent or at very low levels in cereals and particularly in durum wheat (Mellado-Ortega and Hornero-Méndez, 2015).

Unlike kinetics of free and bound phenolic acids and carotenoids, the kinetic of tocopherols reported on Figure 3Findicates a gradual accumulation during the course of maize grain development. A similar pattern of total tocopherols was established by Gutierrez-Gonzalez et al. (2013) in oat seeds. According to the results of Falk et al. (2004) in developing barley kernels, tocopherols reach a maximum level at milk stage and remain stable until final harvest time. In rice, the total tocopherols in immature grains is about 2-fold higher than in mature ones (Lin and Lai, 2011).

Altogether, data describing the time course of *F. graminearum* infection and reporting the evolution of phytochemical levels in grains, provide evidence that the main antioxidant metabolites *F. graminearum* is likely to encounter when the production of mycotoxin starts *in planta* are free phenolic acids such as chlorogenic acid and bound ferulic and diferulic phenolic acids. Although present in lower concentrations, xantophylls, carotenes and benzoxazinoids, which show a high fungal toxicity, could also interfere with the fungus. Additional information on the impact antioxidant phytochemicals could exert on accumulation of mycotoxins in grains is provided by the results of recent studies that attempted to link plant resistance to *Fusarium* and antioxidant content of cereal grains (Siranidou et al., 2002; Bollina et al., 2011; Picot et al., 2013; Atanasova-Penichon et al., 2014).

RELATION BETWEEN RESISTANCE TO *FUSARIUM* SPP. AND ANTIOXIDANT CONTENT IN CEREALS

Host resistance is one of the primary traits that can be used as a control measure, and its manipulation is recognized as one of the best economic and ecological strategies to reduce damage caused by *Fusarium* (Bai and Shaner, 2004). Several authors argued that the use of cereal genotypes resistant to *Fusarium* infection (Champeil et al., 2004) and mycotoxin accumulation (Boutigny et al., 2008) is one of the most promising ways to reduce or prevent contamination. Combined with genetic approaches and the detection of QTL linked with FHB, GER, or FER resistance, biochemical ones aiming at deciphering the chemical mechanisms plants use to fight against *F. graminearum* and reduce toxin production hold great potential for assisting breeding programs (Gauthier et al., 2015). Most of the biochemical approaches that have addressed cereal resistance to *Fusarium* spp. and have been published to date, are based on a comparative analysis of the metabolite composition of resistant and susceptible cultivars, challenged or not with*Fusarium*. Targeted analytical tools and, more recently, metabolomics strategies were implemented. However, while these approaches can provide interesting insights that need to be further validated through genetic studies, they cannot allow conclusive evidence on the involvement of metabolite(s) or group(s) of metabolites in resistance. Indeed, the experimental designs frequently considered a set of limited and genetically unrelated genotypes and very rarely near isogenic lines, as done in the study of Gunnaiah et al. (2012). Moreover, data delivered through metabolomics approaches require to be considered with caution since differences in metabolic profiles of the studied genotypes may actually be confounding with the effects resulting of environment, cultivation practices and developmental stage. Lastly, it should be borne in mind that chemical identification remains a significant bottleneck in plant metabolomics studies and that most of the proposed identification are putative ones.

A number of studies focusing on phenolic acids supports the assumption that, in cereals, cell-wall-bound ferulic acid together with its dehydrodimers and free chlorogenic acid could be key components of the chemical defense against toxigenic *Fusarium* species (Siranidou et al., 2002; Atanasova-Penichon et al., 2012; Sampietro et al., 2013). Bily et al. (2003) highlighted that ferulic acid and its dehydrodimers in maize act as resistance factors to *F. graminearum* through type I resistance (resistance to initial penetration) and type II resistance (resistance to propagation due to a lower degradability of the cell wall). This hypothesis was further corroborated by Picot et al. (2013), Atanasova-Penichon and Richard-Forget (2014) and Atanasova-Penichon et al. (2014) who revealed the highest concentrations of ferulic acid, ferulic acid dehydrodimers and chlorogenic acid in immature grains of the more resistant varieties of a panel of maize genotypes with different susceptibility to FER or GER. Correlations between levels of GER resistance and phenolic acid contents in maize grains are reported on Figures 4A–C. When addressing maize resistance to GER, mechanisms resistance of silk that can slow down the process of infection need also to be addressed. Cao et al. (2011) have investigated the role of hydroxycinnamic acids in silk resistance and observed that, unlike data gathered in Figure 4, high concentrations in hydroxycinnamic acids were not related with a delayed progression of *F. graminearum* through silks. In wheat, positive relations between both free and cell wall bound phenolic acid levels and wheat resistance to FHB were reported by Siranidou et al. (2002). By the same reasoning, Choo et al. (2015) hypothesized that the high level of black barley resistance to FHB is linked to its richness in phenolic compounds. In addition to phenolic acids but with less conclusive evidence, many other phenylpropanoid compounds have been suggested to contribute to the chemical defense to FHB, GER, or FER. This potential contribution was mainly highlighted through comparative metabolomic profiling of grains issued from resistant and susceptible genotypes, challenged or not with toxigenic*Fusarium* strains (Hamzehzarghani et al., 2005, 2008; Browne and Brindle, 2007; Paranidharan et al., 2008; Bollina et al., 2010, 2011; Kumaraswamy et al., 2011a,b; Gunnaiah et al.,

2012; Cajka et al., 2014; Chamarthi et al., 2014;Gunnaiah and Kushalappa, 2014). A large set of constitutive as well as inducible defense metabolites potentially related to *Fusarium* resistance was highlighted in the afore-mentioned studies. Among these metabolites, phenylpropanoids (approximately 180 compounds), including flavonoids and non-flavonoids, represent more than 50% of the total reported metabolites (Gauthier et al., 2015). Among these 180 phenylpropanoid candidates, more than 56% are putatively assigned as flavonoids that encompasse anthocyanins, flavones, flavonols, flavanones, flavanols, isoflavones, isoflavanones, isoflavonols, and chalcones. The remaining 44% is mainly composed by phenolic acids and derivatives, including benzoic and cinnamic ones.

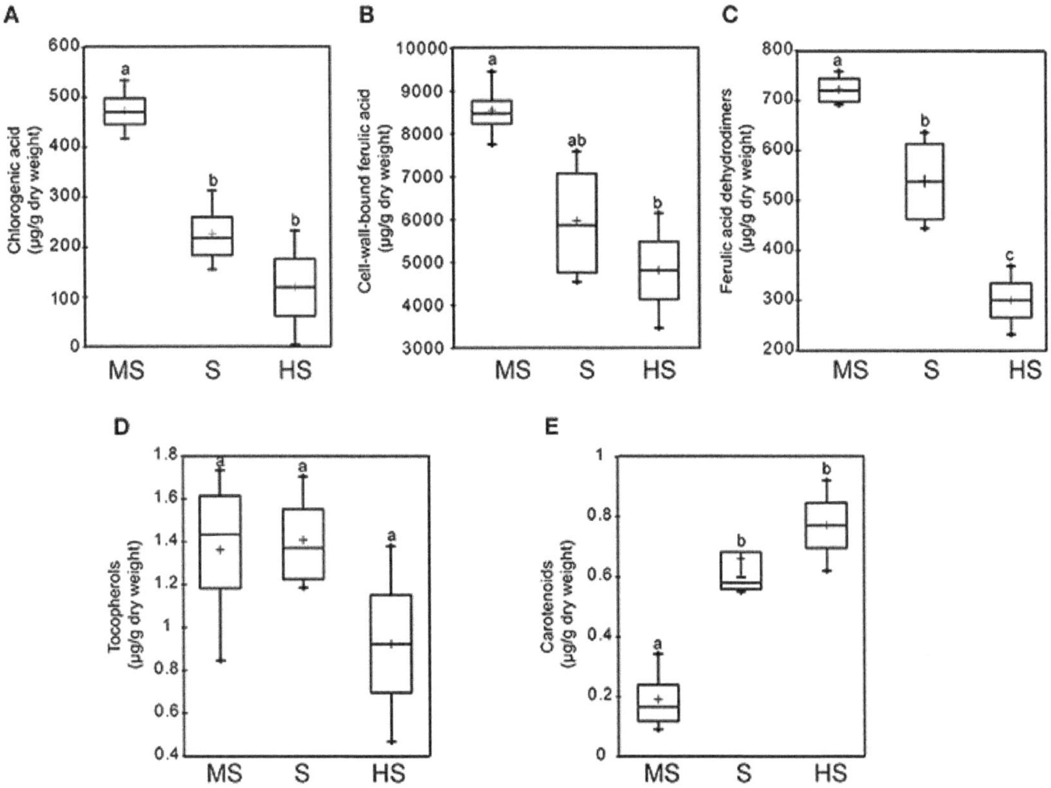

Figure 4. Levels of some antioxidant metabolites in immature maize grains from three groups of varieties differing in their susceptibility to GER: highly susceptible (HS, $n = 3$), susceptible (S, $n = 4$), and moderately susceptible (MS, $n = 3$). (A) chlorogenic acid; (B) cell-wall-bound ferulic acid; (C) ferulic acid dehydrodimers; (D) tocopherols and (E) carotenoids. Groups with different letters are significantly different. (Duncan, $\alpha = 0.05$; Atanasova-Penichon et al., 2012; Atanasova-Penichon and Richard-Forget, 2014).

Indeed, the role of phenylpropanoids in disease resistance has been the subject of intensive research (Treutter, 2006). In response to pathogen infection, they are released from the cell wall or massively synthesized by the plant accumulating rapidly at the site of infection (Nicholson and Hammerschidt, 1992). The main role ascribed to these compounds in plant defense mechanisms results from their antioxidant properties (Dykes and Rooney, 2007; Agati et al., 2012), that allow them to quench reactive oxygen species (ROS), generated by both the pathogen and the plant during infection. In addition, phenolic compounds operate in defense response through direct interference with the fungus, or through the reinforcement of plant structural components to act as a mechanical barrier against the pathogen (Siranidou et al., 2002; Treutter, 2006). Phenolic compounds such as flavonoids can also protect plant cell wall integrity upon fungal infection by inhibiting the activity of several plant cell wall degrading enzymes secreted by fungal pathogens to penetrate plant tissues (Treutter, 2005).

Concerning carotenoids and tocopherols, they are rarely regarded in comparative metabolomic studies due to their lipophilicity and the requirement of a specific extraction protocol and analytical equipment. While compilation of the metabolomic studies reported above results in a list of about 30 terpenoids (Gauthier et al., 2015), no carotenoid or tocopherol were among these terpenoid candidates. Targeted approaches aiming at relating lipophilic antioxidant composition of grains and resistance to *Fusarium* have also been implemented and showed positive or negative correlations, depending on the addressed group of compounds, carotenoids or tocopherols. Thus, based on the use of a set of maize genotypes with moderate to high susceptibility to GER, the experimentations we performed in our laboratory indicated higher levels of carotenoids (lutein + zeaxanthin + β carotene) in immature grains of the more susceptible genotypes (Figure 4E). A similar trend but not statistically significant was also observed between the level of FER resistance and carotenoid contents in maize grains (Picot et al., 2013). Accordingly, positive correlations between the levels of lutein and DON accumulation in durum wheat cultivars were reported by Delgado et al. (2014). As regard to tocopherols, while an absence of correlation between resistance levels of maize and their tocopherol contents in immature grain was observed by Picot et al. (2013), the data of Iqbal et al. (2014) showed the existence of a negative correlation between the concentrations of tocopherols and aflatoxins in rice cultivars. Interestingly, the report of Boba et al. (2011), based on the use of transgenic flax overproducing carotenoids, indicates that a general level of lipophilic antioxidants rather than the content of any particular compound is the most important factor in resistance to *F. culmorum* infection. The main role ascribed to carotenoids and tocols in plant/*Fusarium* interactions directly results from their ability to quench the free radicals produced by plant cells (the so-called "oxidative-burst") as a first response to the fungal pathogen attack (Boba et al., 2011; Gutierrez-Gonzalez et al., 2013). Moreover, carotenoids are directly linked to abscisic acid, which level in wheat and barley was shown to increase after *F. graminearum* or *F. culmorum*inoculation (Gunnaiah et al., 2012; Kumaraswamy et al., 2012; Petti et al., 2012). In fact, abscisic acid is an apocarotenoid synthetized from the cleavage of carotenoids (Tan et al., 1997). 9-cis-epoxycarotenoid dioxygenases cleave 11,12 double bonds of the cis isomers of violaxanthin and neoxanthin to form the C15 product xanthin, the first committed and key regulatory step in the abscisic acid biosynthesis pathway (Sreenivasulu et al., 2010). Abscisic acid is well known for its roles in orchestrating stress response as well as grain maturation in plants. In addition, its role in resistance of wheat to FHB has been linked to a regulatory effect on callose deposition in the transition zone between the spikelet's rachilla and rachis. This was shown to contribute to the type II resistance (Kang and Buchenauer, 2000;Flors et al., 2005). Besides, the involvement of abscisic acid in FHB resistance has also been ascribed to its negative interaction with the signaling ethylene pathway (Flors et al., 2005) since, according to Chen et al. (2009), *F. graminearum* can exploit ethylene signaling to enhance colonization in wheat tissues. Lastly, the possibility that abscisic acid could limit *F. graminearum* penetration through its control of stomatal aperture cannot be omitted (Mauch-Mani and Mauch, 2005).

As regards to benzoxazinoids, while several reports have indicated their ability to inhibit fungal activities linked with FER and GER (Miller et al., 1996; Glenn et al., 2001; Etzerodt et al., 2015), very few studies have investigated their concentration in cereals in relation to *Fusarium* sensitivity. One study (Søltoft et al., 2008) has revealed positive correlations between the susceptibility of wheat to FHB and the concentrations of some benzoxazinoid derivatives, suggesting that the capacity of wheat to produce these secondary metabolites could contribute to resistance mechanisms. However, as emphasized above, the results of Søltoft et al. (2008) based on the use of a set of unrelated germplasms are not sufficient to draw conclusive evidences on the involvement of benzoxazinoids in FHB resistance.

CONCLUSION

Cereal diseases caused by pathogenic and toxigenic *Fusarium* species are responsible for major economic damage worldwide. Hence, the developments of sustainable strategies to avoid *Fusarium* and mycotoxin contamination have been the issue of intense research over past years and decades and a broad consensus has emerged to acknowledge that the use of FHB, GER or FER resistant genotypes is one of the primary pillars of any disease management programs. However, to date, knowledge of the complex mechanisms governing cereal resistance remains insufficient, and selection for resistant genotypes is still challenging.

Considering the available data on the interactions between antioxidant metabolites in grains and *Fusarium* species, we can assume that some of these compounds could significantly contribute to the protection of grains against toxigenic*Fusaria* and mycotoxin accumulation. Five main classes of antioxidant metabolites, phenolic acids, flavonoids, carotenoids, tocopherols and benzoxazinoids, have been evidenced for the pivotal role they could play in FHB, GER, or FER resistance. A first shared argument in favor of the involvement of phenolics, carotenoids and tocopherols is linked to their ability to quench reactive oxygen species, thus protecting biological cells. In addition, tocopherols and carotenoids have the capacity to scavenge lipid peroxyl free radical and therefore to stop the chain propagation of the lipid peroxidation cycle (Das and Roychoudhury, 2014). A second shared argument rests on the fungal toxicity exhibited by cereal antioxidant metabolites. Indeed, as demonstrated by the present review, there are numerous studies illustrating the efficiency of phenolic compounds, carotenoids, tocopherols and even benzoxazinoids to restrain the growth of toxigenic *Fusaria* and their production of toxins. Lastly, phenolic compounds are known to participate to the reinforcement of plant structures and contribute therefore to the establishment of a physical barrier against fungal infection.

However, while involvement of antioxidant metabolites in resistance mechanism to *Fusarium* spp. has been highly suggested, this involvement is far from being elucidated. One major challenge for the coming years will be to obtain conclusive proofs. Even though the genetic architecture underlying the synthesis and regulation of secondary metabolites in cereals is extremely complex, pieces of evidence can certainly come from extensive genetic and functional genomic studies.

AUTHOR CONTRIBUTIONS

VA and FR have made substantial, direct and intellectual contribution to the work. CB has made intellectual contribution. All authors approved this work for publication.

CONFLICT OF INTEREST STATEMENT

The authors declare that the research was conducted in the absence of any commercial or financial relationships that could be construed as a potential conflict of interest.

ACKNOWLEDGMENTS

We greatly acknowledge Dr N. Ponts for her valuable comments.

REFERENCES

1. Adom, K. K., and Liu, R. H. (2002). Antioxidant activity of grains. *J. Agric. Food Chem.* 50, 6182–6187. doi: 10.1021/jf0205099
2. Agati, G., Azzarello, E., Pollastri, S., and Tattini, M. (2012). Flavonoids as antioxidants in plants: location and functional significance. *Plant Sci.* 196, 67–76. doi: 10.1016/j.plantsci.2012.07.014
3. Alexander, N. J., Proctor, R. H., and McCormick, S. P. (2009). Genes, gene clusters, and biosynthesis of trichothecenes and fumonisins in *Fusarium.Toxin Rev.* 28, 198–215. doi: 10.1080/15569540903092142

4. Alfieri, M., and Redaelli, R. (2015). Oat phenolic content and total antioxidant capacity during grain development. *J. Cereal Sci.* 65, 39–42. doi: 10.1016/j.jcs.2015.05.013

5. Andersson, A. A. M., Dimberg, L., Aman, P., and Landberg, R. (2014). Recent findings on certain bioactive components in whole grain wheat and rye. *J. Cereal Sci.* 59, 294–311. doi: 10.1016/j.jcs.2014.01.003

6. Atanasova-Penichon, V., Bernillon, S., Marchegay, G., Lornac, A., Pinson-Gadais, L., Ponts, N., et al. (2014). Bioguided isolation, characterization, and biotransformation by *Fusarium verticillioides* of maize kernel compounds that inhibit fumonisin production. *Mol. Plant-Microbe Interact.* 27, 1148–1158. doi: 10.1094/MPMI-04-14-0100-R

7. Atanasova-Penichon, V., Pons, S., Pinson-Gadais, L., Picot, A., Marchegay, G., Bonnin-Verdal, M.-N., et al. (2012). Chlorogenic acid and maize Ear Rot Resistance: a dynamic study investigating *Fusarium graminearum* development, deoxynivalenol production, and phenolic acid accumulation. *Mol. Plant-Microbe Interact.* 25, 1605–1616. doi: 10.1094/MPMI-06-12-0153-R

8. Atanasova-Penichon, V., and Richard-Forget, F. (2014). Les phytomicronutriments des céréales: un élément de résistance à la Fusariose et à l'accumulation de mycotoxines. *Innov. Agronom.* 42, 63–76. Available online at: http://prodinra.inra.fr/record/284317

9. Awika, J. M., Rooney, L. W., Wu, X. L., Prior, R. L., and Cisneros-Zevallos, L. (2003). Screening methods to measure antioxidant activity of sorghum (Sorghum bicolor) and sorghum products. *J. Agric. Food Chem.* 51, 6657–6662. doi: 10.1021/jf034790i

10. Bai, G., and Shaner, G. (2004). Management and resistance in wheat and barley to *Fusarium* head blight. *Annu. Rev. Phytopathol.* 42, 135–161. doi: 10.1146/annurev.phyto.42.040803.140340

11. Bakan, B., Bily, A. C., Melcion, D., Cahagnier, B., Regnault-Roger, C., Philogene, B. J. R., et al. (2003). Possible role of plant phenolics in the production of Trichothecenes by *Fusarium graminearum* strains on different fractions of maize kernels. *J. Agric. Food Chem.* 51, 2826–2831. doi: 10.1021/jf020957g

12. Balmer, D., Flors, V., Glauser, G., and Mauch-Mani, B. (2013). Metabolomics of cereals under biotic stress: current knowledge and techniques. *Front. Plant Sci.* 4:82. doi: 10.3389/fpls.2013.00082

13. Beekrum, S., Govinden, R., Padayachee, T., and Odhav, B. (2003). Naturally occurring phenols: a detoxification strategy for fumonisin B1. *Food Addit. Contam.* 20, 490–493. doi: 10.1080/0265203031000098678

14. Bennett, J. W., and Klich, M. (2003). Mycotoxins. *Clin. Microbiol. Rev.* 16:3. doi: 10.1128/CMR.16.3.497-516.2003

15. Bezuidenhout, S. C., Gelderblom, W. C. A., Gorstallman, C. P., Horak, R. M., Marasas, W. F. O., Spiteller, G., et al. (1988). Structure elucidation of the fumonisins, mycotoxins from *Fusarium moniliforme*. *J. Chem. Soc. Comm.* 11, 743–745. doi: 10.1039/c39880000743

16. Bily, A. C., Reid, L. M., and Taylor, J. H. (2003). Dehydrodimers of ferulic acid in maize grain pericarp and aleurone: resistance factors to *Fusarium graminearum*. *Phytopathology.* 93, 712–719. doi: 10.1094/PHYTO.2003.93.6.712

17. Blandino, M., Haidukowski, M., Pascale, M., Plizzari, L., Scudellari, D., and Reyneri, A. (2012). Integrated strategies for the control of *Fusarium* head blight and deoxynivalenol contamination in winter wheat. *Field Crops Res.* 133, 139–149. doi: 10.1016/j.fcr.2012.04.004

18. Boba, A., Kulma, A., Kostyn, K., Starzycki, M., Starzycka, E., and Szopa, J. (2011). The influence of carotenoid biosynthesis modification on the *Fusarium culmorum* and *Fusarium oxysporum* resistance in flax. *Physiol. Mol. Plant Pathol.* 76, 39–47. doi: 10.1016/j.pmpp.2011.06.002

19. Bollina, V., Kumaraswamy, G. K., Kushalappa, A. C., Choo, T. M., Dion, Y., Rioux, S., et al. (2010). Mass spectrometry-based metabolomics application to identify quantitative resistance-related metabolites in barley against *Fusarium* head blight. *Mol. Plant Pathol.* 11, 769–782. doi: 10.1111/j.1364-3703.2010.00643.x

20. Bollina, V., and Kushalappa, A. C. (2011). In vitro inhibition of trichothecene biosynthesis in *Fusarium graminearum* by resistance-related endogenous metabolites identified in barley. *Mycology* 2, 291–296. doi: 10.1080/21501203.2011.604354

21. Bollina, V., Kushalappa, A. C., Choo, T. M., Dion, Y., and Rioux, S. (2011). Identification of metabolites related to mechanisms of resistance in barley against *Fusarium graminearum*, based on mass spectrometry. *Plant Mol. Biol.* 77, 355–370. doi: 10.1007/s11103-011-9815-8

22. Bottalico, A., and Perrone, G. (2002). Toxigenic *Fusarium* species and mycotoxins associated with head blight in small-grain cereals in Europe. *Eur. J. Plant Pathol.* 108, 611–624. doi: 10.1023/A:1020635214971

23. Boutigny, A.-L. (2007). *Etude de l'effet de Composés du Grain de blé dur sur la Régulation de la voie de Biosynthèse des Trichothécènes B Purification de Composés Inhibiteurs, Analyse des Mécanismes Impliqués*. Thèse de doctorat. Université de Bordeaux, France.

24. Boutigny, A.-L., Barreau, C., Atanasova-Penichon, V., Verdal-Bonnin, M.-N., Pinson-Gadais, L., and Richard-Forget, F. (2009). Ferulic acid, an efficient inhibitor of type B trichothecene biosynthesis and Tri gene expression in *Fusarium* liquid cultures. *Mycol. Res.* 113, 746–753. doi: 10.1016/j.mycres.2009.02.010

25. Boutigny, A.-L., Richard-Forget, F., and Barreau, C. (2008). Natural mechanisms for cereal resistance to the accumulation of *Fusarium* trichothecenes.*Eur. J. Plant Pathol.* 121, 411–423. doi: 10.1007/s10658-007-9266-x

26. Boz, H. (2015). Ferulic acid in cereals—a review. *Czech J. Food Sci.* 33:1. doi: 10.17221/401/2014-CJFS

27. Brown, D. W., Butchko, R. A. E., Busman, M., and Proctor, R. H. (2007). The *Fusarium verticillioides* FUM gene cluster encodes a Zn(II)2Cys6 protein that affects FUM gene expression and fumonisin production. *Eukaryot. Cell.* 6, 1210–1218. doi: 10.1128/EC.00400-06

28. Browne, R. A., and Brindle, K. M. (2007). H-1 NMR-

based metabolite profiling as a potential selection tool for breeding passive resistance against*Fusarium* head blight (FHB) in wheat. *Mol. Plant Pathol.* 8, 401–410. doi: 10.1111/j.1364-3703.2007.00400.x

29. Cajka, T., Vaclavikova, M., Dzuman, Z., Vaclavik, L., Ovesna, J., and Hajslova, J. (2014). Rapid LC-MS-based metabolomics method to study the*Fusarium* infection of barley. *J. Sep. Sci.* 37, 912–919. doi: 10.1002/jssc.201301292

30. Cao, A., Reid, L. M., Butrón, A., Malvar, R. A., Souto, X. C., and Santiago, R. (2011). Role of hydroxycinnamic acids in the infection of maize silks by*Fusarium graminearum Schwabe*. *Mol. Plant-Microbe Interact.* 24, 1020–1026. doi: 10.1094/MPMI-03-11-0079

31. Chamarthi, S. K., Kumar, K., Gunnaiah, R., Kushalappa, A. C., Dion, Y., and Choo, T. M. (2014). Identification of *fusarium* head blight resistance related metabolites specific to doubled-haploid lines in barley. *Eur. J. Plant Pathol.* 138, 67–78. doi: 10.1007/s10658-013-0302-8

32. Champeil, A., Doré, T., and Fourbet, J. (2004). *Fusarium* head blight: epidemiological origin of the effects of cultural practices on head blight attacks and the production of mycotoxins by *Fusarium* in wheat grains. *Plant Sci.* 166, 1389–1415. doi: 10.1016/j.plantsci.2004.02.004

33. Chen, X., Steed, A., Travella, S., Keller, B., and Nicholson, P. (2009). *Fusarium graminearum* exploits ethylene signalling to colonize dicotyledonous and monocotyledonous plants. *New Phytol.* 182, 975–983. doi: 10.1111/j.1469-8137.2009.02821.x

34. Chitarrini, G., Nobili, C., Pinzari, F., Antonini, A., De Rossi, P., Del Fiore, A., et al. (2014). Buckwheat achenes antioxidant profile modulates *Aspergillus flavus* growth and aflatoxin production. *Int. J. Food Microbiol.* 189, 1–10. doi: 10.1016/j.ijfoodmicro.2014.07.029

35. Choo, T. M., Vigier, B., Savard, M. E., Blackwell, B., Martin, R., Wang, J. M., et al. (2015). Black barley as a means of mitigating deoxynivalenol contamination. *Crop Sci.* 55, 1096–1103. doi: 10.2135/cropsci2014.05.0405

36. Cushnie, T. P. T., and Lamb, A. J. (2005). Antimicrobial activity of flavonoids. *Int. J. Antimicrob. Agents* 26, 343–356. doi: 10.1016/j.ijantimicag.2005.09.002

37. Das, K., and Roychoudhury, A. (2014). Reactive oxygen species (ROS) and response of antioxidants as ROS-scavengers during environmental stress in plants. *Front. Environ. Sci.* 2:53. doi: 10.3389/fenvs.2014.00053

38. Das, A. K., and Singh, V. (2015). Antioxidative free and bound phenolic constituents in pericarp, germ and endosperm of Indian dent (*Zea mays* var. indentata) and flint *(Zea mays* var. indurata) maize. *J. Funct. Foods.* 13, 363–374. doi: 10.1016/j.jff.2015.01.012

39. De Boevre, M., Landschoot, S., Audenaert, K., Maene, P., Mavungu, D. J. D., Eeckhout, M. D., et al. (2014). Occurrence and within field variability of*Fusarium* mycotoxins and their masked forms in maize crops in Belgium. *World Mycotoxin J.* 7, 91–102. doi: 10.3920/WMJ2013.1608

40. Delgado, R. M., Sulyok, M., Jirsa, O., Spitzer, T., Krska, R., and Polisenska, I. (2014). Relationship between lutein and mycotoxin content in durum wheat. *Food Addit. Contam. Part a-Chem. Anal. Control Expo. Risk Assess.* 31, 1274–1283. doi: 10.1080/19440049.2014.925589

41. Desjardins, A. E. (2007). *Fusarium* mycotoxins chemistry, genetics and biology. *Plant Pathol.* 56:2. doi: 10.1111/j.1365-3059.2006.01505.x

42. Desjardins, A. E., Plattner, R. D., and Spencer, G. F. (1988). Inhibition of trichothecene toxin biosynthesis by naturally-occurring shikimate aromatics.*Phytochemistry* 27, 767–771. doi: 10.1016/0031-9422(88)84090-1

43. Dykes, L., and Rooney, L. W. (2007). Phenolic compounds in cereal grains and their health benefits. *Cereal Foods World.* 52, 105–111. doi: 10.1094/cfw-52-3-0105

44. Edwards, S. G. (2004). Influence of agricultural practices on *fusarium* infection of cereals and subsequent contamination of grain by trichothecene mycotoxins. *Toxicol Lett.* 153, 29–35. doi: 10.1016/j.toxlet.2004.04.022

45. El Modafar, C., Tantaoui, A., and El Boustani, E. (2000). Effect of caffeoylshikimic acid of date palm roots on activity and production of *Fusarium oxysporum* f. sp albedinis cell wall-degrading enzymes. *J. Phytopathol.* 148, 101–108. doi: 10.1046/j.1439-0434.2000.00472.x

46. Etzerodt, T., Maeda, K., Nakajima, Y., Laursen, B., Fomsgaard, I. S., and Kimura, M. (2015). 2,4-Dihydroxy-7-methoxy-2H-1,4-benzoxazin-3(4H)-one (DIMBOA) inhibits trichothecene production by *Fusarium graminearum* through suppression of Tri6 expression. *Int. J. Food Microbiol.* 214, 123–128. doi: 10.1016/j.ijfoodmicro.2015.07.014

47. Falk, J., Krahnstover, A., van der Kooij, T. A. W., Schlensog, M., and Krupinska, K. (2004). Tocopherol and tocotrienol accumulation during development of caryopses from barley (*Hordeum vulgare* L.). *Phytochemistry* 65, 2977–2985. doi: 10.1016/j.phytochem.2004.08.047

48. Ferruz, E., Atanasova-Pénichon, V., Bonnin-Verdal, M.-N., Marchegay, G., Pinson-Gadais, L., Ducos, C., et al. (2016). Effects of phenolic acids on the growth and production of T-2 and HT-2 toxins by *Fusarium langsethiae* and *F. sporotrichioides*. *Molecules* 21:449. doi: 10.3390/molecules21040449

49. Flors, V., Ton, J., Jakab, G., and Mauch-Mani, B. (2005). Abscisic acid and callose: team players in defence against pathogens? *J. Phytopathol.* 153, 377–383. doi: 10.1111/j.1439-0434.2005.00987.x

50. Gauthier, L., Atanasova-Penichon, V., Chéreau, S., and Richard-Forget, F. (2015). Metabolomics to decipher the chemical defense of cereals against*Fusarium graminearum* and deoxynivalenol accumulation. *Int. J. Mol. Sci.* 16:10. doi: 10.3390/ijms161024839

51. Gauthier, L., Bonnin-Verdal, M.-N., Marchegay, G., Pinson-Gadais, L., Ducos, C., Richard-Forget, F., et al. (2016). Fungal biotransformation of chlorogenic and caffeic acids by *Fusarium graminearum*: new insights in the contribution of phenolic acids to resistance to deoxynivalenol accumulation in cereals. *Int. J. Food Microbiol.* 221, 61–68. doi: 10.1016/j.

ijfoodmicro.2016.01.005

52. Glenn, A. E., Hinton, D. M., Yates, I. E., and Bacon, C. W. (2001). Detoxification of corn antimicrobial compounds as the basis for isolating *Fusarium verticillioides* and some other *Fusarium* species from corn. *Appl. Environ. Microbiol.* 67, 2973–2981. doi: 10.1128/AEM.67.7.2973-2981.2001

53. Gorinstein, S., Lojek, A., Ciz, M., Pawelzik, E., Delgado-Licon, E., Medina, O. J., et al. (2008). Comparison of composition and antioxidant capacity of some cereals and pseudocereals. *Int. J. Food Sci. Tech.* 43, 629–637. doi: 10.1111/j.1365-2621.2007.01498.x

54. Guiraud, P., Steiman, R., Seigle-Murandi, F., and Benoitguyod, J. L. (1995). Comparison of the toxicity of various lignin-related phenolic-compounds toward selected fungi Perfecti and fungi Imperfecti. *Ecotoxicol. Environ. Saf.* 32, 29–33. doi: 10.1006/eesa.1995.1081

55. Gunnaiah, R., and Kushalappa, A. C. (2014). Metabolomics deciphers the host resistance mechanisms in wheat cultivar Sumai-3, against trichothecene producing and non-producing isolates of *Fusarium graminearum*. *Plant Physiol. Biochem.* 83, 40–50. doi: 10.1016/j.plaphy.2014.07.002

56. Gunnaiah, R., Kushalappa, A. C., Duggavathi, R., Fox, S., and Somers, D. J. (2012). Integrated metabolo-proteomic approach to decipher the mechanisms by which wheat QTL (Fhb1) contributes to resistance against *Fusarium graminearum*. *PLoS ONE* 7:7. doi: 10.1371/journal.pone.0040695

57. Gutierrez-Gonzalez, J. J., Wise, M. L., and Garvin, D. F. (2013). A developmental profile of tocol accumulation in oat seeds. *J. Cereal Sci.* 57, 79–83. doi: 10.1016/j.jcs.2012.10.001

58. Hamzehzarghani, H., Kushalappa, A. C., Dion, Y., and Rioux, S. (2005). Metabolic profiling and factor analysis to discriminate quantitative resistance in wheat cultivars against *fusarium* head blight. *Physiol. Mol. Plant Pathol.* 66, 119–133. doi: 10.1016/j.pmpp.2005.05.005

59. Hamzehzarghani, H., Paranidharan, V., Abu-Nada, Y., Kushalappa, A. C., Mamer, O., and Somers, D. (2008). Metabolic profiling to discriminate wheat near isogenic lines, with quantitative trait loci at chromosome 2DL, varying in resistance to *fusarium* head blight. *Can. J. Plant Sci.* 88, 789–797. doi: 10.4141/CJPS07209

60. Hazel, C. M., and Patel, S. (2004). Influence of processing on trichothecene levels. *Toxicol. Lett.* 153, 51–59. doi: 10.1016/j.toxlet.2004.04.040

61. Hu, Q. L., Zhang, L. J., Li, Y. N., Ding, Y. J., and Li, F. L. (2010). Purification and anti-fatigue activity of flavonoids from corn silk. *Int. J. Phys. Sci.* 5, 321–326. doi: 10.5897/IJPS

62. Humpf, H.-U., and Voss, K. A. (2004). Effects of thermal food processing on the chemical structure and toxicity of fumonisin mycotoxins. *Mol. Nutr. Food Res.* 48, 255–269. doi: 10.1002/mnfr.200400033

63. Iiyama, K., Lam, T., and Stone, B. A. (1994). Covalent cross-links in the cell wall. *Plant Physiol.* 104, 315–320.

64. PubMed Abstract | Iqbal, S. Z., Mustafa, H. G., Asi, M. R., and Jinap, S. (2014). Variation in vitamin E level and aflatoxins contamination in different rice varieties. *J. Cereal Sci.* 60, 352–355. doi: 10.1016/j.jcs.2014.05.012

65. Irakli, M. N., Samanidou, V. F., Biliaderis, C. G., and Papadoyannis, I. N. (2012). Development and validation of an HPLC-method for determination of free and bound phenolic acids in cereals after solid-phase extraction. *Food Chem.* 134, 1624–1632. doi: 10.1016/j.foodchem.2012.03.046

66. Jilek, M. L., and Bunzel, M. (2013). Dehydrotriferulic and dehydrodiferulic acid profiles of cereal and pseudocereal flours. *Cereal Chem.* 90, 507–514. doi: 10.1094/CCHEM-11-12-0144-R

67. Kang, Z. S., and Buchenauer, H. (2000). Cytology and ultrastructure of the infection of wheat spikes by *Fusarium culmorum*. *Mycol. Res.* 104, 1083–1093. doi: 10.1017/S0953756200002495

68. Karlovsky, P. (2011). Biological detoxification of the mycotoxin deoxynivalenol and its use in genetically engineered crops and feed additives. *Appl. Microbiol. Biotechnol.* 91, 491–504. doi: 10.1007/s00253-011-3401-5

69. Kumaraswamy, G., Bollina, V., Kushalappa, A., Choo, T., Dion, Y., Rioux, S., et al. (2011a). Metabolomics technology to phenotype resistance in barley against *Gibberella zeae*. *Eur. J. Plant Pathol.* 130, 29–43. doi: 10.1007/s10658-010-9729-3

70. Kumaraswamy, G. K., Kushalappa, A. C., Choo, T. M., Dion, Y., and Rioux, S. (2012). Differential metabolic response of barley genotypes, varying in resistance, to trichothecene-producing and -nonproducing (tri5-) isolates of *Fusarium graminearum*. *Plant Pathol.* 61, 509–521. doi: 10.1111/j.1365-3059.2011.02528.x

71. Kumaraswamy, K. G., Kushalappa, A. C., Choo, T. M., Dion, Y., and Rioux, S. (2011b). Mass spectrometry based metabolomics to identify potential biomarkers for resistance in barley against *Fusarium Head Blight* (*Fusarium graminearum*). *J. Chem. Ecol.* 37, 846–856. doi: 10.1007/s10886-011-9989-1

72. Li, W., Wei, C. V., White, P. J., and Beta, T. (2007). High-amylose corn exhibits better antioxidant activity than typical and waxy genotypes. *J. Agric. Food Chem.* 55, 291–298. doi: 10.1021/jf0622432

73. Lin, P. Y., and Lai, H. M. (2011). Bioactive compounds in rice during grain development. *Food Chem.* 127, 86–93. doi: 10.1016/j.foodchem.2010.12.092

74. Loiseau, J., Vu, B. L., Macherel, M. H., and Le Deunff, Y. (2001). Seed lipoxygenases: occurrence and functions. *Seed Sci. Res.* 11, 199–211. doi: 10.1079/SSR200176

75. Maresca, M. (2013). From the gut to the brain: journey and pathophysiological effects of the food-associated trichothecene mycotoxin deoxynivalenol.*Toxins.* 5:4. doi: 10.3390/toxins5040784

76. Martyniuk, S., Stochmal, A., Macías, F. A., Marín, D., and Oleszek, W. (2006). Effects of some benzoxazinoids on *in vitro* growth of *Cephalosporium gramineum* and other fungi pathogenic to cereals and on Cephalosporium stripe of winter wheat. *J. Agric. Food Chem.* 54, 1036–1039. doi: 10.1021/jf050901x

77. Masood, A., Dogra, J. V. V., and Jha, A. K. (1994). The influence of coloring and pungent agents of red chilli (*Capsicum-Annum*) on growth and aflatoxin production by *Aspergillus flavus*. *Lett. Appl. Microbiol.* 18, 184–186. doi: 10.1111/j.1472-765X.1994.tb00841.x

78. Mauch-Mani, B., and Mauch, F. (2005). The role of abscisic acid in plant-pathogen interactions. *Curr. Opin. Plant Biol.* 8, 409–414. doi: 10.1016/j.pbi.2005.05.015

79. McCallum, J. A., and Walker, J. R. L. (1990). Phenolic biosynthesis during grain devlopment in wheat (*Triticum aestivum* L.). Changes in phenylananine-ammonio-lyase activity and soluble phenolic content. *J. Cereal Sci.* 11, 35–49. doi: 10.1016/S0733-5210(09)80179-3

80. McKeehen, J. D., Busch, R., and Fulcher, R. (1999). Evaluation of wheat (*Triticum aestivum* L.) phenolic acids during grain development and their contribution to *Fusarium* resistance. *J. Agric. Food Chem.* 47, 1476–1482. doi: 10.1021/jf980896f

81. Mellado-Ortega, E., and Hornero-Mendez, D. (2016). Carotenoid evolution during short-storage period of durum wheat (*Triticum turgidum* cony. durum) and tritordeum (x Tritordeum Ascherson et Graebner) whole-grain flours. *Food Chem.* 192:714–723. doi: 10.1016/j.foodchem.2015.07.057

82. Mellado-Ortega, E., and Hornero-Méndez, D. (2015). Carotenoid profiling of Hordeum chilense grains: the parental proof for the origin of the high carotenoid content and esterification pattern of tritordeum. *J. Cereal Sci.* 62, 15–21. doi: 10.1016/j.jcs.2014.12.005

83. Merhej, J., Richard-Forget, F., and Barreau, C. (2011). Regulation of trichothecene biosynthesis in *Fusarium*: recent advances and new insights. *Appl. Microbiol. Biotechnol.* 91, 519–528. doi: 10.1007/s00253-011-3397-x

84. Merrill, A. H. Jr., Sullards, M. C., Wang, E., Voss, K. A., and Riley, R. T. (2001). Sphingolipid metabolism: roles in signal transduction and disruption by fumonisins. *Environ. Health Perspect.* 109, 283–289. doi: 10.1289/ehp.01109s2283

85. Mesterhazy, A. (1995). Types and components of resistance to *Fusarium* head blight of wheat. *Plant Breeding* 114, 377–386. doi: 10.1111/j.1439-0523.1995.tb00816.x

86. Mesterhazy, A. (2002). Role of deoxynivalenol in aggressiveness of *Fusarium graminearum* and F. culmorum and in resistance to *Fusarium* head blight.*Eur. J. Plant Pathol.* 108, 675–684. doi: 10.1023/A:1020631114063.

87. Mierziak, J., Kostyn, K., and Kulma, A. (2014). Flavonoids as important molecules of plant interactions with the environment. *Molecules* 19, 16240–16265. doi: 10.3390/molecules191016240

88. Milicevic, D. R., Skrinjar, M., and Baltic, T. (2010). Real and perceived risks for mycotoxin contamination in foods and feeds: challenges for food safety control. *Toxins.* 2:4. doi: 10.3390/toxins2040572

89. Miller, J. D., Fielder, D. A., Dowd, P. F., Norton, R. A., and Collins, F. W. (1996). Isolation of 4-acetyl-benzoxazolin-2-one (4-ABOA) and diferuloylputrescine from an extract of gibberella ear rot-resistant corn that blocks mycotoxin biosynthesis, and the insect toxicity of 4-ABOA and related compounds. *Biochem. Sys. Ecol.* 24, 647–658. doi: 10.1016/S0305-1978(96)00050-6

90. Miller, J. D., Young, J. C., and Sampson, D. R. (1985). Deoxynivalenol and *Fusarium* head blight resistance in spring cereals. *Phytopathology* 113, 359–367. doi: 10.1111/j.1439-0434.1985.tb04837.x

91. Montibus, M., Pinson-Gadais, L., Richard-Forget, F., Barreau, C., and Ponts, N. (2015). Coupling of transcriptional response to oxidative stress and secondary metabolism regulation in filamentous fungi. *Crit. Rev. Microbiol.* 41, 295–308. doi: 10.3109/1040841X.2013.829416

92. Moreau, R. A., Nuñez, A., and Singh, V. (2001). Diferuloylputrescine and *p*-coumaroyl-feruloylputrescine, abundant polyamine conjugates in lipid extracts of maize kernels. *Lipids* 36, 839–844. doi: 10.1007/s11745-001-0793-6

93. Ndolo, V. U., and Beta, T. (2013). Distribution of carotenoids in endosperm, germ, and aleurone fractions of cereal grain kernels. *Food Chem.* 139, 663–671. doi: 10.1016/j.foodchem.2013.01.014

94. Ndolo, V. U., and Beta, T. (2014). Comparative studies on composition and distribution of phenolic acids in cereal grain botanical fractions. *Cereal Chem.*91, 522–530. doi: 10.1094/CCHEM-10-13-0225-R

95. Nicholson, R. L., and Hammerschidt, R. (1992). Phenolic compounds and their role in disease resistance. *Ann. Rev. Phytopathol.* 30, 369–389. doi: 10.1146/annurev.py.30.090192.002101

96. Niemeyer, H. M. (2009). Hydroxamic acids derived from 2-hydroxy-2H-1,4-benzoxazin-3(4H)-one: key defense chemicals of cereals. *J. Agric. Food Chem.*57, 1677–1696. doi: 10.1021/jf8034034

97. Nordkvist, E., and Haggblom, P. (2014). *Fusarium* mycotoxin contamination of cereals and bedding straw at Swedish pig farms. *Anim. Feed Sci. Technol.*198, 231–237. doi: 10.1016/j.anifeedsci.2014.10.002

98. Norton, R. A. (1999). Inhibition of aflatoxin B-1 biosynthesis in *Aspergillus flavus* by anthocyanidins and related flavonoids. *J. Agric. Food Chem.* 47, 1230–1235. doi: 10.1021/jf980995t

99. Palumbo, J. D., O'Keeffe, T. L., and Mahoney, N. E. (2007). Inhibition of ochratoxin—A production and growth of *Aspergillus* species by phenolic antioxidant compounds. *Mycopathologia* 164, 241–248. doi: 10.1007/s11046-007-9057-0

100. Pani, G., Scherm, B., Azara, E., Balmas, V., Jahanshiri, Z., Carta, P., et al. (2014). Natural and natural-like phenolic inhibitors of type B trichothecene *in vitro* production by the wheat (*Triticum* sp.) pathogen *Fusarium culmorum*. *J. Agric. Food Chem.* 62, 4969–4978. doi: 10.1021/jf500647h

101. Paranidharan, V., Abu-Nada, Y., Hamzehzarghani, H., Kushalappa, A. C., Mamer, O., Dion, Y., et al. (2008). Resistance-related metabolites in wheat against *Fusarium graminearum* and the virulence factor deoxynivalenol (DON). *Bot. Botaniq.* 86, 1168–1179. doi: 10.1139/B08-

052

102. Passone, M. A., Ruffino, M., Ponzio, V., Resnik, S., and Etcheverry, M. G. (2009). Postharvest control of peanut *Aspergillus* section *Flavi* populations by a formulation of food-grade antioxidants. *Int. J. Food Microbiol.* 131, 211–217. doi: 10.1016/j.ijfoodmicro.2009.02.023

103. Paul, S. S., Kamra, D. N., Sastry, V. R. B., Sahu, N. P., and Kumar, A. (2003). Effect of phenolic monomers on biomass and hydrolytic enzyme activities of an anaerobic fungus isolated from wild nil gai (*Baselophus tragocamelus*). *Lett. Appl. Microbiol.* 36, 377–381. doi: 10.1046/j.1472-765X.2003.01331.x

104. Petti, C., Reiber, K., Ali, S. S., Berney, M., and Doohan, F. M. (2012). Auxin as a player in the biocontrol of *Fusarium* head blight disease of barley and its potential as a disease control agent. *BMC Plant Biol.* 12:224. doi: 10.1186/1471-2229-12-224

105. Picot, A., Atanasova-Pénichon, V., Pons, S., Marchegay, G., Barreau, C., Pinson-Gadais, L., et al. (2013). Maize kernel antioxidants and their potential involvement in *Fusarium Ear Rot* resistance. *J. Agric. Food Chem.* 61, 3389–3395. doi: 10.1021/jf4006033

106. Picot, A., Barreau, C., Pinson-Gadais, L., Caron, D., Lannou, C., and Richard-Forget, F. (2010). Factors of the *Fusarium verticillioides*-maize environment modulating fumonisin production. *Crit. Rev. Microbiol.* 36, 221–232. doi: 10.3109/10408411003720209

107. Picot, A., Barreau, C., Pinson-Gadais, L., Piraux, F., Caron, D., Lannou, C., et al. (2011). The dent stage of maize kernels is the most conducive for fumonisin biosynthesis under field conditions. *Appl. Environ. Microbiol.* 77, 8382–8390. doi: 10.1128/AEM.05216-11

108. Pihlava, J. M., Nordlund, E., Heinio, R. L., Hietaniemi, V., Lehtinen, P., and Poutanen, K. (2015). Phenolic compounds in wholegrain rye and its fractions.*J. Food Comp. Anal.* 38, 89–97. doi: 10.1016/j.jfca.2014.10.004

109. Pirgozliev, S. R., Edwards, S. G., Hare, M. C., and Jenkinson, P. (2003). Strategies for the control of *Fusarium* head blight in cereals. *Eur. J. Plant Pathol.* 109, 731–742. doi: 10.1023/A:1026034509247

110. Ponts, N. (2005). *Inluence de Stress Oxydatiss sur la Biosynthèse de Mycotoxines de Fusarium* spp. Contaminantes de l'épi de maïs. Thèse de doctorat. Université Bordeaux, France.

111. Ponts, N., Pinson-Gadais, L., Boutigny, A. L., Barreau, C., and Richard-Forget, F. (2011). Cinnamic-derived acids significantly affect *Fusarium graminearum* growth and *in vitro* synthesis of type B trichothecenes. *Phytopathology* 101, 929–934. doi: 10.1094/PHYTO-09-10-0230

112. Proctor, R. H., Brown, D. W., Plattner, R. D., and Desjardins, A. E. (2003). Co-expression of 15 contiguous genes delineates a fumonisin biosynthetic gene cluster in *Gibberella moniliformis*. *Fungal Genet. Biol.* 38, 237–249. doi: 10.1016/S1087-1845(02)00525-X

113. Régnier, T., and Macheix, J.-J. (1996). Changes in wall-bound phenolic acids, phenylalanine and tyrosine ammonia-lyases, and peroxidases in developing durum wheat grains (*Triticum turgidum* L. Var. Durum). *J. Agric. Food Chem.* 44, 1727–1730. doi: 10.1021/jf950607c

114. Reid, L. M., Mather, D. E., and Arnason, J. T. (1992). Changes in phenolic constituents of maize silk infected with *Fusarium graminearum*. *Can. J. Bot.*70, 1697–1702. doi: 10.1139/b92-209

115. Rodríguez-Suárez, C., Mellado-Ortega, E., Hornero-Méndez, D., and Atienza, S. (2014). Increase in transcript accumulation of Psy1 and e-Lcy genes in grain development is associated with differences in seed carotenoid content between durum wheat and tritordeum. *Plant Mol. Biol.* 84, 659–673. doi: 10.1007/s11103-013-0160-y

116. Roleira, F. M. F., Siquet, C., Orrù, E., Garrido, E. M., Garrido, J., Milhazes, N., et al. (2010). Lipophilic phenolic antioxidants: correlation between antioxidant profile, partition coefficients and redox properties. *Bioorg. Med. Chem.* 18, 5816–5825. doi: 10.1016/j.bmc.2010.06.090

117. Salas, M. P., Reynoso, C. M., Celiz, G., Daz, M., and Resnik, S. L. (2012). Efficacy of flavanones obtained from citrus residues to prevent patulin contamination. *Food Res. Int.* 48, 930–934. doi: 10.1016/j.foodres.2012.02.003

118. Samapundo, S., De Meulenaer, B., Osei-Nimoh, D., Lamboni, Y., Debevere, J., and Devlieghere, F. (2007). Can phenolic compounds be used for the protection of corn from fungal invasion and mycotoxin contamination during storage? *Food Microbiol.* 24, 465–473. doi: 10.1016/j.fm.2006.10.003

119. Sampietro, D. A., Fauguel, C. M., Vattuone, M. A., Presello, D. A., and Catalan, C. A. N. (2013). Phenylpropanoids from maize pericarp: resistance factors to kernel infection and fumonisin accumulation by *Fusarium verticillioides*. *Eur. J. Plant Pathol.* 135, 105–113. doi: 10.1007/s10658-012-0069-3

120. Sandmann, G., Römer, S., and Fraser, P. D. (2006). Understanding carotenoid metabolism as a necessity for genetic engineering of crop plants. *Metab. Eng.* 8, 291–302. doi: 10.1016/j.ymben.2006.01.005

121. Santiago, R., Reid, L. M., Arnason, J. T., Zhu, X., Martinez, N., and Malvar, R. A. (2007). Phenolics in maize genotypes differing in susceptibility to*Gibberella* stalk rot (*Fusarium graminearum Schwabe*). *J. Agric. Food Chem.* 55, 5186–5193. doi: 10.1021/jf070641e

122. Santos, L., Kasper, R., Gil-Serna, J., Marin, S., Sanchis, V., and Ramos, A. J. (2010). Effect of Capsicum carotenoids on growth and ochratoxin A production by chilli and paprika *Aspergillus* spp. isolates. *Int. J. Food Microbiol.* 142, 354–359. doi: 10.1016/j.ijfoodmicro.2010.07.018

123. Sanzani, S. M., Schena, L., Nigro, F., De Girolamo, A., and Ippolito, A. (2009). Effect of quercetin and umbelliferone on the transcript level of *Penicillium*expansum genes involved in patulin biosynthesis. *Eur. J. Plant Pathol.* 125, 223–233. doi: 10.1007/s10658-009-9475-6

124. Schatzmayr, G., and Streit, E. (2013). Global occurrence of mycotoxins in the food and feed chain: facts and figures. *World Mycotoxin J.* 6, 213–222. doi: 10.3920/WMJ2013.1572

125. Schroeder, H. W., and Christensen, J. J. (1963). Factors affecting resistance of wheat to scab caused by *Gibberella*

zeae. Phytopathology 53, 831–838.

126. Shewry, P. R., Mitchell, R. A. C., Tosi, P., Wan, Y. F., Underwood, C., Lovegrove, A., et al. (2012). An integrated study of grain development of wheat (cv. Hereward). *J. Cereal Sci.* 56, 21–30. doi: 10.1016/j.jcs.2011.11.007
127. Silva, A. M. S., Weidenborner, M., and Cavaleiro, J. A. S. (1998). Growth control of different *Fusarium* species by selected flavones and flavonoid mixtures. *Mycol. Res.* 102, 638–640. doi: 10.1017/S0953756297005480
128. Siranidou, E., Kang, Z., and Buchenauer, H. (2002). Studies on symptom development, phenolic compounds and morphological defence responses in wheat cultivars differing in resistance to *Fusarium* Head Blight. *J. Phytopathol.* 150, 200–208. doi: 10.1046/j.1439-0434.2002.00738.x
129. Søltoft, M., Jørgensen, L. N., Svensmark, B., and Fomsgaard, I. S. (2008). Benzoxazinoid concentrations show correlation with *Fusarium* Head Blight resistance in Danish wheat varieties. *Biochem. Syst. Ecol.* 36, 245–259. doi: 10.1016/j.bse.2007.10.008
130. Sreenivasulu, N., Radchuk, V., Alawady, A., Borisjuk, L., Weier, D., Staroske, N., et al. (2010). De-regulation of abscisic acid contents causes abnormal endosperm development in the barley mutant seg8. *Plant J.* 64, 589–603. doi: 10.1111/j.1365-313X.2010.04350.x
131. Streit, E., Schatzmayr, G., Tassis, P., Tzika, E., Marin, D., Taranu, I., et al. (2012). Current situation of mycotoxin contamination and co-occurrence in animal feed—focus on Europe. *Toxins* 4:10. doi: 10.3390/toxins4100788
132. Sung, W. S., and Lee, D. G. (2010). Antifungal action of chlorogenic acid against pathogenic fungi, mediated by membrane disruption. *Pure Appl. Chem.*82, 219–226. doi: 10.1351/PAC-CON-09-01-08
133. Tan, B. C., Schwartz, S. H., Zeevaart, J. A. D., and McCarty, D. R. (1997). Genetic control of abscisic acid biosynthesis in maize. *Proc. Natl. Acad. Sci. U.S.A.* 94, 12235–12240. doi: 10.1073/pnas.94.22.12235
134. Teodoro, G. R., Ellepola, K., Seneviratne, C. J., and Koga-Ito, C. Y. (2015). Potential use of phenolic acids as anti-Candida agents: a review. *Front. Microbiol.* 6:1420. doi: 10.3389/fmicb.2015.01420
135. Terzi, V., Tumino, G., Stanca, A. M., and Morcia, C. (2014). Reducing the incidence of cereal head infection and mycotoxins in small grain cereal species. *J. Cereal Sci.* 59, 284–293. doi: 10.1016/j.jcs.2013.10.005
136. Treutter, D. (2005). Significance of flavonoids in plant resistance and enhancement of their biosynthesis. *Plant Biol.* 7, 581–591. doi: 10.1055/s-2005-873009
137. Treutter, D. (2006). Significance of flavonoids in plant resistance: a review. *Environ. Chem. Lett.* 4, 147–157. doi: 10.1007/s10311-006-0068-8
138. Zilic, S., Sukalovic, V. H. T., Dodig, D., Maksimovic, V., Maksimovic, M., and Basic, Z. (2011). Antioxidant activity of small grain cereals caused by phenolics and lipid soluble antioxidants. *J. Cereal Sci.* 54, 417–424. doi: 10.1016/j.jcs.2011.08.006

CHAPTER 18

BIOPHYSICAL, SOCIO-ECONOMIC, AND INSTITUTIONAL CONSTRAINTS FOR PRODUCTION AND FLOW OF CEREALS IN ETHIOPIA

Gemechu Shale Ogato

Keywords: Cereals, ecosystem-based adaptation, Ethiopia, food Security, price, sustainable production

INTRODUCTION

With 85% of the population living in the rural areas and depending on agriculture for livelihood, there is no doubt for the economic importance of the agricultural sector for sustainable development and poverty reduction in Ethiopia. In other words, the agricultural sector accounts for more than 40% of national GDP, 90% of exports, and provides basic needs and income to more than 90% of the poor. Moreover, a better performed agricultural sector has provided growth to the overall economy, improved the food security and reduced poverty in the recent years (Alemu, 2005; Cooper et al., 2008; Ludi, 2009; Diao, 2010).

Ethiopia has been well known for its agricultural development challenge given its large and rapid growing population and limited and deteriorated land resource. Scholars in the field contend that these two factors together have caused extreme land shortage in the highland of Ethiopia, the area most population lives and most agricultural production occurs (IFOAM, 2009; Diao, 2010; Asenso-Okyere & Jemaneh, 2012). For instance, Diao (2010) asserts that population pressure has led to expanded cultivation into forest areas and steep slopes. This creates serious repercussion for the environment, which, together with fluctuation in rainfall, have made agricultural production very vulnerable to weather shock. Odendo et al. (2010) also affirm that soil fertility degradation on smallholder farms has been the fundamental biophysical cause of food insecurity and poverty in sub-Saharan African countries like Ethiopia where most of the people live in rural areas and derive their livelihoods from agriculture.

Citation: Gemechu Shale Ogato. Biophysical, socio-economic, and institutional constraints for production and flow of cereals in Ethiopia, American Journal of Human Ecology, 2014; Vol. 3, No. 3, 51-71. DOI: 10.11634/216796221403571.

The rationale for focusing on cereals in this study is the place of cereals in production and consumption in Ethiopia. In other words, cereals are by far the largest group in terms of their share in area cultivated, output, and consumption. Moreover, cereal production and marketing is the single largest sub-sector within Ethiopia's agriculture. It dominates in terms of its share in rural employment, agricultural land use, and calorie intake, as well as its contribution to national income. In other words, the sub-sector accounts for roughly 60 percent of rural employments, about 73 percent of total cultivated land, more than 40 percent of a typical household's food expenditure, and more than 60 percent of total caloric intake of a typical household in the country (Hassan & Nhemachena, 2008; Diao, 2010; Rashid & Negassa, 2011). Barley, maize, teff, wheat, and sorghum are the most important crops for Ethiopian agriculture. According to Diao (2010), while 64% of agricultural value added comes from crops, more than 70% of crop land is devoted to cereal production.

According to World Bank (2011), Global circulation models predict an increase in temperature of 1.7 to 2.1 degrees Celsius for Ethiopia by 2050. The predictions regarding rainfall for Ethiopia are much less consistent; but increasing unpredictability, together with more frequent and intense patterns of extreme weather, is likely to be the trend. The combination of higher temperatures and more unpredictable rains has negative implications for the length and reliability of the growing season. Parts of Ethiopia have abundant surface water resources, but so far irrigated agriculture is not well developed (World Bank, 2011).

Low levels of socio-economic development, limited infrastructure, lack of institutional capacity and a higher dependency on natural resources result in Ethiopia being highly vulnerable to climate variability and extreme climate events (Cooper et al., 2008; Hassan & Nhemachena, 2008; Colls, et al., 2009; Ludi, 2009; Edame et al., 2011; Shiferaw et al., 2011; USAID, 2011; Asenso-Okyere & Jemaneh, 2012). Moreover, recurrent drought events and political instability in the past have resulted in famines and migration of people. The country is also vulnerable to severe flooding and associated soil erosion of the fertile Ethiopian highlands (Colls, et al., 2009). According to Edame et al. (2011), climate change could lead to increased water stress, decreased biodiversity, damaged ecosystems, rising sea levels, and potentially, to social conflict due to increased competition over limited natural resources.

Food insecurity in Ethiopia is persistently caused by a combination of factors that include recurrent drought; the flooding that has become more frequent in flood prone areas along the main river basins; small land holdings with an average of 0.5 to 2 hectares per household associated with population growth. In addition, factors such as death, job loss, crop damage and animal deaths increase household vulnerability in different regions. The most recent risk phenomenon is the price risk especially of grains which constitute the major diet in Ethiopia. These hazards and shocks increase the vulnerability of both most rural and some of the urban households. Added to this is the economic shock of increasing food prices. Rural-urban migration is a common phenomenon, further encouraged by land shortages in rural areas and the perception that food, health services and jobs are more easily available in urban centres. These major shocks have important implications for the welfare of both urban and rural households (WFP-Ethiopia, 2009).

Scholars of ecosystem-based adaptation in agricultural sector contend that ecosystem-based adaptation practices like organic agriculture encourage the use of local and indigenous farmer knowledge and observation techniques and recognize the critical role of women throughout the entire food chain, as farmers, consumers and mothers. However, their agricultural systems have attracted little attention from research and development and are usually assumed to be inherently unproductive. Nevertheless, smallholder farmers throughout the world have developed a multitude of practices and innovations that should be seen for what they are; the basis for any realistic development – including productivity improve-ments (Cooper et al., 2008; Hassan & Nhemachena, 2008; IFOAM, 2009). A major barrier to food security and accessibility in the least developed countries like Ethiopia is lack of adoption of affordable, productive and resilient farming systems by many of their smallholder farmers. Many of these farmers are relatively poor and unproductive and highly vulnerable to climate change and other biophysical constraints (Cooper et al., 2008; Hassan & Nhemachena, 2008). Scholars of ecosystem-based adaptation in agricultural sector contend that optimizing productivity and building resilience of these farms is critical for improving local food security and accessibility. For instance, IFOAM (2009), attest that local food production in smallholder farms can often be increased significantly through improving the use of locally available resources and agro-ecological methods for soil fertility building and pest prevention. In other words, eco-functional intensification increases productivity and enhances food security especially in challenging environments such as water scarce regions (IFOAM, 2009).

Several studies have been undertaken to assess constraints for production and flow of cereals in Ethiopia. Research in this area has predominantly focused on independent assessment of biophysical, socio-economic or institutional constraints for production and flow of cereals in Ethiopia. Comprehensive studies on the biophysical, socio-economic, and institutional constraints for production and flow of cereals and the interaction among these constraints were given less attention. Hence, this study assessed the bio-physical, socio-economic, and institutional constraints for production and flow of cereals and forward strategic measures to address them in a sustainable manner. Such analysis can contribute to a better understanding of the relevant growth process and, as a consequence, to the design of sustainable agricultural development policy and strategy in Ethiopia.

The study answered the following research questions: 1. what are the bio-physical, socio-economic and institutional constraints associated with cereals' production and flow in Ethiopia? 2. What are the strengths, weaknesses, opportunities, and threats of Ethiopian policy and strategy in agricultural sector in responding to bio-physical, socio-economic, and institutional constraints? 3.

What does the trends of major cereals' annual production in Ethiopia looks like? 4. What does the trends of major cereals' annual prices in Addis Ababa grain market looks like?

The general objective of the study is to assess the bio-physical, socio-economic, and institutional constraints for production and flow of cereals in Ethiopia and forward strategic measures to deal with these constraints. The specific objectives of the study are: To assess bio-physical, socio-economic, and institutional constraints associated with cereals' production and flow in Ethiopia; To analyse the weaknesses, strengths, opportunities, and threats of the current Ethiopian agricultural development policy and strategies in dealing with constraints for production and flow of cereals; To analyse the trends of major cereals' annual production in Ethiopia; and To analyse the trends of major cereals' annual prices in Addis Ababa grain market.

RESEARCH METHODOLOGY

A comprehensive review of the existing literature was made in order to obtain both theoretical insights and secondary data on the review themes. The secondary sources or existing literatures in reference to the review themes conceptualized in the study were carefully selected and consulted for secondary data and information. In other words, robust secondary sources on the review themes in the context of Ethiopia were identified for the purpose of this study. The selected review themes were: the conceptual linkage between ecosystem-based adaptation to climate change and food security in Ethiopia; biophysical, socio-economic, and institutional constraints for production and flow of cereals in Ethiopia; strengths, weaknesses, opportunities, and threats of the current Ethiopian agricultural development policy and strategies in dealing with constraints for cereals' production and flow; trends of major cereals' annual production in Ethiopia (2000-2013); and trends of major cereals' annual prices in Addis Ababa grain market (2005-2010).

Finally the collected secondary data were analysed through both qualitative and quantitative techniques and presented under relevant themes. Qualitative data were analysed through narration and description. On the other hand, quantitative production and price data collected through secondary method of data collection from the Ethiopian Grain Trade Enterprise and Central Statistical Agency of Ethiopia were analyzed in computer with the help of SPSS-Version 20 software and Micro-Soft Excel 2007 version. Accordingly, simple descriptive statistics like frequencies and percentages, mean, and standard deviations were employed. The results of the analyzed data were presented with the help of line graphs, bar graphs, pie charts, and tables.

RESULTS AND DISCUSSION

This section of the seminar paper analyses and discusses the conceptual linkage between ecosystem-based adaptation to climate change and food security in Ethiopia; biophysical, socio-economic, and institutional constraints for production and flow of cereals in Ethiopia; strengths, weaknesses, opportunities, and threats of policies and strategies in the Ethiopian agricultural sector in dealing with constraints for cereals' production and flow; trends of major cereals' annual production in Ethiopia (2000-2013); and trends of major cereals' annual prices in Addis Ababa grain market (2005-2010).

The Conceptual Linkage between Ecosystem-Based Adaptation to Climate change and Food security in Ethiopia

Vulnerability to climatic change may refer to inability among households to engage in strategies to cope with and adjust to extremes such as droughts that form part of current climate variability and which may increase in frequency and/or intensity in future (Eriksen et al., 2007). Climate change is already causing severe problems of drought, flooding and unpredictable weather, creating losses in food production and destroying peoples' livelihoods (Eriksen et al., 2007; Gebre-Egziabher, 2007; Nærstad, 2007; Cotter & Tirado, 2008; Edame et al., 2011). Causes for vulnerability of Ethiopia to climate variability and change include very high dependence on rain-fed agriculture which is very sensitive to climate variability and change, under-development of water resources, low health service coverage, high population growth rate, low economic development level, low adaptive capacity, inadequate road infrastructure in drought prone areas, weak institutions, lack of awareness, etc (Georgis et al., 2009). According to Lamboll et al. (2011), agriculture is a cause of climate change – as it is a major emitter of GHGs – but will also be affected by it. Impacts will not be felt evenly. For instance, smallholders' crops and animals, production and livelihoods will be affected directly and indirectly through off-site impacts and as a result of climate change responses. Moreover, climate change impacts will be greatest where they interact with other shocks, stresses and vulnerabilities (Lamboll et al., 2011).

Vulnerability assessment based on existing information and rapid assessments carried out indicated that the most vulnerable sectors to climate variability and change are Agriculture, Water and Human health. In terms of livelihood approach smallholder rain-fed farmers and pastoralists are found to be the most vulnerable. The arid, semi-arid and the dry sub-humid parts of the country are affected most by drought (Georgis et al., 2009). Climate change worsens the living conditions for many who are already vulnerable, particularly in developing countries because of the lack of assets and adequate insurance coverage.

According to Edame et al. (2011), climate change impacts the four key dimensions of food security – availability, stability, access, and utilization. For instance, availability of agricultural products is affected by climate change directly through its impacts on crop yields, crop pests and diseases, and soil fertility and water-holding properties. It is also affected by climate change indirectly through its impacts on economic growth, income distribution, and agricultural demand. In addition, stability of crop yields and food supplies is negatively affected by variable weather conditions. Physical, economic, and social access to food would be affected negatively by climate change as agricultural production declines, food prices rise, and purchasing power decreases. Last but not least, climate change poses threats to food utilization through effects on human health and the spread of diseases in geographical areas which were previously not affected (Cotter & Tirado, 2008; Ludi, 2009; Edame et al., 2011).

Responses to climate change and climate variability are usually grouped into two main categories: mitigation (addressing causes) and adaptation (addressing effects). In agricultural adaptation there is a need for new technologies and farm-level innovations, but also changes in broader institutional arrangement (eg, greater equity in land ownership, adaptive management in relevant organisations). There is a range of options to generate incremental changes at farm level – eg, adaptation of agricultural practices, adapting livestock, pasture and rangeland management, farm-level climate change mitigation practices, and diversification of species and varieties (Cotter & Tirado, 2008; Ludi, 2009; Lamboll et al., 2011).

Adaptation covers reactive and proactive actions taken on different levels by individuals, communities, private companies and public bodies such as governments. Given the close linkages between climate change and the MDGs, it must also be noted that sustainable poverty reduction is a key adaptation strategy, with regard to food security but also in general (Ludi, 2009). Some of the appropriate adaptation strategies for African smallholder farmers include: crop diversification; using different crop varieties; varying the planting and harvesting dates; increasing the use of irrigation; increasing the use of water and soil conservation techniques; shading and shelter; shortening the length of the growing season; diversifying from farming to non–farming activities; increasing diversification by planting crops that are drought tolerant and/or resistant to temperature stresses; taking full advantage of the available water and making efficient use of it; and growing a variety of crops on the same plot or on different plots, thus reducing the risk of complete crop failure since different crops are affected differently by climate events (see also Box 1).

Box 1: Climate Change Adaptation Strategies for African Smallholder Farmers
List of appropriate adaptation strategies for African smallholder farmers include:

- Intensification of food production by smallholders through better access to improved seed, soil fertility management (eg, fertilizer application) and reliable water supply;
- Improved agricultural water management (AWM) (smallholder irrigation, rainwater harvesting, sustainable extraction of groundwater and other underutilized water resources), conservation agriculture and improved on-farm water use efficiency;
- Shifts towards crop and livestock types/varieties/breeds with greater drought and heat tolerance and improved pest and disease resistance;
- Enterprise diversification towards higher value crops, value adding (processing), off-farm employment, and marketing infrastructure;
- Grain storage improvements (from household to national levels) to ensure security of carryover stocks and access to surpluses;
- Climate forecasting and provision of timely advice to governments, private sector (agro-dealers), extension services and farmers;
- Weather-related crop and livestock insurance;
- A more efficient use of water through drip irrigation and the choice of high yielding and high-value crops;
- Bunds, agroforestry, crop rotation and rainwater harvesting;
- Agricultural diversification such as the integration of livestock and crops (mixed farming);
- Migration to wetter regions (from drier to wetter regions), in pursuit of wetter and more fertile lands; and
- Engagement in off-farm activities

Source: (Cotter and Tirado, 2008; Hassan & Nhemachena, 2008; Ngigi, 2009; Shiferaw et al., 2011).

Ecosystem-based adaptation may be defined as the adaptation policies and measures that take into account the role of ecosystem services in reducing the vulnerability of society to climate change, in a multi-sectoral and multi-scale approach. Moreover, it involves national and regional governments, local communities, private companies and NGOs in addressing the different pressures on ecosystem services, including land use change and climate change, and managing ecosystems to increase the resilience of people and economic sectors to climate change (Vignola et al., 2009). In other words, it integrates the use of biodiversity and ecosystem services into an overall strategy to help people adapt to the adverse impacts of climate change. Furthermore, it includes the sustainable management, conservation and restoration of ecosystems to provide services that help people adapt to both current climate variability, and climate change. For instance, it contributes to reducing vulnerability and increasing resilience to both climate and non-climate risks and provides multiple benefits to society and the environment (Colls, et al., 2009; Ludi, 2009). In a nutshell, the ecosystem approach to climate change adaptation involves the integrated management of land, water and other resources that promote their conservation and sustainable use in an equitable way (Cotter & Tirado, 2008; Ngigi, 2009; Ludi, 2009).

According to Ngigi (2009), one of the important characteristics of an adaptation strategy is that it should reflect the needs and aspirations of the society or community it is meant to benefit. Thus, the most effective mechanisms are flexible and relatively independent of scale. In other words, adaptation efforts must be coordinated across sectors and between agencies, which is a challenge in practice. In most cases, measures will mean a strengthening of existing policies, emphasizing the importance of basing climate change policies on existing coping mechanisms and the necessity of integrating them into national development plans (Ngigi, 2009; Ludi, 2009).

Vignola et al. (2009) contend that ecosystem degradation and vulnerability to climate change should be considered as development issues rather than strictly environmental problems. In other words, as the loss of natural capital and the associated vulnerabilities are a threat for sustainable development, national development policies should integrate ecosystem management and adaptation to climate change (Cotter & Tirado, 2008). For Nærstad (2007) agriculture and environment are inseparable. In other words, preservation of biodiversity and other natural resources is a prerequisite for long term food security and to eradicate hunger and poverty. Despite the aforementioned connection, the loss of forests and biodiversity, and the destruction of other natural resources like watersheds, pastures, soil and mangrove, are causing hunger, poverty and the destruction of livelihoods for millions of people, and are undermining the very basis of life for future generations in least developed countries like Ethiopia (Nærstad, 2007; Cotter & Tirado, 2008).

Agro-biodiversity provides ecosystem services which can contribute to the improvement of agricultural productivity (MEA, 2005; Cotter & Tirado, 2008; Hadgu et al., 2009). These ecosystem services include: yield improvement by intercropping; mixed cropping, and soil fertility enrichment (e.g. soil nitrogen) by perennial plants in agricultural landscapes; and insurance for agricultural production by increasing resilience, i.e. speedy recovery from a disturbance or stress, and decreasing the risk of crop failure (Vernooy & Song, 2004; Hadgu et al., 2009).

The agro-ecosystem approach is a strategy for the integrated management of land, water and living resources that promotes conservation and sustainable use in an equitable way. In other words, the application of the ecosystem approach will help to reach a balance of the three objectives of conservation, sustainable use, and the fair and equitable sharing of the benefits arising out of the utilization of genetic resources. Moreover, an ecosystem approach is based on the application of appropriate scientific methodologies focused on levels of biological organization, which encompass the essential structure, processes, functions and interactions among organisms and their environment. It also recognizes that humans, with their cultural diversity, are an integral component of many ecosystems and requires adaptive management to deal with the complex and dynamic nature of ecosystems and the knowledge or understanding of their functioning (Vernooy & Song, 2004; MEA, 2005; Stroud & Khandelwal, 2006; Cotter & Tirado, 2008; Colls, et al., 2009).

Ecosystem-based Adaptation reduces vulnerability to both climate and non-climate risks and provides multiple economic, social, environmental and cultural benefits, including (Cotter & Tirado, 2008; Ngigi, 2009; Colls, et al., 2009; Ludi, 2009): disaster risk reduction; livelihood sustenance and food security; biodiversity conservation; carbon sequestration; and sustainable water management.

Ecosystem-based agricultural practices like organic agriculture plays vital role for ensuring local and national food security (IFOAM, 2009). Organic agriculture is a production system that sustains the health of soils, ecosystems and people. It relies on ecological processes, biodiversity and cycles adapted to local conditions, rather than the use of inputs with adverse effects. It also combines tradition, innovation and science to benefit the shared environment and promote fair relationships and a good quality of life for all involved (IFOAM, 2009). Some scholars of sustainable agriculture claim that organic farming produces higher crop yields than chemical-intensive farming. It can also lead to environmental benefits such as improved soil fertility, better water retention and resistance to drought (Vernooy & Song, 2004; Stoop & Hart, 2005; Cotter & Tirado, 2008; Clements, 2009; Hadgu et al., 2011).

Among other factors, climate change, increasing population and food security highlight the importance of various ecosystem services and the finite nature of land resources (Lamboll et al., 2011). According to Jones & Thornton (2003), the impacts of climate change on agriculture may add significantly to the development challenges of ensuring food security and reducing poverty. Thompson et al. (2010) contend that the importance of ecological and climatic processes for food production, and sensitivity of African food systems to climate, makes climate change a concern for food security. In other words, the ability to achieve food security has broader implications for development and health, and is thus vitally important for future considerations of international development in the region.

Given the multiple demands placed upon agriculture, a number of potential synergies and trade-offs are emerging between agricultural production and climate change adaptation and mitigation objectives. Where there are trade-offs rather than synergies, governments, farmers and other agricultural innovation system (AIS) actors will need to prioritise actions and climate resilience will be a key factor in future agricultural systems (Lamboll et al., 2011).

Improving food security is recognised within the Sustainable Development Poverty Reduction Paper (SDPRP) of Ethiopia as a central concern of government. In other words, the poverty reduction strategy has 'agricultural development-led industrialisation and food security' as one of its four key pillars or building-blocks. Moreover, food security programmes are acknowledged to be a 'subset of poverty reduction interventions'. The food security strategy in turn claims to address the 'supply and demand side', at national and household level, 'taking into account the diversity of the national economy'. The three pillars of the strategy are: increasing the availability of food through domestic (own) production; ensuring access to food for food-deficit households; and strengthening emergency response capability (Haan et al., 2006).

However, both the poverty reduction strategy and food security strategy fails to recognize the connection between food security and ecosystem-based adaptation to climate change. In other words, ecosystem-based adaptation to climate change is not mainstreamed in the aforementioned strategies. Moreover, the current agricultural production and food security initiatives in the country are heavily dependent on industrially produced chemical fertilizers with many negative environmental impacts (MEA 2005; Hutton, 2010).

According to Gebre-Egziabher (2007), the intensification of agricultural production in Africa and Ethiopia can take place without the use of industrially produced chemical fertilizers. For instance, preparing compost from household and farm waste and using it to raise soil fertility has been found to be as effective as, and in the case of crops by smallholder farmers to be more effective than, using chemical fertilizers to raise agricultural productivity. In other words, eco-functional intensification increases productivity and enhances food security especially in challenging environments such as water scarce regions (MEA 2005; Cotter & Tirado, 2008; IFOAM, 2009; Hutton, 2010).

Adapting to climate change will entail adjustments and changes at every level from community to national and international. Communities must build their resilience, including adopting appropriate technologies while making the most of traditional knowledge, and diversifying their livelihoods to cope with current and future climate stress. Local coping strategies and traditional knowledge need to be used and integrated with government and local interventions. In a nutshell, to enable effective adaptation measures in Ethiopian agricultural sector, governments as well as non-government organizations, must consider integrating climate change in their planning and budgeting in all levels of decision making (Georgis et al., 2009; World Bank, 2011; Mekonnen, 2012).

Bio-physical, Socio-Economic, Institutional Constraints for Production and Flow of Cereals in Ethiopia

Biophysical Constraints for Production and Flow of Cereals in Ethiopia

Geographically, Ethiopia can be subdivided into five agroecological zones, based on moisture and land use: 1) drought-prone highlands with insufficient rainfall; 2) rainfall-sufficient highlands dominated by enset-based farming; 3) rainfall-sufficient highland areas mainly planted with cereal-based crops; 4) generally dry, pastoral lowland areas (bordering on Eritrea); and 5) humid lowland areas further inland that primarily support crop farming (Gebreegziabher et al., 2011).

USAID (2011) also identified three agroecological zones (high rainfall, low rainfall, and pastoralist).

The principal biophysical constraint for production of cereals in Ethiopia is land degradation in the form of soil erosion, gully formation, soil fertility loss and severe soil erosion (Ringler et al., 2010; Ali & Surur, 2012; Schmidt & Tadesse, 2012). According to Shiferaw (2011), the rate of soil erosion is severe in the highlands of Ethiopia. Rapid population growth, cultivation on steep slopes, clearing of vegetation and overgrazing are identified as the main factors that accelerate soil erosion in Ethiopia (Shiferaw, 2011; Ali & Surur, 2012).

Another critical biophysical constraint for the production and flow of cereals in Ethiopia is climate change. In other words, the production and flow of cereals is very much affected biophysically by meteorological variables, including rising temperatures, changing precipitation regimes, and increased atmospheric carbon dioxide levels. Biophysical effects of climate change on production of cereals are claimed to be positive in some agricultural systems and regions, and negative in others, and these effects are claimed to vary through time (Parry et al., 2004; Harmeling et al., 2007; Alcadi et al., 2009). The negative effects are attributable to increasing temperature and decreasing precipitation (Deressa, 2007; Georgis et al., 2009; Ludi, 2009; FDRE, 2010; Ringler et al., 2010; World Bank, 2010; Gebreegziabher et al., 2011; Mekonnen, 2012; Schmidt & Tadesse, 2012; AfDB, 2013).

Climate variability and change have been implicated to have significant impacts on global and regional food production particularly the common staple food crops (cereals) performance in tropical sub-humid climatic zone. For example, the most food insecure regions and most climate change vulnerable regions in Ethiopia are those that experience both the lowest and most variable rainfall patterns (Mekonnen, 2012).

Ringler et al. (2010) assert that climate change affects crop area, yield, and production of cereals in Ethiopia. The negative effects on crop area, yield, and production are attributable to the indirect effects of climate change like changes in soil moisture, land and water condition, change in frequency of fire and pest infect and the distribution of diseases (Mekonnen, 2012).

Changes in rainfall pattern are also likely to lead to severe water shortages and/or flooding. Rising temperatures also will cause shifts in crop growing seasons which affects food security and changes in the distribution of disease vectors putting more people at risk from diseases such as malaria (Gebreegziabher et al., 2011; Mekonnen, 2012). In a nutshell, The direct and indirect effects of climate change on agriculture play out through the economic system, altering prices, production, productivity, food demand, calorie availability, and, ultimately, human well-being (Ringler et al., 2010).

Socio-Economic Constraints for Production and Flow of Cereals in Ethiopia

The production and flow of cereals in Ethiopia is constrained by socio-economic constraints like poor infrastructure (constraining access to both agricultural inputs and markets for outputs), dominance by small-scale resource-poor farmers, shortage of draft power, low level of modern farm inputs; inefficient working habit (less working days per week) of the peasants; substantial increase in food prices; and endemic poverty, limited access to capital and global markets, ecosystem degradation, complex disasters like war and conflicts; population growth (FDRE, 2010; Gebreegziabher et al., 2011; Mekonnen, 2012; MOA, 2012). For instance, the availability of financial services is constraining the capacity of smallholders and emerging commercial farmers to adopt improved agricultural practices. This includes both short term seasonal credit for crop inputs and medium-longer term finance for capital investments (FDRE, 2010).

Institutional Constraints for Production and Flow of Cereals in Ethiopia

The critical institutional constraints for the production and flow of cereals in Ethiopia are: tenure insecurity; weak agriculture research base and extension system; lack of appropriate financial system; imperfect agricultural markets; lack of appropriate pricing and incentive policies; and insufficient information dissemination mechanism (Parry et al., 2004; Tefera, 2009; FDRE, 2010; Ringler et al., 2010; Gebreegziabher et al., 2011; Mekonnen, 2012; MOA, 2012).

There are systemic capacity limitations at all levels and in all of the sectoral institutions, but the problem is most severe at the district level. Capacity limitations include human resources, working premises, equipment, communications, machinery, furniture and other facilities (FDRE, 2010). There are also identified institutional gaps related to sector-wide linkages, relationships and synergies. Specific issues include lack of communication among ministries and between Ministries and CSOs and parastatals; inadequate vertical and horizontal collaboration among research institutes; weak research-extension-farmer linkages; and lack of communication and collaboration with the private sector (FDRE, 2010). In summary, there is interplay among the aforementioned biophysical, socio-economic, and constraints for production and flow of cereals in Ethiopia. For instance, the biophysical effects of climate change on the production and flow of cereals is claimed to induce changes in prices, which is claimed to play out through the economic system as farmers and other market participants adjust autonomously, altering crop mix, input use, production, food demand, food consumption, and trade (FDRE, 2010; Ringler et al., 2010; Gebreegziabher et al., 2011; Mekonnen, 2012).

Strengths, Weaknesses, Opportunities and Threats of Current Ethiopian Agricultural Development Policies and Strategies in Addressing Constraints

The author was able to analyze the strengths and weaknesses of the internal environment (Table 1) and the opportunities and threats for the external environment (Table 1) of the current Ethiopian agricultural development policies and strategies in dealing with constraints of production and flow of cereals.

Table 1: SWOT Analysis for the Current Ethiopian Agricultural Development Policy and Strategies in Dealing with Constraints of Production and Flow of Cereals

Internal Environments	
Strengths: • Government's commitment to improving agricultural productivity; • ensuring food security and gender equality; • inclusion of a significant number of Ecosystem-based Adaptation activities within National Adaptation Programme of Action; • nationally appropriate mitigation actions developed; • increasing irrigated cropland and investing in agricultural research and development; • Major emphasis by the government strategy in agriculture to support the intensification of marketable farm products by both small and large farmers; • Major investments like the construction of farm-to-market roads and area irrigation through multipurpose dams; • measures to improve land tenure security; • reforms to improve the availability of fertilizer and improved seeds, and specialized extension services for differentiated agricultural zones and types of commercial agriculture; • implementation of Sustainable Land Management Program (SLMP); • strong commitment of the government to continued agricultural growth (a three-pillar approach - food security and drought resilience, agricultural productivity, and sustainable land and water management); • Commitment of the government to achieve longer term green growth through focusing on improved land and water management; • Government's support for land registration to clarify land tenure; and • Government's commitment to a major landscape re-greening program.	**Weaknesses:** • There is little explicit mention of climate change in agricultural development policies and strategies; • Policies are generally supportive of agricultural practices that focus on increasing short-term production (eg, expansion of agricultural land, increasing mechanisation, increasing use of fertiliser and other inputs); • The Policies and strategies are generally less supportive of practices which can improve food production, enhance adaptive capacity and address mitigation (eg, restoration of degraded land, improving soil macro and micro nutrients); • Weak market linkages both on the input and output side; • Farmers either cannot afford improved inputs or lack the knowledge to use them; and • Weak linkages between agricultural outputs producers and processors.

External Environments	
Opportunities: • decentralisation of decision making on climate change adaptation; • new investment in agricultural development; • existence of the climate resilient green economy strategy (CRGE) and growth and transformation plan (GTP); • development and promotion of agricultural and pastoral innovations by agricultural research and development initiatives; • huge potential in agricultural sector to take advantage of mitigation opportunities through adopting improved land, water, agriculture and agro-forestry management practices; • existence of numerous indigenous land management practices; • Existence of national policies and strategies protecting indigenous knowledge and rights; and • Presence of traditional administration and social institutions.	**Threats:** • lack of finance; • land use conflict and community opposition; • lack of information about the costs and benefits of ecosystem-based adaptation measures; • insufficient funding to properly implement National adaptation plan of actions (NAPAs); • insufficient packaging and storing; • inability of Ethiopian products to meet international market standards; • restrictive trade regulations; • limited availability and usage of irrigation and improved seed and fertilizer; • poor land management practices resulting in severe land degradation; • government-controlled cooperatives managing the input supply system; • low agricultural investment and productivity; • insufficient access to credit by smallholder farmers; • heavy dependence on rainfed agriculture (highly erratic, and most rain falls intensively); • ever-growing human population and food demand; • high exposure of agricultural sector to climate variability, and in particular to drought and related heat stress; • limited national policies and strategies to promote, develop, conserve and use at greater scale indigenous knowledge; • absence of accurate or reliable information database about the extent and location of the past and present natural forest and woody vegetation cover in Ethiopia; • absence of appropriate land-use classification and land-use policy; • lack of appropriate pricing and incentive policies; and • insufficient information dissemination scheme.

Sources: (Cooper et al., 2008; Colls, et al., 2009; Georgis et al., 2009 ; Kato et al., 2009; World Bank, 2010; Lamboll et al., 2011; Shiferaw et al., 2011; USAID, 2011; World Bank, 2011; Ali and Surur, 2012; Schmidt & Tadesse, 2012).

Trends of Major Cereals' Annual Production in Ethiopia (2000-2013)

For the purpose of this study, seven major cereals in Ethiopia were selected and the trends of their mean production area in hectares, mean estimated total production in quintals (1 quintal=100 Kgs), and mean yield in quintals/hectare were analysed.

With regard to area production, Teff has scored the largest mean area production (2411586 hectares). This was followed by 1703866 hectares, 1508176 hectares, 1398004 hectares, 983831 hectares, 873361 hectares, and 34042 hectares for Maize, Sorghum, Wheat, Barely, Millet and Oat respectively (see figure 1).

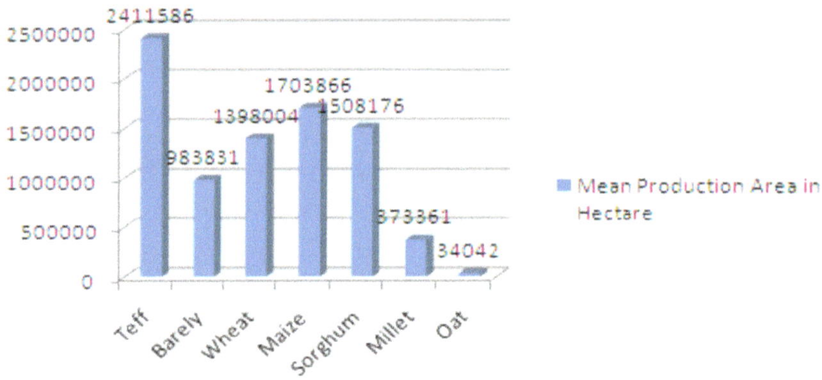

Figure 1: Mean production area in hectares for major cereals in Ethiopia (2000-2013),Source: Computed from CSA data by author].

With regard to total production, Maize has scored the largest mean total production (38756460 quintals). This was followed by 26431205 quintals, 25514706 quintals, 23659564 quintals, 13677339 quintals, 4838731 quintals, and 431223 quintals for Teff, Sorghum, Wheat, Barely, Millet and Oat respectively (see figure 2)

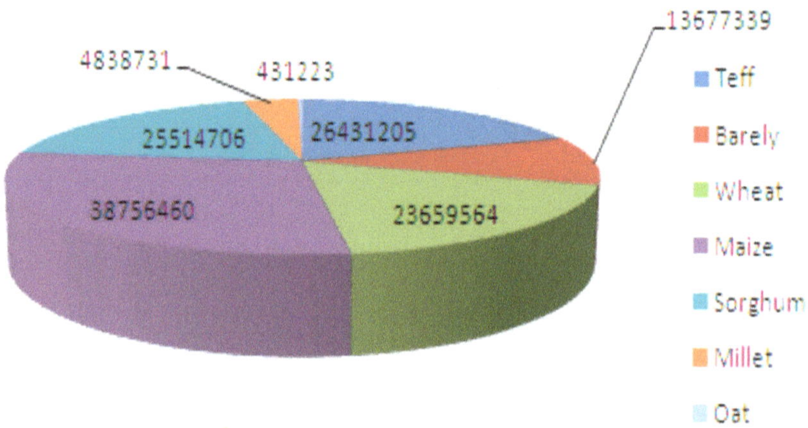

Figure 2: Mean estimated total production in quintals for major cereals in Ethiopia (2000-2013), Source: Computed from CSA data by author.

With regard to yield, Maize has scored the largest mean yield (22 quintal/hectare). This was followed by 17 quintal/hectare, 17 quintal/hectare, 14 quintal/hectare, 13 quintal/hectare, 13 quintal/hectare and 11 quintal/hectare for Wheat, Sorghum, Barely, Millet, Oat, and Teff respectively (see figure 3).

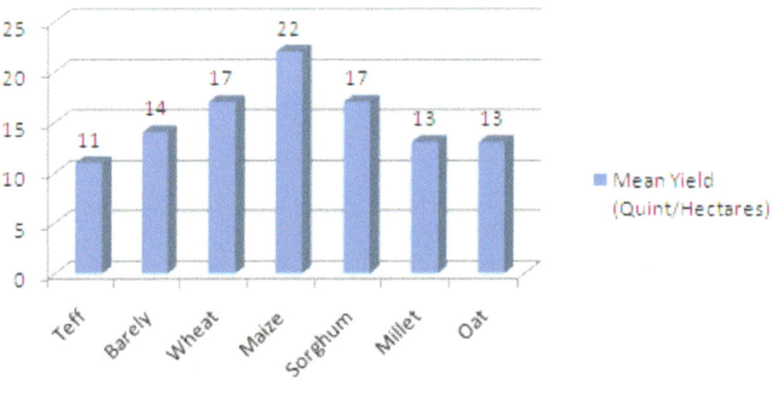

Figure 3: Mean estimated Yield for major cereals in Ethiopia (2000-2013), Source: Computed from CSA data by author.

Ethiopia has experienced large increases in agricultural output in recent years in part due to increased sector support (e.g., the extension system) and in part due to increased cultivation areas (e.g., over the last five years, the area under cereal crop cultivation increased by 63 percent), as opposed to significant productivity increases. However, for Ethiopia to continue the agriculture sector growth required to reduce poverty and meet the country's ever-growing food demand in the coming years, new sources of growth must be found, and greater attention must be placed on productivity enhancement (Georgis et al., 2009; USAID, 2011).

Cereal production and marketing plays a vital role in Ethiopia's economy (Rashid & Negassa, 2011). However, production and flow of cereals in Ethiopia is constrained by biophysical, socio-economic, and institutional constraints (Deressa, 2007; Georgis et al., 2009; Ringler et al., 2010; Gebreegziabher et al., 2011; Mekonnen, 2012; Schmidt & Tadesse, 2012). For instance, Harmeling et al. (2007) assert that climatic conditions are the predominant factors affecting agricultural productivity, and changes in these conditions may have different severe effects. To avert this situation, Edame et al. (2011) contend that developing climate-smart agriculture is crucial to achieving future food security and climate change goals. Georgis et al. (2009) contend that developing drought resistant, early maturing and heat tolerant crop species and varieties is of paramount importance to cope to adapt to climate change in agricultural sector of Ethiopia. Furthermore, IFOAM (2009) asserts that organic agriculture builds soil fertility and structure and is more efficient in using resources, including well adapted local resources such as local seed varieties, which are often limited in areas with extreme poverty and food insecurity. In other words, the eco-functional intensification of organic agriculture systems is claimed to increase productivity and enhance food security especially in challenging environments (IFOAM, 2009).

Production and value chain of four cereals in Ethiopia are discussed hereunder. The four cereals identified for this purpose are Maize, Wheat, Sorghum and Millet.

Maize Production and Value Chain in Ethiopia

The findings of this study indicate that Maize is second to teff in terms of area, but it ranks first among cereals in production. Previous study by USAID & COMPETE (2010) also confirmed the same. Because of its large geographical coverage and the scale of production, maize plays a critical role in the food security of the country and particularly the poor. It is a widely grown crop in Ethiopia; although the extent of cultivation varies, maize is cultivated in some 74 important maize producing administrative zones, of which 20 contribute almost 80% of the national production (Georgis et al., 2009; USAID & COMPETE, 2010).

The largest surplus producers of maize in Ethiopia previously identified by USAID & COMPETE (2010) were: West Gojam (11.3%), Jimma (7.9%), East Shoa (7.8%), East Wellega (6.4%), West Wellega (5.5%), Illubabor (4.3%), Arsi (4.3%), West Shoa (4.2%), East Hararghe (3.5%), Agewawi (3.4%), West Hararghe (2.9%), and Sidama (2.9%). The other important maize producing zones are West Arssi, East Gojam, North Gondar, Horo Gudru, Bale, South Gondar, Silti and Kelem. These zones produce about 80% of national maize production (USAID & COMPETE, 2010).

Maize constitutes about 24%-31% of national cereal consumption and is mainly covered by domestic production. Maize is consumed as a staple food in different forms, including: *injera* (alone or mixed with teff), porridge, and bread. It is also consumed roasted or boiled (especially at green stage). Moreover, it brewed into *tella, araki* and other local drinks (USAID & COMPETE, 2010).

The maize value chain that connects producers and consumers throughout Ethiopia has three key actors: assemblers, wholesalers and retailers. As mentioned earlier the producer assembler wholesaler retailers consumer channel is the most important in terms of the magnitude of the marketed maize that flows from producers to consumers (USAID & COMPETE, 2010).

Previous studies confirmed that production and flow of cereals is affected by biophysical, socio-economic and institutional factors (Deressa, 2007; Georgis et al., 2009; Gebreegziabher et al., 2011; Mekonnen, 2012; MOA, 2012). This is specifically true for Maize. The major constraints previously identified by USAID & COMPETE (2010) were: Shortage of certified maize seeds varieties suitable for different agroecological zones; Inadequate supply and marketing of maize seed; Nonexistence of export market and agro-industries that use maize as a raw material and diversify the utilization of the crop; Inadequate transport infrastructure and high cost of transport; Lack of reliable and timely market information; Inadequate access to credit facilities by grain traders; Lack of storage and marketing facilities in both the surplus producing and the consumption centers; Lack of universally acceptable and enforceable quality standards that reduce transaction costs; and Weak producers' organizations and limited participation in maize seed production, distribution and marketing of members' output.

Wheat Production and Value Chain in Ethiopia

The findings of our study indicate that wheat is the fourth major cereal produced in Ethiopia both in terms of production area and total annual production, but it ranks second together with Sorghum among cereals in yield. Unlike other cereal crops, wheat is an important industrial crop. In other words, it is used as an input for the country's modern food processing factories (USAID & COMPETE, 2010).

The most important wheat producing areas in Ethiopia previously identified by USAID & COMPETE (2010) were: Arsi with (14.7%) share of total wheat production, Bale (13.2%), North Shoa (8.3%), West Shoa (8.1%), East Shoa (6.6%), East Gojam (5.9%), South Wello (5.0%), West Arssi (4.0%), South West Shoa (3.5%), Southern Tigray (3.0%), South Gondar (2.8%), Hadiya (2.8%), and West Gojam. These zones produce more than 85% of national wheat production (USAID & COMPETE, 2010).

The wheat value chain that connects producers and consumers throughout Ethiopia has three key actors: grain wholesalers, processor, and bakery. The producer Grain wholesaler Processor Bakery Consumer channel is the most important in terms of the magnitude of the marketed wheat that flows from producers to consumers (USAID & COMPETE, 2010).

Previous studies confirmed that production and flow of cereals is affected by biophysical, socio-economic and institutional factors (Deressa, 2007; Georgis et al., 2009; Gebreegziabher et al., 2011; Mekonnen, 2012; MOA, 2012). This is specifically true for Wheat also. The constraints in the value chain of wheat in Ethiopia previously identified by USAID & COMPETE (2010) were: Weak seed production and distribution, lack of participation by private firms and farmers' organizations in the production and distribution of improved wheat varieties; High seed cost due to high transport and handling costs; Inadequate coordination between research, seed multiplication and extension; Lack of market information for traders, producers and farmers' organizations; Lack of access to appropriate storage and marketing facilities and infrastructure; Non-existent of forward contractual agreement between producers and millers; Lack of access to bank credit; and Inadequate road infrastructure and high cost of transferring wheat from surplus areas to consumption centres.

Sorghum Production and Value Chain in Ethiopia

The findings of our study indicate that Sorghum is the third major cereal produced in Ethiopia both in terms of production area and total annual production, but it ranks second together with Wheat among cereals in yield.

The most important sorghum producing areas previously identified by USAID & COMPETE (2010) were: North Gondar with 10.6% share in national sorghum output, North Shoa (9.5%), East Hararghe (7.9%), West Hararghe (7.2%), West Shoa (4.8%), South Wello (4.3%), Jimma (4.2%), and North West Tigray (4.0%). Other important sorghum areas include North Wello, East Wellega, Illubabor, South Tigray, Central Tigray and West Tigray, each contributing 2-3% of national production.

In Ethiopia a major part of sorghum is produced for human consumption and the production is largely based on small-scale farming (Georgis et al., 2009; USAID & COMPETE, 2010). Sorghum is consumed in various forms, including *enjera*, porridge and local drinks such as *tella* and *arekie.* The principal use of sorghum, however, is enjera, prepared alone or mixed with teff. Consumption of sorghum is partly determined by the availability and price of teff in the market. There are different varieties of sorghum in the market, but the varieties most demanded are the white and yellow sorghum that are mostly in North Gondar, South Wello, East and West Hararghe, North Shoa, and Southern Tigray (USAID & COMPETE, 2010).

The Sorghum value chain that connects producers and consumers throughout Ethiopia has three key actors: assemblers, wholesalers and retailers. The producer assembler wholesaler retailers consumer channel is the most important in terms of the magnitude of the marketed Sorghum that flows from producers to consumers (USAID & COMPETE, 2010).

Previous studies confirmed that production and flow of cereals is affected by biophysical, socio-economic and institutional factors (Deressa, 2007; Georgis et al., 2009; Gebreegziabher et al., 2011; Mekonnen, 2012; MOA, 2012). This is specifically true for Sorghum also. The most important constraints in the value chain of sorghum in Ethiopia previously identified by USAID & COMPETE (2010) were: Serious problems of Striga, a parasitic weed, in many parts of the country; Non-acceptance of some of the released varieties by farmers for various reasons, including susceptibility to bird attacks and low biomass; Lack of improved seed varieties to meet the demand of farmers, because of the limited government capacity to multiply and distribute certified sorghum seed; Lack of strong linkage among research, extension system and producers; Lack of access to market information; High transport and transaction costs; Lack of storage and market infrastructure; Lack of access to bank credit by traders; Weak organization of producers; and Lack of diversified use of the crop apart from subsistence.

Millet Production and Value Chain in Ethiopia

The findings of this study indicate that Millet is the sixth major cereal produced in Ethiopia both in terms of production area and total annual production, but it ranks fourth together with Oat among cereals in yield. According to USAID & COMPETE (2010), Finger millet (a variety of millet produced in Ethiopia) production is insignificant; it accounts for only 3-3.5% of total cereal production.

The most important millet producing areas in Ethiopia previously identified by USAID & COMPETE (2010) were: West Gojam (14.8%), North Gondar (10.8%), North West Tigray (10.6%), West Welega (10.6%), Agewawi (10.2%), and South Gondar (8.9%) that together have some 66% share of national millet production.

Finger millet has high nutritional value and is consumed in different ways, including as staple food (porridge and *enjera*) and as local brews (*tella* and *araki*). About 41% of the millet supply is used as direct food and the balance for other uses, including local brews.

The finger millet value chain that connects producers and consumers throughout Ethiopia has two key actors: wholesalers and retailers. The producer wholesaler retailers consumer channel is the most important in terms of the magnitude of the marketed that producers to consumers (USAID & 2010).

A Case Study on the Trends of Major Cereals' Annual Prices in Addis Ababa Grain Market (2005-2010)

For the purpose of this study, five major cereals in Addis Ababa market were selected and the trends of their mean annual prices were analysed.

Teff white scored the largest mean annual price (584 birr/quintal). This was followed by 536 birr/quintal, 436 birr/quintal, 409 birr/quintal, 381 birr/quintal, 374 birr/quintal, 369 birr/quintal, 334 birr/quintal, 328 birr/quintal, and 253 birr/quintal for Teff mixed, Teff red, Sorghum white, Barely white, Sorghum mixed, Wheat white, Wheat mixed, Barely mixed, and maize respectively (see figure 4, figure 5, figure 6, figure 7, figure 8, and figure 9).

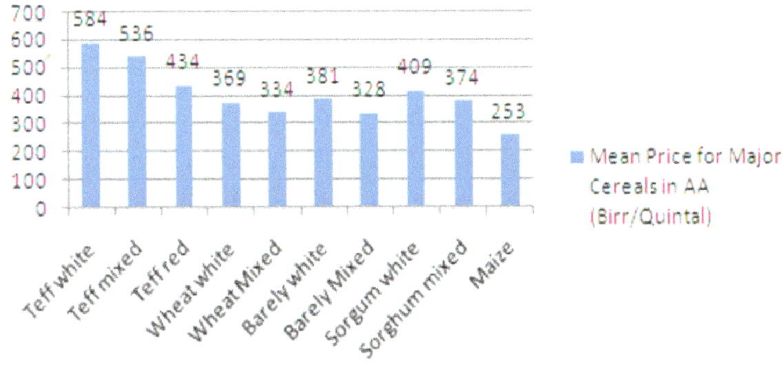

Figure 4: Mean price for major cereals in Addis Ababa market (2005-2010). Source: Computed from EGTE data by author.

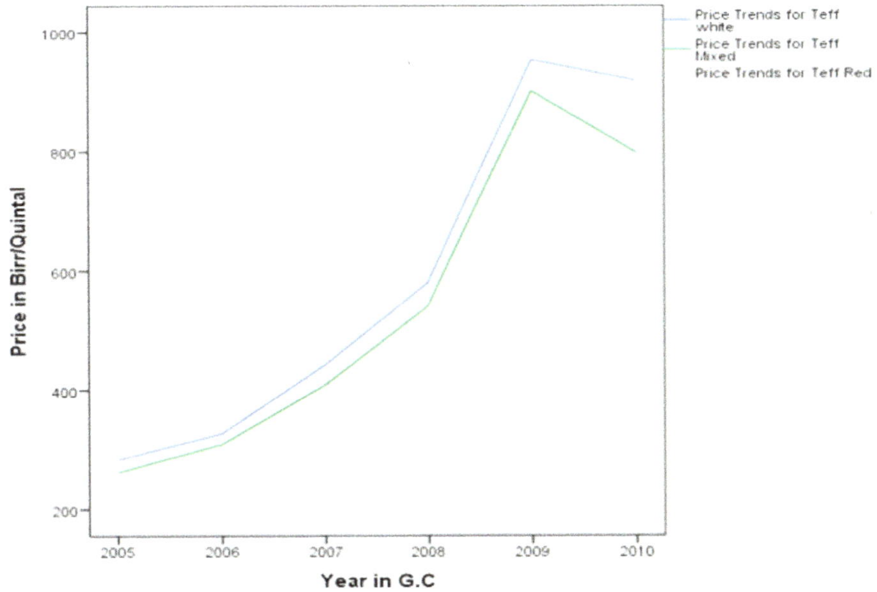

Figure 5: Trend of mean price for Teff in Addis Ababa market (2005-2010). Source: Computed from EGTE data by author.

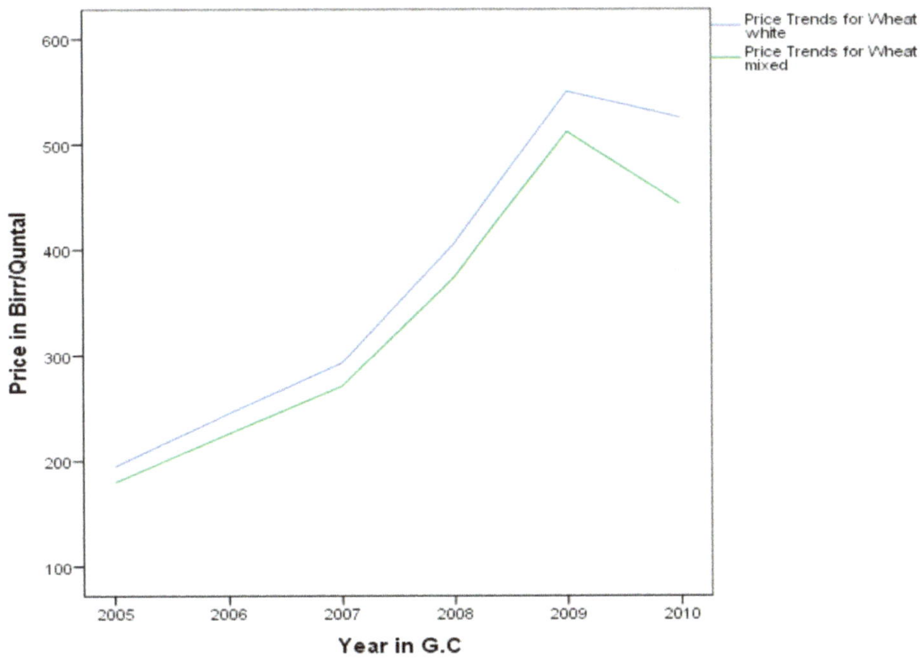

Figure 6: Trend of mean price for Wheat in Addis Ababa market (2005-2010). Source: Computed from EGTE data by author.

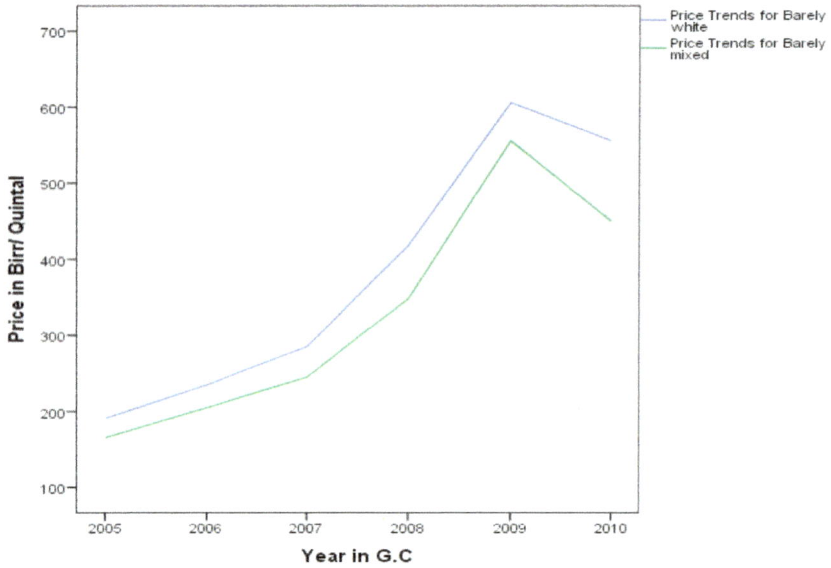

Figure 7: Trend of mean price for Barely in Addis Ababa market (2005-2010). Source: Computed from EGTE data by author.

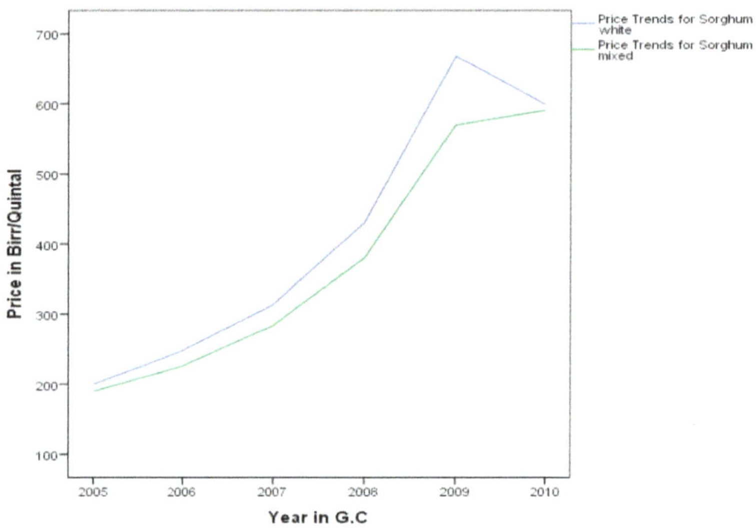

Figure 8: Trend of mean price for Sorghum in Addis Ababa market (2005-2010), Source: Computed from EGTE data by author.

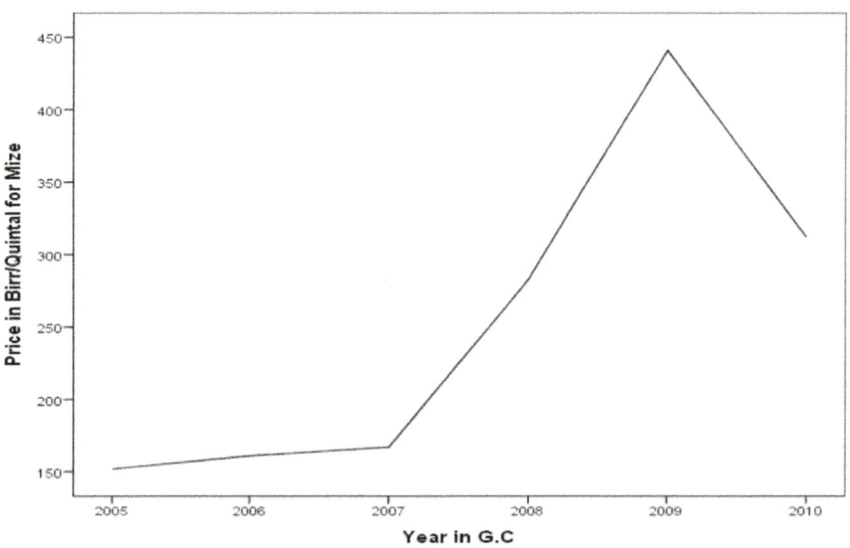

Figure 9: Trend of mean price for Maize in Addis Ababa market (2005-2010). Source: Computed from EGTE data by author.

Rising food prices in Ethiopia has been the outcome of monetary policy misalignment, the balance of payment problems resulting from sharp increases in fuel prices, as well as overestimated cereal production. For instance, according to an IFPRI-EDRI study, official estimates of cereal production were around 30 percent higher (Rashid, 2010). Another case in point is the impact of inflation as one key element that has resulted in increased food insecurity in urban areas of Ethiopia.

Worako (2012) contends that there has been unprecedented high rate of inflation in Ethiopia during 2005-2008 mainly driven by food price inflation mainly of cereal prices. Moreover, the puzzling price trend may account for an over-estimation of cereal production (Worako, 2012). The findings of this study reveal that there are increasing trends in prices of major cereals (Teff, Sorghum, Barely, Wheat, and maize in Addis Ababa grain market between 2005 and 2009. However, decreasing trends in prices of cereals in Addis Ababa grain market were observed between 2009 and 2010 due to increase in production and government's intervention in marketing system.

Though this study could not get reliable data for recent years (2011, 2012, and 2013), previous study by WFP-Ethiopia (2009) asserts that the prices of cereals increased by more than 100% since mid-2005 when the country faced a spiral of price increases. Worako (2012) asserts that the food price in Ethiopia is not only high but also relatively more volatile. As result, this abnormal food price surge put millions of rural and urban net food buyer at risk. This is specifically true for Addis Ababa city.

The puzzle is that the increase in inflation in the recent years coincided with relatively favourable harvests, whereas in the past inflation in Ethiopia had typically been associated with agricultural supply shocks due to droughts (Deressa, 2007; Georgis et al., 2009; Gebreegziabher et al., 2011; Mekonnen, 2012; Schmidt & Tadesse, 2012). Moreover, malfunctioning market at different levels and lack of adequate increase in food production due to structural factors are considered as major causes. Although agricultural production has shown drastic growth as stated in the government's official statistics, high cost of transportation limited its movement from food surplus to deficit area is considered as underlining cause (Worako, 2012). These aforementioned major and underlining causes have seriously affected the food security situation of Addis Ababa city.

According to Worako (2012), instead of stimulating economic growth, inflationary pressure in Ethiopia seems to be on the verge of distorting the allocation of resources and is likely to be a deterrent to undertaking productive investments. Moreover, consumers in general and civil servants in particular are at verge of crisis (Worako, 2012).

If the cause of the inflation is clearly determined, it will be trouble-free to manage. For instance, if the cause of inflation is demand-pull which is initiated by high aggregate demand due to high money supply, it can be managed through monetary policy by increasing interest rate which helps to increase saving and decrease consumption, decreasing money supply, imposing tax can be good instruments. On the other hand, if it is known that the cause of inflation is cost-push which is manifested by low aggregate supply due to scarcity of factors of production, high prices of inputs, low production can be equilibrated by reducing input prices through subsidizing and so on. If it is resulted from malfunctioning marketing system, reform in market institutions, building market infrastructure, and enhancing competition are partly means for improving the situation (Rashid & Negassa, 2011; Worako, 2012).

High fuel prices make agricultural production more expensive by raising the cost of fertilizers, irrigation, and transportation (Worako, 2012). Food security depends on availability of food, access to food, and utilization of food (Edame et al., 2011). Climate variability directly affects agricultural production, as agriculture is inherently sensitive to climate conditions and is one of the most vulnerable sectors to the risks and impacts of global climate change. Many factors impact the type of policies implemented at a national level (such as domestic politics, redistribution of land/wealth, exchange rates, and trade issues, etc.). According to Edame et al. (2011), climate variability should be factored into these policies, as these policies can impact the availability of staple foods, for example, by providing incentives to grow crops appropriate for the climate conditions.

To address the current crisis in cereal markets in Ethiopia in general and Addis Ababa in particular, Worako (2012) suggests Prudent macroeconomic and fiscal policy management; Deeper understanding of current marketing structure and designing alternative and competitive marketing system, mainly for wholesale markets; Investing more on agricultural production and productivity, or pursuing more substantive policy on food production and marketing (more incentive and support for private sector to engage in agricultural production); and Promoting diversification in staples consumption(changes in food culture).

Factors that determine whether people will have access to sufficient food through markets are considered to include: income-generating capacity, amount of remuneration received for products and goods sold or labour and services rendered, and the ratio of the cost of a minimum daily food basket to the average daily income (Edame et al., 2011; Worako, 2012). In other words, if climate change creates other more urgent claims on public resources, support for food distribution schemes may decline, with consequent increases in the incidence of food insecurity, hunger and famine related deaths (Edame et al., 2011; Rashid & Negassa, 2011). Moreover, climate impacts on income-earning opportunities can affect the ability to buy food, and a change in climate or climate extremes may affect the availability of certain food products, which may influence their price. Furthermore changes in the demand for seasonal agricultural labour, caused by changes in production practices in response to climate change, can affect income-generating capacity positively or negatively. For instance, mechanization may decrease the need for seasonal labour in many places, and labour demands are often reduced when crops fail, mostly owing to such factors as drought, flood, frost or pest outbreaks, which can be influenced by climate (Edame et al., 2011).

The price effects from climate change on key cereals is that food prices increase for all staple crops because climate change acts as an additional stressor on the already tightening price outlook (Gebremedhin & Hoekstra, 2007; Ringler et al., 2010; Edame et al., 2011).

In a nutshell, the productive areas of Ethiopia suffer from low capacity and market constraints, preventing these areas from living up to their productivity potential (Rashid &

Negassa, 2011; USAID, 2011). Ringler et al. (2010) attest that the substantial increase in food prices has the potential to slow growth in calorie consumption, with both direct price impacts and reductions in real incomes for poor consumers who spend a large share of their income on food. In other words, if current food price situation is not circumvented, it is expected to be the major source of welfare deterioration for the urban and rural poor (net food buyers) (Worako, 2012). This is particularly true for urban hubs like Addis Ababa (WFP-Ethiopia, 2009).

CONCLUSION

In a nutshell, this study concludes that there is strong conceptual linkage between eco-system based adaptation to climate change and food security in Ethiopia; there is interplay among bio-physical, socio-economic, and institutional constraints for production and flow of cereals in Ethiopia; there are strengths, weaknesses, opportunities, and threats of the current Ethiopian agricultural development policies and strategies in addressing constraints for production and flow of cereals; there are increasing trends of major cereals' mean annual production in Ethiopia; and the food security situation of Addis Ababa is very much affected by increase of mean annual prices of major cereals in Addis Ababa grain market.

RECOMMENDATIONS

While the policy responses by the government of Ethiopia to adapt to climate change and improve the production and flow of cereals are commendable, this study the following strategic measures to address these issues on sustainable basis:

- Ecoagriculture should be adopted in agricultural sector of Ethiopia as a mosaic arrangement of land uses can yield diversified income sources and conserve biodiversity while providing ecosystem services that sustain agricultural productivity, such as pollination, water purification and soil fertility enhancement;
- Appropriate organic matter management should be employed since it is the foundation for high-quality soil and a more sustainable and thriving agriculture;
- Stronger research systems capable of improving the resistance of crops and animals to biotic stresses, as well as investments in irrigation and water management;
- There should be shifts in policy and institutional arrangements of Integrated watershed management, which broadens the freshwater outlook to address both direct and indirect green and blue water functions in a landscape;
- The relevant United Nations agencies in collaboration with Government organizations and other partners should strengthen support to improve the supply of basic services such as water, sanitation, and health facilities in Addis Ababa and other urban areas;
- The Government market stabilization program should be reviewed so that it can effectively contribute to food security of the urban poor as access to food remains a problem and also stabilize the ever increasing cereal prices;
- An urban monitoring system should be established so as to capture any deteriorating food insecurity conditions in Addis Ababa and other urban areas;
- Any efforts to increase the resilience of Ethiopian agriculture in the face of climate change must involve the development of improved crop varieties and animal breeds, as well as more prudent and integrated management of crops, animals and the natural resource base that sustain their production, while providing other vital services for people and the environment;
- Since small farmers and rural communities are the starting point for efforts to adapt to climate change, the problems and solutions should be defined with their direct and active participation;
- Agroecological farming should be promoted through government support in the form of credit for organic fertiliser and new seed varieties provided for smallholders at relatively low cost;
- Appropriate policy measures and institutional support should strengthen Ethiopian smallholder farmers' capacity to adapt to climate change;
- Sustainable, alternative livelihood options (including building communities' capacity to manage sustainable natural resources and promoting agro-forestry and agro-ecology) in rural communities should be promoted since it is a relatively low-cost and complementary approach to adaptation;
- The development community should encourage the generation of innovations at the local level, accompanied by a framework for evaluating experiments and a political and legal space to transform the lessons learned into large-scale initiatives to reduce hunger and poverty;
- Technological innovations need to target productivity increases and be adapted to the emerging challenges facing food production and producers (especially smallholders);
- The link between climate change and agriculture needs to be exploited through triple-win strategies that improve climate change mitigation and adaptation and agricultural productivity;
- The principle of collective action like building a strong, cohesive network to facilitate adaptation through the community and by individuals should be employed in ecosystem-based adaptation in agricultural sector of Ethiopia;
- Any intervention must recognize that women and men adapt differently to new technologies or production systems;
- Policy direction for climate change mitigation and improved sustained food security and agricultural productivity must among other things include improved land management, adjustment of planting dates, and introduction of new crop varieties, while the mitigation options include improved energy efficiency and crop yields, and land management techniques to increase carbon storage;
- Information and training of farmers in organic farming methods should be an important part of the agricultural policy and practice for African governments and donors;
- Short-term plans to address food insecurity, provide

access to water resources, or encourage economic growth must be placed in the context of future climate change, to ensure that short-term activities in a particular area do not increase vulnerability to climate change in the long term; and

- Climate change urgently needs to be assessed at the level of the household, so that poor and vulnerable people dependent on agriculture can be appropriately targeted in research and development activities whose objective is poverty alleviation.

ACKNOWLEDGEMENT

I would like to thank Gete Zeleke (PhD) who was a supervisor for a seminar paper written as a partial fulfilment of the requirements for PhD in Environmental Planning at Ethiopian Institute of Architecture, Building Constriction and City Development (EiABC), Addis Ababa University, Ethiopia. I am also thankful to staff members of Ethiopian Grain Trade Enterprise and Central Statistical Agency of Ethiopia for their full cooperation in providing necessary data during data collection process. The sources of funding for the author were Ambo University; and Ethiopian Institute of Architecture, Building Construction and City Development, Addis Ababa University of the Ethiopian Ministry of Education which also deserve great appreciation for their financial support.

REFERENCES

1. African Development Bank (AfDB).2013. African Development Report 2012: Towards Green Growth in Africa. Tunis-Belvedere, Tunisia: AfDB.
2. Alcadi, R., Mathur, S. and Rémy, P. 2009. Research and innovation for smallholder farmers in the context of climate change. Discussion Paper for Round Table 3. Rome: IFAD.
3. Alemu, Z.G. 2005. Causes of instability in cereal production in Ethiopia. Working Paper. Department of Agricultural Economics Faculty of Natural and Agricultural Sciences at the University of the Free State.
4. Ali, M., and Surur, K. 2012. Soil and water conservation management through indigenous and traditional practices in Ethiopia: A case study. Ethiopian Journal of Environmental Studies and Management (EJESM), Vol. 5 No. 4.
5. Asenso-Okyere, K., and Jemaneh, S. 2012. Increasing agricultural Productivity and enhancing food security in Africa: New Challenges and Opportunities. Synopsis of an international conference. Washington, DC: International Food Policy Research Institute (IFPRI).
6. Binswanger-Mkhize, H. 2008. Food security in Africa: Towards a support strategy for Australia. Australia: AusAID.
7. Butler, L. M. and Mazur, R. E. 2007. Principles and processes for enhancing sustainable rural livelihoods: Collaborative learning in Uganda. International Journal of Sustainable Development & World Ecology 14 (2007) 604–617.
8. Chambers, R. and Conway,G.R.1991. Sustainable Rural Livelihoods: Practical Concepts for 21st Century. IDS Discussion Paper 296. Institute of Development Studies: London.
9. Clements, R.2009. The Economic Cost of Climate Change in Africa. Pan African Climate Justice Alliance (PACJA).
10. Colls, A., Ash, N., and Ikkala. N. 2009. Ecosystem-based Adaptation: a natural response to climate change. Gland, Switzerland: IUCN.
11. Cooper, P.J.M., Dimes, J. Rao, K.P.C., Shapiro, B., Shiferaw, B., Twomlow, S. 2008. Coping better with current climatic variability in the rain-fed farming systems of sub-Saharan Africa: An essential first step in adapting to future climate change? Agriculture, Ecosystems and Environment, 126: 24–35.
12. Cotter, J. and Tirado, R.2008. Food Security and Climate Change: The answer is biodiversity. A review of scientific publications on climate change adaptation in agriculture. United Kingdom: Greenpeace International.
13. Deressa, T.T. 2007. Measuring the Economic Impact of Climate Change on Ethiopian Agriculture: Ricardian Approach. Policy Research Working Paper 4342. Washington D.C: The World Bank.
14. Diao, X. 2010. Economic Importance of Agriculture for Sustainable Development and Poverty Reduction: The Case Study of Ethiopia. Paris: OECD.
15. Edame, G. E, Ekpenyong, A. B., Fonta, W. M., and EJC, Duru.2011. Climate Change, Food Security and Agricultural Productivity in Africa: Issues and policy directions. International Journal of Humanities and Social Science, Vol. 1, No. 21.
16. Ellis, F.1998. Household Strategies and Rural Livelihood Diversification. The Journal of Development Studies, Vol.35, No.l, pp.1-38.
17. Eriksen, S., Ulsrud, K., Lind, J., and. Muok, B.2007. The urgent need to increase adaptive capacities: evidence from Kenyan drylands. In: Nærstad, A.2007. (ed). Africa can Feed itself. Oslo, Norway: The Development Fund.
18. Federal Democratic Republic of Ethiopia (FDRE). 2010. Ethiopia's Agricultural Sector Policy and Investment Framework (PIF) 2010-2020. Addis Ababa: FDRE.
19. Gebreegziabher, T-B. 2007. Potential of the African Environment for the Intensification of Agricultural Production. In: Nærstad, A.2007. (ed). Africa can feed itself. Oslo, Norway: The Development Fund.
20. Gebreegziabher, Z., Stage, J., Mekonnen, A., and Alemu, A.2011. Climate Change and the Ethiopian Economy: A Computable General Equilibrium Analysis. Discussion Paper Series, Environment for Development.
21. Gebremedhin, B. and Hoekstra, D.2007. Cereal Marketing and Household Market Participation in Ethiopia: The Case of Teff, Wheat and Rice. AAAE Conference Proceedings (2007) 243-252.
22. Georgis, K., Dejene, A., and Malo, M. 2009. Agricultural based Livelihood Systems in Drylands in the Context of Climate Change: Inventory of Adaptation Practices and Technologies of Ethiopia.
23. GeoSAS Consulting Service PLC. 2012. African Gender, Climate Change and Agriculture Support Program

(GCCASP): A program of the NEPAD Planning and Coordinating Agency (NPCA) of the African Union (AU). Ethiopia Consultation Report. Addis Ababa: GeoSAS Consulting Service PLC.

24. Haan, N., Majid, N., and Darcy, J.2006. Review of Emergency Food Security Assessment Practice in Ethiopia. Rome: WFP.

25. Hadgu, K. M., Rossing, Walter. A. H., Kooistra, L., and van Bruggen, A. H. C. 2009. Spatial variation in biodiversity, soil degradation and productivity in agricultural landscapes in the highlands of Tigray, Northern Ethiopia. Food Sec., 1:83–97.

26. Hadgu, K., M., J., M., DP. G., G.S. 2011. Current extent of Evergreen Agriculture and Prospects for Improving Food Security and Environmental Resilience in Ethiopia. International Journal of Agricultural Sciences, Vol. 1(1), pp. 006-016.

27. Harmeling, S., Bals, C., and Windfuhr, M.2007. Climate Change and Food Security in Africa. In: Nærstad, A.2007. (ed). Africa can feed itself. Oslo,Norway: The Development Fund.

28. Hassan, R., and Nhemachena, C.2008. Determinants of African farmers' strategies for adapting to climate change: Multinomial choice analysis. AfJARE, Vol 2, No 1:83-104.

29. Hutton, J.2010. Food Security, Biofuels and Agriculture: Major Drivers of Biodiversity Loss and an Opportunity for Sustainability? MacArthur Foundation Conservation White Paper Series.

30. International Federation of Organic Agriculture Movements (IFOAM). 2009. Organic Agriculture: a Guide to Climate Change and Food Security, IFOAM.

31. Jones, P.G. and Thornton, P. K. 2003. The potential impacts of climate change on maize production in Africa and Latin America in 2055. Global Environmental Change, 13: 51–59.

32. Kato, E., Ringler, C., Yesuf, M., and Bryan, E.2009. Soil and Water Conservation Technologies: A Buffer against Production Risk in the Face of Climate Change? Insights from the Nile Basin in Ethiopia. IFPRI Discussion Paper 00871. New York: IFPRI.

33. Lamboll, R, Nelson, V and Nathaniels, N. 2011. Emerging approaches for responding to climate change in African agricultural advisory services: Challenges, opportunities and recommendations for an AFAAS climate change response strategy. Kampala, Uganda and Accra, Ghana: African Forum for Agricultural Advisory Services (AFAAS) and Forum for Agricultural Research in Africa (FARA) and Forum for Agricultural Research in Africa (FARA).

34. Ludi, E.2009. Climate change, water and food security. ODI Background Notes London: Overseas Development Institute.

35. Magdoff, F., and Es, H.V. 2009. Building soils for better crops: Sustainable soil management. Third edition. Waldorf, USA: SARE and USDA.

36. Mekonnen, A.B. 2012. Analysis of Climate Variability and Its Economic Impact on Agricultural Crops: The Case of Arsi Negele District, Central Rift Valley of Ethiopia. M.sc.Thesis. Hawassa, Ethiopia: Wondo Genet College of Forestry and Natural Resources, Hawassa University.

37. Milder, J. C., McNeely, J. A., Shames, S. A. and Scherr, S. J. 2008. Biofuels and ecoagriculture: can bioenergy production enhance landscape-scale ecosystem conservation and rural livelihoods? Intl. J. of Ag. Sustainability, 6: 105-21.

38. Millennium Ecosystem Assessment (MEA). 2005. Ecosystems and Human Well-being: wetlands and water synthesis. Washington, DC: world resources institute.

39. Ministry of Agriculture (MOA). 2012. Ethiopia Country Programming Paper to End Drought Emergencies in the Horn of Africa. Addis Ababa: MOA.

40. Nærstad, A.2007. (ed). Africa can feed itself. Oslo, Norway: The Development Fund.

41. Ngigi, S. N. 2009. Climate Change Adaptation Strategies: Water Resources Management Options for Smallholder Farming Systems in Sub-Saharan Africa. New York: The MDG Centre for East and Southern Africa, the Earth Institute at Columbia University.

42. Odendo, M., Obare, G., and Salasya, B. 2010. Farmers' Perceptions and Knowledge of Soil Fertility Degradation in Two Contrasting Sites in Western Kenya. Land Degrad. Develop. 21: 557–564.

43. Parry, M.L., Rosenzweig, C., Iglesias, A., Livermore, M., and Fischer, G. 2004. Effects of climate change on global food production under SRES emissions and socio-economic scenarios. Global Environmental Change, 14: 53–67.

44. Pinstrup-Andersen, P.2009. Food security: definition and measurement. Food Sec. (2009) 1:5–7.

45. Postel, S.L.2003. Securing water for people, crops, and ecosystems: New mindset and new priorities. Natural Resources Forum, 27:89-98.

46. Rashid, S.2010. Staple Food Prices in Ethiopia. Addis Ababa: IFPRI

47. Rashid, S. and Negassa, A. 2011. Policies and Performance of Ethiopian Cereal Markets. ESSP II Working Paper 21. Addis Ababa: International Food Policy Research Institute (IFPRI) and Ethiopian Development Research Institute (EDRI).

48. Ringler,C., Zhu T., Cai, X., Koo, J., and Wang, D. 2010. Climate Change Impacts on Food Security in Sub-Saharan Africa: Insights from Comprehensive Climate Change Scenarios. IFPRI Discussion Paper 01042. Washington, DC: IFPRI.

49. Rockstrom, J., Folke, C., Gordon, L., Hatibu, N., Jewitt, G., de Vries, F. P., Rwehumbiza, F., Sally, H. Savenije,H., Schulze R. 2004. A watershed approach to upgrade rainfed agriculture in water scarce regions through Water System Innovations: an integrated research initiative on water for food and rural livelihoods in balance with ecosystem functions. Physics and Chemistry of the Earth, 29: 1109–1118.

50. Sanchez, P. A., Denning, G.L., and Nziguheba, G.2009. The African Green Revolution Moves forward. Food Sec. (2009) 1:37–44.

51. Schmidt, E. and Tadesse, F. 2012. Household and Plot Level Impact of Sustainable Land and Watershed Management (SLWM) Practices in the Blue Nile. ESSP II Working Paper 42. Washington, DC: IFPRI.

52. Shiferaw, A.2011. Estimating soil loss rates for soil conservation planning in the Borena district of South Wollo highlands, Ethiopia. Journal of Sustainable Development in Africa,Volume 13, No.3:87-106.

53. Shiferaw, B., Prasanna, B. M., Hellin, J., and Bänziger, M. 2011. Crops that feed the world: Past successes and future challenges to the role played by maize in global food security. Food Sec., 3:307–327.

54. Shiferaw, A., Hurni, H., Zeleke, G., and Debele, Berhanu. 2011. Role of indigenous knowledge in land management for carbon sequestration & soil-based ecological services in Damot Sore (Wolyita), Southern Ethiopia. Report submitted to International start secretariat (report number 3).

55. Stoop, W. A., and Hart, T. 2005. Research and Development towards Sustainable Agriculture by Resource-Poor Farmers in Sub-Saharan Africa: Some Strategic and Organisational Considerations in Linking Farmer Practical Needs with Policies and Scientific Theories. International journal of agricultural sustainability, Vol. 3, No. 3:206-2016.

56. Stroud, A. and Khandelwal,R.2006. In Search of Substance: 'State of the Art' of Approaches, Strategies and Methods for Improving Natural Resource Management and Livelihoods. African highlands initiative (ahi), working PAPERS # 6.

57. Tefera, M. M. 2009. Causes of rural household food insecurity: A case from kuyu district, central Ethiopia. Journal of Sustainable Development in Africa, Volume 11, No.4:286-304.

58. Thompson, H. E., Berrang-Ford, L., and Ford, J. D. 2010. Climate Change and Food Security in Sub-Saharan Africa: A Systematic Literature Review. Sustainability, 2: 2719-2733.

59. Vernooy, R., and Song, Y.2004. New Approaches to Supporting the Agricultural Biodiversity Important for Sustainable Rural Livelihoods. International Journal of Agricultural Sustainability, Vol. 2, No. 1:55-66.

60. Vignola, R., Locatelli, B., Martinez, C., and Imbach, P.2009. Ecosystem-based adaptation to climate change: what role for policy-makers, society and scientists? Mitig Adapt Strateg Glob Change, 14:691–696.

61. United Nations Framework Convention on Climate Change (UNFCCC). 2007. Impacts, vulnerabilities and adaptation in developing countries. Bonn, Germany: UNFCCC.

62. USAID. 2011. The Feed the Future (FTF) Multi-Year Strategies: Ethiopia FY 2011-2015 multi-year strategy. Washington, D.C: USAID.

63. USAID and Competitiveness and Trade Expansion Program (COMPETE). 2010. Staple foods value Chain analysis: Country report–Ethiopia. Addis Ababa and Washington, D.C: USAID and COMPETE).

64. Worako, T. K.2012. Dynamics of Food Price Trends and Policy Options in Ethiopia. Addis Ababa: Ethiopian Development Research Institute.

65. World Bank. 2010. Economics of Adaptation to Climate Change: Ethiopia. Washington, DC: The World Bank Group.

66. World Bank. 2011. Increased Productivity and Food Security, Enhanced Resilience and Reduced Carbon Emissions for Sustainable Development. Opportunities and Challenges for a Converging Agenda: Country Examples. Washington, DC: World Bank.

67. WFP-Ethiopia.2009. Food Security and Vulnerability in Addis Ababa, Ethiopia: Vulnerability Assessment and Mapping (VAM). Addis Ababa: WFP-Ethiopia.

CHAPTER 19 MARKETING POTENTIAL FOR COMMERCIAL SORGHUM MALT IN NORTHERN GHANA

Boaresa K. Isaac, Bawah Demuyakor, and Joseph A. Awuni

Keywords: Sorghum varieties, commercial sorghum malt, marketing potential for sorghum.

INTRODUCTION

Marketing is the process of planning and executing the conception, pricing, promotion and distribution of ideas, goods and services to create exchanges that satisfy individual and organizational objectives ([2] American Marketing Association, 1964). This definition reveals that marketing involves more than just a particular activity such as sales or promotion. Effective marketing requires that managers recognize the interdependence of these various activities and how they can be combined to develop a marketing program.

Sorghum (Sorghum bicolor) is a tropical cereal grass which is cultivated in southern Africa over 3000 years. Today, sorghum is cultivated across the world in warm climate areas. In terms of volume, it is the world's fifth most important cereal grain, after wheat, maize, rice and barley. Sorghum is still largely a subsistence food crop, but is increasingly becoming the foundation for successful food and beverage industries ([11] Taylor, 2003). In Africa, the major sorghum growing areas run across the West Africa, south of the Sahara to the coast and eastward into Sudan, Ethiopia and Somalia ([6] House, 1995).

In the West Africa sub region, among other uses, sorghum is malted and used to brew various traditional beers ([4] Demuyakor and Ohta, 1991). Beer production from sorghum grain or malt is an economically viable option for countries in the West Africa sub region, since barley cannot be cultivated in tropical Africa ([7] Mugode, 2009).

Malts are produced when cereals are subjected sequentially to steeping, germination and kilning under controlled conditions. In spite of the availability of several cereal types, barley has traditionally been the grain of choice when it comes to malting. In tropical Africa, however, barley cultivation has not seen any success. Thus, industries that use barley malt as their major raw materials have to rely on imports of this grain. This has not been a problem only to the industries concerned, mostly brewing industries, but also to the economies of mostly Tropical African countries. Therefore, some tropical cereals including sorghum have been investigated for their malting properties ([3] Beta, et al, 1995, [5] Dufour et al., 1992).

The potential of sorghum as an alternative substrate for lager beer brewing has been cited in several parts of Africa, particularly in Nigeria and elsewhere ([1] Agu and Palmer, 1998; [8] Owuama, 1997).

Malting is largely a cottage industry, usually performed at home by women; it requires great expertise. Maltsters and brewers from most parts of Ghana are faced with challenges such as shortage of sorghum within certain periods of the year, inadequate sunshine during the raining season, high price of sorghum between March and August, and these challenges affect their production levels. Malting is a tiring and tedious work which brewers have to repeat every time they want to prepare a batch of their brew. Also, the traditional techniques used for malting are ill-suited to markets. The risks for human health are serious and the malt technological quality is uneven when these traditional brewers prepare their own malt. The development of cyanogenic compounds, enterobacteria or moulds can impair the organoleptic qualities and healthfulness of sorghum malt prepared by the traditional maltsters.

Quality requirements for malting sorghum are reasonably strict and are directly related to processing efficiency and product quality in the malting and brewing industries. The major quality criteria for malt are noted to be taste and presence or absence of roots in the malt. The parameters affecting malt quality were perceived to be malt production period, proportions of grain and amount of water entering the malt, presence of pesticide residues in malt grains and age of grain.

Citation: Boaresa, I. K., Demuyakor, B., and Awuni, J. A. (2012). Marketing Potential for Commercial Sorghum Malt in Northern Ghana. Open Science Repository Economics, Online (open-access), e70081901. DOI:10.7392/Economics.70081901.

The problems associated with malting sorghum include high malting losses estimated at 10–30%, against 8–10% for barley. Sorghum malt has high gelatinization temperatures, which limits starch solubilization or hydrolysis by the amylolytic enzymes during mashing; this is a limiting factor in malting sorghum. Sorghum malts have low extract yield or low diastatic power (DP) due to inadequate hydrolytic enzyme activities especially β-amylase. This is attributed to poor endosperm modification caused by inadequate hydrolytic enzyme activities especially β-amylase (Dufour et al., 1992, [9] Palmer, 1989).

Pito brewing is a key off-farm economic activity that provides income for numerous households, particularly women in Ghana. It is estimated that over 10,000 women in Northern Ghana depend on pito brewing as their main source of income ([10] PSI-Sorghum, 2006). It has been observed that most of the pito brewers produce their own malt for brewing. Meanwhile some women entrepreneurs produce a limited amount of sorghum malt for sale to small scale brewers.

Therefore, the objective of this study was to investigate the commercial potential for sorghum malt in the northern region of Ghana and to find out if small scale brewers are willing to purchase industrially prepared sorghum malt.

MATERIALS AND METHODS

Primary data was obtained from the field through the administration of semi-structured questionnaires as well as personal observation.

The sample frame included all pito brewers and maltsters in the four districts under the study. Data collected include their brewing capacity per batch of brew, their ability to expand their business and their willingness to purchase commercial sorghum malt. The brewing capacity was measured using the five (5) litre 'frytol' gallon. The 5 litre 'frytol' gallon is the measuring material used by the local brewers to measure their brew. Nominal data was collected on their ability to expand their business and interest in purchasing commercial sorghum malt. These were tested by the use of closed questions where possible answers were defined in advance for the respondents to choose. These variables were coded 'Yes' and 'No' for the respondents to choose to indicate whether they can expand their business and if they are willing to purchase commercial sorghum malt.

Methods of Data Analysis

The data obtained was analyzed using descriptive statistics such as charts, means, frequency distribution tables and percentages with the aid of statistical package for social sciences (SPSS) version 16 and Microsoft excel 2007.

According to Texas State Auditor's Office, Methodology Manual, Review 5 (1995) [12], descriptive statistics are recommended when the objective of the study is to describe and discuss a data set more generally and conveniently than would be possible using raw data alone.

Descriptive statistics are routinely used in reports which contain a significant amount of qualitative or quantitative data. Descriptive statistics helps in summarizing and supporting assertions of facts. Furthermore thorough understanding of descriptive statistics it is essential for the appropriate and effective use of all normative and cause-and-effect statistical techniques, including hypothesis testing, correlation and regression analysis.

Descriptive statistics is essential for arranging and displaying data, make data much easier to work with, interpret and discuss than raw data. It help examine the tendencies, spread, normality, and reliability of a data set and it include useful techniques for summarizing data in visual form. Therefore it is appropriate to use descriptive statistics to analyze the data collected from these four districts.

RESULTS AND DISCUSSION

Categories of Respondents

Results of the research indicated that 2.3% of the respondents are brewers; that is, people who only brew by purchasing malt, 0.5% were people who retail beer (pito) and 97.2% undertake malting and brew using their own malt.

Table 1: Categories of respondents

Categories	Frequency	Percent
Brewers	5	2.3
Maltster and Brewer	209	97.2
Others	1	0.5
Total	215	100

Age of Respondents and Time Period in Business

Ninety two percent (92%) of the respondents in the local brewing industry in the study area are in the active age group between 20-40 years of age. It can be observe from Table 2 that most of the respondents within the active age group are between 0-20 years of experience in the local brewing industry. This suggests that they will have long time in the business if all things being equal. Malting and brewing is labor intensive, therefore, it requires more energetic people to carry out. As revealed by table 2, there is a good potential for growth of the local brewing industries as it is made up of people within the active age group.

Table 2: Age of respondents and time period in brewing

Time period in business (years)	Age of respondents				
	20-25	26-31	32-37	38-43	44-49
0-5	21	15	0	0	0
6-10	6	21	19	0	0
11-15	0	12	29	19	0
16-20	0	2	29	14	17
21-25	0	0	0	9	6
26-30	0	0	0	2	4

Time Period of Respondents in the Local Brewing Industry and Their Brewing Capacity

The table below shows the present brewing capacity of respondents and their respective number of years of experience in the local brewing industry. It could be seen from Table 3 that the capacity of the respondents between 0-20 years of experience is the highest. Those within 21-30 years of experience have the least brewing capacity. The brewing capacity of the respondents is relatively proportional to their age (physical strength), financial strength and availability of labour. Respondents between 0-20 years of experience are within the active age group and are more energetic in terms of labour, they are also better financially capacitated than those between 21-30 years of experience, which constitute the aged.

Table 3: Time period in brewing and brewing capacity

Time period in business (years)	Brewing capacity in liters (L) / batch			
	0-90 L	91.5-135 L	136.5-180 L	181.5-225 L
0-5	19	34	11	8
6-10	10	28	20	8
11-15	20	12	14	4
16-20	-	5	10	2
21-25	-	2	2	-
26-30	-	4	2	-
Total	49	85	59	22

Respondent Age Group and Their Ability to Expand Their Business

The respondents in the various age groups that have the capability to expand their business are more than those who cannot expand. It is clear from figure 1 that most of the brewers in the active age group have the ability to expand their business. The median age group (32-37) which have the highest brewing capacity also have the greatest potential of expanding their brewing business. The financial capability of the respondents, availability of raw materials (sorghum) and labor determine the size of their business.

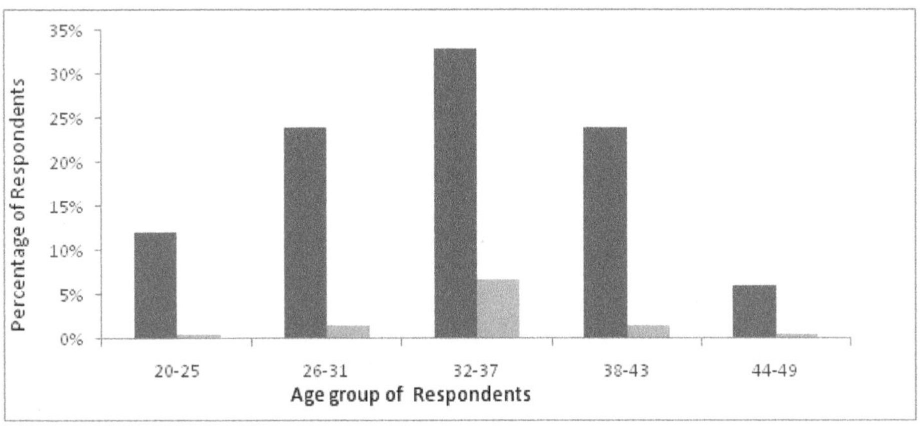

Figure 1: Respondents age and ability to expand their business.

Prospects for Sorghum Malt

The research reveals that out of the 214 brewers interviewed from the four districts, 85% the respondents showed interest in purchasing commercial sorghum malt and 15% respondents were not willing to purchase sorghum malt from firms that may exist to produce and sell sorghum malt. From figure 2 in Tamale, respondents willing and those not willing to purchase commercial malt were 93% and 7% respectively, Krachi-Nchumbru district also recorded 77.6% and 22.4% willing and not willing respectively. In Gushiegu, 90% of the respondents were willing to purchase commercial malt and 10% were not willing. Karaga also recorded 79% against 21% willing and not willing respectively.

Those that were willing to purchase commercial malt are of the view that the existence of firms to produce and sell sorghum malt will help in reducing their work load, as malting and brewing is labour intensive; they also believe that it will reduce their cost of production, which will in turn increase their production capacity. The respondents that did not show interest were of the view that commercial malt may be expensive and the quality may not be guaranteed.

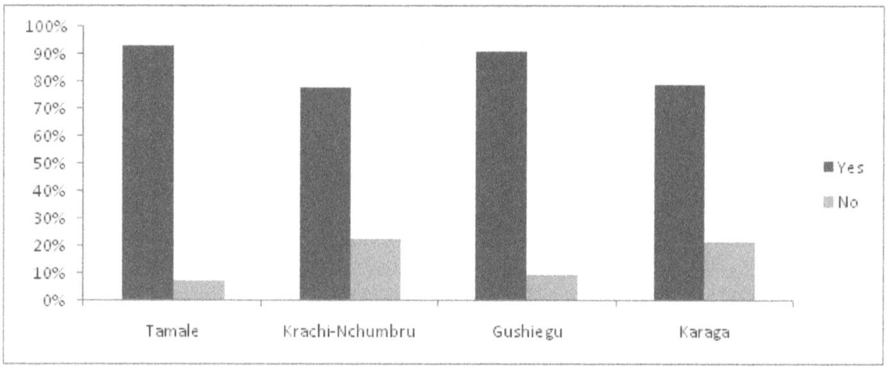

Figure 2: Prospects for sorghum malt in the four districts.

Age of Respondents and Their Willingness to Purchase Sorghum Malt

Figure 3 below showed that the number of respondents that are willing to purchase sorghum malt increases between the age group 20-37 and declines from the age group between 38-49. The figure also showed that out of the 85% of the respondents that were willing to purchase sorghum malt was highest within the age group 32-37 which was the median age group. The lowest limits within the 20-25 and 44-49 age groups are a result of their brewing capacity and financial strength. The age group between 20-25 is those within 0-5 and those between 44-49 are within 21-25 and 26-30 years of experience, as shown in table 3.

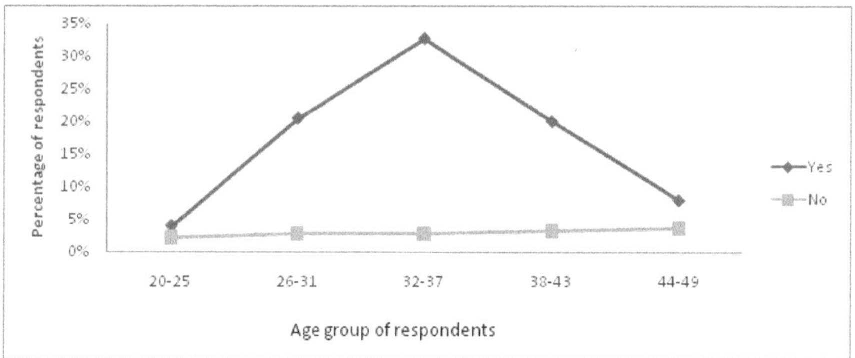

Figure 3: Age group of respondents and marketing prospects for sorghum malt.

CONCLUSION

The marketing prospect for commercial sorghum malt is very promising as revealed by the study. The study indicates that 85% of the respondents were willing to purchase commercial sorghum malt. The existence of firms that will produce sorghum malt for sale to small scale brewers according to the respondents will help in reducing their work load, and it will reduce the cost of production as well; those firms will be available all year round, which will help solve the problem of sorghum shortage and the quality and hygiene of the malt will be assured. Fifteen percent (15%) of the respondents did not show interest in purchasing commercial sorghum malt. They were of the view that, the quality of commercial malt cannot be guaranteed and that the cost of the malt will be expensive as compared to their self-made malt.

There are presently maltsters who supply sorghum malt to brewers, but their numbers and the quality of malt is not adequate, according to the respondents.

The result of the study showed that there is a good market potential for sorghum malt for firms that may exist for selling it. The promotion of malt with consistent quality will allow better use of sorghum in the small scale brewing industry in the Northern region and the Northern part of the Volta region of Ghana. The production of quality malt will generate new jobs, better income, facilitate the work of women and allow the small scale brewing industry to better serve the increasingly demanding rural and urban markets.

REFERENCES

1. A reassessment of sorghum for lager-beer brewing. R.C. Agu, G.H. Palmer (1998). *Bioresource Technology* 66 (3) p. 253-261.
2. *Marketing information and advice on marketing strategies*. American Marketing Association (1964).
3. Malting characteristics of sorghum cultivars. T. Beta, L.W. Rooney, R.D (1995). *Waniska. Cereal Chemistry* . 72 (6) p. 533-538.
4. Malt characteristics of sorghum vulgare varieties from ghana. B. Demuyakor, Y. Ohta (1991). *Journal of the Science of Food and Agriculture* 59 (4) p. 457-462.
5. Sorghum malts for the production of a lager beer. J. P. Dufour, L. Melotte (1992). J*ournal of the American Society of Brewing Chemists* 50 (3) p. 110-119.
6. Sorghum and millet: History, taxonomy and distribution. House, L.R. (1995). In: Dendy, D. A.V. (Ed).*Sorghum and millets chemistry and technology*. America Association of cereal chemist, St, Pauls, MN.Pp. 1-10.
7. *Free amino nitrogen improvement in sorghum malt brewing*. Mugode, L. (2009). MSc dissertation, University of Pretoria, Pretoria.
8. Sorghum: a cereal with lager beer brewing potential. C. I. Owuama (1997). *World Journal of Microbiology and Biotechnology* 13 (3) p. 253-260.
9. Cereals in malting and brewing. Palmer, G.H. (1989). In *Cereal Science and Technology*, ed Palmer, G.H. pp. 61-242. Aberdeen: Aberdeen University Press.
10. *Final Report On The President's Special Initiative On Sorghum*. Ministry Of Trade, Industry, Private Sector Development And Presidential Special Initiatives (PSI) President's Special Initiative (PSI-Sorghum), (2006).
11. *Overview: Importance of sorghum in Africa*. J.R.N. Taylor (2003). Department of Food science, University of Pretoria, South Africa.
12. *Methodology Manual, Review 5, Descriptive Statistics*. Texas State Auditor's Office (1995).

INDEX